THE RUMEN

AND ITS MICROBES

THE RUMEN
AND ITS MICROBES

Robert E. Hungate

DEPARTMENT OF BACTERIOLOGY
and
AGRICULTURAL EXPERIMENT STATION
UNIVERSITY OF CALIFORNIA
DAVIS, CALIFORNIA

1966

ACADEMIC PRESS • New York and London

COPYRIGHT © 1966 BY ACADEMIC PRESS INC.
ALL RIGHTS RESERVED.
NO PART OF THIS BOOK MAY BE REPRODUCED IN ANY FORM,
BY PHOTOSTAT, MICROFILM, OR ANY OTHER MEANS, WITHOUT
WRITTEN PERMISSION FROM THE PUBLISHERS.

ACADEMIC PRESS INC.
111 Fifth Avenue, New York, New York 10003

United Kingdom Edition published by
ACADEMIC PRESS INC. (LONDON) LTD.
Berkeley Square House, London W.1

LIBRARY OF CONGRESS CATALOG CARD NUMBER: 65–26041

PRINTED IN THE UNITED STATES OF AMERICA

Preface

During early experiments on the nutrition and metabolism of termites and their protozoa, I became intrigued with the possibility of describing quantitatively the mutualistic relationship between microbes and host. Later a description seemed possible also for ruminants, and experiments were directed toward quantitative analysis of the rumen. This goal, pursued over the past twenty-five years, has led to this monograph, which will serve as a contribution to the ecology of this important microbial habitat.

Relatively few microbial habitats have been subjected to a thorough quantitative ecological analysis. The rumen fermentation is peculiarly suitable because of its relatively constant and continuous nature and because of the very rapid rates of conversion of organic matter. Although analysis of the ruminant-microbe symbiosis is still far from complete, knowledge is sufficient for formulation of principles and for identification and measurement of important parameters.

The first eight chapters include a description of the rumen and its microbes, their activities, and the extent of these activities. This basic biology provides a framework in which applications to agriculture can be evaluated. These applications are discussed in the last four chapters: host metabolism, variation in the rumen, possible practical applications, and abnormalities in rumen function.

Historical developments have been traced as completely as possible, but the magnitude of the task has prevented its accomplishment in many instances. References are provided to introduce the student to the literature and to encourage independent evaluation of the evidence.

Friends who have assisted by critical reading of portions of the manuscript include A. L. Black, M. P. Bryant, R. T. J. Clarke, R. W. Dougherty, R. J. Moir, K. el Shazly, and D. W. Wright. Their assistance is gratefully acknowledged. To them and to the many other associates and students who have worked with me on many aspects of the rumen, I am deeply indebted, not only for their contributions which are interwoven into the account, but even more for their enthusiasm in this shared endeavor.

Finally, I am most indebted to my wife, Alice, who has perseveringly, patiently, and assiduously helped with the references and has generously encouraged the research and writing over a period of many years.

March, 1966   R. E. HUNGATE

Contents

CHAPTER XII

Abnormalities in the Rumen

CHAPTER I

Introduction

Ruminants, cloven-hoofed mammals of the order *Artiodactyla,* obtain their food by browsing or grazing, subsisting on plant material. Innumerable wild species have been hunted for centuries, and many have been domesticated, including sheep, cattle, goats, camels, llamas, buffalo, reindeer, and caribou.

Peculiarities of ruminants as compared to other mammals must have been recognized by the primitive hunter, long before recorded history. Written evidence that ruminants were distinguished from their mammalian relatives is found in Leviticus, eleventh chapter, third verse: "Whatsoever parteth the hoof, and is clovenfooted, and cheweth the cud among the beasts, that shall ye eat." The term for cud-chewing is rumination; hence ruminants.

Aristotle described the four compartments of the ruminant stomach. Books on natural history prior to 1800 contain numerous descriptions of ruminants and their digestive activities, including experiments to test the rate of passage of material through the alimentary tract (Spallanzani, 1776). Peyer (1685, quoted by Colin, 1886) concluded that a fermentation occurred in the rumen. Tiedemann and Gmelin (1831) demonstrated what they concluded to be acetic and butyric acid in rumen contents.

In 1832, when Karl Sprengel published his "Chemistry for Farmers, Foresters, and Cameralists," decomposition of plant materials in the rumen was known to give rise to volatile substances which at that time were assumed to consist of acetic and butyric acids. Methane and carbon dioxide were later (Popoff, 1875; Reiset, 1863a) recognized as products of rumen activity. The other important product of the rumen fermentation, propionic acid, was not identified with certainty until 1945 (Elsden, 1945a).

It is interesting that the terms "Selbstentmischung" (self-mixing) and "Gährung" (fermentation) were used by Sprengel to denote microbial activity. With the later restriction of fermentation to anaerobic metabolism (Pasteur, 1861), the descriptive term "Selbstentmischung" was lost. Interestingly enough it is a more apt expression for these processes as they are now understood than is Gährung, the term retained. Sprengel reported that plant fibers could undergo a "Selbstentmischung," especially when attacked in association with more readily decomposable materials, but he did not state clearly that fiber decomposition occurred in the stomach of ruminants. He viewed acetic acid as assisting in digestion, rather than as a product of fermentation.

In 1837 Haubner postulated that water was absorbed in the omasum, but erroneously assumed that this concentrated the saliva and made it the effective agent of digestion. Later (Haubner, 1854) he studied fiber utilization. By that time cellulose had been distinguished by its solubility in strong acid, and insolubility in weak acid and alkali, and established as an important constituent of plants. By quantitative analyses of feed and feces, Haubner (1855) showed that a significant quantity of cellulose disappeared during passage of forage through the ruminant. This was soon confirmed (Henneberg and Stohmann, 1860), and the rumen was shown to be the site of the digestion (Wildt, 1874).

The discovery of the rumen protozoa by Gruby and Delafond in 1843 was the first identification of microorganisms in the ruminant stomach. This was only a few years after the rediscovery of microorganisms and the emergence of protozoa as an important subdivision of the animal kingdom. A popular subject for debate at that time was the question, must animal material make up at least a part of the food of all animals?, a forerunner of the recent interest in growth factors of animal origin. Gruby and Delafond suggested that the protozoan fauna explained the ability of ruminants to survive on an exclusively plant diet.

The role of bacteria in the fermentation of plant materials became well known as a result of the genius of Pasteur (1863). It was inferred by Zuntz (1879) that rumen microorganisms fermented fiber anaerobically and thus formed acids and gas. He postulated that the acidic fermentation products were absorbed and oxidized by the host and demonstrated a capacity of the rabbit to oxidize lactate and butyrate (Zuntz and von Mehring, 1883). Zuntz was the first to formulate clearly the "fermentation" hypothesis to explain the mechanism of forage utilization by ruminants. Oxidation of acetic acid had already been demonstrated in mammals by Woehler (1824) and others. Hofmeister (1881) showed that cellulose was digested by rumen contents and also by sheep saliva. The effectiveness of the latter misled him to believe that the saliva itself was responsible for the digestion, but the saliva was obtained from the throat and probably contained numerous bacteria responsible for the observed cellulose digestion, an explanation suggested by von Tappeiner (1884a).

Experimental support for the fermentation hypothesis was quickly provided by von Tappeiner (1884a), a student of Zuntz. Cellulose, incubated with juices of the ruminant itself, showed no digestion, provided bacterial growth was inhibited with mild antiseptics not impeding hydrolysis. In the absence of antiseptics, cellulose incubated with rumen fluid disappeared, and gas (methane or hydrogen and carbon dioxide) and acids (assumed to be acetic and butyric) were formed. The fermentative activities were ascribed to rumen bacteria which occurred in profusion along with the protozoa.

In the light of modern knowledge (and also as suggested by Zuntz, 1913) it is doubtful that "rumen" bacteria were responsible for the cellulose disappearance in von Tappeiner's cultures. Initiation of the fermentation was delayed and activity persisted longer than is common with *in vitro* cultures of rumen bacteria, but his conclusions that bacteria in the rumen were active in cellulose fermentation and that volatile acids were an important product were entirely correct. In accord with Zuntz, von Tappeiner inferred that the acids were absorbed and oxidized by the host to supply its energy needs. Wilsing (1885) showed that the volatile acids were not recovered in the feces. Mallèvre (1891) substantiated Woehler's finding that animals utilized acetic acid, and he extended the observations to propionate. Boycott and Damant (1907) showed that the corrected respiratory quotient of ruminants indicated that fatty acids were oxidized.

In spite of these results the fermentation hypothesis was not widely accepted. The papers of Ellenberger and Hofmeister (1887a) and Henneberg and Stohmann (1885) may explain this rejection. They proposed that, instead of fermenting cellulose, the bacteria digested it to a disaccharide and then to a monose which was absorbed and utilized by the ruminant. Even though no experimental evidence was presented, this scheme appeared plausible. It agreed with the known mechanism for utilization of starch in man, and when Pringsheim (1912) showed that glucose appeared in cellulose cultures inhibited with antiseptics, it was widely accepted that glucose was the final microbial product in the ruminant, even though Zuntz (1913) pointed out that in the absence of antiseptics the sugar demonstrated by Pringsheim was used immediately by the rumen microorganisms. "He [Pringsheim] has shown that the agents of the fermentation secrete an enzyme splitting cellulose to sugar; however, he can only find the sugar when the bacteria are killed with an antiseptic; as long as they are alive the sugar was dissimilated to fatty acids and gases at the moment it was formed" (translation).

Important exceptions were the paper by Markoff (1913), another student of Zuntz, in which rates of fermentation were measured and the studies (Boycott and Damant, 1907; Krogh and Schmidt-Jensen, 1921) in which the energy lost as methane was estimated. During this period numerous investigations of the rumen protozoa were completed by Mangold in Germany (1929, 1943), Dogiel in Russia (1927b), Kofoid and MacLennan in the United States (1930, 1932, 1933), and Usuelli in Italy (1930a,b). The author (Hungate, 1939) developed a fermentation hypothesis to account for the utilization of cellulose in termites, postulated that it applied also in cattle, and, upon investigating the literature, found that the theory had been fully developed for ruminants by Zuntz and von Tappeiner many years before.

Between 1884 and 1944 additional investigators undoubtedly accepted the fermentation hypothesis, but little published information was presented in its support. The stimulus of methods for quantitative separation and identification of the volatile fatty acids was lacking, as were methods for cultivation and isolation of the rumen microbes.

By 1940 the time was ripe for a more widespread appreciation of rumen microbial activities. An Animal Physiology Unit was set up at Cambridge, England, under the leadership of Sir Joseph Barcroft. The group recognized that a fermentation was the basic mechanism involved, and they demonstrated it quantitatively by showing that significant amounts of volatile acids appeared in sheep blood as it circulated through the rumen wall (Barcroft *et al.,* 1944). The study was materially aided by Elsden's adaptation (1946) of partition chromatography to separate small quantities of the lower volatile acids and by Phillipson's development (Phillipson and Innes, 1939) of experimental and surgical techniques permitting access to the ruminant stomach with minimal disturbance of normal function. Rachel McAnally also participated in these early investigations of the Animal Physiology Unit.

Interest burgeoned after the war and led to a surge of investigations which has continued and greatly expanded. The insight into rumen functions resulting from the wartime investigations demonstrated the wisdom of the British in supporting basic research even at times of national stress.

Von Tappeiner's experiments (1884a) stimulated an interest in the cultivation of the rumen microorganisms, and numerous attempts were made between 1900 and 1940 to grow the protozoa (Trier, 1926; Waentig and Gierisch, 1919; Willing, 1933; Knoth, 1928; Westphal, 1934a,b) and the bacteria (Ankersmit, 1905; Henneberg, 1919; Hopffe, 1919; Kreipe, 1927; Steiger, 1904; Neubauer, 1905; Arnaudi, 1931; Orla-Jensen, 1919; Pochon, 1935. In the case of the bacteria these were unsuccessful except for *Streptococcus bovis* (Orla-Jensen, 1919).

Efforts to cultivate the protozoa had been generally unsuccessful, though Poljansky and Strelkow (1934) obtained clones in the rumen and Margolin (1930) reported growth of *in vitro* mass cultures. Margolin's work was not repeated, and the lack of success of similar attempts in other laboratories prevented its receiving the attention which on the surface was merited. Perhaps the failure to grow them for a long time and to provide quantitative information on the numbers and kinds of protozoa cultured made it difficult to appreciate Margolin's results.

The author obtained *in vitro* cultures of certain rumen protozoa (Hungate, 1942, 1943) and later extended the investigation to include cultivation of the cellulolytic bacteria (Hungate, 1947), adapting methods to their growth and pure culture which permitted application of classical isolation procedures. Quite independently, A. J. Kluyver initiated a study of the rumen bacteria during World War II, with the assistance of Miss A. Kaars Sijpesteijn. It is characteristic of the courage and the scientific interest and perseverance of Professor Kluyver that he and his student tackled so difficult a problem at a time when food shortages and the hazards of an occupying power rendered existence so precarious. They obtained evidence on the nature of some of the important cellulolytic bacteria of the rumen (Sijpesteijn, 1948) and obtained one strain in a pure state.

As a result of these and many other investigations of the various facets of rumen function it is now firmly established that the ruminant and the rumen microbial population exist in a reciprocally beneficial relationship in which many of the plant materials consumed by the mammalian host are digested and fermented by the rumen microbes to form chiefly carbon dioxide, methane, and volatile acids. The gases are excreted by the ruminant, and the acids are absorbed and oxidized. This type of symbiotic relationship is called mutualism, after deBary (1879). It is important to stress that this symbiotic relationship is between the *total microbial population* and the host. Individual rumen species may perform functions of greater or less value to the host. The role of each is not simply with the host, but also with the remaining elements of the total population.

As previously mentioned, the first discoverers of the rumen microbiota suggested that the protozoa were used as food, and evidence was presented for this interpretation. Subsequently the bacteria were included (Hagemann, 1891). According to this view the microorganisms provide the host not only with a source of energy, but also with the proteins and other materials needed for the construction and maintenance of host cells. This utilization of microorganisms as food recalls the numerous plankton feeders of the ocean, but whereas typical plankton feeders have numerous devices for straining the microorganisms out of large volumes of water, the ruminant grows the "plankton" continuously in tremendous numbers in a small volume and then harvests them. In a sense the ruminant can be termed a plankton feeder, but its adaptations go much further since it utilizes not only the bodies of the microorganisms, but also some of the waste products formed during their growth.

The zoologist Hegner placed the ruminants at the top of the evolutionary tree of mammals, evaluating their digestive specializations as more advanced than the brain of man. Although we optimistically deny this
misanthropic attitude, it at least emphasizes an appreciation of the unique
arrangements and highly specialized nature of the ruminant digestive
mechanism.

The favorable conditions provided by the ruminant permit growth of
many species of bacteria and protozoa and of myriads of individuals; the
concentration of microorganisms in the rumen is as great as in any other
natural habitat. To understand this complex microbiota each species
should be studied in pure culture. These studies must describe the isolated
organisms and measure their activities in order to relate microbe and host
in quantitative fashion (Hungate, 1960). It is necessary also to know how
the host supplies them with food, absorbs their fermentation products,
and regulates conditions in the rumen to support continuous microbial
growth. Finally, the interactions between host and microbes must be
analyzed by quantitative study of events in the rumen itself. Understanding of normal functions provides a basis for examining disturbances leading to malfunction.

Utilization of forages with the aid of symbiotic microorganisms is not
confined to ruminants (Zuntz, 1891; Elsden *et al.,* 1946; Moir *et al.,*
1956). In the horse and its relatives, the cecum is much enlarged relative
to other parts of the alimentary tract, and there occurs in it an extensive
fermentation of cellulose and other materials by bacteria and protozoa. The
protozoa differ from those of the rumen and are assigned to a different taxonomic group. Also in rodents, lagomorphs, elephants, and probably other
mammals there is an active cecal fermentation, and in some such as field
mice and hamsters there is a stomach fermentation. The elephant cecum
contains numerous round worms (*Nemathelminthes*) and also some gi-
gantic protozoa, *Elephantophilus.*

The Australian herbivorous marsupials utilize forage with the aid of
abundant bacteria and unique protozoa in the enlarged stomach (Moir
et al., 1954, 1956), and they survive under quite adverse conditions of
forage and water (Storr, 1964). The stomach, though not divided into
highly specialized divisions as is the stomach of ruminants, shows an esophageal groove and compartmentation and supports a dense population of
bacteria and protozoa which bear the same relationship to the marsupial
host as that exhibited in ruminants. The completeness of digestion of fiber
is less than in ruminants, but more than in rabbits (Calaby, 1958). The
lack of an organ resembling the omasum may be explained by the fact
that the forage is ground extremely fine as it is consumed.

Evidence has been obtained recently of a microbial fermentation in the stomach of *Cololus* monkeys in Africa (Drawert *et al.,* 1962; Kuhn, 1964) and in the *Langur* monkeys of India.

An important characteristic of ruminants, monkeys, some rodents, and marsupials is the occurrence of the microbial fermentation prior to the gastric and duodenal activity in which the bodies of the microorganisms are digested. In cecal digestion of fiber, the microbial waste products can be absorbed, but the bodies of the microorganisms (Zuntz, 1891) are voided in the feces. A habit of coprophagy among rabbits and rodents permits recovery of this microbial protoplasm, but in the horse, elephant, and their relatives the microorganisms themselves are not utilized.

CHAPTER II

The Rumen Bacteria

The rumen bacteria are adapted to live at acidities between pH 5.5 and 7.0, in the absence of oxygen, at a temperature of 39–40°C, in the presence of moderate concentrations of fermentation products, and at the expense of the ingesta provided by the ruminant. The steady supply of food and continuous removal of fermentation products and food residues maintain relatively constant conditions in which an extremely dense population develops.

A. Diversity of Kinds

The constancy of rumen conditions, as compared, for example, with soil, diminishes the number of possible natural niches for microorganisms. Constant conditions might appear suited for the selection of a few especially well-fitted microbial types which would outgrow others and predominate. Certain bacterial types do occur almost universally in the rumen and may occasionally constitute a large proportion of the cultivated colonies, but the diversity in the types of rumen bacteria is striking.

There are several possible explanations for this diversity. The feed of ruminants is complex, containing carbohydrates, proteins, fats, numerous other organic compounds, and minerals. Two avenues of adaptation are open. Organisms may become narrowly adapted (highly specialized), compete for a few of the foods, and occupy a limited niche, or become widely adapted and capable of using many nutrients. Examples of both types and of intermediates occur in the rumen.

A second factor selects for diversity in the rumen population. This can be expressed as *selection for maximum biochemical work*. During fermentation, the energy-containing components (chiefly carbohydrate) in ruminant feeds are converted into microbial cells and into microbial wastes, i.e., carbon dioxide, methane, and acetic, propionic, and butyric acids (Table II-1). Some of these microbial wastes are food for the host.

The hypothesis of selection for maximum biochemical work postulates that in a system such as the rumen, open to invasion by innumerable microbes, those organisms accomplishing the most growth will survive. Since growth is limited by the quantity of available food, efficiency in transformation of food into cells carries survival value. During ruminal conversion of carbohydrates to acetic, propionic, and butyric acids, carbon dioxide, and methane, the cell yield is greater than by other theoretically possible biochemical pathways for carbohydrate fermentation. Pure fermentations—such as the homo- and heterolactic fermentations, the acetic-butyric types, or formic-acetic types—are replaced by types more efficient in the production of new cells. Among the innumerable theoretically possible biochemical reactions leading to cell formation, certain combinations produce more cells than do others. These combinations include many individual reactions. The conditions necessary for each reaction differ from conditions favoring others. A single kind of cell cannot contain the diver-

Table II-1

RUMEN FERMENTATION PRODUCTS[a,b]

Experiment	Acetic acid	Propionic acid	Butyric acid	Carbon dioxide	Methane
1	20.5	6.9	5.8	22.0	9.0
2	20.0	7.3	4.7	16.1	6.6
Average	20.3	7.1	5.3	18.6	7.8

[a] All values in μmoles per hour per gram of bovine rumen contents.
[b] From Hungate *et al.* (1961).

sity of conditions and enzymes needed for all the individual reactions in the combination yielding maximum cell growth. There are limits to the biocatalytic capacities of a single cell. Maximum total growth requires a complex population in which are included many more pathways for biochemical work than can be accommodated in a single cell. Part of the diversity in the rumen bacteria may be explained in this way.

A third factor increasing the complexity of the rumen microbiota is variation with time in the metabolic type occupying a particular niche (Margherita *et al.,* 1964). In theory it might seem that, of all possible variants, one would be best fitted to occupy the niche and would displace all others, but this "most-fitted" type becomes itself a part of the niche! It composes an element of a new environment (niche) differing from the original by the presence of the selected organism. This leads to selection of still another type. The changes with time in the strains of a particular bacterial species in a single ruminant can be explained in this way. The validity of the explanation must await experimental test. The fact is that there can be variation in the detailed characteristics of strains of the same species isolated at different times, even from the same animal (Margherita and Hungate, 1963).

The great majority of the rumen bacteria in forage-fed animals are obligately anaerobic. Euryoxic types capable of living in oxygen tensions ranging from none to atmospheric are relatively few. If the environment is chiefly anaerobic, synthesis during cell growth of the machinery for utilization of the scarce oxygen may be of negative survival value.

Although anaerobic conditions prevail within the mass of the rumen contents, preventing aerobic growth, the gas above the rumen digesta often contains oxygen, and probably some aerobic bacterial growth occurs. The extent of this has not been estimated. Because abundant bacteria may be ingested with the feed (Gutierrez, 1953), growth estimates of aerobes must be based on careful quantitative measurement of the aerobic bacteria in the feed and in the rumen. Those aerobic forms cultivated from the rumen (Stellmach-Helwig, 1961) which require high levels of carbon dioxide may be normal residents. They are few in number.

B. Distribution

Because the complex of conditions characteristic of the rumen habitat is not commonly encountered outside of warm-blooded hosts, the rumen bacteria are limited in their distribution. Bacteria closely related to certain rumen types have been found in the cecum of the rabbit (Hall, 1952; Brown and Moore, 1960) and porcupine (McBee, personal communication), in the feces of a wide variety of mammals (Seeley and Dain, 1960; Brown and Moore, 1960), and even in the human mouth (MacDonald and Madlener, 1957). Further study of other habitats, especially in the alimentary tracts of mammals, will almost certainly disclose additional bacteria similar to those of the rumen. Not all the rumen bacteria are widely distributed, since some types do not develop in isolated calves (Bryant and Small, 1960).

Although the easily identified rumen protozoa have not been commonly reported from other habitats, one common species in the horse cecum has been found in the rumen (Jameson, 1925). The differences between the rumen microbiota and those in other habitats probably exceed the similarities. The protozoa in the marsupial *Setonix brachyurus* differ markedly from those in the rumen and in the horse cecum, and preliminary estimates of the rodent cecal flora (McBee, personal communication) disclose many important bacteria not found in the rumen.

Rumen bacteria are not specifically required for production of the gases and volatile fatty acids characteristic of the rumen. During incubation of grass with water, volatile fatty acids form (Barnett and Duncan, 1953), presumably due to action of nonrumen bacteria.

There is some evidence of a specificity within ruminants as to the nutritional characteristics of the rumen bacteria. Cellulolytic bacteria tentatively identified as *Bacteroides succinogenes* from African antelope species did not grow as well on bovine rumen fluid as on homologous fluid (Hungate *et al.,* 1959). The cellulolytic bacteria from sheep on teff hay grew better in the homologous rumen fluid than in media containing rumen fluid from sheep on a high protein diet, and the converse also was observed (Gilchrist and Clark, 1957). Large oval microbial forms (see Fig. II-1B,C, and D) are common in the sheep rumen, but are not reported from cattle. The microbiota of the tiny African suni is microscopically quite different from the population in cattle.

Some of these differences stem from differences in food, others from differences in the other environmental factors maintained by the host.

Relatively little is known about the distribution of microorganisms within the rumen. The stratification of the digesta suggests that also the bacteria are not evenly distributed. They are more abundant in the whole digesta, including solids, than in the liquid left when solids float to the top of freshly drawn contents, presumably because many are attached to the solids particles. The fermentation rate when feed is added to the whole contents, including solids, is about four times the rate when feed is added to the liquid alone. This is in part due to lack of protozoa in the liquid (they settle to the bottom), but the difference is too great to be accounted for solely by the protozoa. The bacteria tend to be associated with the solids, and in animals on a variety of feeds the bacterial culture count from the dorsal rumen contents is usually higher than from the ventral.

The little available evidence suggests that rumen bacteria are more readily transmitted from one individual ruminant to another than are the protozoa—under conditions in which no protozoa developed in isolated calves, the bacterial flora included many species found in faunated animals (Abou Akkada and Blackburn, 1963; Bryant and Small, 1960; Bryant *et al.,* 1958a), though some were different (Bryant and Small, 1960). Transmission in soil, air, and water of bacteria originating in other ruminants is not excluded in these experiments, and careful studies in which calves are isolated at birth under conditions permitting no bacterial inoculation from other ruminants would clarify the problem of the degree of specificity of the association between the rumen bacteria and their host. Anaerobic bacteria similar to rumen bacteria have been detected in the air of cattle sheds (Hobson, 1963). If they withstand drying, airborne transmission would account for their spread to considerable distances.

C. Morphology

In an animal on a predominantly hay or forage ration the majority of the bacteria are gram-negative. With high grain rations, there is an increased proportion of gram-positive cells. The significance of this shift is not known. Perhaps it reflects the increased numbers of lactobacilli under the more acid conditions usually accompanying high grain rations. No ecological significance has been associated with the gram reaction.

Most rumen bacteria are cocci and short rods occurring in various sizes, but usually within the range of 0.4–1.0 μ in diameter and 1–3 μ long. The great majority can be identified accurately only after axenic culture, but some species with a distinctive morphology can be identified microscopically (Moir and Masson, 1952).

Both sheep (Moir, 1951; Moir and Masson, 1952) and cattle (Pounden and Hibbs, 1949c) contain *Oscillospira guillermondii* (Figs. II-1A, 2A, and 2C). This organism is characterized by motility, iodophily, and large size in comparison with most bacteria. The filaments are transversely partitioned into cells in a fashion similar to the related *Caryophanon* isolated from fresh cow dung. Spherical spores may be formed. The *Oscillospira* in the rumen has not been cultured, and single cells picked with a micromanipulator (Purdom, 1963) have failed to grow. Numbers in sheep range between 2.6×10^4 and 5×10^7 per milliliter (Moir, 1951; Warner, 1962a,b).

The regularly arranged plates often seen in the rumen (Fig. II-1B) are *Lampropedia*. It is probably identical with the *Bacillus merismopedioides* seen by List (1885) and should be designated as *Lampropedia merismopedioides*. Julius Kirchner, working at Davis, isolated this organism from rumen contents of a fistulated heifer. Preliminary enrichment occurred when rumen fluid was left standing for several days in an open Erlenmeyer flask at room temperature. A thick layer of *Lampropedia* accumulated at the surface. Streaking on aerobic plates of rumen fluid agar gave isolated colonies. The *Lampropedia* did not grow under anaerobic conditions and may be obligately aerobic.

It has not been definitely established that *Lampropedia* grows in the rumen. It usually occurs only in small numbers and could conceivably enter with the feed. If it does grow in the rumen it may be at the expense of the limited quantities of swallowed oxygen in the gas overlying the rument contents.

Lampropedia occurs in the alimentary tract of certain species of turtle and in their intestinal nematodes (Schad *et al.*, 1964), and it may serve as food for the worms. The turtle colon in which the nematodes occur is a site of cellulose digestion.

Sheep contain at least two small motile organisms (Moir and Masson, 1952; Smiles and Dobson, 1956) large enough to appear almost like small protozoa, but lacking the characteristic internal structures. One of these types (Oxford, 1955b) is *Selenomonas ruminantium* (Certes) Weynon (Lessel and Breed, 1954). It is a crescentic cell, 0.7–3.0 × 3.5–8.0 μ, with a laterial tuft of flagella, slightly iodophilic with a red-brown reaction. Some of the large rumen selenomonads are serologically similar (Hobson *et al.,* 1962) to the *Selenomonas* cultured from the rumen (Bryant, 1956), to be described later, but others are different (Purdom, 1963). Direct counts in sheep of 2.2 × 10⁷ to 3 × 10⁸ selenomonads per milliliter have been recorded (Warner, 1962b). The long crescents of Moir and Masson (1952) and Smiles and Dobson (1956) (Fig. II-1B) resemble the division forms of selenomonads (Lessel and Breed, 1954), but may be a distinct group.

The other organism was described by Woodcock and Lapage (1913) as being closely related if not identical to *Selenomonas ruminantium.* It was believed by Quin (1943a) to be a yeast, but later found (McGaughey and Sellers, 1948; Westhuizen *et al.,* 1950) to be motile and therefore distinctly a nonyeast type. It is 2.5-3.0 × 4.0-9.0 μ (Fig. II-1B and C) and more strongly iodophilic (red-brown reaction) than *Selenomonas.* Further study is needed to determine its relationship to *Selenomonas.* Organisms in sheep rumen contents resembling the one described by Quin did not react to anti-*Selenomonas* serum (Hobson *et al.,* 1962). It has been reported (Helwig, 1960) that the yeastlike forms disappeared when casein was added to the ration. Attempts to cultivate this organism from single picked cells have been unsuccessful (Purdom, 1963), but Moir, in the author's laboratory, has recently isolated a strain.

Yeasts of the genera *Candida* and *Trichosporon* have been found in small numbers in the bovine rumen and not in the feed (Clarke and di Menna, 1961). Yeasts occur also in sheep and goat rumens (van Uden *et al.,* 1958).

The LC coccus discovered by Elsden and Lewis (1953) and later named *Peptostreptococcus elsdenii* (Gutierrez *et al.,* 1959a) is a distinctive type which has been grown in pure culture. It will be discussed later.

Sarcina bakeri (Mann *et al.,* 1954a) is a euryoxic coccus (Fig. II-2A) occurring singly or in pairs and tetrads. The large diameter of the cells (4 μ) is their distinguishing feature. It was grown in pure culture and shown to ferment glucose. Some large cocci and rods have been reported as characteristic of the rumen when hay is fed (Pounden and Hibbs, 1948a), but an identity with *S. bakeri* has not been established.

Additional morphological types of cocci occur, especially in the sheep

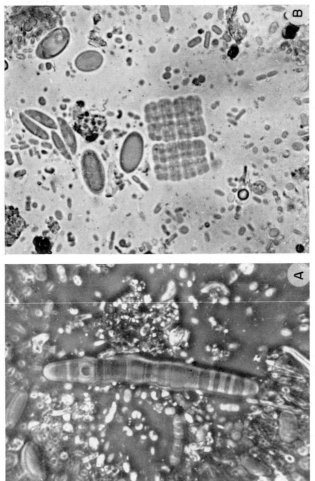

Fig. II-1. A. *Oscillospira guillermondii* from the sheep rumen. Ultraviolet-phase illumination. Note the transverse septa and the spore at the upper end of the filament. After Smiles and Dobson (1956). Magnification 1830 ×. B. *Lampropedia* from the sheep rumen, also long almost crescentic forms, and the oval form of Woodcock and Lapage (1913), also called Quin's oval. Ultraviolet illumination. After Smiles and Dobson (1956). Magnification 1830 ×.

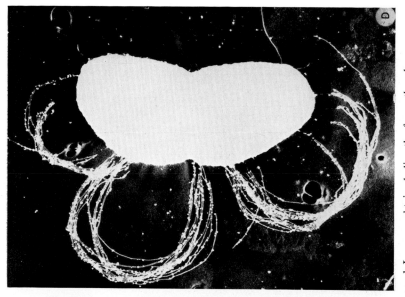

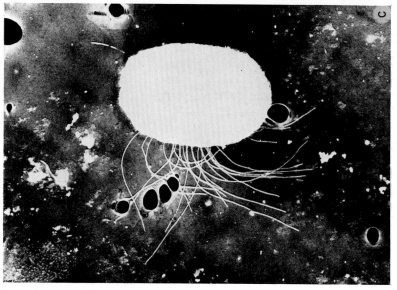

Fig. II-1. (*Cont'd.*) C. The oval form of Woodcock and Lapage, obtained directly from the sheep rumen. Fixed with osmium, washed in distilled water, uranium shadowed. Electron micrograph by J. Pangborn. Magnification: 6800 ×. D. Dividing oval, or possibly *Selenomonas*; same preparation as C. Electron micrograph by J. Pangborn. Magnification: 6800 ×.

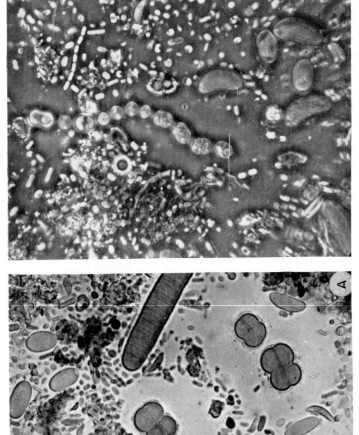

Fig. II-2. A. Large tetracocci resembling *Sarcina bakeri,* and one end of an *Oscillospira* showing transverse septa. Ultraviolet illumination. After Smiles and Dobson (1956). Magnification: 1830 ×. B. Chain of large cocci, selenomonads, and ovals. Ultraviolet illumination. After Smiles and Dobson (1956). Magnification: 1830 ×.

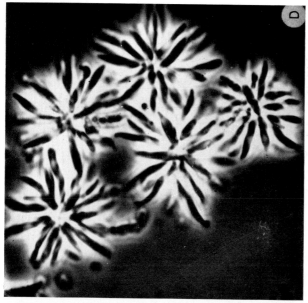

Fig. II-2. C. Two rosettes from the sheep rumen, with oscillospiras and oval forms. Ultraviolet-phase illumination. After Smiles and Dobson (1956). Magnification: 1830 ×. D. A rosette-forming bacterium cultured concurrently with *Ruminococcus albus* on rumen cellulose agar by R. A. Mah. From a culture of *Ophryoscolex purkynei*. Dark-phase contrast. Photograph by R. A. Mah. Magnification: 2200 ×.

rumen, as can be seen in the excellent plates of Smiles and Dobson (1956). Distinctive rosettes of cells (Figs. II-2C and D) are sometimes found in the rumen. These are shown in one of Fiorentini's early sketches (Fiorentini, 1889). Some cultured strains of *Bacteroides succinogenes* show a rosette arrangement (Bryant and Burkey, 1953a), but an identity with rumen rosettes such as those in Fig. II-2C has not been established.

A morphologically distinctive filamentous bacterium intermediate between *Thiothrix* and *Beggiatoa* has been reported as abundant in sheep on rations low in protein (Jamieson and Loftus, 1958). The organism has not been cultured.

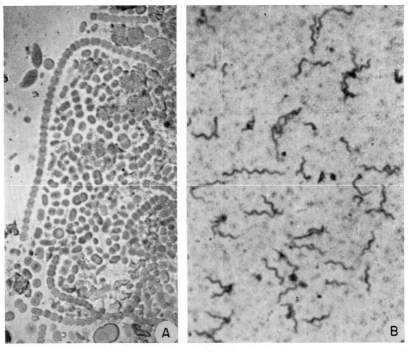

Fig. II-3. A. Chains of cocci from the sheep rumen, with numerous other small coccoid forms. Ultraviolet illumination. After Smiles and Dobson (1956). Magnification: 1830 ×. B. Pure culture of a spirochete from the bovine rumen. Stained smear. Isolated and photographed by M. P. Bryant. Magnification: ca. 2000 ×. C. Gram-stained chain of cocci observed in a direct microscopic examination of bovine rumen bacteria. Magnification: ca. 2000 ×. D. Various large and small bacteria from sheep rumen contents. Ultraviolet-phase illumination. After Smiles and Dobson (1956). Magnification: 1830 ×.

Morphologically distinctive spirochetes (Fig. II-3D) are usually seen in fresh rumen contents. The spiral forms resemble the *Borrelia* which has been isolated from the sheep rumen (Bryant, 1952) (Fig. II-3B). This species does not always occur in large numbers in rumen contents,

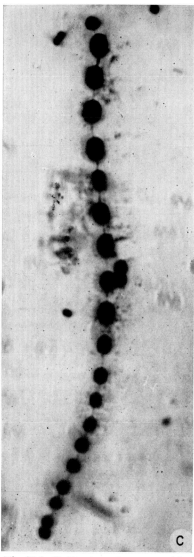

Fig. II-3. C. For legend see page 18.

but it is almost always present. It can penetrate through the agar and produces colonies with diffuse edges. These colonies expand through the agar, becoming less distinct as they spread. They may ultimately completely disappear if the agar is not too concentrated. Microscopic examination shows cells spread evenly throughout the agar.

Another distinctive rumen type is the chain of evenly spaced large cocci (Fig. II-3C and D).

Direct microscopic examination is valuable for checking the results of cultural studies. Some of the large microscopically identifiable forms have not been cultivated, which indicates that, although types performing most of the known rumen conversions have been isolated, additional organisms and reactions remain to be identified.

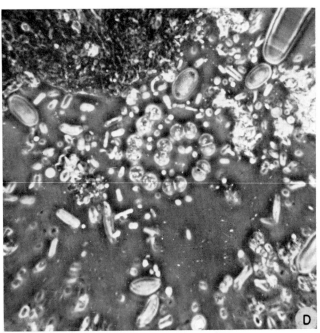

Fig. II-3. D. For legend see page 18.

D. Cultivation

Most rumen bacteria, so similar in size and shape as to be indistinguishable by microscopic examination, require special treatment, either direct, e.g., as with fluorescent antibodies, or indirect, through measurement of their characteristics after isolation as a pure culture. Purity of culture assures that the ascertained characteristics apply to a single strain.

Early attempts to isolate strains of rumen bacteria enjoyed less success than had been experienced in examinations of other habitats. The reason lay chiefly in failure to simulate rumen conditions. Culture procedures were borrowed from medical, dairy, or soil microbiology, without modifications to fit the rumen habitat. Though all the elements of successful techniques for cultivating strict anaerobes were discovered before 1900 (Hungate, 1950), they were not applied to the rumen until much later (Hungate, 1947).

Much of the early interest centered on the ability of rumen bacteria to decompose cellulose. The mass cultures of von Tappeiner (1884a) were the first attempt to grow rumen cellulolytic bacteria. Rumen contents were inoculated into meat-extract media containing cellulose, and, after a lag of a week or so, growth and cellulose digestion commenced. This long lag is not characteristic of mass cultures of rumen cellulolytic bacteria, and, as mentioned in the introductory chapter, the bacteria in von Tappeiner's cultures were probably not the ones predominant in the rumen. This does not invalidate his contribution since the conclusions regarding the nature of the microbial fermentation were correct and the products found, within the limitations of the analytical methods of that time, were the same as those subsequently identified.

Ankersmit (1905) counted cellulolytic bacteria of the rumen by employing the inorganic medium of Omeliansky plus cellulose. He found only low numbers, chiefly sporeformers differing microscopically from the bacteria seen in the rumen, and he concluded that the important rumen cellulolytic bacteria did not grow under the conditions employed. Pochon (1935) similarly obtained chiefly sporeforming cellulolytic bacteria from rumen contents. Lack of organic growth factors in the culture medium employed by these investigators could have been an important factor in their lack of success. Anaerobic sporeforming bacteria have subsequently been isolated in fairly large numbers from the bovine rumen (Hungate, 1957), but only from certain animals. The sporeformer *Bacillus circulans*, in the rumen to the extent of 10^2 per milliliter (Hobson and Purdom, 1959), was probably ingested with the feed.

The first use of rumen fluid in the culture medium was by Hoppfe (1919). She could obtain cellulose digestion by mixed rumen bacteria during several transfers and could show it in the absence of the protozoa, but continued subcultures failed to grow, presumably because the conditions in the tubes were aerobic. When carbon dioxide was used to obtain anaerobiosis, cellulose was rapidly digested, but pure cultures could not be obtained. She postulated that unknown bacteria were concerned with the cellulose digestion and that culture conditions must simulate the rumen more closely than had been accomplished.

Henneberg (1919, 1922), in Germany, abandoned attempts at cultivation and relied upon microscopic observation to detect bacteria attacking plant fibers. He employed Bredemann's practice of staining with iodine to disclose bacteria containing starch or other reserves giving a dark color (iodophily). Iodophilic bacteria occupying holes in plant cell walls were identified as cellulose digesters. These were probably similar to the iodophilic cocci observed (Kreipe, 1927) on lettuce leaves added to rumen contents.

Baker (1943; Baker and Harriss, 1947), in England, also used microscopic techniques and iodophily to identify the amylolytic and cellulolytic organisms in the rumen and cecum of various mammals, employing polarized light to detect attack on cellulose (Fig. II-8B). His results confirmed and extended Henneberg's postulate that iodophily in bacteria attached to plant fibers was evidence of cellulose attack. Modern culture results confirm the conclusions of Baker and of Henneberg that iodophilic cocci play an important role in rumen cellulose digestion, but have disclosed also important noniodophilic cellulolytic forms (Hungate, 1950), e.g., *Bacteroides succinogenes*.

Lack of success of early attempts to cultivate rumen cellulolytic bacteria stemmed from (1) failure to supply the important nutritional factors in rumen fluid, including carbon dioxide, *n*-valeric acid, and the C_4 and C_5 branched-chain fatty acids (Bryant and Doetsch, 1954a), (2) lack of sufficient anaerobiosis, (3) use of too high concentrations of cellulose, (4) use of types of cellulose not readily attacked, and (5) the confusion resulting from the rather general lack of success in obtaining pure cultures of anaerobic cellulolytic bacteria from any habitats, with a few exceptions (Khouvine, 1926).

Lorraine Gall (Gall *et al.*, 1947) made one of the first modern attempts to cultivate the rumen bacteria. Operating on the correct assumption that the inorganic culture media of Omeliansky did not supply the necessary nutrients, she used commercially prepared peptones and other extracts in a phosphate-buffered medium, obtained culture counts from liquid dilutions, and found very high numbers of bacteria per gram of rumen contents. Although the unusually high counts initially reported (as high as 10^{12}) have not been generally confirmed in later studies, she cultured a number of typical rumen bacteria and stimulated great interest in rumen microbiology among American students of ruminant nutrition. Culture counts were made on the rumen contents from animals on various feeds (Gall *et al.*, 1949a, 1949b, 1951; Gall and Huhtanen, 1951; Huhtanen *et al.*, 1949). Later, bicarbonate was added to the medium (Huhtanen *et al.*, 1952, 1954a). Groups of bacteria were distinguished, but were not characterized in detail (Huhtanen and Gall, 1953a,b). It is doubtful that many of the pure cultures were cellulolytic since the medium did not contain the branched- and straight-chain fatty acids subsequently shown to be required by the more important cellulolytic bacteria (Bryant and Doetsch, 1954b; Bryant and Robinson, 1961a).

Another method for pure culture of the rumen bacteria (Hungate, 1947, 1950) employed rumen fluid as one-third of the medium. Factors such as substrate, carbon dioxide-bicarbonate buffer, a reducing agent, and exclusion of oxygen were designed to duplicate *in vitro* the factors important in the rumen. When rumen fluid media were employed for initial isolation, twenty out of ninety strains from the rumen required rumen fluid in the medium (Wegner and Foster, 1960). The methods and media have been improved and extended (Bryant and Burkey, 1953a; Bryant and Robinson, 1961b; P. H. Smith and Hungate, 1958; Kistner, 1960; Bryant, 1963) and are now used rather widely. Some of the features will be discussed in the following section.

Sijpesteijn (1948) paid attention to the conditions in the rumen habitat by providing a culture medium with rumen fluid, bicarbonate buffer, and strictly anaerobic conditions. She succeeded in pure culturing one of the important cellulolytic cocci of the rumen and named it *Ruminococcus flavefaciens*.

Recently several strains of rumen bacteria (Hungate, 1963a; Hobson and Smith, 1963; Hobson, 1965a,b) have been grown in continuous cultures.

1. THE CULTURE MEDIUM

Two types of media can be employed to cultivate bacteria (Hungate, 1962). One simulates the particular niche an organism is postulated to occupy in nature. Such media were used successfully by Winogradsky and Beijerinck to isolate soil bacteria. They are designated selective, enrichment, or niche-simulating media. Ideally an enrichment medium should support only one kind of bacterium; in practice this cannot often be achieved, but the desired organism becomes relatively more abundant. Selective media give quick information on the bacteria for which they are designed, but cannot disclose the total population.

The other type is the habitat-simulating medium (Hungate, 1962). It is designed to grow as many kinds of bacteria as possible and to reveal bacteria occupying many niches. No single medium suitable to grow every important bacterium in the rumen has yet been developed, but some may approach this ideal.

a. Media Simulating the Total Habitat

In the habitat-simulating media (Hungate, 1960; Bryant and Burkey, 1953a,b; Bryant and Robinson, 1961b; Bryant, 1963) rumen fluid or its components are included to supply essential accessory nutrients, with various carbohydrates as energy source. The necessity for materials in rumen fluid has been tested by comparing culture counts from rumen contents inoculated into media with and without rumen fluid. The count with rumen fluid is nearly always significantly higher (King and Smith, 1955; Gilroy and Hueter, 1957; Hobson and Mann, 1961; McNeill *et al.*, 1954) than with other more common media (Heald *et al.*, 1953). Efficacy of other media used without a control rumen fluid medium has been difficult to assess (M. K. Wilson and Briggs 1954, 1955; S. M. Wilson, 1953) because of uncertainty whether similar bacteria were concerned. Such media have usually given much lower counts (Munch-Peterson and Boundy, 1963). Failure of an extensive comparison to show a significant difference in media with and without rumen fluid (Munch-Peterson and Boundy, 1964) may be due to inadequate technique, as pure cultures could not be obtained. Also, the statistical comparison apparently did not take account of the dilution showing growth.

It might seem that a medium of undiluted rumen fluid plus added carbohydrate should support growth of the most bacteria. Though rumen fluid contains the nutrients it also contains inhibitory microbial wastes. In a favorable medium the concentration of wastes is reduced, but nutrients are not diluted to a limiting level.

Rumen contents can be obtained from abbatoir animals, through a fistula, or by stomach tube. The fluid can be separated by straining rumen contents through cheese cloth lined with cotton, or, if fermentation gas has floated solids to the top, the underlying fluid can be pipetted out with little disturbance of the surface mat.

The rumen bacteria are difficult to separate by filtration. Centrifugation of rumen fluid for an hour at a relative centrifugal force of 18,000 g is needed to sediment them, and even at this force the upper portion of the sediment is not well packed. Bacteria are not particularly troublesome in the medium, except that their dead bodies may be a handicap when the culture is examined microscopically.

Sterile rumen fluid can be obtained by filtration through millipore filters. The fluid is first centrifuged to remove bacteria, then equilibrated with oxygen-free hydrogen gas to remove carbon dioxide and raise the pH. The increased alkalinity causes a precipitate to form. This is centrifuged off under hydrogen. The supernate is drawn by vacuum through a millipore filter of 0.45-μ porosity. To insure complete sterility a second filtration has proven advisable. For culture of rumen holotrichs, filtered rumen fluid possesses some nutritional superiority over the autoclaved material.

Inorganic salts are added to provide balanced nutrients and to make the total mixture isosmotic with rumen fluid. The exact percentages of the various inorganic salts in the medium are not critical. Most formulas for a balanced mineral solution meet the requirements, provided enough bicarbonate is included. The medium used originally (Hungate, 1947) consisted of: 2 parts rumen fluid; 1 part salt solution containing 0.3% KH_2PO_4, 0.6% NaCl, 0.3% $(NH_4)_2SO_4$, 0.03% $CaCl_2$, and 0.03% $MgSO_4$; 1 part 0.3% KH_2PO_4; and 2 parts of distilled water containing the energy-yielding nutrients. Resazurin (purple) to a final concentration of 0.0001% was included later to detect oxidation of the medium. At pH 6.7 the E_0 of resorufin, the pink reduction product of resazurin, is −0.042 (Twigg, 1945).

Resorufin reduces to a colorless compound which readily reoxidizes to the pink resorufin, but the oxidation does not proceed back to resazurin. Sterile sodium bicarbonate (0.5% final concentration) and a reducing agent are added prior to inoculation. The medium is equilibrated with carbon dioxide or a carbon dioxide-hydrogen mixture (Bryant and Robinson, 1961b), and the gas is used throughout the procedure to exclude oxygen. The carbon dioxide-bicarbonate buffer provides an initial pH of about 6.7. The concentration of bicarbonate is slightly higher than that in the rumen. Rumen gas ordinarily contains about 70% carbon dioxide and 20% methane (Lungwitz, 1891; Reiset, 1868b).

The energy-containing substrates to be included in the habitat-simulating medium must replace the fermentable components of the plant material ingested by the host. Some of these are soluble carbohydrates and can be supplied as sugars or plant extracts. Others such as cellulose and hemicellulose usually are insoluble and do not readily lend themselves to the preparation of culture media, but organisms attacking them can grow on their constituent saccharides. Many feed constituents are chemically or physically combined within the insoluble plant material and not easily separable into pure components which can be used as substrate in studying the organisms that attack them. Success in culturing all types may depend on provision of all substrates normally present in the rumen.

As detailed knowledge of the nutritional requirements of individual types of rumen bacteria accumulates (Bryant and Robinson, 1962b) and nutrients are identified (Bryant and Doetsch, 1954a; Caldwell *et al.,* 1962; Bentley *et al.,* 1954a, 1955; MacLeod and Murray, 1956; Wegner and Foster, 1960), media of known composition can be supplied (Bryant, 1963). It should ultimately be possible to replace the rumen fluid with chemically defined nutrients. This will eliminate the variability inherent in the use of rumen fluid and permit more exact comparisons by investigators in different laboratories. The exact purposes of the study will determine the advisability of replacing rumen fluid by defined ingredients. If it is desired to detect as many kinds of bacteria as possible, a rumen fluid medium should be used as a check on the nutritional adequacy of other media. Its use is particularly applicable when new organisms are sought.

An advantage of a habitat-simulating medium is that, in addition to increasing the chances of success in isolation, it assures that the growth requirements, ultimately identified, will be ecologically significant.

b. Media Simulating Individual Niches

These media (Kistner, 1960; Hamlin and Hungate, 1956; Hungate, 1947) are employed for isolation of restricted groups of bacteria and possess the advantage that numerous undesired types are eliminated at the stage of the initial culture. Success depends on the specificity for the desired niche. Highly selective media cannot be designed for all niches, particularly during initial studies. Use of rumen fluid diminishes the selectivity of the medium. For this reason, as required nutrients of rumen species become known it will be advantageous to substitute them for rumen fluid.

Most niche-simulating media used to grow rumen bacteria depend on selection by the energy-containing food. The following have been used: cellulose (Hungate, 1947), starch (Hamlin and Hungate, 1956), lactate (Elsden and Lewis, 1953; Kistner et al., 1962), hydrogen and carbon dioxide (P. H. Smith and Hungate, 1958), casein (Blackburn and Hobson, 1960b), saponin (Gutierrez et al., 1958), glycerol (Hobson and Mann 1961), and xylan (Hobson and Purdom, 1961). Wolin et al. (1959) have used ammonia as the sole nitrogen source in selecting nonfastidious strains of Streptococcus bovis, and urea (Gibbons and McCarthy, 1957) and casein (Blackburn and Hobson, 1960b) have been used to select bacteria hydrolyzing these materials.

2. TECHNIQUE FOR ANAEROBIOSIS

a. Preparation of the Medium

In order to maintain conditions as anaerobic as possible during preparation of the medium, a stream of the oxygen-free gas, usually carbon dioxide, is passed into the container, which should be a round-bottom boiling flask capable of resisting autoclave pressures if the total medium is sterilized in it. Ordinarily it is most convenient to dispense the medium into culture tubes before sterilization. This is again performed in a manner avoiding exposure of the medium to air, by passing oxygen-free gas into the large container of medium and a second stream into the culture tube to be filled.

It is undesirable to have too great a flow of gas into the culture tube since a rapid stream sets up air currents at the tip of the capillary. A flow of about 4 ml of gas per second is usually satisfactory. The gas is used to displace oxygen from the gas phase, not the liquid. The latter is freed of oxygen by boiling the flask of medium while the oxygen-free gas is being passed into the container.

A 3-inch No. 19-gauge hypodermic needle bent so that about $2\frac{1}{2}$ inches extend down into the container, the sleeve portion hanging over the side, has been found useful for introducing the gas. Metal needles have the advantage over glass of stability against mechanical shocks and rapid heating and cooling. The needle can be cotton-plugged and connected to the carbon dioxide source with rubber tubing or attached to a sterile 1-ml syringe barrel plugged with cotton and with the flange at the open end cut off to permit insertion into the rubber tube delivering the gas. Mixtures of nitrogen, hydrogen, and carbon dioxide may be employed. The best method for freeing the gas of oxygen is to pass it over hot copper turnings or wire heated to 350°C and reduced with hydrogen. Care must be taken to displace all oxygen gas from the copper container before hydrogen is passed through, to avoid explosions. Escape of excessive hydrogen into the laboratory should be avoided by working under a hood or by igniting the gas issuing from the needle when it is not being used to flush.

Straight-sided culture tubes (16 $\times$ 150 mm) taking a No. 0 rubber stopper have been used. On insertion, the stopper is placed loosely in the open end with the gassing needle extending down along one side. As the needle is withdrawn the stopper is slipped into place and then seated firmly with a twisting motion. Pushing the stopper should be avoided as it increases the severity of cuts if the glass breaks. It is a good idea to tilt the tube so that the medium moistens the area around the stopper and thus facilitates firm insertion. The most common error is to allow air to enter the tube as the gassing needle is withdrawn and the stopper inserted.

For the cultivation of some rumen bacteria the extreme gassing precautions are unnecessary. It is sufficient to use media exposed to air and add a reducing agent to each culture tube just before it is inoculated, gassed, and stoppered. Sometimes the gassing of the tubes while they are open can be omitted for a short interval during transfer, provided the cultures are gassed well immediately after the transfer and then stoppered. Because a bacterium grows in an initial dilution series in which no great pains are taken to exclude oxygen, it should not be concluded that similar methods will be adequate for growing it in pure culture. Fastidious strains may grow in the mixed culture because the medium is reduced by accompanying bacteria less sensitive to oxygen. For pure cultivation of such rumen bacteria the oxygen in the culture tube must be removed completely; cysteine, sodium sulfide, or some other reducing reagent must be used.

The methanogenic rumen bacteria have been most difficult to grow. The key to their cultivation has been the use of rumen fluid in the medium, hydrogen and carbon dioxide as substrate, and redox potential as low as or lower than -0.35 volts to initiate growth (P. H. Smith and Hungate, 1958). The isolation of the methane bacteria was of particular interest because methane was an important product of the rumen fermentation for which previously no causal organism could be found.

It has been found recently that various types of rubber differ in their permeability to oxygen. Success with the stoppered roll-tube method has been achieved almost exclusively with the black-rubber stoppers routinely supplied in the United States. The red-rubber stoppers commonly used in the United Kingdom permit sufficient diffusion of oxygen into culture tubes to inhibit growth (Hungate, 1963a). They also do not protect stored reducing agents against oxidation. This factor may explain some of the difficulties of investigators attempting to use the roll-tube method.

Butyl rubber stoppers have been found recently to be less permeable to oxygen, hydrogen, and carbon dioxide than are the common black stoppers. A sterile hypodermic needle (1-inch, 21-gauge) attached to a 1-ml Luer-Lok syringe is inserted through the recessed stoppers (A. H. Thomas catalog No. 8820-B) to add bicarbonate, cysteine, sulfide, sugars, and growth factors to sterile tubes without risking contamination by opening them. This operation is quicker for many purposes than is transfer with a pipette, and it is especially suited to quantitative dilution procedures or routine transfers of liquid cultures. Narrow neck tubes and #00 stoppers are used.

For autoclaving, the stoppered tubes—containing usually 4.5 ml of medium minus the volume of the solutions of sugar, bicarbonate, and reducing agent to be added later—are arranged in a flat-bottom test tube rack, and the rack is then placed in a press. A lid covers the stoppers and prevents their blowing out when the pressure drops during cooling. The covered racks supplied by Labtool Specialties Co. of Ypsilanti, Michigan, are quite useful for holding the tubes during sterilization. After sterilization, tubes of medium can be stored with relatively little oxidation until used, provided the stoppers are impermeable to oxygen. It is advantageous to cover them to avoid contamination on the stoppers.

Before use, the tubes of medium are placed in a shallow boiling water bath to melt the agar and then kept in a 45°C-water bath until inoculated. If sugar is to be included in the medium, it is added as a 5 or 10% solution, usually to give a final concentration of 0.025 to 0.2%. Low concentrations of sugar permit more colonies in a tube to be counted (Bryant and Robinson, 1961b). They do not spread as much. Cysteine, sodium sulfide, and sodium bicarbonate are added as 0.35 ml of a sterile solution prepared by mixing 5 ml of 10% bicarbonate with 1 ml of 3% cysteine and 1 ml of 3% $Na_2S \cdot 9H_2O$. While the tubes are open, they are kept anaerobic by passing gas through them.

When carbon dioxide is used, the final bicarbonate concentration should be 0.5% to give a pH of about 6.7. Bicarbonate concentration can be reduced proportionally if the carbon dioxide is diluted. In tests for acid production from sugars it is advisable to decrease the bicarbonate to 0.1% and use 5% carbon dioxide in nitrogen to reduce the buffering capacity of the medium. Solutions of sodium sulfide must be stored under hydrogen or nitrogen gas. Carbon dioxide is a stronger acid than hydrogen sulfide and produces hydrogen sulfide from sodium sulfide. The hydrogen sulfide may be lost when the tube is opened.

b. Inoculation and Transfer

The inoculum is added from a pipette or hypodermic syringe, and the tubes are tilted to mix thoroughly. If the tubes are shaken, small air bubbles form and may be included in the medium when it solidifies; this increases the difficulty in distinguishing colonies. The inoculated tubes are rotated and tipped in ice water until the agar has gelled. It is best to spread the medium evenly on the wall of the tube when accurate counts are desired. For other purposes it is sometimes desirable to leave a portion of the agar as a small slant in the butt of the tube. The agar layer over the remaining surface is thinner and, in the case of cellulose agar, small cellulolytic colonies can be more easily detected than in thick layers. Occasionally when conditions are not sufficiently anaerobic, colonies of bacteria develop in the slant, but not in the thin agar. A tube roller is commercially available from McBee Laboratories, Bozeman, Montana.

After solidification of the agar, the cultures are incubated at 39–40°C.

One of the advantages of anaerobic cultures prepared as above is that no special incubators are required. The tubes can be removed for examination at any time. Various gas mixtures can be used, either as substrate, buffer, or as inert filler to exclude air. When suitable stoppers are used, the tubes remain anaerobic for very long periods, as judged by continued reduction of resazurin when that indicator is included in the medium. Most rumen bacteria can be grown by these methods, whereas with techniques employing petri plates and anaerobic jars the more anaerobic rumen bacteria have not been grown.

Special precautions are required with bacteria that produce much fermentation gas. The gas may blow out the stopper if too much substrate is provided. Less medium or a smaller percentage of substrate avoids this.

Pathogenic bacteria are more of a problem, and the author has not attempted to adapt the procedure to their growth. The fact that the medium covers the inside of the stopper makes contamination of the surroundings possible when the stopper is withdrawn. This disadvantage can be overcome by using tubes with a constricted neck and rotating horizontally, or by rotating the inoculated melted agar in the ice water on a slant that deposits the thin agar layer over the lower two-thirds of the tube without contacting the stopper.

When it is desired to preserve anaerobiosis in an opened culture tube, a bent capillary carrying oxygen-free gas is inserted into the mouth of the tube and the stopper laid upside down on a clean surface. The desired colony is withdrawn with a microspatula (the flattened end of a platinum-iridium inoculating needle) or a bent drawn-out sterile glass capillary (Pasteur pipette) attached to a rubber mouth tube. The glass capillary permits cleaner picking of a colony with less contamination by adjacent material than is possible with the microspatula, but it is slower and for many purposes is not necessary. It has been found inadvisable to use Nichrome inoculating needles or microspatulas because they oxidize the medium when inserted immediately after flame sterilization. The glass capillary used for picking colonies is bent at a right angle about 7 inches from the end attached to the rubber tubing. It is scratched with an edge of broken porcelain about $\frac{1}{4}$-inch distal from the bend and usually breaks squarely across at the scratch. The capillary is flamed lightly and inserted into the culture tube without touching the agar; the tip is then placed over the desired colony and forced into the agar as gentle mouth suction is applied to the rubber tube to draw the colony into the capillary. It is often advisable to work under the dissecting microscope when small colonies are to be picked.

E. Total Numbers of Rumen Bacteria

1. CULTURE COUNTS

Appropriate dilution of rumen contents and inoculation of known quantities into culture media yield a value called the culture count. It is not necessarily identical with the number of cells in the rumen contents, in fact it is usually considerably less, but if counts on different samples are made with the same method the results can disclose relative differences in numbers.

If culture counts of the total bacterial population are desired, a habitat-simulating medium is employed. The cells in a particular niche are estimated with a selective medium. Media may be liquid or solid. For most purposes solid media are preferable because a single culture can yield a statistically more significant estimate than can be obtained with a liquid culture, but there are exceptions, e.g., with some cellulolytic bacteria which cannot digest cellulose dispersed in an agar medium.

During dilution of rumen contents for culture counts, it is necessary to observe the precautions employed in culturing. Many of the rumen bacteria can be killed by oxygen (Giesecke, 1960a) and by exposure to unfavorable pH or temperature.

There is an interesting inverse relationship between the concentration (i.e., numbers per milliliter) of bacteria and the fermentation rate in a given animal fed on hay (Walter, 1952; Bryant and Robinson, 1961b) (Fig. II-4). The culture count per milliliter is higher just before feeding

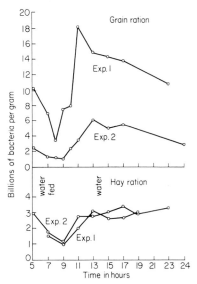

Fig. II-4. Total bacterial colony counts from samples of ruminal contents collected at various times of day from a cow fed alfalfa hay or grain once a day. From Bryant and Robinson (1961b).

than several hours after feeding (McAnally, 1943), but the fermentation rate is low due to decreased available substrate. During feeding the count of bacteria per milliliter diminishes because of dilution by feed and saliva (Kistner *et al.,* 1962), but the fermentation rate is high because of the increased substrate. The count per milliliter ultimately increases after feeding (Helwig, 1960).

A micromethod for cultural enumeration of rumen bacteria has been described (Claypool *et al.,* 1961).

2. DIRECT COUNTS

This is the most reliable method for determining the total number of bacteria in the rumen. It is also the easiest and quickest. It measures dead cells as well as living, and any unknown types which cannot be grown. The weakness of the direct count is that it may not reflect accurately the bacteria attached to particles of digesta.

Because rumen bacteria are so numerous, the direct count must be made on a dilution of the rumen contents. Cells in the dilution may be killed with formalin and stained with a dilute solution of gentian violet to render them more easily visible. The actual counting can be done with various counting chambers designed for enumeration of bacteria (van der Wath, 1948b; Warner, 1962a), or, less precisely, a known quantity of the dilution can be spread over a measured area of a slide and dried, fixed, and stained by any chosen technique (Hungate, 1957). The bacteria in two strips at right angles across the smear are counted. Quantitation is achieved by knowing the diameter of the microscope field, the volume of the diluted material counted, and the area over which it is spread. Nigrosine staining has been used to make the bacteria more easily visible (Moir, 1951).

There is occasionally some uncertainty whether or not stained particles are bacteria, but the great majority can be distinguished with confidence from debris (Warner, 1962a). Bacteria far outnumber other particles.

Use of fluorescent-labeled specific antisera permits counts of serologically specific types (Hobson *et al.,* 1955, 1958; Margherita *et al.,* 1964; Hobson and Mann, 1957), and success with the Neufeld technique has been reported (Macpherson, 1953).

Microscopic examination is the only direct method for determining the numbers of cells attached to the rumen solids. Extensive qualitative studies of the particles were made by Baker. The method (Baker, 1945) has been used to detect morphological types of bacteria attacking cellulose, digestion of the latter being disclosed by the decrease in birefringence.

An estimate of the relative numbers of bacteria in rumen solids as compared to liquid can be obtained by measuring the fermentation rate when excess substrate is added (McAnally, 1943; Coop, 1949; Hungate, 1956a). With high concentration of substrate the cells ferment at their maximum rate, and the total rate can be used as a rough measure of the total number of cells. For many purposes this provides a more practicable method than direct or culture counts. It has been used to estimate the relative numbers of microbial cells in rumen liquid and rumen solids, with results as shown in Fig. II-5. The solids show a much greater activity than the liquid, which indicates that more microorganisms are associated with it.

The direct counts obtained by a number of investigators are shown in Table II-2, and for comparison some culture counts are given.

An especially careful study of problems in the direct counting of rumen microorganisms has been completed by Warner (1962a). Bacteria freed from particles of solids by blending were more than twice as numerous as the bacteria in the rumen fluid, which agrees with results of the relative rate of fermentation of liquids and solids with excess substrate.

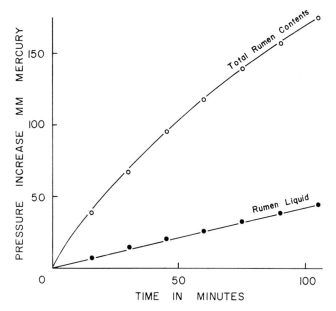

Fig. II-5. Relative rates of fermentation by rumen liquid and total rumen contents in the presence of added alfalfa hay. Ten grams of rumen material, 20 ml of balanced salt solution, and one gm of ground alfalfa hay in each vessel. Temperature: 39°C. Vessel volume: 110 ml.

The total count, including both liquid and solids, appeared to bear a fairly constant relationship to the count on the fluid alone, but errors of as much as 150% were encountered with spiral forms and gram-positive rods.

These methods have been used to complete excellent studies on the incidence of various rumen species at various times of day and under differing feeding regimes (Warner, 1962b). Much information can be ob-

Table II-2

NUMBERS OF BACTERIA IN THE RUMEN

Animal	Ration	Direct microscopic count $\times 10^{-9}$	Culture count $\times 10^{-9}$	Reference
Sheep	Test rations for effects of cobalt	30–56	—	Gall et al. (1949b)
Sheep	Chaffed oaten hay, casein, and starch to give:			
	24 gm N	29	—	Moir and Williams (1950)
	62 gm N	40	—	Moir and Williams (1950)
	82 gm N	55	—	Moir and Williams (1950)
	100 gm N	55	—	Moir and Williams (1950)
	122 gm N	61	—	Moir and Williams (1950)
Sheep	Pasture at various seasons	32–88	—	Moir (1951)
Sheep	Chaff + various concentrates	18–44	—	Williams and Moir (1951)
Cattle	Hay, straw, and concentrates in several combinations	14–25	1.3–6.5	Bryant and Burkey (1953b)
Sheep	Hay, protein, and starch in various combinations	27–58	—	Williams et al. (1953)
Year-old steers	Hay + 4 lb grain	—	1.01–1.35	King and Smith (1955)
Pregnant range cows	2/3 timothy hay, 1/3 concentrates	8–47	0.008–0.6	Hungate (1957)
	8/9 timothy hay, 1/9 concentrates	7–29	0.002–0.2	Hungate (1957)
Ewes, Rahmani	Chopped wheat straw + 65 gm various protein equivalents	4–61	—	Shehata (1958)
Steers	Feed-lot bloat ration	—	0.2–10	Gutierrez et al. (1959)
Calves, no protozoa	Hay-grain mixture	—	13.6	Bryant and Small (1960)
Calves, protozoa	Same hay-grain mixture	—	6.0	Bryant and Small (1960)
Steers	Lush ladino clover pasture	—	5.8–8.0	Bryant et al. (1960)
Steers	Feed-lot bloat ration	—	1.6–23	Bryant et al. (1961)
Cow	4/5 alfalfa hay, 1/5 concentrates	—	4.0–12.0	Bryant and Robinson (1961b)
Sheep	Pasture	88	—	Warner (1962a)

Sheep	400 gm lucerne chaff + 400 gm wheaten chaff once daily	11.8-24.5	—	Warner (1962b)
Sheep	1 kg mixed lucerne hay	—	0.07-0.84	Kistner et al. (1962)
	1 kg lucerne hay stalks	—	0.06-0.3	Kistner et al. (1962)
	1 kg lucerne hay leaves	—	0.13-0.97	Kistner et al. (1962)
Mature sheep	Hay and gluten, before feeding	24	—	Nottle (1956)
	Hay and gluten, 3 hr after feeding	41	—	Nottle (1956)
	Hay and casein, before feeding	49	—	Nottle (1956)
	Hay and casein, after feeding	38	—	Nottle (1956)
Sheep	Oaten and lucerne chaff + 20% sheep cubes fed in various fashions:			
	1 daily feeding	32	—	Moir and Somers (1957)
	2 daily feedings	28	—	Moir and Somers (1957)
	4 daily feedings	32	—	Moir and Somers (1957)
	chaff A.M., concentrates P.M.	33	—	Moir and Somers (1957)
	concentrates A.M., chaff P.M.	37	—	Moir and Somers (1957)
Cows	Alfalfa hay *ad libitum*	10.3-12.4	0.33-0.9	Maki and Foster (1957)
	15 lb grain + alfalfa hay	14.8	1.8	Maki and Foster (1957)
	12 lb grain, 25 lb alfalfa meal	22.2	12.7	Maki and Foster (1957)
	15 lb grain, 27 lb alfalfa meal	28.2	20.0	Maki and Foster (1957)
	25 lb grain, 9 lb alfalfa meal	31.4	22.9	Maki and Foster (1957)

tained, and in less time, by these methods than with pure culture techniques, but they cannot be applied to bacteria with no distinctive morphology.

F. The Kinds of Rumen Bacteria

By means of the enrichment and habitat-simulating techniques a great many rumen bacteria have been isolated. They can be grouped according to the type of substrate presumably attacked in the rumen, though this assignment is tentative and in many cases unsupported by experimental evidence beyond demonstration that the organism can attack the substrate in pure culture. This organization is followed, at the risk of erroneous interpretations, in order to emphasize the ecological problems concerned in rumen microbiology and the need to ascertain the food actually used in the rumen.

1. CELLULOSE DIGESTERS

a. Cultivation

Because of the long history of unsuccessful attempts to cultivate anaerobic cellulolytic bacteria it is important to take every precaution to insure that strains cited as pure are actually so. Since the suspended cellulose in agar cultures makes it difficult to detect contaminating colonies, it is advisable, after several subcultures made by picking a cellulolytic colony to a cellulose-agar dilution series, to subculture in a cellobiose medium in which all colonies can be readily detected. If they all appear alike and further subculture of a colony in high dilution into cellobiose medium again shows colonies all of the same sort, and if these are cellulolytic when inoculated back into cellulose medium, it can reasonably be concluded that the culture is pure, provided no evidence to the contrary is observed.

Cellulolytic bacteria can be picked from primary dilutions in sugar media with less chance of contamination (Bryant and Burkey, 1953a), but a number of colonies must be picked and tested on liquid cellulose medium to identify those attacking cellulose.

If a cellulolytic bacterium grows only in association with a contaminant, mixed colonies would appear in cellulose agar and they would all be alike. This condition can be detected, even though the morphology of the two organisms is identical, by determining the numbers of cellulolytic colonies in serial dilution tubes. If a single type is responsible for the colony, the numbers of colonies will decrease in proportion to the dilution, whereas if more than one is concerned the cultivated numbers diminish much more rapidly than expected from the dilution.

A cellulose suspension suitable for agar cultures to grow certain kinds of rumen cellulolytic bacteria can be prepared by wet grinding 3% (w/v) filter paper or absorbent cotton cellulose in a pebble mill. The required amount of grinding varies with the particular mill and the loading. Grinding should continue until a fine state of division is reached and the suspension has become viscous, but should be stopped before viscosity decreases and copper-reducing substances appear. Flint pebbles are better than porcelain balls. With the latter, fine bits of porcelain break off and the suspension is milky even in the absence of cellulose. Cellulose prepared with porcelain-ball grinding has not supported growth of cellulose decomposers which grew with pebble-milled cellulose.

Several weeks of incubation may be required for development in higher dilution tubes of zones of cellulose digestion visible to the naked eye. Unless stoppers impermeable to oxygen are used, the cultures may not remain sufficiently anaerobic during this time to permit colony development.

b. Rod-Shaped Cellulolytic Bacteria

Three kinds of rod-shaped cellulolytic bacteria can be abundant in the rumen and not in the feed. They are *Bacteroides succinogenes, Butyrivibrio fibrisolvens,* and *Clostridium lochheadii.* In addition *Clostridium longisporum, Cillobacterium cellulosolvens,* and an acetigenic rod have been found more rarely.

(1) *Bacteroides succinogenes. Bacteroides succinogenes* was the first important cellulolytic rumen bacterium to be isolated in pure culture (Hungate, 1947, 1950). It has been found in sheep, cattle, and several wild African antelopes [eland, kongoni, and Grant's gazelle (Hungate *et al.,* 1959)]. The greatest recorded numbers were in the animal from which pure cultures were first obtained, a steer killed in the abbatoir at Austin, Texas. The culture count was in excess of one billion per milliliter. Since this figure has never been equaled in any subsequent counts on rumen cellulose agar the author would be inclined to ascribe it to error, except that the colonies were relatively free of contaminating bacteria—a situation not usually encountered when the primary culture medium is cellulose.

On a nonselective glucose-cellobiose rumen fluid medium the culture count of *B. succinogenes* was 0.5×10^8 per milliliter, about 8.6% of the total culture count (Bryant and Burkey, 1953b). Straw and hay were important parts of the ration of the cattle studied.

A difference between isolated strains has been noted. The original strains (Hungate, 1947) were isolated from rumen fluid cellulose-agar dilution tubes in which zones of cellulose digestion appeared. Later isolates from rumen fluid sugar agar (Bryant and Burkey, 1953b) showed cellulose digestion when inoculated into a liquid rumen fluid cellulose medium, but not in cellulose agar. This difference may be due to an inability of some strains to move through agar.

Bacteroides succinogenes is a gram-negative rod, 0.3–0.4 μ in diameter and 1–2 μ long when growing in cellulose agar. In cellobiose rumen fluid agar the cells are somewhat larger, crescentic in shape, and often show more intense staining at the poles of the cells (Hungate, 1950), or they may be lemon-shaped. Old cells swell and lyse readily.

Growth in cellulose agar is distinctive. There is no discrete colony. The cells migrate through the agar and are most abundant at the periphery of the clearing, in close proximity to the undigested cellulose (Hungate, 1950). In the center of the clearing are weakly staining oval cells which appear to be degeneration stages of the vegetative cells. In old colonies only these ghost cells are seen. The central portion of old colonies gives no growth when inoculated into fresh medium, which confirms that the spherical forms are nonviable.

The fact that *Bacteroides succinogenes* is able to migrate through agar has suggested an affinity with the myxobacteria (Hungate, 1950). This is born out by the similarity in the pattern of fermentation products of *Cytophaga succinicans* (R. L. Anderson and Ordal, 1961) and *Cytophaga fermentans* (Bachmann, 1955), except that the latter decarboxylates succinic acid to propionate.

When subcultures of *Bacteroides succinogenes* are made from the primary isolation cellulose-agar tubes, they are often contaminated with the spirochete *Borrelia* sp. (Bryant, 1952). Both organisms are able to move through the agar. The *Borrelia* accumulates in the same region as *Bacteroides succinogenes* since it is there that the concentration of sugar from cellulose digestion is greatest. It has not been possible to obtain pure cultures of *Bacteroides succinogenes* from such a mixed culture by subcultivating in cellulose agar. The *Borrelia* can move through the agar and invade the *Bacteroides succinogenes* colonies by the time the latter can be detected. Primary isolation on sugar media usually avoids this problem.

The cellulose in autoclaved hay is digested by *B. succinogenes* to almost the extent that hay is digested in the rumen (Table II-3).

Bacteroides succinogenes strains ferment glucose and cellobiose in addition to cellulose. Pectin (Bryant and Doetsch, 1954b) and starch (Hungate, 1950) are fermented by some strains, but xylans are not (Dehority, 1965). One or several of a few other carbohydrates such as maltose, starch, trehalose, and lactose may be fermented. Growth on sugar is rapid, colonies appearing in 24 hours.

A requirement of *B. succinogenes* for mixed branched- and straight-chain fatty acids was discovered by Bryant (Bryant and Doetsch, 1954a).

The straight-chain acid can be from 5 to 8 carbons in length, or 14, 15, or 16 (Wegner and Foster, 1963), and the branched-chain can be iso-butyric or 2-methylbutyric acid. These acids occur in small concentrations in rumen fluid (Table VII-6) and account in part for the requirement for rumen fluid. They have been shown by el-Shazly (1952a) to result from fermentation of certain amino acids.

Other nutritional requirements (Bryant *et al.*, 1959) include biotin (Scott and Dehority, 1965), *p*-aminobenzoic acid, ammonia, and sulfide or cysteine. The ammonia requirement is strict, being expressed even in the presence of numerous amino acids (Bryant and Robinson, 1961a). The nutritional requirements indicate an ability to synthesize most of the monomers used in growth.

Table II-3

DIGESTIBILITY OF THE WATER-INSOLUBLE FRACTION OF HAY BY
Bacteroides succinogenes[a]

	Alfalfa hay		Poor grass hay	
Strain	Percent digestibility of insoluble fraction	Percent total digestibility and solubility	Percent digestibility of insoluble fraction	Percent total digestibility and solubility
8/5–52–3	23	54	35	54
68d	32	60	25	50
7/5–67–3	24	55	38	58

[a] 40 mg of sterilized ground hay in rumen fluid medium was inoculated with a pure culture. After growth, the hay particles were allowed to settle, and the supernate with suspended bacteria was drawn off. The particles were washed by resuspension in water and sedimentation. Similar procedures with uninoculated controls gave the weight of water-insoluble material. This was 24.2 mg for the poor grass hay and 21.6 mg for the alfalfa hay. Initial samples were air dried. Oven-dry weight was approximately 36 mg.

Sugar does not accumulate in the medium of old cultures in which *B. succinogenes* has ceased to grow. This again is reminiscent of the myxo-bacteria; *Cytophaga* produces only a very small quantity of reducing sugars (Fahraeus, 1944). The fermentation products are chiefly acetic and succinic acids, with variable amounts of formic acid (Hungate, 1950; Bryant and Doetsch, 1954b). Carbon dioxide is fixed in the fermentation, presumably as the carboxyl group in succinate, and strains will not grow unless carbon dioxide is provided in the medium.

The ratios in which fermentation products may be formed by *B. succinogenes* are shown in Table II-4.

Succinic acid is the most important product of cellulose fermentation by *B. succinogenes,* 70–80% of the recovered carbon being found in this material. The carbon recovered accounts for (Table II-4) about 65% of the cellulose used. Similar low carbon recoveries, usually accounting for about 70% of the carbon in the substrate fermented, have been encountered in most analyses of the fermentation products of the cellulolytic bacteria isolated from the rumen and other sources. The cells formed may account for part of the missing carbon (Hungate, 1963a).

(2) *Butyrivibrio fibrisolvens. Butyrivibrio fibrisolvens,* originally described as the less actively cellulolytic rod (Hungate, 1950), was later studied thoroughly and named by Bryant and Small (1956a). Cellulose around colonies of this organism is not always cleared to the same extent

Table II-4

PRODUCTS OF CELLULOSE FERMENTATION BY *Bacteroides succinogenes*[a]

Experi- ment	Cellulose fermented (as glucose)	Formic acid	Acetic acid	Succinic acid	CO_2 utilized	Reference
1	0.554	—	0.28	0.458	0.125	Hungate (1950)
2	0.531	—	0.208	0.424	0.148	Hungate (1950)
3	1.72	0.23	0.86	0.89	0.69	Bryant and Doetsch (1954b)
4	1.78	0.21	0.61	1.26	0.83	Bryant and Doetsch (1954b)
5	2.73	0.28	1.00	2.46	—	Bryant and Doetsch (1954b)

[a] All values in mmoles.

as around colonies of *Bacteroides succinogenes* and some of the cellulolytic cocci. Some strains show rapid and complete cellulose disgestion, but for many strains it is rather slight. In agreement with this, α-celluose and hemicellulose, fermented *in vitro* by mixed rumen bacteria and uniformly labeled with carbon-14, yield a relatively low percentage of label in butyric acid as compared to acetic and propionic (Bath and Head, 1961). The majority of *Butyrivibrio* isolates obtained on a nonselective sugar medium are noncellulolytic, with other characteristics similar to those described for the cellulolytic types (Bryant and Small, 1956a). The capacity for cellulose digestion can vary with time in a single strain (Bryant and Small, 1956a).

It is doubtful that butyrivibrios actively decompose cellulose in the rumen. Other forage constituents such as soluble carbohydrates are probably the chief substrate. However, the most abundant cellulolytic bacteria isolated on cellulose rumen fluid agar from starving zebu cattle on very poor forage at Archer's Post, Kenya, were representatives of *Butyrivibrio* (Margherita and Hungate, 1963). Some of them were quite actively cellulolytic, whereas others were only slightly so. Also in sheep fed poor quality teff hay, cellulolytic butyrivibrios predominate (Gouws and Kistner, 1965).

The fiber-digesting ability of pure cultures of cellulolytic butyrivibrios, tested by incubating them with a known quantity of hay and determining the loss in weight of the insoluble components, shows variations (Table II-5). Species which do not digest pebble-milled filter paper digest little of the fiber in hay.

Table II-5
WEIGHT OF THE WATER-INSOLUBLE ALFALFA-HAY
FRACTION DIGESTED BY *Butyrivibrio* STRAINS[a]

Strain number	Water-insoluble material digested (%)	Total material soluble and digested (%)
9/7–18	29	55
28–T	29	55
7/5–42–3	1	38
8/3–44	27	54
9/5–43–1	31	56
8/3–52–1	6	40
9/7–57–1	4	39
7/5–69–2	4	39
8/4–46–2	0	37

[a] 40 mg of ground alfalfa hay (oven-dry weight = 36 mg) in 10 ml of rumen fluid broth. Water-insoluble material in controls = 22.7 mg. Inoculated tubes incubated for 1 month.

Colonies of *Butyrivibrio* in cellulose-agar medium vary in shape from simple lens-shaped to triangular or highly branched colonies. Sometimes the colonies are compound lenticular, sometimes spherical with a smooth or uneven edge. A few strains form branched rhizoidal colonies. Colonies in sugar agar also show a diversity in form. In both cellulose and sugar agar, a discrete outer envelope forms around some colonies and they can be lifted out whole. In some strains, colonies in sugar agar are lenticular after 24 hours of incubation and then at 48 hours are crescentic in shape. Examination of the crescentic colonies shows that the lenticular space formerly occupied is filled with fluid except in the lower part where the cells collect in a crescent-shaped mass. The probable explanation is that capsular material at first surrounds the cells and causes them to occupy a considerable space in the agar. The capsule then liquefies, which allows the cells to settle on the lower side of the lens and gives the colony a crescentic shape. A similar phenomenon has been noted for *Klebsiella* (Read *et al.,* 1956).

Cell shapes in *Butyrivibrio fibrisolvens* also vary from one strain to another. The diameter varies between 0.4 and 0.8 μ. The length is usually 1.5–3.0 μ. Some cells taper to a point at both ends, while some are joined in chains which may be fairly straight or a coiled spiral. The cells are gram-negative, possess a single polar flagellum, and are usually motile, though the latter is not evident in all fresh mounts. Motility in wet mounts is best observed if the energy source in the medium is in low concentration. The water of syneresis of stab-inoculated slant cultures is excellent for detecting motility (Bryant and Small, 1956a). Many strains form mucoid extracellular material (Gutierrez *et al.,* 1959b), and some show a distinct capsule. Cells are not usually iodophilic, though an iodophilic rod described as *Bacteroides amylogenes* exhibits many features of butyrivibrios and may belong to the genus *Butyrivibrio*. Strains are killed by exposure to oxygen (Giesecke, 1960a).

Butyrivibrio can grow rapidly on a wide variety of sugars (with a good deal of variation from strain to strain), usually attacks starch, and can grow on feed ingredients in the absence of other nutrients (Margherita and Hungate, 1963), though some strains need constituents of rumen fluid. Alfalfa hay extract plus rumen fluid gives the greatest yield of *Butyrivibrio* cells (Margherita and Hungate, 1963). Carbon dioxide is not required in the medium (Bryant and Small, 1956a), except in some investigations (Gill and King, 1958). Biotin, folic acid, and pyridoxal are required, and histidine, isoleucine, methionine, lysine, cysteine, leucine, tyrosine, and valine stimulated growth of one strain, but much of the assimilated nitrogen came from ammonia (Gill and King, 1958). Studies of several strains (Bryant and Robinson, 1962b) disclosed much variability in the nutritional requirements for rumen fluid, ammonia, casein, acetate, and volatile fatty acids. Purines and pyrimidines were not required. Growth of a rumen strain of *Butyrivibrio* was increased by mucin (Gordon and Moore, 1961).

The principal fermentation products are carbon dioxide, hydrogen, ethanol, and acetic, butyric, formic, and lactic acids. If rumen fluid is included in the medium there is usually an uptake of acetic acid rather than a production. It is presumably converted into butyric acid. In media initially containing no acetic acid, some accumulates as a final product of the fermentation. The acetokinase and phosphotransacetylase found in mixed rumen bacteria (Van Campen and Matrone, 1964a,b) may reside in the butyrivibrios.

Production of propionate has been observed in a small percentage of strains (Lee and Moore, 1959). They are able to reduce lactate to propionate. This suggests an acrylic pathway.

Over a period of years the author has isolated a number of cellulolylic strains of *Butyrivibrio* on a cellulose medium. Their characteristics are generally similar to noncellulolytic strains, except that most of the cellulolytic strains fermented D-xylose, whereas most of the noncellulolytic strains did not. Pectinolysis was also more common in the cellulolytic strains. Characteristics of butyrivibrios isolated from a nonselective medium are shown in Table II-6. Only one of these twelve strains was cellulolytic. The strains producing no lactate are assigned to a new species, *Butyrivibrio alactacidigens* n. sp., with characteristics as shown in Table II-9.

The variability in the sugar fermentation pattern of *Butyrivibrio* is borne out by serological analyses for common antigens (Margherita and Hungate, 1963; Margherita *et al.*, 1964). Strains isolated from the same animal at different times show serological differences, as do strains isolated from widely separated geographic areas. There are enough common antigens to indicate that the strains are related, but the available evidence indicates that the group as a whole is quite heterogeneous.

(3) *Sporeforming Cellulolytic Rods. Clostridium lochheadii* was first detected (Hungate, 1957) in a group of cows used for a study of the effect of salt on the rate of consumption of concentrate. It occurred in numbers as high as 4×10^6 per milliliter of rumen contents. It was isolated also from a steer in Mississippi fed on a coarse-hay ration and from a pellet prepared by a veterinary supply house and advertised as containing the bacteria active in the rumen. When tested for cellulolytic bacteria the pellet showed only strains of *C. lochheadii*. It is not as common as *Bacteroides succinogenes* and some of the cellulolytic cocci and is usually not noticed in cellulose rumen agar dilution series.

The morphology of these bacteria is distinctive (Fig. II-6A and B). Colonies in cellulose agar show a clearing of the cellulose in a surrounding area 1- or 2-mm wide, or if the organisms have gained access to the surface they spread rapidly and digest the cellulose in the thin agar underlying the growth. The advancing edge of the colony may show fingerlike extensions. The colony itself is irregular in shape, the inside usually less dense than the periphery and containing very few vegetative cells. Recently isolated strains show abundant spores within the small colonies. As the colony grows, the spores in the center disappear, and spores are found in abundance in a thin peripheral shell underlying the vegetative cells at the surface of the loosely knit colony.

After strains of this species have been cultured and transferred in the laboratory for a year or two the spores become less and less prominent. All strains have ultimately formed no spores, and the cultures have finally

II. THE RUMEN BACTERIA

Table II-6
CHARACTERISTICS OF NONSELECTED STRAINS OF *Butyrivibrio*

Characteristics	Strains											
	SW1-4-2	LW5-2	SW1-7-2	SW1-6-1	LW1-6-1	L2-2	L2-1	S8-2-2	S8-2-1	S7-6	S7-4	S7-1
Fermentation products												
Hydrogen	7	6	156	9	22	132	53	16	11	35	68	18.3
Ethanol	4	7	22	20	9	0	0	0	0	0	16	0
Formate	153	159	27	28	170	147	96	200	229	241	135	64
Acetate	29	−34	10	16	53	23	33	3	−75	−12	−12	14
Butyrate	92	62	75	53	92	99	82	84	87	95	139	63
Lactate	20	90	0	0	0	0	0	87	68	100	53	218
Succinate	0	0	0	0	0	0	0	0	0	−	−	−
Substrates fermented												
D-xylose	−	−	−	±	−	−	−	+	−	+	+	+
L-Arabinose	−	−	±	−	±	+	−	−	−	±	+	−
Dextrose	+	−	+	+	+	+	+	+	+	+	+	+
Fructose	+	±	+	+	+	+	+	+	+	+	+	+
Galactose	−	±	+	+	+	+	+	+	−	+	+	−
Mannose	−	−	−	+	+	+	+	+	+	+	+	+
Rhamnose	−	−	+	+	+	+	+	+	−	±	+	+
Mannitol	+	+	−	−	−	−	−	−	−	+	+	+
Esculin	+	+	+	+	+	+	+	−	−	±	+	+
Salicin	+	+	−	+	+	+	+	+	+	±	+	+
Sucrose	−	−	−	−	+	+	+	+	+	+	+	+
Lactose	−	+	−	+	+	+	+	+	+	+	+	+
Trehalose	+	+	−	+	−	−	+	+	±	+	+	+
Cellobiose	+	+	−	−	+	+	+	+	+	+	+	+
Maltose	−	+	−	+	+	+	+	+	+	+	+	+
Melezitose	+	−	−	−	+	−	−	±	−	+	+	+
Dextrin	±	±	±	+	+	−	+	−	−	+	+	+

	1	2	3	4	5	6	7	8	9	10	11	12
Starch	+	+	+	+	+	+	+	−	±	+	−	+
Inulin	+	+	+	+	+	+	+	+	+	+	+	−
Xylan	+	+	+	+	+	+	+	−	−	+	+	+
Pectin	+	+	±	+	−	−	+	−	−	+	+	−
Gum arabic	+	+	±	−	−	−	−	−	−	±	±	−
Cellulose	−	+	−	−	−	−	−	−	−	−	+	−
Other characteristics												
Indole production	−	−	±	−	−	−	−	−	−	−	±	−
Voges-Proskauer	−	−	±	−	−	−	−	−	−	±	−	−
H₂S production	−	−	+	−	±	+	−	−	−	+	−	−
Nitrate reduction	−	−	+	−	−	−	−	−	−	−	−	−
Gelatin liquefaction	−	−	+	−	−	−	−	−	−	±	−	−
Litmus milk	coag. lique-fied	coag. lique-fied	−	coag. lique-fied	coag.	coag. lique-fied	coag. pepto-nized	coag.	coag.	coag. pepto-nized	coag. coag.	coag. coag.

failed to grow when subcultured. Whether the spores germinate prematurely or are digested by enzymes has not been ascertained.

A very potent proteinase was produced by several tested strains. Litmus milk was completely peptonized (i.e., the casein digested) within 48 hours after inoculation. Very few bacterial cells could be detected in the cleared liquid.

The vegetative cells of this strain vary widely in diameter, with the larger ones presumably precursors of spindle-shaped giant clostridial cells within which the spores form. The large vegetative cells have a diameter of 0.7 μ and a length of 7 μ. The clostridial forms (Fig. II-6B) are 1.2–1.7 × 7 μ, and the spores are 1.0–1.5 × 2–3 μ (Fig. II-6A). Vegetative cells are gram-positive (Fig. II-6A).

Clostridium lochheadii is the most rapid cellulose digester isolated from the rumen, though certain strains of *Bacteroides succinogenes* and *Ruminococcus* are almost as rapid. When a piece of agar from an old nonviable culture tube in which all cellulose has digested is inoculated into a fresh tube of cellulose agar, some continuing digestion of cellulose is occasionally observed around an agar chunk of the inoculum, due to the transferred cellulase.

Clostridium lochheadii does not require rumen fluid in the culture medium and grows well on extracts of hay or mixed ruminant feeds. Cultures tested on alfalfa and grass hays digest between 22 and 50% of the insoluble materials with an average of 40%. This corresponds to a net solubilization of as much as 68% of the alfalfa hay, 35% being soluble prior to action by the bacteria.

The sugars fermented are glucose, cellobiose, maltose, sucrose, starch, salicin, and cellulose. Of eight strains, four fermented fructose, and one of six strains fermented L-xylose. Not fermented are D-xylose, L-arabinose, D-arabinose, galactose, mannose, rhamnose, trehalose, lactose, raffinose, inulin, and esculin. One tested strain fermented glycerol.

On fermentable-disaccharide media C. *lochheadii* forms so much slime that liquid tubes containing 0.5% sugar can occasionally be inverted without spilling the contents. The organisms are killed by oxygen, though it has not been determined whether this includes the spores or only the vegetative cells.

The relative quantities of fermentation products of pure cultures are carbon dioxide 2.55, hydrogen 2.76, acetic acid 0.92, formic acid 0.52, butyric acid 0.87, and ethanol 0.54. The hydrogen and carbon dioxide cause a considerable pressure to develop in the culture tubes. Little succinic or lactic acid is formed.

Another sporeformer, *Clostridium longisporum,* was first observed as a very orange colony in a rumen fluid cellulose-agar tube inoculated from

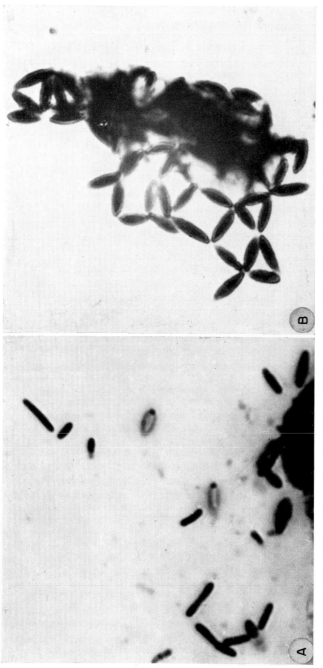

Fig. II-6. A. Vegetative cells, a few clostridia, and two spores of *Clostridium lochheadii*. Gram stain of a 24-hour rumen fluid glucose-agar culture. Magnification: 1750 ×. From Hungate (1957). B. Clostridial forms of *C. lochheadii*. Magnification: 1750 ×. From Hungate (1957).

one of the same group of animals in which *Clostridium lochheadii* was found in large numbers. It is gram-positive and produces a yellow to brilliant orange pigment which makes easy the recognition of this species. It has been found in numbers up to 10^5 per milliliter in a Jersey cow on an alfalfa hay ration, but is never consistently present nor one of the more abundant bacteria in the rumen.

In addition to the colony color, the morphology of the cells and spores is distinctive (Fig. II-7). The vegetative cells are even more variable in

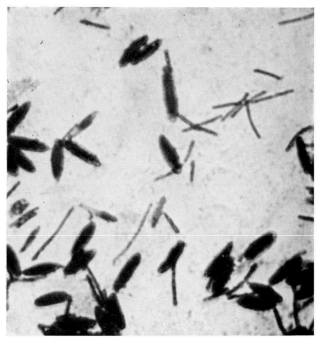

Fig. II-7. Vegetative and clostridial cells of *Clostridium longisporum*. Gram stain. Magnification: 2350 ×. From Hungate (1957).

size than are those of *C. lochheadii,* both in diameter and in length. The larger vegetative cells are 1 μ in diameter and 7 to 15 μ long. They enlarge into clostridial cells which are 2–3 μ in diameter and in which are formed long spores, sometimes slightly curved, 1 μ in diameter by 3–6 μ long (Hungate, 1957). Both the vegetative cells and the clostridia are actively motile, presumably by means of peritrichous flagella, though the location has not been tested.

Strains of *C. longisporum* have also proven difficult to maintain indefinitely in pure culture, though by transferring infrequently and keeping the cultures in cellulose agar some strains have been maintained for several years. Cellulose digestion is complete, but not as rapid as in the case of *C. lochheadii*. Much slime is formed when *C. longisporum* is grown in sugar broth.

Rumen fluid is not required as a nutrient. Yeast extract or nutrient broth supports growth if a carbohydrate source is also supplied. The principal fermentation products are hydrogen, carbon dioxide, acetic acid, formic acid, and ethanol. No butyric, propionic, lactic, or succinic acid was found with the original isolate. Strains obtained subsequently form also propionate (Lester, personal communication).

Unidentified sporeformers occur in numbers up to 10^9 per milliliter in sheep fed on teff hay (Kistner, 1965).

(4) *Cillobacterium cellulosolvens*. *Cillobacterium cellulosolvens* (Bryant *et al.*, 1958c) is an anaerobic, peritrichous, gram-positive rod which was found initially in a single culture tube from an animal pastured on clover and later by the author in an animal on an alfalfa hay ration. The dilutions in which the organism occurred were 5×10^8 and 1×10^8, respectively. This bacterium is of considerable interest because lactic acid is its chief fermentation product. Its numbers are not ordinarily sufficient to make it quantitatively significant.

(5) *Acetigenic Rod*. A curved rod which forms hydrogen, ethanol, acetate, and presumably carbon dioxide was isolated as a single cellulolytic colony developing in a dilution tube inoculated with rumen contents from one of the starving zebu cattle at Archer's Post, Kenya (previously mentioned), and was the only cellulolytic colony in high dilution that was not *Butyrivibrio fibrisolvens*. It is not a rapid cellulose digester, but the cellulose is very completely digested. Serologically it is unrelated to *Butyrivibrio* (Margherita and Hungate, 1963). It fermented mannitol, lactose, xylan, pectin, cellulose, cellobiose, maltose, sucrose, fructose, glucose, and L-arabinose, but not D-xylose, galactose, mannose, rhamnose, esculin, salicin, trehalose, and starch or inulin; it was nitrite negative, hydrogen sulfide negative, indole negative, inactive on litmus milk, and did not liquefy gelatin.

(6) *Cellulomonos fimi*. This organism has been found in Japanese cattle in numbers as high as 10^7 per milliliter, as detected by direct counts after treatment with fluorescent antibody (Higuchi *et al.*, 1962a,b). The organism is euryoxic, grows anaerobically in the presence of rumen fluid, and is believed to play a role in cellulose digestion in the rumen.

c. The Cellulolytic Cocci

These bacteria, identified microscopically by Henneberg (1922), make up a distinctive group of rumen cellulose digesters (Figs. II-8A, C, and

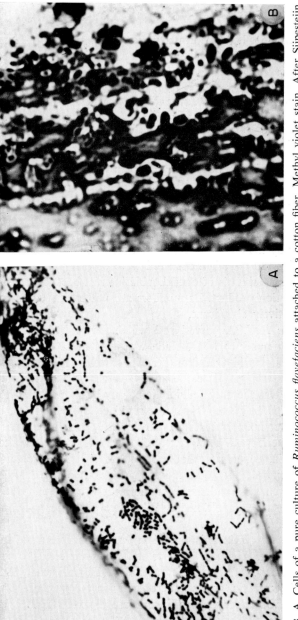

Fig. II-8. A. Cells of a pure culture of *Ruminococcus flavefaciens* attached to a cotton fiber. Methyl violet stain. After Sijpesteijn (1949). Magnification: 900 ×. B. Grass blade after 48 hours in the rumen of a fistulated sheep. Fluorescence microscopy. The undigested plant material and the bacterial cells are white, the digested cavities dark. After Baker and Nasr (1947). Magnification: 1350 ×.

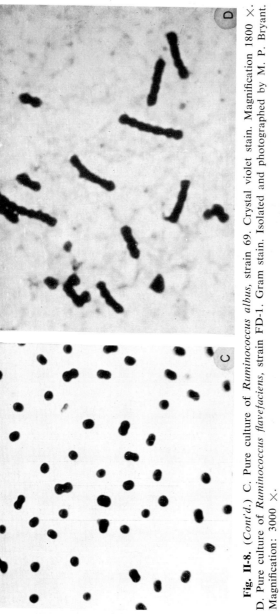

Fig. II-8. (*Cont'd.*) C. Pure culture of *Ruminococcus albus*, strain 69. Crystal violet stain. Magnification 1800 ×. D. Pure culture of *Ruminococcus flavefaciens*, strain FD-1. Gram stain. Isolated and photographed by M. P. Bryant. Magnification: 3000 ×.

D) characterized by coccoid morphology and a cell diameter of 0.8–1.0 μ. They have been grown in pure culture by numerous workers (Hungate, 1947, 1950, 1957; Sijpesteijn, 1948, 1949, 1951; Bryant and Robinson, 1961a; Bryant et al., 1958a) and compose as much as 84% of the cellulolytic bacteria developing in a cellulose-agar medium (Kistner et al., 1962). There is much variation among individual strains (Hungate, 1957). They range from gram-negative to gram-variable, single cells to chained filaments, with the color of colonies varying from white to yellow and on some media almost orange. The features common to most strains are the ability to digest cellulose and xylan (Bryant et al., 1958a) and to utilize cellobiose and the inability to utilize glucose and most of the other mono- and disaccharides. Glucose, D-xylose, esculin, fructose, sucrose, lactose, or L-arabinose is occasionally used. Cellulose digestion is rapid with most strains, but hardly perceptible with others, and for a few may be entirely absent. Strains which digest alcohol-insoluble xylan to an alcohol-soluble form are not necessarily able to utilize the soluble material for energy, although some can (Dehority, 1965).

Ruminococci are killed by oxygen (Ayers, 1958; Giesecke, 1960a).

Digestion of the insoluble components of hay ranges between 24 and 51% (Table II-7).

Since gram negativeness is relatively rare among cocci, this characteristic was investigated thoroughly with the type strain for Ruminococcus albus (Hungate, 1957). An 8-hour culture in rumen fluid cellobiose broth was gram-negative, though the color was darker than is often the case with gram-negative bacteria. The culture was transferred to a second broth tube, and after 12 hours of incubation the organisms in this sec-

Table II-7

DIGESTION OF VARIOUS FRACTIONS OF ALFALFA HAY AND
POOR GRASS HAY BY STRAINS OF Ruminococcus[a,b]

Fraction	Alfalfa hay		Poor quality grass hay	
	Amount in control (mg)	Percent digested	Amount in control (mg)	Percent digested
Soluble in benzene-alcohol	1.0	43	0.9	37
Soluble in H_2O at 85°C	2.2	62	1.3	17
Soluble in 2% H_2SO_4	8.3	50	6.8	21
Soluble in 70% H_2SO_4	8.1	55	10.9	28
Insoluble in the above	3.7	41	3.6	17
Total dry weight	23.3	51	23.5	24

[a] From Hungate (1957).
[b] Each value is the average for four different strains of Ruminococcus.

ond tube were stained in parallel (i.e., on the same slide) with *Escherichia coli* and *Streptococcus pyogenes.* The smears of *E. coli* and *R. albus* overlapped slighty to give an area with both species in one microscopic field. The cells of *Ruminococcus* appeared slightly darker than those of *E. coli.* However, in a second mixed smear simply stained with safranine, the cells showed exactly the same difference in the depth of the color. These results show conclusively that the type strain for this species was gram-negative.

Because of the variability of the cocci in so many characteristics, it is extremely difficult to determine species. The two named species represent a mode around which strains are clustered. *Ruminococcus flavefaciens* (Sijpesteijn, 1951; Hungate, 1957) (Figs. II-8A and D) forms yellow colonies (though this is variable), grows in chains of cells, tends to be gram-positive, and is usually less rapidly cellulolytic. It forms simple lens-shaped colonies in cellulose agar or may form a complex colony composed of portions of lenses joined at various angles. Reducing sugars may accumulate in old cultures containing more cellulose than can be fermented, and cellobiose is presumably formed (Ayers, 1958; Hungate, 1950). The cells may be iodophilic due to storage of reserve polysaccharide and are reported to contain a high concentration of diaminopimelic acid (Synge, 1953). Succinic acid is one of the most important fermentation products, and significant quantities of acetic and formic acid are formed. Small quantities of hydrogen, ethanol, and lactic acid are produced. The hydrogen cannot be detected except by analysis, the quantity produced being insufficient to split the agar. Evidence has been obtained that glucose is metabolized intracellularly, but cells are impermeable to it. Cellobiose is split intracellularly to glucose and α-D-glucose-1-P (Ayers, 1959).

The other species, *R. albus,* (Fig. II-8C) is characterized by white colonies, cells usually single or in short chains. It is gram-negative and rapidly cellulolytic. The colonies in cellulose agar usually appear as multiple lenses joined by narrow isthmuses to give a much-branched series of small discs. It forms very little succinic acid, but produces hydrogen and ethanol in significant quantities, and some strains form a fair amount of lactic acid.

The rumen cellulolytic cocci require isovalerate and isobutyrate (Allison *et al.,* 1958, 1962b). Leucine cannot replace isovalerate as a growth factor for *R. flavefaciens.* Isovalerate was not used by one *R. albus* strain (Allison *et al.,* 1962b), but 2-methylbutyrate was used. There is some uncertainty (Allison *et al.,* 1962b) of the purity of the isovalerate in the earlier nutritional studies (Allison *et al.,* 1959), and some 2-methylbutyrate may have been present. The isomers are difficult to separate. Many strains assigned to *R. albus* require (Bryant and Robinson, 1961c; Hungate, 1963a) isovaleric, isobutyric, and 2-methylbutyric acids. Additional unidentified substances in rumen fluid were stimulatory. All require biotin (Bryant and Robinson, 1961c; Scott and Dehority, 1965), and biotin increases cellulose digestion by a washed-cell suspension of rumen bacteria (Dehority *et al.,* 1960) in which ruminococci are prominent (Dehority, 1963). Many need *p*-aminobenzoic acid and pyridoxamine, a few require thiamine, riboflavin, folic acid, or pantothenic acid, and a few strains are stimulated by acetic acid (Bryant and Robinson, 1961c). *Ruminococcus flavefaciens* is stimulated by methionine, but B_{12} can substitute for it (Scott and Dehority, 1965).

Ammonia is essential to most strains, and they use it in preference to amino acids (Bryant, 1963). On cellobiose the RAM strain stored iodophilic polysaccharide to the extent of 30% of the cell dry weight and in continuous cultures stored 25% of the substrate in the form of cell material (Hungate, 1963a). It produces some yellow pigment, but this is not usually evident in the colonies, and a mutant with little yellow pigment has been obtained. Another mutant digests cellulose less completely than does the parent strain.

In isolating *R. albus* from cellulose-agar cultures the *Borrelia* previously mentioned is frequently encountered. Fortunately, in the case of *R. albus* the colony is at some distance from the zone of cellulose digestion in which the *Borrelia* accumulate, and it is consequently possible to separate them more easily than is the case with *Bacteroides succinogenes.*

The variability among the cellulolytic cocci is so great that further work is needed to determine the proper classification of many of the strains. Tests of several strains for cross agglutinations failed to indicate common antigens.

d. Distribution of Cellulolytic Rumen Bacteria

The rumen cellulolytic bacteria are widely distributed geographically and show an adaptability to different animal species.

Ruminococci have been isolated from cattle and sheep in Holland, England, South Africa, and several geographic localities in the United States, and from the cecum of the guinea pig (Hungate *et al.,* 1964) and rabbit (Hall, 1952).

Bacteroides succinogenes has been found in cattle in Maryland, Texas, Washington, and California in the United States, and in England, Kenya, and the Union of South Africa. It has been found in sheep and several African antelopes (Hungate *et al.,* 1959). It is not certain that the strains in all these hosts are identical, but the colonies in cellulose agar are quite similar and they are probably related.

e. Cellulase Production

Difficulties in culturing anaerobic cellulolytic bacteria led to many misconceptions of their nature. They reputedly lost their ability to digest cellulose when grown on sugar media. Although some cellulolytic rumen bacteria cannot ferment glucose (Hungate, 1950, 1957; Bryant and Burkey, 1953b), all tested thus far can use cellobiose and retain their cellulose-digesting capacity through many consecutive subcultures on sugar media. Over a period of several weeks in a continuous culture on cellobiose with a 10-hour cycle time, the cells retain their cellulose-digesting ability.

The idea that pure cultures lost the ability to digest cellulose originated because cultures in which cellulose was digested were impure, and, after inoculation into solid sugar media, the noncellulose digesters were picked. A few strains of cellulolytic rumen bacteria vary with time in their ability to digest cellulose (Bryant and Small, 1956a), but the capacity can be either gained or lost, unless it is retained by carrying the culture on a cellulose medium.

Although no cellulolytic bacteria produce during growth significantly more sugar than can be used (with the possible exception of *Clostridium lochheadii*), cellulose digestion continues in some cultures of cellulolytic cocci and butyrivibrios after growth has ceased, and sugar accumulates (Hungate, 1950). Sugar does not accumulate in *Bacteroides succinogenes* cultures.

The completeness of digestion of cellulose varies among the different strains. This has already been mentioned in connection with cellulose-agar cultures of each type. It is also evident when strains are inoculated into rumen broth-cellulose cultures. Some digest all the cellulose, whereas others do not attack it sufficiently to cause a noticeable diminution in the volume of the cellulose resting in the bottom of the culture. It cannot necessarily be concluded from lack of a volume change that no cellulose digestion occurs, since hydrolysis may not cause a corresponding reduction in the space occupied by the cellulose strands. Dry weight determinations bear out the observation that most of the cellulose is not attacked when there is no visible decrease in the volume occupied by the settled cellulose.

In pure cultures of cellulolytic *Ruminococcus albus* growing in cellulose agar there is a very sharp line of demarcation between the area cleared of cellulose and that in which the cellulose is unattacked. This is interpreted as indicating that the cellulase is quite firmly attached to the cellulose and unable to diffuse until the substrate has been digested. If the enzyme possessed only a slight affinity for the cellulose it would presumably cause a diffuse zone of cellulose digestion or even accomplish digestion rather uniformly throughout the agar culture.

In general, the strains of bacteria showing the most rapid and complete digestion of cellulose in rumen cellulose agar are the ones which digest hay most completely. Bacteria showing no marked digestion of pebble-milled cellulose in pure cultures usually fail to digest significant quantities of the cellulose in the natural feed.

In testing the digestibility of the hay by the bacteria it is important to use a small percentage of hay to prevent the fermentation products from reaching inhibitory concentrations (Lee and Moore, 1959). Most of the cellulolytic strains cease growth at a pH of 5.5, and even with a fairly high buffering capacity (0.1% K_2HPO_4-KH_2PO_4 and 0.5% bicarbonate) 1% of fermented cellulose drops the pH to inhibitory levels. The rate of digestion of a small concentration of cellulose can be quite high in some pure cultures.

Before pure cultures of rumen cellulolytic bacteria were obtained it was often postulated that fiber digestion in the rumen required the combined action of a number of microorganisms. Lack of the essential contributions by accompanying forms was used to explain the failure of cellulose digestion by the pure cultures. This explanation is in part correct. The rumen cellulolytic bacteria do depend on other rumen bacteria for the synthesis of nutrients such as short-chain fatty acids, but there is no evidence that cellulose digestion *per se* requires action of more than one species. Cellulase is completely synthesized in one cell.

Plant fiber is a complex of many materials. Single bacterial species may not synthesize enzymes attacking all the components. Mixtures of rumen bacteria, each singly incapable of digesting plant fiber, might be able to digest hay if allowed to act together. This has been tested in a preliminary fashion by taking strains of different bacteria (none of them cellulolytic) isolated at random from a high dilution of rumen contents and exposing them singly and together to hay as a substrate. The extent of fiber digestion was little more than in the uninoculated control. Instead of synergism in the splitting of individual chemical bonds in ruminant feeds, it would seem more plausible that action by enzymes of one species on one type of bond might expose other types susceptible to attack by enzymes of other species.

The species able to attack the cellulose in hay attack the hemicellulose to about the same extent (Hungate, 1957) (Table II-7). The enzymes concerned may be the same for both, but until pure enzymes have been isolated exact tests cannot be performed.

Enzymes attacking carboxymethylcellulose have been obtained from rumen contents, and some of their characteristics have been studied

(King, 1956) (see Chapter X). They can be demonstrated in solution in extracellular rumen fluid (Meiske *et al.*, 1958).

Although the capacity for cellulose digestion is retained during growth on sugar media, it is not necessarily expressed. Strains inoculated into cellulose agar which also contains 0.1% cellobiose form detectable colonies with no visible digestion of cellulose as compared to parallel cultures with no sugar (Fig. II-9). Later the cellulose is digested, after the sugar concentration has fallen. As little as 0.07% cellobiose is partially inhibitory, and a concentration of 0.27% stops most of the cellulase activity.

The sugar does not inhibit production of cellulase by *Ruminococcus albus,* but does inhibit activity (W. Smith and Hungate, unpublished experiments). This inhibition explains the failure of sugars to accumulate to any great extent in growing cultures of cellulose digesters and the lack of cellulose digestion in media containing fermentable sugar.

Continuous cultures of *Ruminococcus albus* strain RAM on cellobiose have been used to obtain a cellulase. The crude enzyme in the culture is most active at pH's between 5.8 and 6.3. When cellulose and cellulase are placed in a dialysis bag immersed in a large volume of balanced salts solution, about 60% (40 mg) of the pebble-milled cellulose is digested, the enzyme retaining activity over a period of a week or more. It would be of interest to determine whether the cellulase-inhibiting material in the extract of sericea forage (Smart *et al.,* 1961) is a sugar.

f. Attachment of Cellulolytic Bacteria to Plant Particles

The etching out in cellulosic plant walls of holes, "Frassbetten," exactly accommodating the bacterium as observed by Henneberg (1922) and Baker and Martin (1938) (Fig. II-8B) indicates that at least some of the rumen cellulolytic bacteria are attached to the plant particles in the digesta. The extent to which this is characteristic of all rumen cellulolytic bacteria is not known, and, as mentioned earlier, bacteria cultured from rumen fluid free of plant particles include about the same proportion of cellulose digesters, as do bacteria cultured from the solids.

Information on the importance of attachment of rumen cellulolytic bacteria to the substrate has been provided by observations on the difference in colony development in the presence and absence of accompanying bacteria.

A pure culture of *Ruminococcus* diluted into cellulose media digests cellulose in all dilutions containing viable cells, whether the medium be liquid or solid. Similar digestion in all tubes occurs when rumen digesta with its myriads of bacteria, both cellulolytic and not, are diluted into *liquid* cellulose media. In contrast, dilutions of rumen contents into solid cellulose agar show clearing of cellulose in low dilutions due to *Bacte-*

roides succinogenes strains which move through the agar and occasionally to *Clostridium lochheadii*, but no clearings typical of *Ruminococcus* or *Butyrivibrio* are observed. In the intermediate dilutions of 10^3–10^5 no cel-

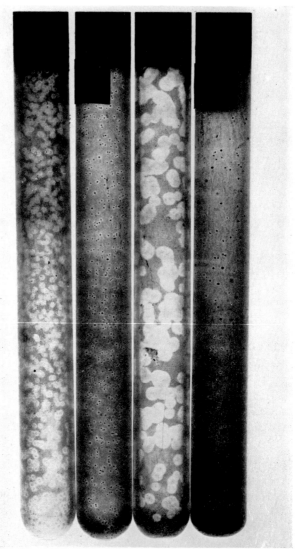

Fig. II-9. Inhibitory effect of cellobiose on cellulose digestion in rumen fluid cellulose agar cultures. The two tubes on the left were inoculated with equal quantities of a suspension of *Butyrivibrio*, strain N1C11, and the two on the right with *Butyrivibrio* strain 8/3 42. For each pair the tube on the right received 0.1% cellobiose. Decreased opacity indicates cellulose digestion. Actual size.

lulolytic zones occur, and the cellulose is not visibly digested after months of incubation. However at still higher dilutions colonies of *Ruminococcus* and *Butyrivibrio* appear, with characteristic clearing of the cellulose around them, but they never become very large. Inhibition of cellulose digestion in the lower dilution tubes is due to the numerous noncellulolytic bacteria. They capture the products from cellulase, and the cells elaborating the enzyme do not receive a sufficient cellobiose yield to support growth and continued cellulose digestion. In liquid the cellulolytic cells move to a preferred position adjacent to the substrate, which enables them to obtain a sufficient yield from their cellulase to permit continued growth even in the presence of numerous noncellulolytic neighbors. Agar prevents this movement.

Zones of cellulose digestion by *Bacteroides succinogenes* in low agar dilutions are possible because these cells can migrate in the solid medium to lie immediately adjacent to the cellulose. This explanation probably does not apply to cellulose digestion by *Clostridium lochheadii* in low dilutions. The cells of this species exhibit some ability to grow through agar and do not form colonies with a discrete surface. However the cells do not reach the cellulose, which indicates that digestion by means of an extracellular enzyme can be successful in spite of accompanying forms. The explanation appears to be that *C. lochheadii* secretes enough cellulase to supply itself as well as competitors with sugar.

g. Numbers of Cellulose-Digesting Bacteria

Dilution cultures into rumen fluid cellulose-agar media have been used to obtain culture counts (Hungate, 1950, 1957; Kistner, 1960; Kistner *et al.,* 1962). The counts were usually on the order of 10^6 to 10^7 per milliliter, and about 5% of the total colony count. Using a habitat-simulating sugar medium for initial isolation of colonies with later testing on liquid cellulose media, Bryant (Bryant and Burkey, 1953b) obtained total cellulolytic culture counts of $0.5–5 \times 10^8$ per milliliter, with cellulolytics constituting 15.6% of the total culture count of forage-fed cattle. The difference from the counts on cellulose agar may reflect inability of many cellulolytic strains to digest cellulose in an agar medium (Bryant and Burkey, 1953b).

2. THE STARCH DIGESTERS

A number of rumen cellulolytic bacteria are also amylolytic, e.g., *Clostridium lochheadii,* some strains of *Bacteroides succinogenes,* and most strains of *Butyrivibrio fibrisolvens.* The noncellulolytic species—*Streptococcus bovis, Bacteroides amylophilus, Bacteroides ruminicola, Succini-*

monas amylolytica, and *Selenomonas ruminantium*—include many starch-digesting strains.

a. Streptococcus bovis

Streptococcus bovis (Orla-Jensen, 1919) is a homofermentative streptoccus common in the rumen (Kreipe, 1927). It does not occur in ruminant feeds (Higginbottom and Wheater, 1954). The orange streptococci growing on potato slices (Steiger, 1904) may have been *S. bovis.* Characteristics of many rumen strains have been described (Macpherson, 1953; Higginbottom and Wheater, 1954; Perry *et al.,* 1955; Hungate, 1957; Giesecke, 1960b; Seeley and Dain, 1960). Rumen strains differ serologically (Macpherson, 1953) from those isolated from infections in man and domestic animals. They belong in Lancefield's group D streptococci (Shattock, 1949; Macpherson, 1953), but this relationship is revealed only by certain methods of antigen preparation. Only a few strains stimulate production of D antibodies in rabbits.

Most rumen strains are reported as α-hemolytic, due to H_2O_2 production (Shattock, 1949), but the author did not observe any effects on blood agar by most of the rumen strains studied (Hungate, 1957). Much dextran is formed from sucrose. Characteristics of strains as determined by various authors are shown in Table II-8. Rhamnose, mannitol, xylan, gum arabic, and cellulose are not fermented, and most strains do not ferment mannitol.

Because it does not require a very low oxidation-reduction potential and grows rapidly *S. bovis* is one of the most commonly isolated of the rumen bacteria (Perry and Briggs, 1955).

It occurs also in feces of ruminants, horses, and pigs, and in the horse cecum (Alexander *et al.,* 1952).

Streptococcus bovis digests heated starch very rapidly by means of an α-amylase (Hobson and Macpherson, 1952), but raw starch grains are attacked only after a lag period (Higginbottom and Wheater, 1954). In rumen fluid starch-agar tubes inoculated with suitable dilutions of rumen contents and incubated at $39°C$, *S. bovis* colonies almost invariably appear first, closely followed by *Butyrivibrio.* In rumen fluid glucose agar, surface colonies are as much as 1 mm in diameter after only 6–8 hours of incubation. This early appearance is due to two factors: rapid growth and the abundant capsular material which greatly increases the volume occupied by the colony. This rapid growth on sugar media is the easiest means to distinguish *S. bovis* from other rumen bacteria.

On initial isolation the rumen strains of this species are sensitive to oxygen (Hungate, 1957; Wolin and Weinberg, 1960; Giesecke, 1960a,b). Later, this sensitivity diminishes, and the cells grow aerobically, according to most authors, though more vitamins are needed under aerobic conditions (Ford *et al.,* 1958), and growth is better with added Na$_2$S (Niven *et al.,* 1948).

In the initial dilution series from the rumen, colonies of *S. bovis* may be white, yellow, or orange in color. In the author's experience this char-

Table II-8

SUGAR FERMENTATION CHARACTERISTICS OF
Streptococcus bovis AS REPORTED BY VARIOUS AUTHORS

	Author					
Compound	McPherson (1953)	Higginbottom and Wheater (1954)	Perry *et al.* (1955)	Hungate (1957)	Seeley and Dain (1960)	Giesecke (1960b)
D-Xylose				−	−	−
L-Arabinose		+	10/81[a]	11/20	46/95	4/5
Dextrose	+	+		+		+
Fructose						+
Galactose	+					+
Mannose						+
Mannitol	−	−	34/81	−	11/95	1/5
Glycerol	±	−	16/81	−	1/95	
Esculin		+		+		4/5
Salicin				+		+
Sucrose	±	+	+		+	+
Lactose	+	+		+	+	+
Trehalose	±		21/81	15/27	65/95	3/5
Cellobiose		+		+		+
Maltose	+	+				+
Melezitose						
Dextrin	+	+				4/5
Starch	+	+	+	+		+
Inulin	±	+	+	+	91/95	+
Raffinose			+	+	90/95	+
Pectin		+				

[a] The denominator indicates the number of strains tested and the numerator the number of strains positive on test.

acteristic is not constant, and stock cultures usually are white. The variability in color is not due to culture conditions, since colonies differing markedly in color may grow side by side in the same tube. In contrast, two orange strains of *S. bovis* (Giesecke, 1960b) exhibited the color constantly, but it did not correlate with any other characteristics.

Surface colonies are flat and mucoid and tend to run down the wall of the tube after a couple of days of incubation. They often are iridescent. Subsurface colonies are lenticular and large initially in comparison with most other species.

Examination of colonies of *S. bovis* grown on starch or glucose and mounted under the microscope without suspending fluid discloses cells very evenly dispersed, with no Brownian motion. They are held in place and spaced by the capsular material. The capsular material contains, in decreasing order of concentration, glucose, galactose, rhamnose, and uronic acid (Bailey and Oxford, 1959; Hobson and Macpherson, 1954).

On sucrose (Niven *et al.,* 1946; Oxford, 1958c) *S. bovis* produces large quantities of a dextran (Niven *et al.,* 1941) with α-1:6 glucosidic linkages (isomaltose) and very little branching (Bailey, 1959a). A molecular weight of 90,000 has been proposed (Greenwood, 1954). Synthesis is catalyzed by a constitutive transglycosidase, dextransucrase (Bailey, 1959b). All strains produce dextran when grown on sucrose (Barnes *et al.,* 1961), but the amount varies. As much as 87% of the anhydroglucose in 8.5% sucrose solution was synthesized into dextran (Bailey and Oxford, 1958a).

Streptococcus bovis forms an α-galactosidase (Bailey, 1962, 1963; Bailey and Bourne, 1961) which splits melibiose, raffinose, stachyose, verbascose, and digalactosyl glycerol, but does not split galactose from the chloroplast lipid. *Streptococcus bovis* extracts also contain sucrose phosphorylase, and occasionally isomaltase (Bailey, 1963). Amylolytic rumen cocci isolated by Williams and Doetsch (1960) formed a β-1 : 4 galactosidase hydrolyzing galactomannan. The lactic dehydrogenase of *S. bovis* requires fructose-1,6-diphosphate for its activation (Wolin, 1964).

Out of twenty-five strains of rumen group D streptococci (probably *S. bovis*), seven deposited an intracellular iodophilic material which did not appear to be identical with starch, amylose, amylopectin, or glycogen (Hobson and Mann, 1955). The formation occurred only at pH 5–6. Iodophily has been observed in cells around starch grains (Higginbottom and Wheater 1954). This may be similar to the iodophily observed by van der Wath (1948a) in rumen cocci attacking starch. The starch phosphorylase of *S. bovis* (Ushijima and McBee, 1957) may digest the intracellular polysaccharide.

The author's analyses of products of glucose utilization have disclosed 8% of the substrate in the form of cells and 80–85% in DL-lactic acid. Giesecke (1960b) found a little acetate, formate, and carbon dioxide as additional products.

The culture count for *S. bovis* in the rumen averages about 10^7 per milliliter, and in most animals on a wide variety of rations the count deviates less than for most rumen bacterial species (Higginbottom and Wheater, 1954; Krogh, 1959; Hungate, 1957). Numbers increase for 2 hours after the host is fed and then diminish. The numbers in the abomasum and small intestine are small, indicating digestion; the concentration again rises in the cecum and large intestine (Higginbottom and Wheater, 1954).

Under certain conditions of changed feed, as when alfalfa hay-fed animals are suddenly given excess grain, the numbers of *S. bovis* increase almost explosively (Hungate *et al.*, 1952; Krogh, 1959) and can cause fatal indigestion (see Chapter XII) due to the high concentration of lactic acid produced. In rumen fluid glucose medium supplemented with yeast extract and peptone the division time is 20 minutes.

Since starch is digested it might be expected that the numbers of *S. bovis* in animals on high grain rations would be very high. This appears to be true occasionally (Bauman and Foster, 1956; Gutierrez *et al.*, 1959a; Gall and Huhtanen, 1951; Phillipson, 1952c), but in most animals well adapted to a grain ration the concentration differs little from that in hay-fed animals (Hungate, 1957; Bryant *et al.*, 1961; Perry *et al.*, 1955). In adapted animals a new microbial balance develops in which *S. bovis* is not exceptionally abundant and in which the protozoa, especially *Entodinium,* are prominent. Since *S. bovis* can be ingested by *Entodinium* (Gutierrez and Davis, 1959) and presumably serve as food, its concentration in the rumen may not reflect its actual importance. Ability to grow rapidly would permit a large production of cells through rapid turnover of a small population.

In animals on a high grain ration the turnover rate of lactate in the rumen is higher than in hay-fed animals (Jayasuriya and Hungate, 1959), which suggests that *S. bovis* was more active, and Gutierrez *et al.* (1959a) found *S. bovis* counts in these same animals as high as 5×10^9. However the lactate turnover still accounted for only about one-sixth of the substrate converted, and at a later period the count of *S. bovis* in these same animals was not high (Bryant *et al.*, 1961). On certain starchy rations which have been subjected to a heat treatment lactic acid accumulates in fairly high concentrations (Phillipson, 1952c), possibly due to *S. bovis.*

The nutrition of *S. bovis* is especially interesting because it differs so much from the streptococci growing in other habitats. There is no requirement for peptides and amino acids [except arginine (Prescott and Stutts, 1955; Prescott *et al.*, 1959)], but carbon dioxide (Prescott and Stutts, 1955; Dain *et al.*, 1956) is needed for growth initiation, though carbon dioxide is produced after growth starts (Prescott *et al.*, 1957). The nutritional characteristics fit the rumen habitat in which amino acids are in short supply though carbon dioxide is abundant. Biotin is required, (Niven *et al.*, 1948; Barnes *et al.*, 1961), thiamine is stimulatory, and some strains need pantothenic acid, nicotinic acid (Niven *et al.*, 1948), or purines, and a few amino acids. Arginine supports the best growth of any single amino acid (Barnes *et al.*, 1961)—but this is somewhat of an anomaly since resting cell suspensions do not metabolize it, and ammonia gives slightly better growth when used as the sole nitrogen source.

Ammonia can serve as the sole source of nitrogen for some strains (Wolin *et al.*, 1959; Prescott *et al.*, 1959), yet many investigators report stimulation of growth by amino acids (Paul, 1961), purines, yeast extract, and casein digest. Stimulation by substances showing some of the characteristics of peptides has been observed (Paul, 1961), and in some strains (Bryant and Robinson, 1961a) amino acids are assimilated in preference to ammonia.

These variable results can be explained by strain differences (Barnes *et al.*, 1961) and the amount of growth. The strains capable of growth with ammonia as the sole source of nitrogen were selected by a medium containing only ammonia as a nitrogen source. In other cases the initial isolation medium has selected for other nutritional capacities. Even in cases in which mixtures of amino acids and complex growth substances are stimulatory, carbon dioxide can be essential (Wright, 1960c). Radioactive carbon in C^{14}-labeled carbon dioxide was incorporated chiefly into aspartic acid, then glutamic, threonine, and some into the purines. Aspartic acid labeled with C^{14} was not incorporated to any extent unless Tween 80 was added to decrease the membrane barrier. Rumen fluid can be stimulatory even in the presence of amino acids (Paul, 1961).

It is uncertain whether the nutritional factors of ecological importance to *S. bovis* have yet been detected. Long incubation times (72 hours) (Barnes *et al.*, 1961) suggest that the growth rate was less than would be expected from the division time of 20 minutes observed with a medium containing rumen fluid, peptone, and glucose. The fact that the numbers of *S. bovis* in grain-fed animals are often no greater than in hay-fed animals suggests that a study of the nutrients in rumen fluid from these habitats might disclose interesting ecological relationships. Limitation of *S. bovis* to 2 ATP molecules per hexose molecule may explain its ultimate lack of competitive success in grain-fed ruminants (see Chapter VI).

The amylolytic coccus studied by Provost and Doetsch (1960) is almost identical with *S. bovis* in the pattern of fermented sugars and in being somewhat heat resistant. It differs in other respects and has been classified as *Peptostreptococcus*. It is gram-negative, though gram-positive granules were observed in young cultures. Lactic acid is the chief fermentation product, with some acetic and formic acids. Litmus milk is digested.

b. *Bacteroides amylophilus*

Bacteroides amylophilus (Hamlin and Hungate, 1956) is a starch-digesting bacterium resembling the cellulolytic cocci in being unable to attack glucose. Starch and the di- and oligosaccharides derived from it are the only carbohydrate substrates fermented. The cells attach to starch grains during digestion (Fig. II-10A). The characteristics of this species clearly indicate decomposition of starch as the niche occupied in the rumen.

The occurrence of *B. amylophilus* is sporadic (Hamlin and Hungate, 1956), but when present it may be the predominant starch digester and constitute 10% of the total bacteria in the rumen (Bryant and Robinson, 1961b; Blackburn and Hobson, 1962).

The cells are irregular in shape and do not remain viable for a long period without transfer. Fermentation products are formic, acetic, and succinic acids, with possibly some lactate. Carbon dioxide is required, but rumen fluid is not. It grows readily on maltose-yeast extract-peptone or even on maltose in a mineral medium and assimilates ammonia. Amino acids are used to a limited extent in the presence of ammonia.

c. *Bacteroides ruminicola*

Bacteroides ruminicola (Bryant *et al.,* 1958b) (Fig. II-10B) includes strains which digest starch and others which are inactive on this substrate. It attacks many more sugars than does *B. amylophilus* and is encountered in cattle fed almost any ration (Bryant *et al.,* 1958b). The fermentation products are similar to those of other rumen *Bacteroides,* succinic, acetic, and formic acids being important, with carbon dioxide fixed. The individual strains assigned to this species (Bryant *et al.,* 1958b) show variation in the sugars fermented and in other characters, which suggests that more than one species is represented.

Bacteroides ruminicola constitutes 6 to 19% of the colonies counted in glucose-cellobiose rumen fluid agar cultures inoculated from animals on various feeds (Bryant *et al.,* 1958b) and appears to be relatively more important in animals receiving low starch rations, constituting 64% of the cultivable starch digesters in an animal fed wheat straw, but only 10% of the starch digesters in an animal fed solely on a grain mixture.

Its nitrogen nutrition is peculiar. Enzymatically hydrolyzed casein is superior to acid-hydrolyzed casein (Bryant and Robinson, 1963), presumably due to availability of peptides (Pittman and Bryant, 1963). As a nitrogen source, ammonia is superior to other tested substances, except peptides of three or more amino acids (Pittman and Bryant, 1964).

Heme or a related compound is the nutritional factor in rumen fluid re-

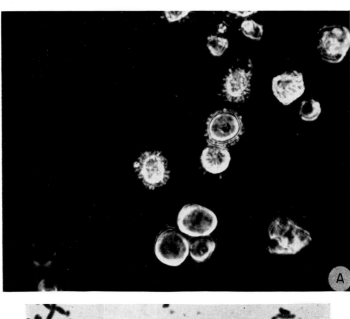

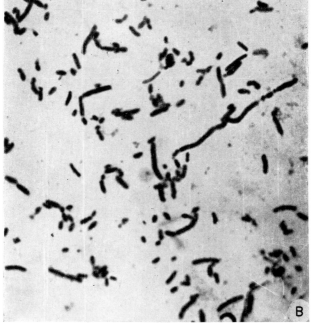

Fig. II-10. A. Cells of *Bacteroides amylophilus* attached to starch grains. Light-phase contrast. After Hamlin and Hungate (1956). Magnification: 970 ×. B. Cells of *Bacteroides ruminicola*. Stained smear. Pure cultured and photographed by M. P. Bryant. Magnification: ca. 2000 ×.

quired for growth of the *ruminicola* variety (Caldwell *et al.,* 1962). The requirement is met by deuteroheme, mesoheme, zinc protoheme, manganese protoheme, hemoglobin, catalase, peroxidase, protoporphyrin IX, deuteroporphyrin, mesoporphyrin, hematoporphyrin, coproporphyrinogen, or uroporphyrinogen. The conversions of plant pigments in the rumen have been little studied, though a few examinations (Davidson, 1954) support the conclusion from visual observations that the pigments are degraded, and in some cases intermediates such as phylloerythrin have been demonstrated (Quin *et al.,* 1935).

The heme is required for synthesis of cytochrome *b* and a cytochrome similar to *o* in spectrum but lacking oxidase activity. These are found also in var. *brevis* which can synthesize the necessary heme from simple foods. Cytochrome b and flavoprotein function in the reduction of fumarate to succinate (White *et al.,* 1962a,b) in a reaction which may yield high-energy phosphate.

This discovery of cytochromes in rumen bacteria is of great interest. It may assist in explaining cell yields under anaerobic conditions greater than those accounted for by previously known pathways of ATP formation.

d. Succinimonas amylolytica

Succinimonas amylolytica (Bryant *et al.,* 1958b) is another starch-digesting type found in the rumen, but not in as large numbers as the various species of *Bacteroides*. It is a short rod with a single polar flagellum. The fermentation products resemble those of *Bacteroides* except that a trace of propionic acid is formed, but no formic acid. Carbon dioxide is required for growth, as is expected in anaerobic species producing succinic acid as an important fermentation product. This species was relatively abundant in one steer on a high grain ration, but not in a second animal (Bryant *et al.,* 1961), and it was not a prominent form in animals on alfalfa silage, alfalfa hay, or wheat straw. It ferments only starch and hydrolysis products of starch and appears to occupy a niche comparable to that of *Bacteroides amylophilus*.

e. Selenomonas ruminantium

Selenomonas ruminantium (Bryant, 1956) (Fig. II-11) is the other starch-digesting bacterium isolated from the rumen, though not all strains are amylolytic. It has been observed by direct microscopic examination in sheep in numbers of $1.5–428 \times 10^6$ per milliliter (Reichl, 1960; Warner, 1962a,b). The crescentic shape and lateral flagella of the large phase of this genus have been mentioned in the section of this chapter on morphology. Many sugars are fermented, which suggests that *Selenomonas* maintains itself in the rumen under a wide variety of feeding conditions.

A salient feature is the production of hydrogen sulfide from media containing cysteine. Another is a lower final pH in culture media (4.3) than is characteristic of most bacteria except *Streptococcus bovis* (4.0) and species of *Lactobacillus*.

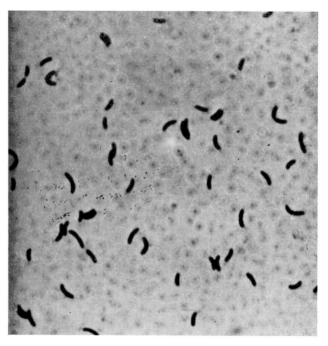

Fig. II-11. Cells of *Selenomonas ruminantium*. Gram stain. Pure cultured and photographed by M. P. Bryant. Magnification: ca. 1000 ×.

The fermentation products (Bryant, 1956; Hobson and Mann, 1961) vary among strains, but all produce acetic and propionic acid. Some carbon dioxide is formed, but no hydrogen. Butyric acid, succinic acid, lactic acid, and formic acid are other identified products, but not all are produced by all strains. No volatile alcohol was detected. The succinic pathway is concerned in the substrate conversion to propionic acid (Paynter, 1962). Strains of the variety *lactilytica* do not combine with antiserum against the variety *ruminantium* (Hobson *et al.,* 1962). These differences warrant their separation into two species: *Selenomonas ruminantium, sensu strictu,* to include only the nonlactate fermenters, and *Selenomonas lactilytica* n.sp., to include those fermenting lactate, with characteristics as shown in Table II-9. When grown on glucose *S. lactilytica* strains produce some lactate.

Acetate stimulated growth of *Selenomonas* on a yeast extract-casitone-glycerol medium (Hobson *et al.*, 1963). Propionate, butyrate, and branched-chain fatty acids inhibited growth. Independent studies on other strains given glucose (Bryant and Robinson, 1962b) showed requirements for volatile fatty acids and stimulation by casein hydrolyzate.

Selenomonas ruminantium was not found in very large numbers in animals on a variety of rations (Bryant, 1956). It constituted 4% of the total isolates as a maximum and in some cases was not among the predominant bacteria. In another investigation it occurred in dilutions as high as 10^9 (Hobson and Mann, 1961). Large forms, identified microscopically, were more abundant in sheep fed on green or dry alfalfa than in those fed ground grain (Reichl, 1960). In steers fed cracked corn and urea *Seleno-monas* constituted 20–40% of the total colony count (3×10^{10} per gram), an abundance confirmed by direct microscopic examination.

f. Enzymatic Breakdown of Starch

The amylolytic enzymes of the rumen bacteria presumably resemble those in other organisms. They have not been identified except for the α-amylase of *S. bovis* (Hobson and Macpherson, 1952).

The question of utilization of intermediate digestion products by organisms unable to attack the polysaccharide arises with starch as with cellulose. There is more evidence for release of large molecule intermediates from starch. The proportion of amylolytic to nonamylolytic bacteria in the rumen of animals on a ration consisting largely of grain could provide some clue as to the extent to which intermediates are formed. Bryant *et al.* (1961) do not report this in detail, but *Selenomonas, Bacteroides ruminicola, Butyrivibrio,* and *Succinimonas,* genera containing many starch digesters, constituted 56–60% of the isolates from steers on a feed-lot bloat ration. *Succinivibrio,* able to use dextrin but not starch, constituted 13% of the isolated strains from one animal. The numbers of dextrin-using nonamylolytic bacteria may have been higher and the starch digesters fewer than these percentages since some strains of *Butyrivibrio, Selenomonas,* and *Eubacterium* can ferment dextrin but not starch. These considerations suggest that consumption of dextrin or other oligosaccharide intermediates from starch digestion by nonamylolytic cells may be quantitatively significant. This was believed by Baker (1942) to be important because when starch was fed a great many cells showed an increase in iodophilic polysaccharide, more than he thought could be directly concerned with starch digestion. The relative abundance of *Peptostreptococcus elsdenii* in high-grain-fed animals (Gutierrez *et al.,* 1959a) could be due in part to the use of split products from the starch. Glucose can be detected in the rumen contents of ruminants on a starch ration (Ryan, 1964a,b; Waldo and Schultz, 1956; Shinozaki and Sugawara, 1958), and the concentration increases when the pH drops to 6.0 or lower (Ryan, 1964b).

The ability of *P. elsdenii* and some *Selenomonas* strains to use lactate provides an alternative hypothesis to account for their numbers in association with starch fermenters.

3. HEMICELLULOSE DIGESTERS

Hemicellulose, i.e., plant carbohydrate insoluble in water but soluble in dilute acid or alkali, constitutes a large percentage of the forage consumed by ruminants. It undergoes digestion in the rumen to about the same extent as cellulose (Heald, 1953). Pure hemicellulose is almost completely digested in the rumen (McAnally, 1942). Xylan was hydrolyzed by extracts from frozen stored rumen contents (Pazur *et al.,* 1957), and reducing sugars formed, including xylose. Optimal pH was 6.0–7.0. As much as 40% of added xylan was digested.

Xylan is attacked by strains of *Eubacterium, Bacteroides ruminicola, Bacteroides amylogenes, Butyrivibrio fibrisolvens, Ruminococcus flavefaciens* and *Ruminococcus albus.* Gum arabic is fermented by *Bacteroides amylogenes* and a few strains of *Bacteroides ruminicola.* Much additional study is needed on the fermentation of individual components of the hemicellulose of forage. A microbial enzyme capable of hydrolyzing glucuronides, most active at pH 6.1, has been obtained from the rumen (Karuniratnam and Levvy, 1951) and purified (Marsh, 1954, 1955). It is specific for β-D-glucosyl pyranuronic acids.

Just as *Ruminococcus* can use cellulose yet cannot ferment glucose, except for a few strains, so also many *Ruminococcus* strains which ferment xylan are unable to ferment xylose. These cellulolytic strains all ferment cellobiose, and it is probable that where xylan but not xylose is fermented, xylobiose is the product of digestion that is attacked. Xylobiose, xylotriose, and new di-, tri-, and tetrasaccharides from xylan have been found in the hydrolyzates of purified pentosans subjected to the digestive action of a toluene-treated suspension of a pure culture of *Butyrivibrio* (Howard *et al.,* 1960). Cells grown on insoluble substrates contained intracellular enzymes attacking the substrate, but the enzymes from cells grown on soluble substrates could be obtained from the medium. Since an excess of solid substrate was used, the enzymes were possibly adsorbed on it. Xylobiase was active against xylobiose and oligosaccharides up to 6 xylose units. The enzyme split a single xylose from the oligosaccharides. Cellobiose was not split. The optimum pH was 6.8. With concentrated xylobiose some transfer of xylose to form oligosaccharides was noted. Xylanase produced chiefly xylobiose, at an optimum pH of 5.6. It showed some activity on oligosaccharides, but none on xylobiose.

The ability to digest hemicelluloses is characteristic of all cellulolytic strains. This is shown for *Ruminococcus* in Table II-7 (Hungate, 1957). The hemicellulose is digested to about the same extent as the cellulose.

Pectin is rapidly fermented by mixed rumen bacteria (Howard, 1961) and also by some of the rumen protozoa. It is fermented by *Lachnospira multiparus, Bacteroides succinogenes, Bacteroides ruminicola, Butyrivibrio fibrisolvens,* and *Succinivibrio dextrinosolvens.* A pectin methylesterase splits off methanol, which is metabolized slowly, and does not accumulate in a concentration greater than 4 μg per milliliter (Howard, 1961). In experiments with bacteria from the rumen of animals fed on hay and grass there were formed per hundred μmoles of pectin (as galacturonic acid) 114 μmoles acetic and 38 μmoles propionic acid. The bacteria from animals fed hay and concentrate produced, in μmoles, 102 acetic, 36 propionic, 4 butyric, and 1 lactic acid. In each case approximately 35% of the substrate was recovered as intracellular polysaccharide, calculated as glucose.

It has been repeatedly observed that *Lachnospira multiparus* (Bryant and Small, 1956b) is one of the most abundant bacteria in the rumen of a fistulated heifer fed alfalfa hay. It is also abundant in animals fed ladino clover (Bryant *et al.,* 1960b). In agar cultures it shows a diffuse thready or woolly appearance due to growth through the agar as long filaments of cells. Whether a similar penetration of plant material occurs in the rumen has not been determined, but it has been noted that a greater proportion of *Lachnospira* colonies are obtained from the solids fraction of rumen digesta than from the liquid. Since this species readily ferments pectin in pure culture, pectin may be an important substrate in the rumen. The spread in agar may reflect an ability to penetrate the pectic middle lamella of plant cells. These organisms are gram-positive and possess a single lateral flagellum (Leifson, 1960).

Hemicellulose fermenters isolated from the rumen include some euryoxic xylose-fermenting bacteria (Heald, 1952), *Bacteroides amylogenes* (Doetsch *et al.,* 1957) which digested pentosan and the streptococcus using galactomannan (Williams and Doetsch, 1960).

Two types of xylan-fermenting bacteria were isolated from the sheep rumen (Hobson and Purdom, 1961). One resembled *Butyrivibrio fibrisolvens* and the other was a coccoid- to rod-shaped bacterium which produced formic, acetic, propionic, and succinic acids. A butyrivibrio capable of digesting xylan was almost identical with *Bacteroides amylogenes* except that it did not store iodophilic reserves (Butterworth *et al.,* 1960).

A rod which occurred in numbers up to 10^8 per milliliter was isolated from the rumen of sheep fed wheaten hay (D. J. Walker, 1961). It actively fermented a hemicellulose prepared from the hay and was assigned to the genus *Bacteroides,* though it differed from described species in several important characteristics. It did not form succinic acid, and only acetic and lactic acids were detected as fermentation products, with traces of butyric acid when grown on xylose.

Succinivibrio dextrinosolvens (Bryant and Small, 1956b) (Fig. II-12) attacks dextrin, pectin, and a few sugars. From 100 mmoles of glucose it forms 37 mmoles of formate, 42 mmoles of acetate, 82 mmoles of succinate and 50 mmoles of lactate (Bryant and Small, 1956b; Scardovi, 1963). The EMP glycolytic enzymes have been demonstrated in this species (Scardovi, 1963), as well as a Co^{++}-dependent enzyme which catalyzes the carboxylation of phosphoenolpyruvate to oxalacetate, ATP being formed. A dextran-fermenting *Lactobacillus bifidus* strain has been isolated from the rumen (Clarke, 1959).

4. FERMENTERS OF SUGAR

All of the polysaccharide-digesting bacteria can also utilize mono-or disaccharides. Since these bacteria are abundant in the rumen they play a role also in the fermentation of simple sugars. It would be interesting to know whether the activity of polysaccharidases in the rumen is inhibited during the period in which sugars are present. It could be the mechanism decreasing fiber utilization when the feed contains much sugar.

Alfalfa hay contains about 35% cold-water-soluble material of which one-fourth to one-fifth is sugar. Red clover contains xylose, glucose, fructose, sucrose, a fructosyl fructose, a fructosyl glucose, and an isomer of raffinose (Bailey, 1958a,b). White clover contains the same sugars. Rye grass contains 1% glucose, 1% fructose, 9% sucrose, and 19% fructan (Thomas, 1960).

A few bacteria isolated in the author's laboratory from the rumen of cattle fed alfalfa hay grew in a water extract of the hay, but they were unable to ferment glucose. They could ferment some of the other carbohydrates tested.

Pure cultures of some bacteria isolated from the rumen cannot digest polysaccharides. Presumably in the rumen also they are unable to utilize polyoses and depend on mono- and disaccharides. Since these sugars are available only during short periods after food is ingested, the microorganisms depending on them would be handicapped during nonfeeding periods unless small concentrations of sugars are maintained by hydrolysis of the polysaccharides by other microbes. A lack of continuous supply of sugar

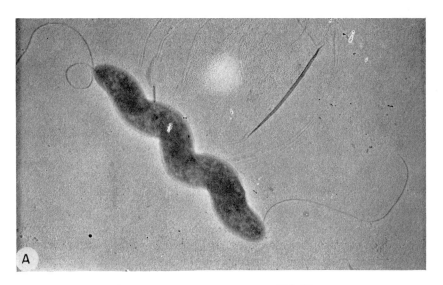

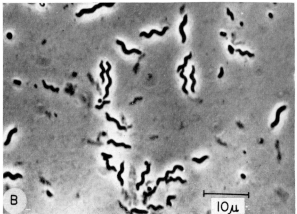

Fig. II-12. A. Electron micrograph of a cell of *Succinivibrio dextrinosolvens* strain S-4-2. Pure cultured by Hannelore Storz and photographed by J. Pangborn. Magnification: 13,500 ×. B. Cells of the same pure culture of *Succinivibrio dextrinosolvens*. Magnification: 1200 ×.

Characteristics of strain S-4-2 were: fermentation products—ethanol, acetate, lactate, and succinate; substrates fermented—L-arabinose, glucose, fructose, mannose, mannitol, sucrose, lactose, cellobiose, and dextrin (±); substrates not fermented—D-xylose, galactose, rhamnose, esculin, salicin, trehalose, maltose, melezitose, starch, inulin, xylan, pectin, and cellulose; indole production negative; nitrate reduced; Voges-Proskauer negative; H₂S production negative; gelatin not liquified.

may be partially compensated by rapid sugar uptake and conversion into storage polysaccharide (Thomas, 1960).

Some sugar utilizers do not store iodophilic polysaccharide. *Borrelia,* which cannot digest the insoluble fractions of the hay and does not form polysaccharide reserves, may be at a disadvantage in the rumen during periods after the sugars have been assimilated and only the insoluble components are available. During the ruminating and rest periods which constitute two-thirds of the day, the bacteria able to digest the fiber would have the advantage and would outstrip the species able to use only sugars. Even if the animal grazed more or less continuously, sugar utilizers such as *Borrelia,* in competition with cells storing polysaccharide, would be at a disadvantage.

Although intermediate sugars can hardly be an important digestion product absorbed by the host, because they occur in such low concentrations that the great distance from the mass of digesta to the rumen wall precludes any quantitatively important diffusion to the absorptive surface of the rumen, glucose or other sugars in the immediate vicinity of digesting polysaccharides in the rumen might supply energy for a significantly large number of nonfiber-digesting bacteria mixed intimately with the fibrous material undergoing digestion. An increase in copper-reducing materials in the rumen with increase in fermentative activity has been found, but the reducing materials have not been identified.

Inspection of Table II-9 shows that a great many of the noncellulolytic rumen bacteria can utilize cellobiose. A β-glucosidase enzyme has been demonstrated in the bodies of mixed rumen bacteria (Conchie, 1954). In view of the possible occurrence of cellobiose in the extracellular fluid during cellulose digestion, this capacity of many of the nonfibrolytic bacteria may have ecological significance.

A number of species of *Lactobacillus* have been found in the rumen. They are an important component of the rumen flora of young calves and on occasion are found in large numbers in the adult rumen. Many are homofermentative (Bryant, 1959). Lactobacilli can grow under rather acid conditions and can be enriched by supplying complex nutrients plus a fermentable carbohydrate at a pH around 5.0. It is probable that acid conditions in the rumen are responsible for the marked development of lactobacilli sometimes encountered. The nutritional requirements of rumen strains tend to be simpler than for other lactobacilli.

Milk which enters the rumen of the young calf undergoes a lactic type of fermentation which leads to an acidity in which growth of most anaerobes is inhibited, but in which lactobacilli can develop. Certain strains have been found chiefly in calves with a rumen pH of 5.7.

In experiments in which a high rumen acidity was induced by sudden feeding of large amounts of grain to cattle (Hungate *et al.,* 1952), the burst of *Streptococcus bovis* which led to marked acidity (pH 4.5) was followed by a great reduction of streptococci and development of a very abundant population of lactobacilli. In such animals, high numbers of lactobacilli indicate a relatively high acidity of the rumen, pH 5–6, during the period preceding observation. Lactobacilli may be abundant also in the rumen of cattle grazed on ladino clover (Bryant *et al.,* 1960) in which the rumen was not particularly acid. The increase in gram-positive rods in animals on a high grain ration may be due in part to an abundance of lactobacilli, as the rumens of these animals are often very acid.

Several species of lactobacilli have been identified in rumen contents. These include *Lactobacillus lactis* (Mann and Oxford, 1954), *Lacobacillus bifidus* (Krogh, 1961b; Phillipson *et al.,* 1962; Bauman and Foster, 1956; Clarke, 1959) of which one strain was ureolytic (Gibbons and Doetsch, 1959), *Lactobacillus brevis* (Briggs, 1953; Jensen *et al.,* 1956; Krogh, 1961b), *Lactobacillus acidophilus* (Jensen *et al.,* 1956; Mann and Oxford, 1955; Perry and Briggs, 1957), *Lactobacillus buchneri* (Jensen *et al.,* 1956; Krogh, 1961b), *Lactobacillus casei* (Jensen *et al.,* 1956; Perry and Briggs, 1957; Krogh, 1961b), *Lactobacillus fermenti* (Jensen *et al.,* 1956; Mann and Oxford, 1955; Perry and Briggs, 1957), *Lactobacillus plantarum* (Jensen *et al.,* 1956; Perry and Briggs, 1957; Krogh, 1961b), and *Lactobacillus cellobiosus* (Krogh, 1961b).

In the animals studied by Jensen *et al.* (1956) the lactobacilli were recovered from the 10^1 to the 10^7 dilutions.

Strains of *L. bifidus* produce a dextranase which hydrolyzes some of the dextrans formed in the rumen (Bailey and Clarke, 1959). The enzyme is an α-1,6-glucosidase which hydrolyzes dextran to isomaltotriose and higher polymers of short chain length. In addition an intracellular α-1,6-glucosidase splits these lower polymeric compounds to glucose (Bailey and Roberton, 1962). Splitting occurs at the nonreducing end of the molecule.

Eubacterium ruminantium (Bryant, 1959) is a gram-variable, nonmotile rod which produces carbon dioxide, acetate, formate, and lactic and butyric acids from various sugars. Strains within the group differ in fermentable sugar pattern and in some other characteristics.

5. Bacteria Utilizing Acids

Utilization of lactate by *Selenomonas* has been mentioned previously. Unknown rumen bacteria must decompose succinate since it does not accumulate in the rumen even though produced by many bacterial types. Many rumen bacteria must also be able to attack formate. Some of these bacteria capable of attacking acids in the rumen have been identified, but it is probable that this capability is characteristic also for additional known and unknown types.

a. Lactate-Utilizing Bacteria

The first of these bacteria to be found in the rumen was isolated by Johns (1948). He isolated *Veillonella alcalescens* [=*Veillonella gazogenes*, =*Micrococcus lactilyticus* (Foubert and Douglas, 1948)] from sheep and showed that it fermented lactate to propionate, acetate, hydrogen, and carbon dioxide. It was a particularly interesting organism because a C_4 dicarboxylic acid was the precursor of propionic acid (Johns, 1951b), a feature later found characteristic also of the propionibacteria. Many of the metabolic characteristics of this organism have been studied (McCormick *et al.*, 1962a,b; Rogosa, 1964; Rogosa and Bishop, 1964a,b). Failure to ferment glucose is explained by a lack of hexokinose (Rogosa *et al.*, 1965).

Veillonella alcalescens has been found in cattle, but the numbers are small (Gutierrez, 1953; and confirmed by Rouf in the author's laboratory), and it is doubtful that this bacterium is quantitatively significant in the bovine rumen. It has been found also in other ruminants in numbers varying between 10 and 10^6 per milliliter (Ogimoto and Suto, 1963). On occasion, large numbers of an organism identified as *Corynebacterium enzymicum* (Gyllenberg and Lampila, 1955) appear to occupy a similar ecological niche. *Propionibacterium acnes* in the rumen (Gutierrez, 1953) may be derived from the feed.

A filamentous bacterium first discovered in sheep (Elsden and Lewis, 1953; Lewis and Elsden, 1955; Elsden *et al.*, 1956) and called the LC (large coccus), later named *Peptostreptococcus elsdenii* (Gutierrez *et al.*, 1959a), may be an important fermenter of lactate in cattle during adaptation to a high grain ration conducive to feed-lot bloat. In feed-lot cattle it occurs in conjunction with *Streptococcus bovis* and other bacteria which ferment the starch to lactate. The lactate is attacked by *P. elsdenii*, with the production of acetic, propionic, butyric, valeric, and caproic acids, carbon dioxide, and a trace of hydrogen. Several amino acids are fermented. After cattle have been adapted to the feed-lot bloat ration (Jacobson and Lindahl, 1955) *P. elsdenii* and *S. bovis* may not remain as prominent as during the period of adaptation.

Peptostreptococcus elsdenii shows the acrylic pathway of propionate formation (Ladd, 1959; Ladd and Walker, 1959; Baldwin and Milligan, 1965; Baldwin *et al.*, 1962, 1963, 1965) and contains a high percentage of flavin (Peel, 1955). Both *P. elsdenii* and *Veillonella alcalescens* contain ferredoxin, and in both it serves as a hydrogen acceptor in the breakdown of pyruvate. Hydrogenase catalyzes the oxidation of reduced ferredoxin by forming hydrogen (Valentine and Wolfe, 1963). Vitamin K compounds are concerned in the metabolism of both these species (Gibbons and Engel, 1964).

Peptostreptococcus elsdenii was abundant in the rumen of young calves, (Hobson *et al.*, 1958), but as they grew older the concentration decreased. A number of serological types were isolated, their proportions fluctuating with time. In some examinations certain types were entirely absent, which indicates a variability in serotypes comparable to that in *Butyrivibrio* and *Ruminococcus*.

Only 40% of single filaments morphologically similar to *P. elsdenii* showed growth when picked to culture media (Purdom, 1963), which suggests that not all cells with this morphology are alike. This may explain the failure to obtain growth in lactate medium of this morphological type from cultures of *Ophryoscolex* (Mah, 1964).

A lactate-fermenting strain of *Butyribacterium* has been isolated from the bovine rumen, but it was found only rarely (Clarke, 1964b). It may be related to *Butyribacterium limosum* Barker and Haas (Moore and Cato, 1965).

b. *Vibrio succinogenes*

Vibrio succinogenes (Wolin *et al.*, 1961; Jacobs and Wolin, 1963a,b) provides an interesting example of an organism utilizing intermediary fermentation products. Hydrogen or formate is oxidized, with stoichiometric reduction of malate or fumarate to succinate or of nitrate to ammonia. Cytochromes b and c are present and presumably function in a phosphorylation during the electron transfer from the reducing to the oxidizing substrate.

Growth with fumarate, malate, or nitrate occurs in the complete absence of oxygen. With hydrogen the organism can grow on oxygen, if the latter is kept at very low concentrations. This microaerophily is explained by accumulation of toxic concentrations of hydrogen peroxide. At low oxygen concentration a peroxidase prevents peroxide accumulation, but is inadequate for the disposition of the peroxide formed at high oxygen concentrations.

The ecology of this organism is obscure. It has been demonstrated in numbers as high as 10^5 per milliliter, but a more suitable selective medium might disclose larger numbers. To the extent that hydrogen is transferred to fumarate to form succinate, the activity might be of significance in increasing the ratio of propionate, but it is doubtful that the reactions of *V. succinogenes* are quantitatively important in the rumen. Perhaps the concentration of malate, fumarate, or nitrate limits growth.

Even if unimportant, the organism is of interest as an example of a species able to exist only in natural habitats characterized by a vigorous mixed fermentation, since it requires continuous availability of at least two transitory fermentation products. Its capacity for stoichiometric reduction of nitrate to ammonia, shown in resting cell suspensions, is an exceptional metabolic feature since, though many bacteria can reduce nitrate to ammonia for growth purposes, few can utilize the electron transfer to ammonia as part of a reaction providing energy to the cell.

c. Oxalate

An euryoxic bacterium decomposing oxalate has been demonstrated in the rumen (O'Halloran, 1962), but its numbers have not been determined, which makes difficult the assessment of its importance in the decomposition of rumen oxalate.

6. METHANOGENIC BACTERIA

As indicated earlier, these have been the most difficult of the rumen bacteria to isolate in pure culture (P. H. Smith and Hungate, 1958). The methane bacteria are unusually sensitive to oxygen, and growth is initiated only when the oxidation-reduction potential of the medium is low, in the case of the rumen methane bacteria about −0.35 volts at pH 7.0. This low potential was first obtained by using hydrogen and a palladium catalyst (Mylroie and Hungate, 1954). Later (P. H. Smith and Hungate, 1958) a little sodium pyruvate was included in the medium, and it was inoculated with *Escherichia coli*. After 24 hours of incubation the *E. coli* was killed by heating the tubes in a boiling water bath. The inoculum of methane bacteria was then injected through the rubber stopper without opening the tube. As little as 0.8 μl of oxygen in a culture tube delays the development of methanogenic colonies, and 6 μl is completely inhibitory. In practice it has been found extremely difficult to avoid entrance of these amounts of oxygen if the tubes are opened.

Reduction of the medium with a solution of sodium sulfide and cysteine assists in obtaining the low oxidation-reduction potential needed for growth of the rumen methane bacteria, and with it the *E. coli* can be omitted. If primary dilution cultures of rumen contents on sugar are gassed with 80% hydrogen-20% carbon dioxide in place of pure carbon dioxide, pale yellow to colorless colonies of the methanogenic bacteria can sometimes be seen. Their incidence correlates with the production of methane in the cultures.

The chief organism producing methane in the rumen is *Methanobacterium ruminantium* (P. H. Smith and Hungate, 1958). It is gram-positive, coccoid- to rod-shaped, single or in chains, with the ends somewhat tapered. It is nonmotile. Colonies often have a slightly yellow color. The cells are surrounded by a thin capsule.

Methanobacterium ruminantium requires rumen fluid in the medium as a source of branched- and straight-chain volatile fatty acids and acetate as well as heme and other growth factors (Bryant, 1965). Ether-soluble and ether-insoluble materials in rumen fluid are also needed. It grows with hydrogen and carbon dioxide as an energy source and converts these materials to methane and water. Of numerous other substances tested, formate is the only one which appears to be utilized.

Although methanogenic bacteria which attack acetate, propionate, and butyrate have been isolated from a number of natural sources, including the rumen (Nelson *et al.,* 1958; Oppermann *et al.,* 1957), these substrates are not converted to methane to any great extent in the rumen (Oppermann *et al.,* 1961), and these methanogenic types occur only in low numbers. This is fortunate, since such a conversion would seriously deplete the quantity of fermentation products oxidizable by the ruminant. The problem is really to determine what are the factors which *prevent* conversion of the host-valuable volatile fatty acids to the useless methane and carbon dioxide.

7. PROTEOLYTIC BACTERIA

The rumen fermentation is primarily saccharoclastic rather than proteoclastic, since the carbohydrates in forages usually preponderate. Proteolytic bacteria have been cultured, as will be discussed in Chapter VII, and proteinases have been found in a number of saccharoclastic species. There have been relatively few nonsaccharoclastic proteolytic bacteria found in rumen contents, but one euryoxic strain has been isolated and the proteinase studied (W. G. Hunt and Moore, 1958). Its ecologic significance has not been estimated. Some of the sporeformers common in soil are primarily proteolytic and have been reported in the rumen in numbers between 10^4 and 10^7. Proteolytic nonsporeformers such as *Proteus, Corynebacterium,* and *Micrococcus* were found, but their origin was not ascertained, and it is uncertain whether they are an important element in the rumen microbiota.

The author has found proteolytic nonsporeforming bacteria in dilutions of rumen contents varying between 10^4 and 2×10^7. Some of these strains digest the cells of the other bacteria and reduce the optical density of a suspension of rumen bacteria in rumen fluid from an initial value of 0.630 to a final 0.470. These bacteria do not utilize sugars. They may live upon other rumen bacteria, and become relatively more abundant during periods when substrates for other bacteria are in short supply, e.g., just before a daily feeding. The rise in the ammonia concentration in the rumen contents before feeding (Moir and Somers, 1957) may be due to actions of these cytoclastic types.

8. LIPOLYTIC BACTERIA

Mixed rumen bacteria (Garton *et al.,* 1958, 1961; Wright, 1961b) hydrolyzed fats into glycerol and fatty acids. The fermentation of glycerol by *Veillonella alcalescens* (Johns, 1953) suggests that this organism may be concerned with fat metabolism in the rumen.

Motile rods with a single polar flagellum were isolated (Hobson and Mann, 1961) from the rumen on a medium containing emulsified linseed oil. Bacteria hydrolyzing the oil occurred in numbers as high as 10^9 per milliliter. The fermentation products included acetic, propionic, butyric, and succinic acids (Hobson and Mann, 1961), but no formic or lactic acid. No large quantities of gas were produced, though small amounts may have been formed or consumed since the gases were not determined with precision. Hydrogen sulfide was formed. Of a number of sugars and derivatives tested, only glycerol, fructose, and ribose were fermented. Two of the three strains isolated produced abundant slime. Old cells ultimately disintegrated, leaving a mass of granules.

The characteristics of these strains do not fit any of the described rumen bacteria. They are gram-negative rods with a polar flagellum. The paucity of sugars fermented separates them from *Butyrivibrio,* the morphology from *Selenomonas*. The fermentation products and the few sugars fermented cause them to resemble *Succinimonas,* but the morphology differs, and the sugars fermented are not the same. The name *Anaerovibrio lipolytica,* n. gen., n. sp., is proposed for these organisms.

9. SUMMARY OF RUMEN BACTERIA SPECIES

A summary of the characteristics of rumen bacteria has been prepared in Table II-9.

Acetic acid is formed by more rumen species than is any other single fermentation product. This agrees with its preponderant production in the rumen. Propionate, which is usually second in abundance, is formed by only a fourth of the species, and the same is true of butyrate. Of the species producing propionic acid, only *Selenomonas* is quite abundant. The quantities of propionate formed in the rumen appear too large to be ascribed to the species shown as propionic acid producers in the table. *Butyrivibrio* is the only butyrate producer occuring in large numbers in most rumens and may be abundant enough to account for the quantities of butyrate formed.

Pure cultures of many species of rumen bacteria produce hydrogen, ethanol, and formic, lactic, and succinic acids, yet these substances do not

ordinarily appear as final products in the rumen. If formed, they are further converted, the formate presumably to carbon dioxide and hydrogen,
the succinate to propionic acid and carbon dioxide. *Butyrivibrio* may use
hydrogen in the reduction of acetate, but otherwise the only species known
to utilize hydrogen actively is *Methanobacterium*. The details of some of
these conversions and their significance in the rumen will be discussed in
Chapter VI.

G. Nutrition

The nutritional requirements of the rumen bacteria are of considerable
practical interest since much of the ruminant feed nourishes the rumen
microbes and is only indirectly destined for the host.

Examination of the nutritional requirements early disclosed a need by
some of the cellulolytic bacteria for materials in rumen fluid (Hungate,
1950). As mentioned for individual strains, some of these materials have
been identified as the C_4 to C_6 branched- and straight-chain fatty acids
(Bryant and Doetsch, 1954a; Bryant and Robinson, 1961c).

Ammonia is essential for some rumen bacteria and stimulatory for
others. In many instances it is preferentially assimilated in the presence of
amino acids. *Bacteroides succinogenes* requires Mg^{++}, Ca^{++}, K^+, Na^+,
and PO_4^{3-} (Bryant *et al.*, 1959), requirements which are probably common to a great many rumen bacteria.

Biotin, *p*-aminobenzoic acid or folic acid, thiamine, pyridoxine, and
pantothenic acid are among the vitamins required by some species. Relatively few of the rumen bacteria can grow on a mineral medium with
carbohydrate as the only organic substrate, though strains of *Bacteroides
amylophilus* can grow under these conditions (Blackburn and Hobson,
1962). The requirement of some strains of *Bacteroides ruminicola* for
precursors of heme (Caldwell *et al.*, 1962) stands in striking contrast with
the ability of other strains to synthesize it.

Nutrient requirements of a number of species are shown in Table II-10
(Bryant and Robinson, 1962b).

In nutritional characteristics, as in most others, the rumen bacteria are
quite variable. They have evolved in many directions with independent
selection for particular traits. Some strains grouped within a species on
the basis of one characteristic differ markedly in others, including nutrition. For this reason the description of nutritional requirements of a given
species cannot be precise. The total variability is not as great, however, as
might be expected, because many of the nutritional requirements are common to several species.

The variability in rumen species precludes a precise analysis of the ru-

Table II-9A

CHARACTERISTICS OF RUMEN BACTERIAL SPECIES

Organism	Shape of cells	Dimensions (μ)	Motility	Sporulation	Capsule	Niche
1. *Bacteroides succinogenes*	Rods	0.3–0.4 by 1–2 (larger on sugar media)	None	None	None	Attacks resistant cellulose
2. *Ruminococcus flavefaciens*	Cocci	0.8–1.0	None	None	Abundant	Fiber digestion
3. *Ruminococcus albus*	Cocci	0.8–2.0	None	None	Abundant	Fiber digestion
4. *Bacteroides amylophilus*	Rods to irregular	0.9–1.6 by 1.6–4.0	None	None	None	Starch digestion
5. *Succinimonas amylolytica*	Coccoid to rod	1.0–1.5 by 1.2–3.0	Single polar flagellum	None	None	Starch digestion
6. *Veillonella alcalescens*	Cocci	0.3–0.6	None	None	None	Lactate fermenter
7. *Methanobacterium ruminantium*	Curved rod	0.7–0.8 to 1.8	None	None	Slight	Methane production
8. *Anaerovibrio lipolytica*	Rods	0.4 by 1.2–3.6	Single polar flagellum	None	Much slime	Lipolytic
9. *Peptostreptococcus elsdenii*	Cocci in chains	1.2–2.4	None	None	None	Lactate fermenter
10. *Clostridium lochheadii*	Rods	0.7–1.7 by 2.0–6.0	None	Spores, 1.0–1.5 2–3.5	Yes	Cellulose digester poor forage
11. *Clostridium longisporum*	Rods	1.0 by 7–15 or 2.3 by 7.0	Motile	Spores, 1 by 3–6	Yes	Unknown
12. *Borrelia* sp.	Spirochete	0.3–0.5 by 4.0–7.0	Motile	None	None	Unknown

13. *Lachnospira multiparus*	Curved rod	0.4–0.6 by 2.0–4.0	Motile by single lateral flagellum	None	None	Pectin digester
14. *Cillobacterium cellulosolvens*	Coccoid to rod	0.5–0.7 by 1.0–2.0	Peritrichous flagella	None	None	Cellulose digestion
15. *Butyrivibrio fibrisolvens*	Curved rod	0.4–0.6 by 2.0–5.0	Single polar flagellum	None	Yes	Starch digestion to widely adapted
16. *Butyrivibrio alactacidigens*	Curved rod	0.5–1.0 by 1.5–8.0	Single polar flagellum	None	—	Starch digestion to widely adapted
17. *Bacteroides ruminicola*	Coccoid to rod to irregular	0.8–1.0 by 0.8–30.0	None	None	Present	Widely adapted
18. *Selenomonas ruminantium*	Crescentic	0.8–2.5 by 2.0–7.0	By tuft of lateral flagella	None	—	Widely adapted
19. *Selenomonas lactilytica*	Crescentic	0.4–0.6 by 1.8–3.0	1–4 flagella tuft of lateral flagella variously located	None	—	Lactate fermenter or various
20. *Succinivibrio dextrinosolvens*	Spiral	0.3–0.5 by 1.0–1.5	Single polar flagellum	None	—	Dextrin fermenter
21. *Streptococcus bovis*	Cocci	0.7–0.9	None	None	Abundant	Starch digestion to various
22. *Eubacterium ruminantium*	Coccoid to rod	0.4–0.7 by 0.7–1.5	None	None	None	Sugars, xylan
23. *Sarcina bakeri*	Cocci	1.0–4.0	None	None	—	Unknown
24. Lactobacilli	Rods	0.7–1.0 by 1–6	None	None	—	Widely adapted under acid conditions

Table II-9B

FERMENTATION PRODUCTS OF RUMEN BACTERIAL SPECIES

Organism	Formate	Acetate	Propionate	Butyrate	Higher acids	Lactate	Succinate	Ethanol	Carbon dioxide	Hydrogen	Methane	Indole production	Gelatin liquefaction	Nitrate reduction	Voges-Proskauer	H_2S production	Final pH in cultures
1. *Bacteroides succinogenes*	+	+	—	—	—	—	++	—	used	—	—	—	—	—		—	
2. *Ruminococcus flavefaciens*	±	+	—	—	—	±	++	±	used	±	—	—	—	—		—	5.0–5.6
3. *Ruminococcus albus*	+	+	—	—	—	±	±	+	+	+	—	—	2/7	—		—	5.1–5.5
4. *Bacteroides amylophilus*	+	+	—	—	—	±	++	+	used	—	—	—	±	—		—	5.5–5.7
5. *Succinimonas amylolytica*	—	+	+	—	—	—	++	—	used	—	—	—	—	—		—	5.2–5.8
6. *Veillonella alcalescens*	—	+	+	—	—	—	—	—	+	+	—	—	'	+		+	
7. *Methanobacterium ruminantium*	—	—	—	—	—	—	—	—	used	used	+						
8. *Anaerovibrio lipolytica*	—	+	++	trace	+	—	+	—		—	—	—	—	—		+	
9. *Peptostreptococcus elsdenii*	—	+ or used	+	+	+	±	—	—	+	+	—	—	—	—	—	+	4.5–5
10. *Clostridium lochheadii*	+	+	—	+	—	±	—	+	+	+	—	—	+	—	—	—	
11. *Clostridium longisporum*	+	+	—	—	—	±	—	+	+	±	—	—					

																	pH
12. *Borrelia* sp	+	+	−	−	−	−	+	−	used	−	−	−	−	±		±	5.5
13. *Lachnospira multiparus*	++	+	−	−	−	+	−	+	+	+	−	−	−	−	−	−	4.7-5.3
14. *Cillobacterium cellulosolvens*	±	±	−	−	−	+++	±	−		+	−	−	−	−	−	−	4.8
15. *Butyrivibrio fibrisolvens*	++	+ or used	−	+	−	+	−	+	+	+	−	−	−	±	±	±	4.6-5.6
16. *Butyrivibrio alactacidigens*	+	+	−	++	−	−	−	−	+	+	−	−	−	±	−	±	
17. *Bacteroides ruminicola*	+	+	−	−	−	−	++	−	used	−	−	−	±	−	±	±	4.6-5.7
18. *Selenomonas ruminantium*	±	+	++	±	−	±	±	−		−	−	−	−	±	+	+	4.3
19. *Selenomonas lactilytica*	−	++	+	−	−	±	+				−	−	−	±	−	+	
20. *Succinivibrio dextrinosolvens*	+	++	−	−	−	+++	++	−	used	−	−	−	−	−	−	−	4.8
21. *Streptococcus bovis*	±	±	−	−	−	−	−	±	±	−	−	−	−	−	−		4.0-4.5
22. *Eubacterium ruminantium*	+	+	−	+	−	+	−	−	+	−	−	−	−	−	−	−	5.0-5.5

Table II-9C

FERMENTATION OF SUBSTRATES BY RUMEN BACTERIAL SPECIES[a]

Organism	D-Xylose	L-Arabinose	Glucose	Fructose	Galactose	Mannose	Rhamnose	Sucrose	Lactose	Maltose	Cellobiose	Trehalose	Raffinose	Melezitose	Dextrin	Starch	Inulin	Xylan	Pectin	Gum arabic	Cellulose	Lactate	Glycerol	Mannitol	Esculin	Salicin
1. *Bacteroides succinogenes*	−	−	+	−	−	−		−	±	±	+	±	−		±	±	−	−	+	−	+		−	−	−	−
2. *Ruminococcus flavefaciens*	−	±	±	−	−	−	−	±	±	−	+	−	−		−	−	−	+	−	−	+		−	−	±	−
3. *Ruminococcus albus*	−	±	±	±	−	±	−	±	±	±	+	−	−		−	−	−	+	−		+		−	±	±	−
4. *Bacteroides amylophilus*	−	−	−	−	−	−		−	−	+	−	−	−		−	+	−				−		−	−	−	−
5. *Succinimonas amylolytica*	−	−	+	−	−	−		−	−	+	−	−	−		+	+	−				−	−	−	−	−	−
6. *Veillonella alcalescens*	−	−	−	−	−	−		−	−	−	−	−	−		−	−	−				−	+	−	−	−	−
7. *Methanobacterium ruminantium*	−	−	−			−		−	−	−	−	−	−	−	−	−	−	−				−	−	−		
8. *Anaerovibrio lipolytica*	−	−	−	+	−	−		−	−	−	−	−	−		−	−	−	−			−	−	+	+	−	−
9. *Peptostreptococcus elsdenii*	−	−	+	+	−	−		±	−	+	−	−	−		−	−	−	−			−	+	±	+	−	−
10. *Clostridium lochheadii*	−	−	+	±	−	+	−	+	+	+	+	−	−		−	+	−	−			+		−		−	+
11. *Clostridium longisporum*	−	−	+	+	+	+	±	+	+	+	+	+	−		−	−	+	−			+		−	±	+	+

12. *Borrelia* sp.	±	+		+	−	−	−	−		−	−		−	−		+	−	+
13. *Lachnospira multiparus*	−	−	∓	±	+	−	−	−	∓	−	−		−	∓	−	+	+	∓
14. *Cillobacterium cellulosolvens*	−	+	+	+	+	+	−	+		±	−	∓	+	−	+	+	+	+
15. *Butyrivibrio fibrisolvens*	±	+	+	+	+	±	±	∓	∓	+	+	+	∓	−	∓	±	±	∓
16. *Butyrivibrio alactacidigens*	−	∓	+	+	+	+	+	±	±	±	±	±	±	−	+	+	+	+
17. *Bacteroides ruminicola*	±	+	+	+	+	∓		±	∓	+	±	±		−	−	+	∓	∓
18. *Selenomonas ruminantium*	+	+	+	+	+	+	±	±	∓	±	±	±		−	∓	+	∓	∓
19. *Selenomonas lactilytica*	∓	+	+	+	−	−	−	+	±	±	∓	±	±	+	−	+	+	+
20. *Succinivibrio dextrinosolvens*	+	∓	±	+	±	−	+	+	+	∓	+	+		−	±	+	∓	∓
21. *Streptococcus bovis*	−	∓	+	+	−	−	+	∓	∓	±	−	+			−	∓	∓	+
22. *Eubacterium ruminantium*	±	+	+	+	+	±	±	±		+	−	−	±		∓	+	−	∓

[a] Key: ∓ indicates few strains utilize; ± indicates most strains utilize.

Table II-10

NUTRITIONAL REQUIREMENTS OF VARIOUS RUMEN BACTERIA[a]

Organism	C_4 to C_5 volatile fatty acids	CO_2	Acetate	NH_3	Amino acids	Amino acid assimilation	Vitamins
1. *Bacteroides succinogenes*	E	E	–	E	–	Fair	Biotin E, PAB E to –
2. *Ruminococcus flavefaciens*	E	E	+	E	+	Poor	Biotin E, folic acid or PAB E, pyridoxamine E to –, thiamine + or –, riboflavin + or –
3. *Ruminococcus albus*	E	E	±	E	+	Poor	Biotin E, PAB E, pyridoxamine E to –
4. *Bacteroides amylophilus*	–	E	–	E	–	Poor	None
5. *Succinimonas amylolytica*	E	E	E	E	–		
7. *Methanobacterium ruminantium*	E	E	+	E			
9. *Peptostreptococcus elsdenii*	–	–	+	–	E	Good	
12. *Borrelia sp.*	E to ?	+					
13. *Lachnospira multiparus*	–		+++	–	–	Good	
15. *Butyrivibrio fibrisolvens*	E to –	+ to –	++ to –	E to –	E to –	Fair to poor	Biotin E, folic acid E, pyridoxal E
17. *Bacteroides ruminicola*	E to –	+ to –	E to –	+ to –	–, Methionine E	Poor	
18. *Selenomonas ruminantium*	E to –	+ to –	+	–	++ to –	Good	

20. *Succinivibrio dextrinosolvens*	−	+	−	+++	E	Good	
21. *Streptococcus bovis*	+ to −	+	+ to −		E to −, arginine +	Good	Biotin E, thiamine +, pantothenic acid ±, Nicotinic acid +
22. *Eubacterium ruminantium*	E	−	E		−	Poor	
24. Lactobacilli	−	±	−	−			

[a] Key: E equals essential in some media; + indicates degree of stimulation; − indicates not stimulatory

men microcosm in terms of numbers of individual cells of specified characteristics. The concept of a species is essential in attempting to describe the rumen population, but for many practical purposes, the time required for valid assignment of isolated strains to species prevents description of any particular rumen in terms of discrete species. It is necessary to consider fewer characteristics than would be required for species identification, and although with these few many strains can be tentatively assigned to species with a fair reliability, particularly after long acquaintance with rumen organisms, others require extended study for proper classification. This necessity for limitation in the characteristics studied in any particular experiment emphasizes the desirability of selecting for study those microbes or microbial characteristics most pertinent to the problem investigated. Care in designing the selective medium employed will increase the potential significance of the results obtained.

A technique for replica subculturing of rumen bacterial colonies would greatly facilitate their detailed study. Their sensitivity to oxygen and failure to grow on plates incubated according to current anaerobic methods makes usual replica plating unsuitable. If a method could be developed, much more extensive information on the nutritional requirements of the rumen bacteria could be obtained.

Analysis of the mixed rumen microbes on the basis of enzyme concentrations and patterns (Baldwin, 1965; Baldwin and Palmquist, 1965) or other chracteristics holds promise as a means of obtaining desired information on the population, without the necessity for isolating and characterizing pure cultures. Recently Suto and Minato at the Institute of Animal Health, Kodaira-shi, Tokyo, Japan, have obtained encouraging results in separating types of rumen bacteria from rumen liquid by the countercurrent distribution method (Albertson and Baird, 1962). The method involves repeated fractionation of the mixed rumen bacteria between two aqueous phases, one containing dextran and the other polyethylene glycol. Further development of this method for separating rumen bacteria will be of great interest.

CHAPTER III

The Rumen Protozoa

As discovered by Leeuwenhoek in 1675, when infusions are prepared by grinding parts of plants with water, various protozoa appear after a day or two of incubation. The most common of these move by means of many small hairs, cilia, which project out from the surface of the body and beat with wavelike contractions which propel the animal. These protozoa are called ciliates, after their method of locomotion, or infusoria, after the hay infusions in which they so commonly appear.

In ages past, when progenitors of modern ruminants consumed grass and water, ciliates similarly appeared in stomachs containing this natural infusion. Through evolution, during millions of years they became specifically adapted to their habitat (Dogiel, 1947).

When Gruby and Delafond (1843) first saw the rumen protozoa they noticed that some of them ingested parts of plant cells. They also observed that, whereas the protozoa in the rumen and reticulum were viable and active, those in the omasum and abomasum were immobile and disintegrating, and the duodenum showed no traces of the bodies of the protozoa (Günther, 1899). It was concluded that the digested protozoa composed part of the food of the host. Since at that time the relative advantages of animal versus plant food was a topic of lively speculation, the protozoa were greeted as an unexpected source of animal food in the supposedly herbivorous ruminant.

For a period of almost 90 years after the discovery of the rumen protozoa, reports and investigations were concerned with their specificity for this habitat. The question was answered conclusively when means were devised for artificially freeing the ruminant of its protozoa. Goats or sheep were defaunated by introducing hydrochloric acid into the rumen (Günther, 1899), by starving (Mangold and Schmitt-Kramer, 1927; Mangold and Usuelli, 1930; Warner, 1962b), by feeding milk (Dogiel and Winogradowa-Fedorowa, 1930), by administration of copper sulfate to starved animals (Becker et al., 1929), by overfeeding (Warner, 1962b), or by various combinations of these factors (Winogradow et al., 1930).

Earlier reports (Günther, 1899) that animals gained the protozoa by consuming hay were based on the appearance in hay infusions of ciliates superficially resembling some of the rumen protozoa (Mangold and Lenkeit, 1931). It was thought the rumen protozoa survived as cysts in the hay (Certes, 1889; Eberlein, 1895; Liebetanz, 1910), but this was later convincingly disproved (Strelkow *et al.*, 1933). Animals remained protozoa-free for long periods if strictly isolated from other ruminants (Becker and Hsiung, 1929).

The rumen ciliates have evolved into a highly specialized group fitted to survive only in the rumen or closely related habitats. They are anaerobic, can ferment constituents of plant materials for energy, and can grow at rumen temperatures in the presence of billions of accompanying bacteria. An important feature of their adaptation to an anaerobic habitat is the ready storage of polysaccharide (Certes, 1889).

A. Morphology and Classification

Most of the rumen protozoa are ciliates, but a few species of small flagellates are regularly encountered. The flagellates are not numerous in the adult ruminant, counts of 2×10^3–3×10^4 being reported in sheep (Warner, 1962b), and because of their small size they do not compose a very large fraction of the bulk of the rumen protozoa. In calves with a rumen near neutrality, the flagellates *Monocercomonas ruminantium* Levine 1961 ($=Trichomonas ruminantium$ Braune 1914) and *Callimastix frontalis* (Braune, 1914) are fairly abundant during the brief period before the ciliates develop (Eadie, 1962a), and they may become quite numerous in animals without ciliates. The flagellates probably are not restricted to the rumen. Other flagellates in the rumen include (Jensen and Hammond, 1964) *Chilomastix,* two species of *Tetratrichomonas, Pentatrichomonas hominis,* and *Monocercomonas bovis.*

Ciliates are asymmetrical, a feature especially evident in the entodiniomorphs, with anterior, posterior, dorsal, ventral, right, and left surfaces all different. Two chief groups of ciliates, holotrichs and entodiniomorphs, occur in the rumen. The family Isotrichidae, consisting of the genera *Isotricha* and *Dasytricha,* is in the order Trichostomatida in the subclass Holotrichia (Honigberg *et al.,* 1964). *Bütschlia parva* Schuberg 1888 is occasionally found (Clarke, 1964a)—order Gymostomatida, subclass Holotrichia. Other rare holotrichs, *Blepharocorys bovis* (Dogiel, 1926a), *Charon ventriculi* (Jameson, 1925), and *Charon equi* (Schumacher, 1915; Clarke, 1964a), have been found in the rumen.

The family Ophryoscolecidae is in the order Entodiniomorpha, subclass Spirotrichia. These ciliates were earlier called oligotrichs, but entodiniomorphs accords with the present classification and is more descriptive.

The Ophryoscolecidae inhabit chiefly the alimentary tract of ruminants. One has been reported from the intestine and cecum of the gorilla (Reichenow, 1920) and chimpanzee (Brumpt and Joyeux, 1912). Rumen genera include *Entodinium, Diplodinium, Epidinium, Ophryoscolex, Caloscolex,* and *Opisthotrichum* (Noirot-Timothée, 1960). The genus *Diplodinium* has been variously divided into subgenera.

The holotrichs possess cilia over the entire body surface, each inserted singly and not fused with others except in the region of the mouth. In the entodiniomorphs most of the body surface is without cilia. In *Entodinium* the cilia are restricted to a single fused ciliary band which encircles the mouth. These cilia tend to join and have been called membranelles, but the combinations beating in unison are not constant, as in membranelles, and the cilia should therefore be called syncilia (Gelei and Sebestyen, 1932; Noirot-Timothée, 1956a). The remaining genera have in addition a band which is displaced various distances from the mouth and encircles part or most of the cell. This is the dorsal ciliary band. The structure of the entodiniomorphs has been studied in detail (Sharp, 1914; Bretschneider, 1934; Noirot-Timothée, 1960).

The holotrichs superficially resemble in their structure and appearance the free-living paramecia. The entodiniomorphs are among the most complex of ciliates. They are paralleled by the cycloposthiid protozoa in the cecum of the horse and its relatives, by protozoa in the stomach of Australian marsupials, and are related to a few groups of free-living marine protozoa.

1. THE HOLOTRICHS

The detailed structure of *Isotricha* has been described (Campbell, 1929; Noirot-Timothée, 1958c). Two species of *Isotricha, Isotricha prostoma* Stein 1858 (Fig. III-1) and *Isotricha intestinalis* Stein 1858 (Fig. III-2) occur in the rumen. They differ somewhat in shape (Schuberg, 1888), with *I. intestinalis* (97–131 $\mu \times$ 68–87 μ) relatively wider than *I. prostoma* (80–160 $\mu \times$ 53–100 μ). In *I. prostoma* the mouth is located at the end opposite the leading or anterior end. This location has elicited speculation (Schuberg, 1888) as to what is actually the anterior end. In *I. intestinalis* the mouth is on one side of the cell equidistant between the posterior end and the middle.

There is only one species of *Dasytricha, Dasytricha ruminantium* Schuberg 1888, (Fig. III-3). It has the mouth at the posterior end. It is smaller than the isotrichs and commonly occurs in greater numbers in the rumen. *Charon* and *Bütschlia* occasionally occur. Diagrams of these species according to Clarke (1964c) are shown in Fig. III-4.

The holotrichs swim more rapidly than the entodiniomorphs. Because

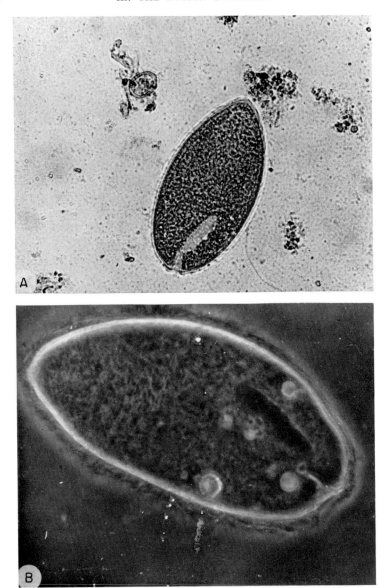

Fig. III-1. A. Photomicrograph of a living cell of *Isotricha prostoma*. Macronu-cleus and cytostome. Direct illumination. Magnification: ca. 375 ×. B. Photomicro-graph by R. A. Mah of a living cell of *Isotricha prostoma*. Macronucleus, cytostome, vacuoles, and cilia covering entire surface. Dark-phase contrast. Magnification: ca. 800 ×.

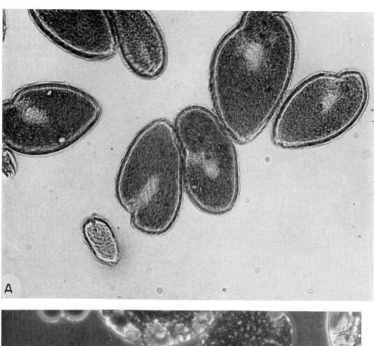

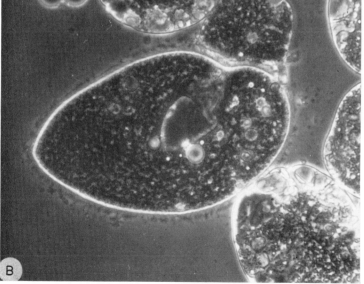

Fig. III-2. A. Photomicrograph of living cells of *Isotricha intestinalis,* one *Isotricha prostoma,* one dasytrich, and one *Entodinium.* Macronucleus and lateral cytostome of *I. intestinalis.* Magnification: ca. 375 ×. B. Photomicrograph by R. A. Mah of living cell of *Isotricha intestinalis.* Macronucleus, cytostome, and surface cilia. Dark-phase contrast. Magnification: ca. 800 ×.

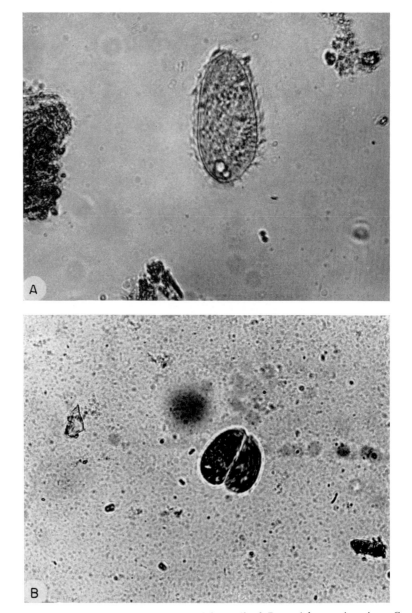

Fig. III-3. A. Photomicrograph of a living cell of *Dasytricha ruminantium*. Contractile vacuole and surface cilia. Direct illumination. Magnification: ca. 625 ×. B. Conjugating *Dasytricha ruminantium*. Magnification: ca. 375 ×.

of this and their large size, the isotrichs are often the most conspicuous protozoa when freshly removed rumen contents are examined under the microscope. The holotrichs maintain motility longer in the presence of oxygen than do most of the other rumen protozoa, and concentrated suspensions stay viable for long periods in the presence of some air (Eberlein, 1895). Low concentrations do not survive long unless anaerobic conditions are maintained. Clarke and Hungate (1966) have noted a correlation between methane production and health of *Dasytricha* in test-tube cultures. The holotrichs are often the first protozoa to become established in the rumen of young lambs (Eadie, 1962a).

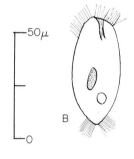

Fig. III-4. A. *Bütschlia parva*. Camera lucida drawing by R. T. J. Clarke. Magnification: ca. 564 ×. B. *Charon equi*. Camera lucida drawing by R. T. J. Clarke. Magnification: ca. 564 ×.

The opacity of the holotrichs is greatly affected by their state of carbohydrate nutrition. They readily assimilate carbohydrate and store what was originally called glycogen (Certes, 1889), more recently identified as amylopectin (Oxford, 1951), which greatly increases the opacity and causes a black reaction when replete cells are stained with iodine. The starch deposited in the ectoplasm is indistinguishable chemically from plant amylopectin (Forsyth and Hirst, 1953). The ectoplasm-endoplasm boundary of *Isotricha* and the fibers positioning the nucleus have been described (Noirot-Timothée, 1958c).

Conjugating *Dasytricha* are sometimes observed (Fig. III-3B) and on occasion may be very numerous (Warner, 1962b).

2. THE ENTODINIOMORPHS

The entodiniomorphs lack the abundant cilia covering the surface of the holotrichs, but have evolved the highly specialized bands of syncilia which function both in locomotion and food ingestion. The vibrating movements of the syncilia can easily be seen under the microscope if the animals are actively moving, and details can be distinguished when movement is slowed by unfavorable conditions. The cilia around the mouth or cytostome are arranged in a clockwise direction (viewed from the anterior end) down into the cytostome (Noirot-Timothée, 1956b). These adoral cilia are the only ones in *Entodinium*. The ciliary bands can be retracted by contraction of retrociliary fibers (Sharp, 1914; Noirot-Timothée 1958a). Retraction is the reaction to unfavorable conditions.

The macronucleus is on the dorsal side. It can be stained by adding a little 0.5% methyl green or carmine in 5% acetic acid to a fresh suspension of the protozoa. The micronucleus, which does not stain readily with the above dyes, lies adjacent to the macronucleus, often in an indentation of the latter. The shape and position of the macro- and micronucleus are important in identifying species. In entodinia the micronucleus is on the inside of the macronucleus, on the outside in the other genera. A migration of the cell and nucleus in an evolutionary transition from *Entodinium* to *Diplodinium* has been postulated (Lubinsky, 1957b).

a. Entodinium

The more common entodinia in domesticated ruminants include *Entodinium bursa* Stein 1858, the type species. It was initially described as the largest entodinial species, and subsequent studies have failed to disclose any that exceed it in size. It differs from many of the other entodinia in ingesting plant material and smaller protozoa.

The descriptions of Stein (1858a), Schuberg (1888), and Dogiel (1927b) indicate that the oral cilia are displaced to one side of the animal, as in Fig. III-5C, and that the endoplasm is usually filled with food particles. The figures of Dogiel (1927b) and Noirot-Timothée resemble Fig. III-5C. The illustrations of Becker and Talbott (1927) and Schuberg (1888) appear to be of one of the smaller entodinia, though the description fits *E. bursa*. In body shape, *E. bursa* (Fig. III-5C) resembles *Diplodinium* more than *Entodinium*. It also ingests plant cell wall material. The groove on the dorsal side is at about the position of the dorsal ciliary band of *Diplodinium,* which increases the resemblance. The appearance suggests that *E. bursa* is a modified *Diplodinium* or a transition step in the evolution from *Entodinium*. It is quite distinct from the other entodinia.

Entodinium caudatum Stein 1858 often exhibits the distinctive appearance shown in Fig. III-5B and F. It is extremely common in many individuals, but missing in some geographic areas. As with so many of the rumen protozoa, the cuticular spines are extremely variable. A single entodinium can give rise to a clone culture (Hungate, 1943) in which typical cells of *E. caudatum* are mixed with individuals resembling two other species. Because of this inconstancy in spination many of the entodinia described as separate species on the basis of spination may simply be variants of *E. caudatum*. The plasticity of the spines in all entodiniomorph species studied in clone culture indicates that spination is a poor taxonomic character.

Some species of *Entodinium* in the rumen of cattle, sheep, and goats are diagrammed in Fig. III-6. Numerous additional species have been described, but until the variability has been tested by examination of clone cultures, excessive splitting appears undesirable. For descriptions of additional species in domesticated and wild ruminants the monograph of Dogiel (1927b) is probably the best single reference.

b. Diplodinium

The remaining genera of entodiniomorphs possess a curved dorsal band of cilia in addition to the adoral band. In diplodinia the dorsal band is located near the anterior end. Diplodinia are compressed laterally and in a fresh mount will usually be seen lying on the right or left side. The nucleus lies dorsally and often has a distinctive shape.

The characteristics of chief importance in distinguishing the species of *Diplodinium* are the size of the cell; position and shape of the macronucleus and micronucleus; number and location of contractile vacuoles; size, number, and position of the skeletal plates; and spination. Of these, spination is the most variable. It is more suited to describe different forms of the same species than to distinguish species. When distinctive spination accompanies other traits it is extremely useful for quick identification of living specimens. Size is also quite variable.

Dogiel (1927b) distinguished four subgenera in the genus *Diplodinium*. These were *Anoplodinium, Eudiplodinium, Polyplastron,* and *Ostracodinium*. Kofoid and MacLennan (1932) created six additional genera and considered each form with a different spination to be a separate species. A clone culture can give rise to several of their species (Wertheim, 1935; Clarke, 1964c), and not all their genera are distinguishable (Noirot-Timothée, 1960). The Dogiel (1927b) classification into forms seems preferable (Wertheim, 1935; Christl, 1958) to the extreme splitting (Kofoid and MacLennan, 1930, 1932, 1933) into species. Noirot-Timothée (1960) has recommended subdivision of *Diplodinium* into six subgenera, with a system which provides distinct genera and reduces splitting.

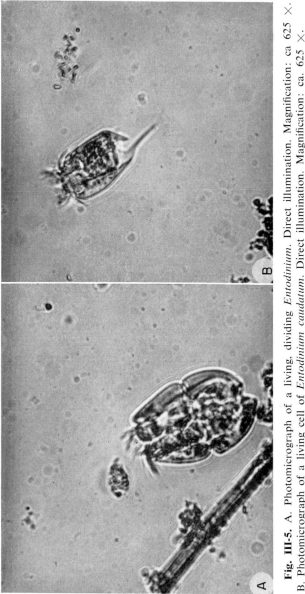

Fig. III-5. A. Photomicrograph of a living, dividing *Entodinium*. Direct illumination. Magnification: ca 625 ×.
B. Photomicrograph of a living cell of *Entodinium caudatum*. Direct illumination. Magnification: ca. 625 ×.

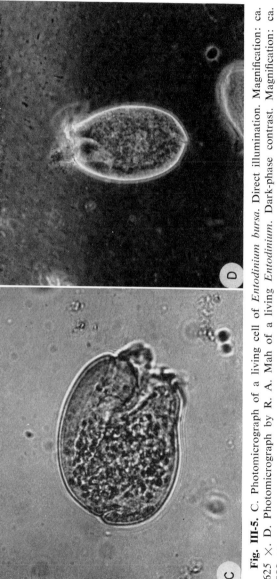

Fig. III-5. C. Photomicrograph of a living cell of *Entodinium bursa*. Direct illumination. Magnification: ca. 625 ×. D. Photomicrograph by R. A. Mah of a living *Entodinium*. Dark-phase contrast. Magnification: ca. 800 ×.

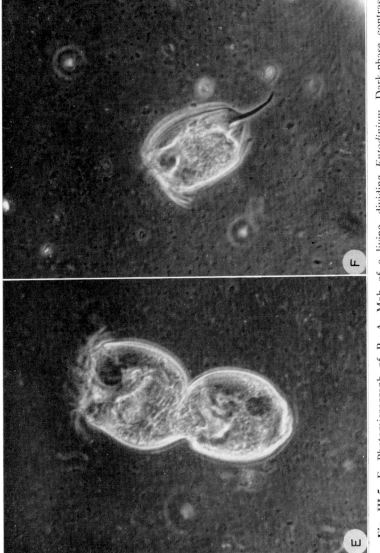

Fig. III-5. E. Photomicrograph of R. A. Mah of a living, dividing *Entodinium*. Dark-phase contrast. Magnification: ca. 800 ×. F. Photomicrograph by R. A. Mah of *Entodinium caudatum*. Dark-phase contrast. Magnification: ca. 800 ×.

The subgenus *Anoplodinium* Dogiel must be replaced by the subgenus *Diplodinium*, since rules of nomenclature require that one subgenus bear the generic name. Noirot-Timothée has adopted two of the genera of Kofoid and MacLennan to give the subgenera *Diplodinium, Eudiplodinium, Polyplastron, Elytroplastron, Ostracodinium,* and *Enoploplastron.* These subgeneric names are often used in place of the generic, but if any ambiguity is involved it is desirable to use the generic name followed by the subgenus in parenthesis.

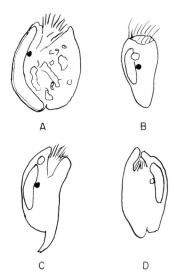

A   B

C   D

Fig. III-6. A. *Entodinium longinucleatum* Dogiel. Redrawn from Dogiel (1927b). Magnification: ca. 440 ×. B. *Entodinium minimum* Schuberg. Redrawn from Dogiel (1927b). Magnification: ca. 520 ×. C. *Entodinium rostratum* Fiorentini. Redrawn from Dogiel (1927b). Magnification: ca. 520 ×. D. *Entodinium elongatum* Dogiel, from the left side. Redrawn from Dogiel (1927b). Magnification: ca. 520 ×.

The subgenus *Diplodinium* is distinguished by the lack of skeletal plates. In his initial description of the rumen ciliates, Stein mentioned *Entodinium dentatum* as having six spines of equal length. He did not distinguish between the protozoa with one or with two ciliary bands. The organism with the six equal spines could have possessed both oral and dorsal ciliature. Schuberg (1888) recognized and named *Diplodinium* as having two ciliary zones, but he did not name a type species nor illustrate any diplodinia. Eberlein (1895) showed drawings of two protozoa, each with six equal-length spines, one with a single zone of membranelles (*Entodinium*) and the other with two (*Diplodinium*). In his extensive examinations of rumen protozoa Dogiel (1927b) never encountered the six-spined *Entodinium*. The author has never observed it. On the assumption that only *Diplodinium* exhibits the six-spined condition orignally described by Stein, *Diplodinium dentatum* [=*Entodinium dentatum* Stein, =*Diplodinium denticulatum* Fiorentini, the difference being the length of the spines (Eberlein, 1895)] is the proper designation for the organism shown in Fig. III-9D. It is a small species, consumes cellulose and presumably digests it (Hungate, 1943), and exhibits great variation in the spination of offspring from a single cell (Hungate, 1943), a clone from a six-spined cell showing individuals with one, two, or no spines, but none like the original type.

Diplodinium posterovesiculatum has been found (Dogiel, 1927b) in cattle, as have *Diplodinium crista-galli, Diplodinium psittaceum,* and *Diplodinium elongatum* (Dogiel, 1927b). All are rare. Diagrams are shown in Fig. III-7.

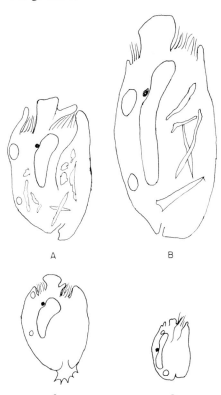

Fig. III-7. A. *Diplodinium psittaceum* Dogiel. Redrawn from Dogiel (1927b). Magnification: ca. 300 ×. B. *Diplodinium elongatum* Dogiel. Redrawn from Dogiel (1927b). Magnification: ca. 300 ×. C. *Diplodinium crista-galli* Dogiel. Redrawn from Dogiel (1927b). Magnification: ca. 300 ×. D. *Diplodinium posterovesiculatum* Dogiel. Redrawn from Dogiel (1927b). Magnification: ca. 300 ×.

The subgenus *Eudiplodinium* possesses either one or two skeletal plates on the right side, which angle from a point near the mouth caudally and dorsally along the macronucleus and diminish in width.

Eudiplodinium neglectum Dogiel 1925(a) (=*Eremoplastron* Kofoid and MacLennan 1932) is one of the most common species and exhibits tremendous variability. Dogiel (1927b) recognized eight different forms.

A clone can give rise to several forms (Clarke, 1964c), each classified as separate species according to Kofoid and MacLennan. In size the individuals vary from *Eudiplodinium neglectum* form *bovis,* 78–100 × 40–54 μ, common in cattle, to the form *giganteum* (Dogiel, 1927b), 330 × 174 μ, found in two African antelopes. A diagram of *E. neglectum* from a clone culture is shown in Fig. III-8B. The protozoa in Figs. III-9A and E resemble *E. neglectum,* but cannot be identified with certainty because from the left side the skeletal plates do not show. Figure III-10C is probably *E. neglectum,* from the left side, the skeletal plate being visible through the cell, with the double appearance caused by poor focus.

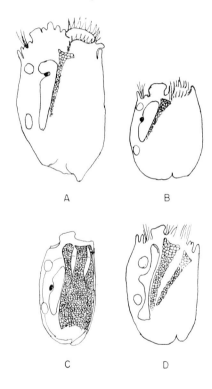

A   B

C   D

Fig. III-8. A. Drawing of a cell of *Eudiplodinium maggii* Fiorentini. Magnification: ca. 264 ×. B. Drawing of a cell of *Eudiplodinium neglectum* Dogiel. Magnification: ca. 330 ×. C. *Enoploplastron triloricatum* (Dogiel) Kofoid and MacLennan. After Noirot-Timothée (1960). Magnification: ca. 264 ×. D. *Eudiplodinium medium* Awerinzew and Mutafowa. Magnification: ca. 99 ×.

Eudiplodinium maggii Fiorentini 1889, frequently encountered in cattle, is easily identified if the macronucleus is stained. It has a characteristic indentation for the micronucleus (Fig. III-8A). The smaller *Eudiplodinium bursa* is almost identical with *E. maggii,* and several authors (Eberlein, 1895; Sharp, 1914) have questioned the separation into two species. The author has observed *Eudiplodinium neglectum* which appear identical with Fiorentini's (1889) drawing of *Diplodinium bursa,* particu-

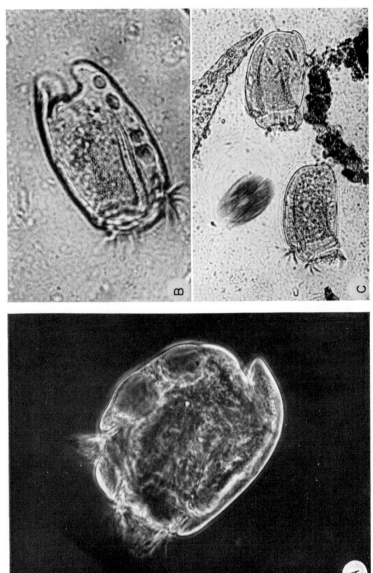

Fig. III-9. A. Living cell of a *Diplodinium* from the left side, showing two very large contractile vacuoles. Dark-phase contrast. Photomicrograph by R. A. Mah. Magnification: ca. 800 ×. B. Living cell of *Ostracodinium*, with the broad skeletal plate barely visible and with two caudal lobes. Viewed from the right side. Note adoral and dorsal ciliary bands. Direct illumination. Magnification 625 ×. C. Living ostracodinia, from the left side. The two curled edges of the large skeletal plate are visible in each cell. Direct illumination. Magnification: ca. 275 ×.

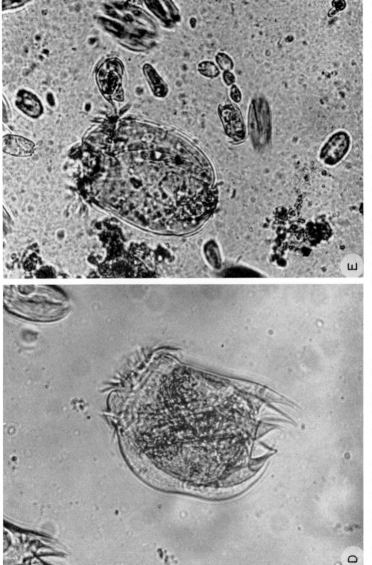

Fig. III-9. D. Living cell of *Diplodinium dentatum* (Stein) Schuberg, from the right side. Direct illumination. Magnification: ca. 625 ×. E. Living *Diplodinium*, possibly *Eudiplodinium neglectum*, from the left side. Some entodinia. Direct illumination. Magnification: ca. 625 ×.

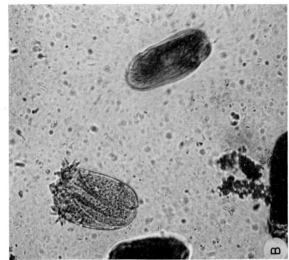

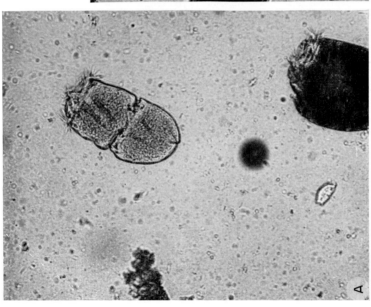

Fig. III-10. A. Living dividing cell of *Eudiplodinium affine* Dogiel, from the right side. Two parallel skeletal plates visible. Direct illumination. Magnification: ca. 275 ×. B. Living cell of *Eudiplodinium affine*, from the right side, and two cells that are probably *Isotricha prostoma*. Direct illumination. Magnification: ca. 275 ×.

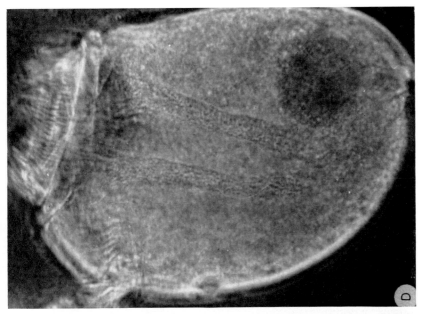

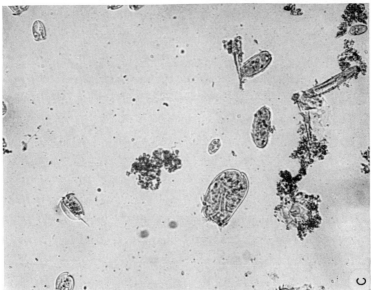

Fig. III-10. C. Living cell of *Eudiplodinium neglectum*, from the left side. Direct illumination. Magnification: ca. 130 ×. D. Living cell, probably *Polyplastron multivesiculatum* Dogiel, from the right side. Two parallel skeletal plates. Magnification: ca. 800 ×. Photomicrograph by R. A. Mah. Dark-phase contrast.

larly the drawing from the dorsal view, and is inclined to believe that *Diplodinium neglectum* is a synonym of *Diplodinium bursa.*

Eudiplodinium medium Awerinzew and Mutafowa 1914 is one of the largest rumen protozoa, the discoverers giving dimensions of 172–272 × 136–170 μ. The macronucleus is depressed at two points on the dorsal side adjacent to the contractile vacuoles (Rees, 1931) (Fig. III-8D). This species, among others, may occasionally ingest other protozoa. The large cell in Fig. III-12B is probably *E. medium,* but without staining the nucleus it is difficult to be sure that it is not *Polyplastron multivesiculatum.*

Eudiplodinium affine Dogiel 1927 (b) resembles *Polyplastron multivesiculatum* and *Eudiplodinium medium* in having two skeletal plates, but it is smaller. A cell of this species, relatively devoid of food particles, is shown in Fig. III-10B, and a dividing cell is shown in Fig. III-10A.

Eudiplodinium rostratum Fiorentini 1889 (Fig. III-14D) is fairly common in cattle (Dogiel, 1927b; Clarke, 1964c). This species has been interpreted as a recently divided *Epidinium ecaudatum* form *caudatum* (Eberlein, 1895). Lack of a skeletal plate in *E. rostratum* excludes it from the epidinia (Clarke, 1964a), and makes it *Diplodinium.*

Polyplastron multivesiculatum Dogiel 1927(b) is the only species of this subgenus in domestic ruminants. The subgenus is distinguished from the others by occurrence of three skeletal plates on the left side of the animal (Figs. III-11A and B). These are often difficult to see in well-fed individuals. The opacity of the cell obstructs examination and the visibility is not improved by staining with iodine, since such a cell is full of amylopectin reserves. In starved specimens the plates can be distinguished, joined by a band which stains more intensely than the rest of the cuticle, but less intensely than the plates. However, the left plates are much less apparent than those on the right. In Figs. III-11A and B are shown sketches of *P. multivesiculatum* as seen by the author. In the author's specimen, no definite middle longitudinal plate on the left side could be detected. The cells shown in Figs. III-10D and III-12A and C are probably *P. multivesiculatum.*

The subgenus *Elytroplastron* Kofoid and MacLennan 1932 is represented by a single species, *Elytroplastron bubali* Dogiel 1928 (=*Elytroplastron longitergum* Hsiung 1931). This was regarded by Dogiel as *Polyplastron,* and it may be related, but the skeletal plate on the left dorsal side is so much better developed than in *Polyplastron* that a separate subgenus seems warranted. *Elytroplastron bubali* occurs in the camel rumen (Dogiel, 1928) and has been reported from Chinese sheep and cattle (Hsiung, 1931, 1932) as well as American cattle (Becker, 1933), though in the latter case the size of the protozoa suggests *Polyplastron multivesi-*

culatum. Diagrams of the plates as seen from the right and left sides are shown in Figs. III-11C and D.

Ostracodinium Dogiel 1927(b) is separated from other subgenera by the possession of a massive skeletal plate. In *Ostracodinium obtusum* Dogiel 1927(b) the plate is relatively flat and covers about half of the right side of the cell. In *Ostracodinium gracile* it is rolled under at the edges. The skeletal plate shows some variability in that it may contain longitudinal slits. Two sketches of ostracodinia are shown in Figs. III-13A and B. The cells in Figs. III-9B and C are *Ostracodinium.*

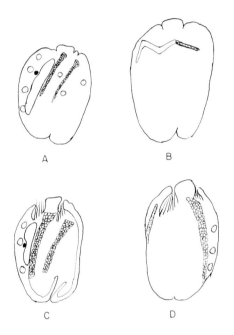

A   B   C   D

Fig. III-11. A. *Polyplastron multivesiculatum,* from the right side, showing two skeletal plates. Magnification: ca. 179 ×. B. *Polyplastron multivesiculatum,* from the left side, showing the indistinct skeletal plates. Magnification: ca. 179 ×. C. *Elytroplastron bubali* (Dogiel) Kofoid and MacLennan, from the right side. Redrawn from Dogiel (1928). Magnification: ca. 220 ×. D. *Elytroplastron bubali,* from the left side, showing the third and fourth skeletal plates. Redrawn from Dogiel (1928). Magnification: ca. 220 ×.

In the last subgenus, *Enoploplastron* Kofoid and MacLennan 1932 (Fig. III-8C), there are three skeletal plates, two of which lie in much the same position as those in *Diplodinium medium,* with the third on the ventral side.

c. Epidinium

The genus *Epidinium* Crawley 1923 includes cells with the dorsal ciliary band displaced posteriorly, as compared to *Diplodinium,* and the body more nearly cone-shaped or cylindrical. Species of this genus were originally included (Fiorentini, 1889) in *Diplodinium.* The many forms, described at various times as separate species, appear (Sharp, 1914) to

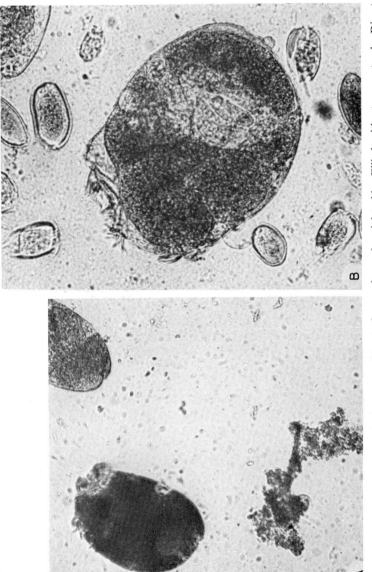

Fig. III-12. A. Probably *Polyplastron multivesiculatum*, from the right side. Filled with storage starch. Direct illumination. Magnification: ca. 275 ×. B. Probably *Eudiplodinium medium*, from the right side. It appears to have ingested another protozoon. Direct illumination. Magnification: ca. 275 ×.

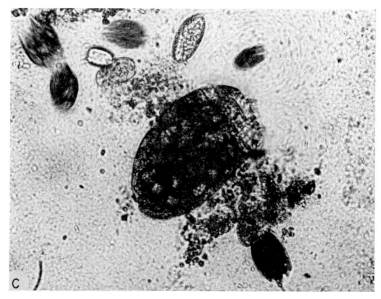

Fig. III-12. C. Probably *Polyplastron multivesiculatum,* from the right side. Filled with ingested starch grains. Note the swirls of bacteria set up by the adoral ciliary band. Direct illumination. Magnification: ca. 275 ×.

belong to one species, *Diplodinium ecaudatum* Fiorentini 1889. This was transferred to *Epidinium* by Crawley (1923). The morphology has been studied in some detail (Sharp, 1914; Bretschneider, 1934; Noirot-Timothée, 1960). The forms *ecaudatum, caudatum,* and *cattanei* (Fiorentini, 1889), and *bicaudatum, tricaudatum,* and *quadricaudatum* (Sharp, 1914) are variants of *Epidinium ecaudatum* (Fiorentini emend. Sharp) Crawley.

In laboratory cultures this genus shows much variability. A photomicrograph of *E. ecaudatum* form *cattanei* is reproduced in Figs. III-14A and C. In the latter the protozoon is filled with starch grains. Form *caudatum* is shown in Fig. III-14B. The motorium described in this species (Sharp, 1914) consists of episkeletal fibers rather than a neural coordinating mechanism (Noirot-Timothée, 1960, p. 637). Numerous other fibers are arranged in a fashion suggesting a coordinating function, but experimental evidence for such a role is lacking.

d. Ophryoscolex

Ophryoscolex Stein 1858 is the most complex of the genera in domestic ruminants. The dorsal ciliary band is displaced posteriorly about one-third of the length of the body and is increased in length to encircle three-fourths of the circumference (Fig. III-15). Several species have been described, but it seems possible that they are forms of *Ophryoscolex purkynei*. Clone cultures of this species (Mah, 1964) show cells indistinguishable from *Ophryoscolex inermis,* and in view of the great variability of the spines it seems possible that the morphological variants represent forms rather than species. An instance in which the two daughter cells from a single division of *Ophryoscolex* showed a difference in spination

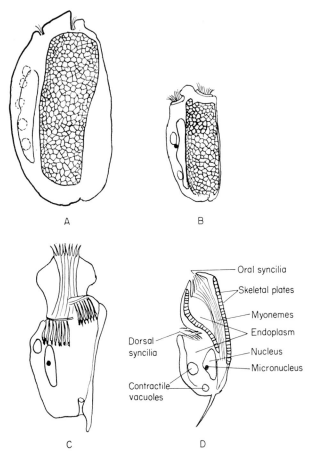

Fig. III-13. A. *Ostracodinium obtusum* Dogiel, from the right side. Redrawn from Dogiel (1927b). Magnification: ca. 440 ×. B. *Ostracodinium gracile* Dogiel, from right side. Redrawn from Noirot-Timothée (1960). Magnification: ca. 264 ×. C. *Caloscolex camelinus* Dogiel, from the right side. Oral cilia and dorsal band. Redrawn from Dogiel (1927b). Magnification: ca. 370 ×. D. *Opisthotrichum janus* Dogiel, from the right side. Redrawn from Dogiel (1927b). Magnification: ca. 330 ×.

has been recorded (Dogiel and Fedorowa, 1926). Nice photographs of *Ophryoscolex caudatus* have been obtained (Ocariz, 1963).

Two other genera of entodiniomorphs have not been reported from domestic ruminants. *Caloscolex* Dogiel 1926(b) occurs in the camel, and *Opisthotrichum* Buisson 1923(a) in some of the African antelopes. Diagrams are shown in Figs. III-13C and D.

The ciliated bands in all the entodiniomorphs are inserted at the base of folds of the body surface. The extended cilia can be retracted by means of fibers (Noirot-Timothée, 1958a) attached to the grooves and folds bearing the bands, and in the tightly retracted condition they show no motion. As the fibers relax, the cilia may be seen beating within the space bounded by the adjacent folds or lips and are gradually exposed and extended as locomotion resumes. In the retracted condition the cilia and cytostome are obliterated as far as detection by microscopic observation of live material is concerned, and identification of species is difficult. The beginning student of the rumen protozoa is advised to examine fresh active preparations until he becomes familiar with the appearance of the various genera in both the active and retracted state.

A Golgi apparatus in the entodiniomorphs has been demonstrated (Noirot-Timothée, 1957).

Division forms are often noted in the entodiniomorphs. Ferber and Winogradowa-Fedorowa (1929) estimated that 7% were in a state of division. Division is preceded by development of a transverse constriction in the middle of the body (Figs. III-10A and III-5A and E) which gradually deepens until the daughter cells separate. A new mouth and membranelles form as division proceeds, these becoming part of the posterior daughter cell, the anterior daughter retaining the old structures. As soon as the membranelles of the new cell begin to form, they start beating, even though not exposed to the external milieu.

Most of the rumen protozoa contain one or more contractile vacuoles (Figs. III-1B, III-3A, III-9A and B, III-14D, III-15D), but these are almost never seen to contract, i.e., to disappear (MacLennan, 1933). The time for a contractile vacuole pulsatory cycle in *Ophryoscolex caudatus* ranged between 15 and 45 minutes, for *Epidinium caudatum* between 1 and 60 minutes, for *Diplodinium maggii* from 2 minutes to an undetermined period, and for *Polyplastron multivesiculatum* between 1 and 10 minutes (MacLennan, 1933). The osmotic pressure in the medium used in these studies may have been lower than is normal for the rumen. Other investigators have not detected contractions. The contractile vacuoles form from lipid granules (MacLennan, 1933; Noirot-Timothée, 1957) and the vacuole itself contains much lipid (Kraschenninikow, 1929).

The central portion of entodiniomorphs is a digestive sac which com-

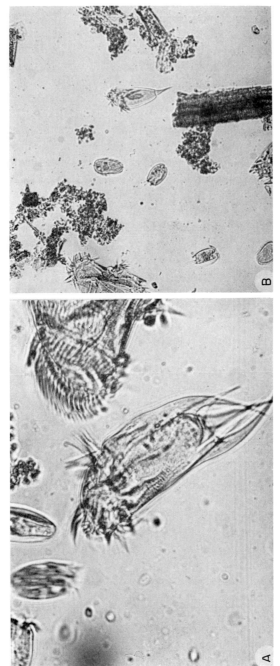

Fig. III-14. A. Living cell of *Epidinium ecaudatum* form *cattanei* (Fiorentini) Sharp, and anterior portion of an *Ophryoscolex*. Direct illumination. Magnification: ca. 594 ×. B. *Epidinium ecaudatum* form *caudatum*, with several entodinia, a dasytrich, and part of an *Ophryoscolex*. Direct illumination. Magnification: ca. 124 ×.

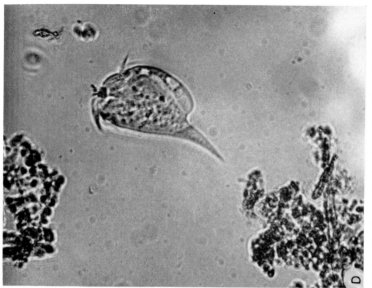

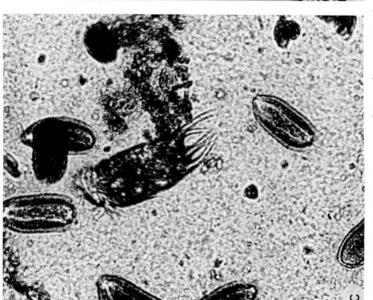

Fig. III-14. C. *Epidinium ecaudatum* form *cattanei*, filled with ingested starch grains. Several dasytrichs. Note masses of bacteria. Direct illumination. Magnification: ca. 275 ×. D. *Diplodinium rostratum* Fiorentini. Direct illumination. Magnification: ca. 625 ×.

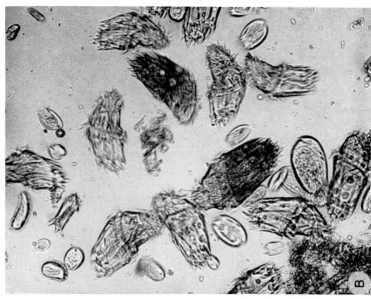

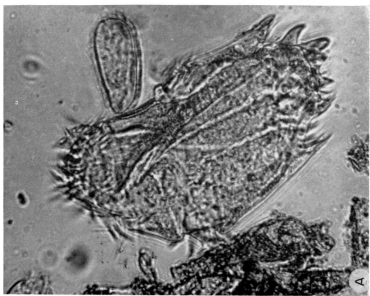

Fig. III-15. A. A living cell of *Ophryoscolex purkynei* Stein, from the right ventral aspect. Note anteriorly forked ventral skeletal plate. Direct illumination. Magnification: ca. 625 ×. B. *Ophryoscolex* and some entodinia. Direct illumination. Magnification: ca. 130 ×.

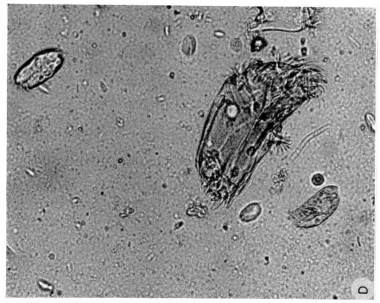

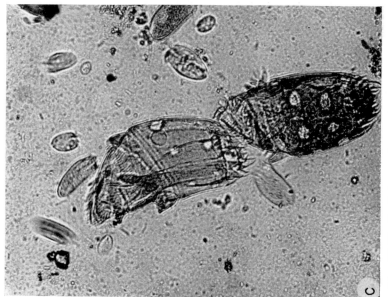

Fig. III-15. C. *Ophryoscolex purkynei*, from the left ventral aspect, showing skeletal plate and contractile vacuoles. Direct illumination. Magnification: ca. 275 ×. D. *Ophryoscolex* showing adoral and dorsal cilia. Direct illumination. Magnification: ca. 275 ×.

municates anteriorly with the cytostome and posteriorly with the anus through which undigested food residues, sometimes large pieces (Dogiel and Fedorowa, 1925), can discharge. It is acid in reaction and contains copper-reducing substances when starch or cellulose have been ingested (Weineck, 1934). Around the digestive sac are one or two endoplasmic layers. The endoplasm is delimited from the ectoplasm by a membrane (Noirot-Timothée, 1956e).

The ectoplasm contains the nucleus (Günther, 1900), the skeletal plates, contractile vacuoles (Bretschneider, 1931), and areas in which amylopectin is deposited. In *Ostracodinium gracile* the amylopectin granules in the ectoplasm are smaller than those in the endoplasm (Noirot-Timothée, 1960, p. 617).

The skeletal plates are not continuous solid structures (Roth and Shigenaka, 1964). When stained with iodine they show a latticelike arrangement, with the brown-staining plate itself the continuous part of the lattice and the black-staining amylopectin (Eadie *et al.,* 1963) in the interstices. The black iodophilic material disappears completely during starvation. The lattice part of the skeletal plate is relatively little affected by starvation, though the reddish-brown color sometimes seems less intense. The black-staining reserve polysaccharides are deposited first in the anterior and posterior lobes of the cell (Hungate, 1943) and then in other ectoplasmic regions, including the lacunae within the skeletal plates (MacLennan, 1934). In animals having access to abundant carbohydrate, polysaccharide deposition may extend to almost all portions of the cell.

The cuticle consists (Noirot-Timothée, 1956d) of a double layer of fine irregular fibrils, the outer longitudinal and the inner transverse. The various projections or spines are extensions of the cuticle.

B. Transfaunation

Young animals are faunated when the mother licks and grooms them and they swallow some of the saliva and digesta (Becker and Hsiung, 1929; Mangold and Radeff, 1930; de Waele and Genie, 1943b). Rumination brings also large numbers of the protozoa into the cow's mouth and is a factor in transfaunation (Strelkow *et al.,* 1933). Faunation may take place also when protozoa left in the food by one animal are consumed by another. Two rumen genera, *Entodinium* and *Diplodinium,* have been found (Fantham, 1922) on the plants in a sheep pasture.

Protozoa have been detected in the stomach of young calves within a week after birth. Inoculation is frequent, but the protozoa do not become immediately established (Fiorentini, 1890; Eadie *et al.,* 1959) because the rumen is too acid. The acidity results from the lactic fermentation of the milk or starter ration (Eberlein, 1895; Mangold and Usuelli, 1930) that escapes from the esophageal groove into the rumen (see Chapter IV). The protozoa cannot tolerate acidity (Purser and Moir, 1959). As forage begins to be consumed, the pH rises, due to less fermentable feed and an increase in the alkaline salivary secretion, and conditions become favorable for establishment of a ciliate population. When the pH of the rumen reaches 6.0, entodinia may become established (Eadie, 1962a; Bryant and Small, 1960), and at 6.5 the holotrichs and higher ophryoscolecids begin to appear. If the rumen has reached a pH of 6.5 before inoculation, all types are established immediately (Eadie, 1962a).

Young ruminants separated from faunated animals soon after birth may require artificial inoculation. Rumen contents for inoculation can be obtained from a fistulated animal, by stomach tube, or by "catching the cud" (Brugnone, 1809; Pounden and Hibbs, 1949b). They are fed to the calf, to inoculate both protozoa and bacteria. If the calf is not separated from other ruminants, inoculation is of doubtful value. With calves that have been stringently isolated or that show a tendency to scour, it may be helpful to inoculate them when forage begins to be consumed. Since inoculation is easily accomplished and is not known to have harmful effects, it may well be a first resort when digestive disturbances occur.

Chance reinoculation of isolated animals occurs more readily than might be expected. Sheep must be separated from each other by a gap of 5 ft. (Eadie, 1962a), and cattle by a greater distance.

Defaunation provides not only animals free of protozoa, but also a means to obtain a rumen with desired types. The rumen can be populated from a clone derived from a single cell (Poljansky and Strelkow, 1934) or with known mixed species (Howard, 1959a; Eadie and Oxford, 1957). Such animals provide an easy and reliable source of masses of the selected protozoa (Eadie, 1962a).

Defaunation tests the essentiality of the rumen protozoa for the host. In early experiments no differences in growth of faunated and defaunated ruminants were observed (Becker *et al.*, 1929; Becker and Everett, 1930; Winogradow *et al.*, 1930; Falaschini, 1934), though in one instance (Usuelli and Fiorini, 1938) faunated sheep grew slightly better. The faunated animals usually exhibited a smoother coat. The total results were interpreted as showing that the protozoa were not essential. This does not mean that they play no significant role. Essential is not equivalent to important. The relatively large size and abundance of the protozoa make them an important element in the rumen.

In the defaunated ruminant the functions of the protozoa may be taken over by bacteria. The observed drop in bacterial numbers following faunation (Eadie, 1962a; Bryant and Small, 1960) supports this interpretation. The question then arises, which performs these functions best, the bacteria or the protozoa? If no superiority in the defaunated animals can be demonstrated, it must be concluded that the protozoa aid the host to the same extent as the bacteria and are thus important, even if not essential. The fermentation gas bubbles which arise from a mass of washed protozoa separated from the rumen provide visible evidence of their activity. The protozoa contain 10–20% of the rumen nitrogen in hay-fed sheep (Ferber, 1929b), and at least 40% in some grain-fed cattle, as found in the author's laboratory. From protozoal counts and the rate of fermentation per cell, Clarke (1964c) estimates the protozoal activity at 30–60% of the total.

In recent experiments with faunated and defaunated lambs reared over a long period (Abou Akkada and el-Shazly, 1964, 1965) the weight gain in faunated animals was significantly greater than in the defaunated, which is in agreement with the tendency recorded earlier (Usuelli and Fiorini, 1938). This may reflect the greater digestibility and biological value of the protozoa as compared to the bacteria.

C. Distribution of the Protozoa

The specificity of the protozoan species for particular ruminant hosts has been studied to some extent. Little experimental transfaunation has been attempted, though some of the cattle rumen protozoa survive in goats (Dogiel and Winogradowa-Fedorowa, 1930). The same kinds and numbers of protozoa do not occur in all hosts of the same species on identical rations (Eadie, 1962b). Predation and intracellular parasites (Jìrovec, 1933; Lubinsky, 1955a,b, 1957d) may be concerned. Some protozoan species are found in a number of host species, but others are restricted. The chief investigations have been those of Dogiel (1927b) and Noirot-Timothée (1956a, 1959b) who studied many of the African antelopes, of Kofoid and MacLennan (1930, 1932, 1933) who examined *Bos indicus* in India and Ceylon and the wild *Bos gaurus* in India (Kofoid and Christensen, 1934), and of Hsiung who examined sheep and cattle in China.

It has been noted that the rumen fauna of a red deer in a zoo differed markedly from the fauna of a wild individual of the same species (Kubikova, 1935), with the inference that the zoo animal's fauna was determined by neighboring ruminants. Food differences could also be concerned in this effect. A summary of the entodiniomorphs in domestic ruminants is given in Table III-1. The relative numbers of ciliates in cattle are shown in Table III-2 (Clarke, 1964a).

Table III-1
OCCURRENCE OF ENTODINIOMORPHS IN CATTLE AND SHEEP

Species	Cattle			Sheep		
	Russia[a]	China[b]	New Zealand[c]	China[d]	Russia[e]	Scotland[f]
Entodinium simplex	+	+		+		
Entodinium elongatum	+					
Entodinium bursa (= *vorax*)	+	+		+		
Entodinium longinucleatum	+	+	+	+		
Entodinium minimum	+	+		+	+	
Entodinium rostratum	+	+	+		+	
Entodinium bicarinatum	+	+				
Entodinium furca	+			+		
Entodinium loboso-spinosum	+			+	+	
Entodinium ovinum			+	+		
Entodinium caudatum	+	+		+	+	+
Entodinium triacum	+			+	+	
Diplodinium posterovesiculatum	+	+	+			
Diplodinium polygonale		+				
Diplodinium dentatum	+	+		+		
Diplodinium crista-galli	+					
Diplodinium psittaceum	+	+				
Diplodinium elongatum	+	+				
Eudiplodinium neglectum	+	+	+	+	+	
Eudiplodinium rostratum	+	+	+			
Eudiplodinium maggii	+	+	+	+	+	+
Eudiplodinium affine	+	+		+	+	+
Eudiplodinium medium	+	+		+	+	+
Polyplastron multivesiculatum	+	+		+		+
Ostracodinium obtusum	+	+	+	+		
Octracodinium dentatum	+	+				
Ostracodinium gracile	+	+		+		
Ostracodinium uncinucliatum		+				
Elytroplastron bubalidis	+	+				
Enoploplastron triloricatum	+	+		+	+	+
Epidinium ecaudatum	+	+	+	+		+
Ophryoscolex purkynei	+	+		+		+

[a] From Dogiel (1927b, 1928).
[b] From Hsiung (1932).
[c] From Clarke (1964a).
[d] From Hsiung (1931).
[e] From Dogiel (1927b).
[f] From Eadie (1957, 1962a).

Table III-2

APPROXIMATE PERCENTAGE GENERIC COMPOSITION OF CILIATES
IN *Bos indicus*, *Bos gaurus*, AND NEW ZEALAND CATTLE[a]

Genus	*Bos indicus*[b]	*Bos gaurus*[c]	N.Z. cattle (mean of 4 cows)[a]
Entodinium	38	52	30
Eodinium	+[d]	+	5
Eremoplastron	6	+	24
Epidinium	1	3	15
Eudiplodinium	3	3	2
Metadinium	2	2	+
Ostracodinium	3	6	2
Diplodinium	9	12	4
Elytroplastron	1	−[e]	−
Ophryoscolex	+	−	−
Polyplastron	−	−	−
Holotrichs	38	21	17

[a] From Clarke (1964a).
[b] From Kofoid and MacLennan (1930, 1932, 1933).
[c] From Kofoid and Christenson (1934).
[d] Plus (+) indicates less than 1%.
[e] Minus (−) indicates not present.

D. Microscopic Examination and Counting

Direct counts obtained by microscopic examination are extremely useful. There is probably no test disclosing the approximate nutritional health of a ruminant as easily and quickly as does the microscopic examination for the protozoa (Ferber, 1929b). Particularly in clinical veterinary medicine, it can be extremely helpful in preliminary diagnosis of alimentary dysfunction. Most of the protozoan species can be identified microscopically in fresh mounts, and total counts as well as counts of the principal genera and species can be obtained (Warner, 1962a,b).

The counting chamber must be deeper than for the bacteria, to accommodate the thicker bodies under the cover slip. A Sedgwick-Rafter counting cell has been used (Purser and Moir, 1959), and careful study using the similar McMaster-type cell (Boyne *et al.*, 1957) has established that counts involve a 10% error if the slide contains five protozoa per field and seventy-five fields are counted. This precision is adequate since the variations in numbers of protozoa in the rumen from one time to another are considerably greater than 10%.

One of the problems in counting the protozoa is the speed with which they settle to the bottom of the suspending medium. The suspension must be vigorously agitated during transfers and sampling. Suspension in 30–40% glycerol diminishes settling (Boyne *et al.,* 1957) of killed protozoa.

As a simple counting method, requiring little apparatus, rumen contents can be diluted to a concentration of 250 to 500 protozoa per milliliter and a 0.1-ml sample counted under the dissecting microscope. The protozoa are drawn singly into a capillary pipette with an orifice slightly larger than the largest protozoon. Gentle suction is applied by means of a rubber mouth tube attached to the pipette.

A method for determining both concentration and total number of protozoa is to count a sample, add water to the rumen, wait till it has mixed, and again count a sample. From the amount of water added and the decrease in the count, an estimate of the total protozoa has been obtained which agrees reasonably well with the number obtained by weighing the rumen contents after slaughter and counting the protozoa in a small sample (Winogradowa-Fedorawa and Winogradow, 1929).

The protozoa are usually so numerous that casual microscopic observation of rumen contents under low power discloses many within a single field. In rumen contents allowed to cool gradually in the laboratory, the protozoa settle and are visible as a white sediment in the bottom of the flask. They sometimes collect inside the flask at points where water drops have been left on the outside surface, perhaps as a result of the lower temperature.

Direct counts of protozoa obtained by various workers are shown in Table III-3.

Because of the great disparity in size between the rumen protozoa and bacteria, counts are not particularly useful for evaluating their relative importance. Mechanical means have been employed to separate the protozoa from the bacteria, and the diaminopimelic acid content of the bacteria has been measured (Weller *et al.,* 1962). Analysis for diaminopimelic acid in the total contents gives a measure of the bacteria. In sheep on wheaten-lucerne hay, the protozoal nitrogen amounted to 5–17% of the total microbial nitrogen at various times after feeding, and in sheep on lucerne hay the value was 3–39%. As mentioned previously, the protozoa may contain more than 40% of the nitrogen in the rumen of cattle on an 82% grain ration.

E. Nutrition and Growth

Cultures of rumen entodiniomorphs have been grown over an extended period (Hungate, 1942, 1943; Coleman, 1960b; Gutierrez and Davis, 1962a; Sayama, 1953; Mah, 1964; Clarke, 1963) without addition of tissues or extracts from the host itself. Types of food commonly consumed

Table III-3

NUMBERS OF PROTOZOA IN THE RUMEN

Animal	Ration	Protozoa per milliliter	Reference
Goat	Hay and grain	1.7–2.9 × 10⁵ total	Winogradowa-Fedorowa
Cattle	Hay and grain	0.7–1.1 × 10⁵ total	and Winogradow (1929)
Cattle	Hay and concentrate	2.7 × 10³ dasytrichs, 0.3–0.4 × 10³ isotrichs	Gutierrez (1955)
Sheep and goats		0.8–1.0 × 10⁶ total	Ferber (1928)
Sheep	Corn-alfalfa	1–4 × 10⁶ total	van der Wath and Myburgh (1941)
	Corn-wheat straw	0.7–0.9 × 10⁶ total	van der Wath and Myburgh (1941)
	Pasture	0.1–0.5 × 10⁶ total	van der Wath and Myburgh (1941)
Sheep	Oaten chaff, alfalfa and linseed oilmeal	2.3–7.8 × 10⁵ total	Purser and Moir (1959)
Sheep	Roughage and cubes at 9:00 A.M. only	1.15 × 10⁶ total	Moir and Somers (1956)
	Cubes at 9:00 A.M., roughage at 5:00 P.M.	2.26 × 10⁶ total	Moir and Somers (1956)
	½ ration at 9:00 A.M., ½ at 5:00 P.M.	2.26 × 10⁶ total	Moir and Somers (1956)
	Roughage at 9:00 A.M., cubes at 5:00 P.M.	2.34 × 10⁶ total	Moir and Somers (1956)
	¼ ration at 9:00 A.M., 11:00 A.M., 1:00 P.M., and 3:00 P.M.	3.14 × 10⁶ total	Moir and Somers (1956)
Sheep	—	0.6–3.3 × 10⁵ total	Boyne et al. (1957)
Sheep	—	1.2 × 10⁴ dasytrichs	Eadie and Oxford (1957)
Sheep	—	0.7–7.3 × 10⁵ total	Reichl (1960)
Cattle	Fresh alfalfa	5 × 10² ophryoscolex	Mah (1962)
Cattle	Timothy hay and concentrate	2.7 × 10⁴ diplodinia, 10⁴ isotrichs, and 0.9 × 10⁴ dasytrichs	Hungate (1957)
Sheep	NaOH-treated oat straw, oat chaff, alfalfa chaff, starch, molasses, and sucrose	10⁶ entodinia, 3.6 × 10⁴ diplodinia, 5.7 × 10⁴ epidinia, and 5.4 × 10⁴ dasytrichs	Purser (1961b)
Sheep	Pasture	1.7–2.2 × 10⁵ entodinia	Warner (1962a)

NUMBERS OF PROTOZOA IN THE RUMEN—*Continued*

Animal	Ration	Protozoa per milliliter	Reference
Sheep	400 gm alfalfa chaff, 400 gm wheaten chaff	1.63–5.9×10^5 entodinia, 3×10^2–1.1×10^4 isotrichs, 1–4.8×10^4 dasytrichs, 0–3.8×10^4 diplodinia, and 0–2.7×10^4 epidinia	Warner (1962b)
Cattle	Fresh red clover	1.9–19.7×10^5 total, 0.4–6.0×10^5 epidinia	Clarke (1964c)
	Clover hay	1.1–4×10^5 total	Clarke (1964c)

by the ruminant have served as food in the *in vitro* cultures, e.g., cellulose, cereal grains, and dried forage. The type of forage is important in culturing some species.

The protozoa do not require any complex organic nutrients synthesized by the host, and rumen fluid is not essential. In some studies the requirement for rumen fluid may reflect the need for more bicarbonate than is provided in the salt solution.

Maintenance of anaerobic conditions is very important in culturing the protozoa. This refers to the concentration of oxygen in the immediate vicinity of the protozoa. A beaker of rumen contents can be brought into the laboratory, exposed to air for several hours, and upon examination found to contain numerous active protozoa. It cannot be concluded from observations of this sort that the protozoa survive in air. Such rumen contents show a darkening which marks the depth to which the oxygen has penetrated, usually only a few millimeters. The rumen contents below that level are as anaerobic as in the rumen.

The slowness of penetration of oxygen into rumen contents may explain in part the ability of the rumen protozoa to survive exposure to oxygen during transfer between animals. Oxygen is slowly consumed by rumen contents (Broberg, 1957d). The chemical reactions involved are not known, but the darkening suggests an oxidase giving rise to melanins.

The protozoa are quite sensitive to acidity (Purser and Moir, 1959). Numbers correlate with minimum pH. At pH 5.3 the count was 3×10^5 per milliliter; at 5.6, 4.7×10^5, and at 5.9, 6.2×10^5, chiefly *Entodinium*. *Entodinium* is somewhat more resistant to acid than are the holotrichs (Abou Akkada *et al.,* 1959), but all protozoa are rapidly killed at very high rumen acidities (Hungate *et al.,* 1952). In controlled pH cultures

(Quinn *et al.,* 1962) the protozoa did not survive extended exposure to acidities outside the pH range of 5.5 to 8.0

1. CARBOHYDRATE SUBSTRATES USED BY THE PROTOZOA

Most of the tests on carbohydrate utilization have been made with protozoa separated mechanically from the rumen bacteria and other digesta. Sedimentation can be assisted by feeding the substrates used for starch storage by the holotrichs; this increases their specific gravity (Heald *et al.,* 1952). Subsequently these cells must be starved to exhaust the reserves before test carbohydrates are added. Another method is to add galactose to the rumen contents (Eadie and Oxford, 1955). Galactose is fermented by the bacteria, but not by the protozoa, except *Dasytricha.* The gas from the bacterial fermentation floats the digesta solids to the surface, and the protozoa on the bottom can be drawn from beneath and washed relatively free of bacteria by differential sedimentation. Cellulose and starch can be employed to increase the specific gravity of the entodiniomorphs.

Migration in an electric field has been employed (Masson *et al.,* 1952) to separate the protozoa from the other rumen solids. To some extent it separates the different genera.

Dasytricha sticks to the sides of a separatory funnel containing rumen liquid to which glucose has been added (Gutierrez, 1955) and remains there after the rest of the contents are drained out. By adding a small portion of balanced salt solution and shaking vigorously, the dasytrichs can be dislodged from the wall and washed to yield a pure suspension.

For obtaining enzyme preparations the holotrichs can be lysed by exposure for 2 hours to saturated solutions of indole, skatole, and related compounds (Eadie and Oxford, 1954; Eadie *et al.,* 1956; Bailey and Howard, 1962). *Ophryoscolex* is also susceptible. It is of interest that these compounds cause marked pleomorphism in *Escherichia coli* (Cavallo and D'Onofrio 1956). Holotrichs can be killed also by anionic detergents (Oxford, 1951), mannose (Oxford, 1958b), and glucosamine (Masson and Oxford, 1951; Oxford, 1959). The other entodiniomorphs do not lyse easily, but *Diplodinium* can be lysed by an antifoam agent (Quinn *et al.,* 1962). The endoplasm of the entodiniomorphs can be extruded by grinding in an abrasive with a mortar and pestle (Hungate, 1942).

a. The Holotrichs

The ecological niche of the holotrichs is the rapid assimilation of soluble sugars (Oxford, 1951). The sugars in feeds such as fresh red clover may constitute 1.34% of the wet weight and 7.35% of the dry weight (Bailey, 1958b). The protozoa from a given volume of rumen contents use sucrose more rapidly than do the bacteria (Thomas, 1960).

As much as 82% of the sugar absorbed by the holotrichs can be stored as starch (Gutierrez, 1955; Thomas, 1960). The storage process itself requires energy, and the fermentation rate with sugar is four to five times the rate without sugar (Gutierrez, 1955). The endogenous rate falls to about half its initial value following starvation for 24 hours. Starch storage from sugar evens out the rumen fermentation and diminishes the burst of postfeeding fermentative activity by storing part of the substrate as starch, to be used later when soluble carbohydrates have been exhausted.

Table III-4

RELATIVE RATES OF FERMENTATION BY HOLOTRICH GENERA[a]

Substrate	Rate of fermentation, as percent of the glucose rate	
	Dasytricha	Isotricha
Glucose	100	100
Fructose	100	100
Galactose	71	0
Maltose	33	0
Cellobiose	100	0
Sucrose	100	100
Raffinose	100	100
Xylose	0	0
Esculin	40	4
Amygdalin	88	13
Arbutin	60	28
Salicin	91	25
Rice-starch grains	0	25
Holotrich starch		
Untreated grains	0	3
Defatted grains	0	3
Precipitated	0	1
Bacterial levan	Attacked	Attacked

[a] After Howard (1959a) and Heald and Oxford (1953).

Many sugars can be utilized. Use of the test carbohydrate is disclosed by deposition of reserve starch within the protozoa. This technique was developed by Oxford (1951, 1955c) and his collaborators (Masson and Oxford, 1951; Heald and Oxford, 1953; Heald *et al.,* 1952) at the Rowett Research Institute. Subsequent analyses (Howard, 1959a; Gutierrez, 1955) have compared the fermentation rate of a test substance with glucose. The results are shown in Table III-4. Starch storage by the holotrichs was maximal 2 to 4 hours after the host was fed (Oxford, 1951).

Dasytricha contains enzymes (Howard, 1959c) hydrolyzing sucrose, maltose (Bailey and Howard, 1963a,b), cellobiose, esculin, arbutin, salicin, and o-nitro-ϕ-β-glucoside. *Isotricha* has more invertase (Christie and Porteous, 1957) and less of the other enzymes. The sucrases from *Dasytricha* and *Isotricha* differ in kinetic properties (Carnie and Porteous, 1959).

Isotricha is unable to use maltose, yet can attack starch, whereas the reverse holds for *Dasytricha*. *Dasytricha* can hydrolyze maltose, but cannot store starch from it (Mould and Thomas, 1958). Both *Dasytricha* and *Isotricha* contain an α-amylase (Mould and Thomas, 1958), equally abundant in both, but with differing isoelectric points and speeds of electrophoretic migration. These amylases may be concerned with digestion of the stored amylopectin. A starch phosphorylase was also demonstrated. The exact biochemical pathways of synthesis and breakdown of starch have not been identified.

The holotrichs are active and viable for a considerable time after reserves can no longer be demonstrated with iodine. Materials other than amylopectin may serve as a source of energy. The loss of nitrogen from holotrichs fermenting glucose (Heald and Oxford, 1953) may indicate an active metabolism of nitrogen, though it is also possible that cell contents were extruded due to the pressure of deposition of the large amounts of amylopectin. If too much sugar is administered the cells store starch until they burst. Such bursting has been reported to occur under normal feeding conditions (Warner, 1964a; Clarke, 1965b).

The starch in holotrichs digested in the abomasum and intestine has been estimated to constitute about 1% of the carbohydrate requirement of the host (Heald, 1951b). Although protozoan starch appears chemically to be identical with plant starch (Forsyth and Hirst, 1953; Walker and Hope, 1963, 1964), it is not equivalent insofar as utilization by the homologous protozoon is concerned. *Isotricha* does not ingest its own starch (Howard, 1957b; Sugden and Oxford, 1952), supplied extracellularly, unless it has been thoroughly washed (Eadie and Oxford, 1955), but *Entodinium* consumes isotrich starch avidly (Quinn *et al.,* 1962).

Pectinesterase and polygalacturonase are abundant in *Isotricha* (Abou Akkada and Howard, 1961) and mixed protozoa (Wright, 1960b), less plentiful in *Dasytricha*. The pH optima are high, 8.6–9.0, and since the products of pectin hydrolysis were not fermented by *Isotricha,* the significance of pectin digestion to the holotrichs is obscure.

b. The Entodiniomorphs

The entodiniomorphs can also be mechanically separated from the rumen digesta (Sugden, 1953). The holotrichs can be eliminated first with mannose or glucosamine (Oxford, 1958b, 1959), or a defaunated animal may be prepared with a single species of entodiniomorph (Eadie and Oxford, 1957). *Polyplastron multivesiculatum* (Eadie, 1962b), erroneously identified originally as *Metadinium medium,* was obtained relatively free from other protozoa, and the utilization of various carbohydrates was tested (Sugden, 1953). Glucose caused no visible storage of iodophilic materials and presumably was not fermented. The entodiniomorphs generally appear to have very limited ability to use soluble substrates (Abou Akkada and Howard, 1960; Williams *et al.,* 1960), though Oxford (1959) reported that *Epidinium ecaudatum* var. *caudatum* stores starch to some extent in the presence of glucose and fructose. The relative amounts of storage by holotrichs and entodiniomorphs, at the expense of soluble sugars, can be strikingly demonstrated by giving glucose to a suspension of starved mixed protozoa from the rumen (Fig. III-16). The mixed protozoa in Fig. III-16A were given glucose. *Isotricha* and *Dasytricha* are opaque and swollen with starch, but the entodinia are transparent. In Fig. III-16B starch was fed, and the entodinia (the only types in this preparation) are opaque with ingested grains. They later become opaque because of their own storage starch.

(1) *Utilization of Starch.* Starch is actively ingested and digested by all the entodiniomorphs except some very small species of *Entodinium* (Sugden, 1953). The speed with which starch is ingested is so rapid that when grains are added to a culture nearly all entodiniomorphs contain at least a few starch grains by the time the material can be examined microscopically. Starch ingestion continues until the cells are packed and swollen. Twenty minutes after 10 gm of starch was fed to a sheep, 83% of the grains were within the protozoa, and at 6 hours 94% (Usuelli, 1930a). Small starch grains are ingested more readily than large ones. *Entodinium caudatum* uses soluble starch and maltose, and to a lesser extent cellobiose, sucrose, and glucose (Coleman, 1964).

Entodinium becomes particularly numerous in ruminants fed large quantities of grain. Ninety-five percent of the protozoa in sheep on a corn-alfalfa hay ration were *Entodinium* (van der Wath and Myburgh, 1941). The protozoa, chiefly *Entodinium,* in cattle fed 80% grain—20% hay contained 40% of the total nitrogen in the rumen. The abundance of *Epidinium* in New Zealand sheep fed fresh red clover may be due to the 3–5.5% starch (Bailey, 1958a). This genus was also abundant in cattle on a high starch ration (Bond *et al.,* 1962).

Maltase and α-amylase have been demonstrated in *Entodinium caudatum* (Abou Akkada and Howard, 1960; Howard, 1963a) and *Epidinium ecaudatum* (Bailey, 1958c). The protozoa contain more amylase than do the bacteria (Schlottke, 1936; Mould and Thomas, 1958; Walker and Hope, 1964).

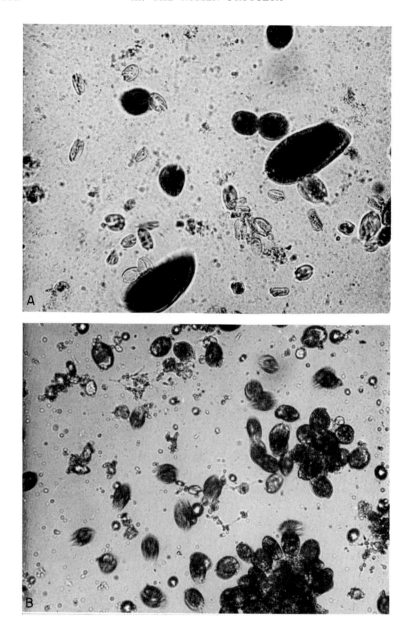

Fig. III-16. A. *Dasytricha ruminantium, Isotricha prostoma,* and entodinia, 30 minutes after glucose was provided. The holotrichs are dark and swollen because of the stored amylopectin granules, whereas the entodinia are transparent. Magnification: ca. 130 ×. B. Entodinia avidly ingesting starch grains from ground wheat. Magnification: ca. 130 ×.

Ingestion of starch does not immediately cause an increase in the metabolism of *Entodinium caudatum,* (Abou Akkada and Howard, 1960) presumably due to a lag in the adsorption of enzyme on substrate. The overall rate of digestion of a starch grain in a protozoon is little greater than that outside (van der Wath and Myburgh, 1941), the digestion in the latter case being accomplished by the bacteria. This can signify that the substrate in both cases is saturated with the enzymes and that the surface area limits the rate. It is not necessarily evidence for participation of the bacteria in the starch digestion in the protozoa. It has been proposed repeatedly (Trier, 1926; van der Wath and Myburgh, 1941; Baker and Harriss, 1947; Sugden, 1953; Quinn, 1962) that digestion in the protozoa is accomplished by bacteria, either ingested with the starch or living symbiotically within the protozoa (McAnally and Phillipson, 1944), but little conclusive evidence on this possibility has been obtained. The penicillin inhibition of growth of *Entodinium* (Coleman, 1960a) and of cellulose utilization by *Polyplastron* (Sugden, 1953) is consistent with this explanation, but could be due to other effects. The starch grains as initially ingested have very few bacteria clinging to them. Cultivable bacteria do not occur to a significant extent within the bodies of the protozoa. This would seem to exclude any important digestion by ingested extracellular bacteria. Those ingested are presumably soon digested. It has been postulated that intracellular symbiotic bacteria digest starch within the protozoa and are so adapted to the intracellular milieu that they cannot grow *in vitro.* This is difficult to examine experimentally.

(2) *Utilization of Cellulose and Other Polysaccharides.* Soon after the protozoa were discovered, the large entodiniomorphs were observed to ingest large quantities of plant materials. Interest centered on their possible capacity to digest cellulose. This was considered the critical factor in the decision as to whether or not the protozoa aided the host.

Because Cleveland's (1923, 1924, 1926) technique of defaunation had been so successful in providing evidence for cellulose digestion by the protozoa in wood-eating termites, defaunation of the ruminant was undertaken as a means to test the capacity for cellulose digestion by the rumen protozoa. Defaunation did not noticeably impair the capacity of goats and sheep to digest cellulose, and it was concluded that the protozoa did not digest cellulose. As discussed previously, the experiment showed that the protozoa were not essential for the digestion of cellulose in these ruminants, but could not prove that the protozoa did not digest cellulose in the rumen, since rumen bacteria capable of cellulose digestion could substitute for any missing cellulolytic protozoa.

The experiments of Cleveland on cellulose digestion by the protozoa in termites are analogous in this respect. As long as a single cellulose-digesting species of *Trichonympha* was left in the termite, cellulose digestion was unimpaired. The protozoan could be any one of three species. All three participated in cellulose digestion in the normal faunated termite, but no one species was essential, provided any other was present.

The initial supposition that all the rumen protozoa behaved alike with respect to the foods used was in part responsible for the confusion over cellulose digestion. Some authors concluded that it was digested, others that it was not, but they did not deal with the same species. Most of the experimental work on cellulose digestion by the rumen protozoa has been with *in vitro* cultures of *Eudiplodinium neglectum* (Hungate, 1942), *Eudiplodinium maggii, Polyplastron multivesiculatum, Diplodinium denticulatum,* and *Entodinium* (Hungate, 1943) and with washed mass cultures of *Polyplastron multivesiculatum* (Sugden, 1953; Abou Akkada *et al.,* 1963).

In the culture work, rumen contents were inoculated into balanced salt solutions containing small concentrations of finely powdered dried grass and finely powdered cellulose. *Eudiplodinium neglectum, Eudiplodinium maggii,* and *Diplodinium dentatum* grew in these cultures, and clones of each species were obtained. Cultures of *Polyplastron multivesiculatum* were obtained with the same medium supplemented with ground wheat grain. In the cultures with grass and cellulose the protozoa ingested large quantities of the cellulose. *Eudiplodinium neglectum* failed to grow if the cellulose was omitted from the medium, but *Eudiplodinium maggii* could survive with dried grass as the only added organic material (Hungate, 1942, 1943). In all these cultures there was an active bacterial fermentation with production of methane, which suggests that *in vitro* bacterial processes were comparable to those of the rumen. Cellulolytic cocci typical of the rumen were demonstrated in these cultures.

The necessity for inclusion of cellulose in the medium could be due to a requirement by the protozoa for the bodies or metabolic products of the bacteria digesting the cellulose. Two lines of evidence indicated that the cellulose was actually digested within the protozoa. Individuals of *E. maggii* were starved until the starch reserves were exhausted and were then fed cellulose. They rapidly ingested it, and within 2 hours iodophilic reserve starch could be seen stored in the posterior and anterior ectoplasmic zones (Hungate, 1943). The number of bacteria in the washed suspensions was too small to account for this digestion.

The second line of evidence was obtained by growing a number of the protozoa *in vitro,* washing them free of accompanying bacteria and debris, and grinding them with quartz sand. The filtered extract was tested for cellulolytic activity. Digestion of the cellulose was inferred from increases in copper-reducing materials after action of the enzymes. The results are summarized in Table III-5.

The absorbent cotton cellulose used in these experiments had been treated in the cold with concentrated hydrochloric acid. This dissolves the amorphous cellulose (Farr and Eckerson, 1934) and leaves the crystalline material. After thorough washing, the dried, HCl-treated cellulose was ground in a mortar and pestle to yield very fine particles of a size easily ingested by the protozoa. For the enzyme experiments a 2% aqueous suspension of this cellulose was pebble-milled and used as substrate. This

Table III-5

Cellulose Digestion by Extracts of *Diplodinium*[a]

	Protozoan extract, plus cellulose	Protozoan extract, no cellulose	Boiled protozoan extract, with cellulose	Extract of digesta, with cellulose	Extract of digesta, boiled, with cellulose
Eudiplodinium maggii					
Experiment 1	+++		±		
Experiment 2	+++	±		−	−
Experiment 3	++	±			
Eudiplodinium neglectum					
Experiment 1	+		−		
Experiment 2	++	±			
Experiment 3	+			−	
Experiment 4	+		−	−	−
Experiment 5	++	±			

[a] After Hungate (1942, 1943). Key: + indicates the approximate magnitude of the copper reduction; ± indicates only a trace of reducing material; − indicates that no reducing material was formed.

treatment decreased the degree of polymerization of the cellulose. It did not increase the copper-reducing power of the cellulose itself or of the liquid. The chemical configuration of the β-1,4-glucosidic linkage in cellulose was unchanged.

These results are consistent with the 50% digestion of cellulose by *Polyplastron* obtained directly from the rumen (Sugden, 1953). Ground cotton cellulose was actively ingested, in preference to the finely ground hay that was also given. *Eudiplodinium medium* ingested cellulose (Westphal, 1934b), but no resulting deposition of polysaccharide reserves was noted.

Reducing sugar accumulates around cellulose ingested by *Diplodinium* (Weineck, 1934). The splitting of cellodextrin by enzymes from *Diplodinium neglectum* (*Eremoplastron bovis*) (Bailey and Clarke, 1963b) is in accord with a capacity for cellulose digestion, though digestion of the latter could not be demonstrated in these experiments. Finally, Abou Akkada *et al.* (1963) demonstrated conclusively a cellulose-digesting ability by *Polyplastron.*

The cultural, enzymatic, and microscopic evidence suggests that cellulose is digested by many entodiniomorphs. Since the biochemistry of cellulose digestion is still somewhat obscure, further experiments on the action of entodioniomorphs on cellulose are highly desirable.

Entodinium, with the possible exception of *Entodinium bursa,* does not digest cellulose (Sugden, 1953; Hungate, 1943), but has been observed to ingest it (Bailey and Clarke, 1963a).

In the author's experience there is no indication of cellulose digestion in *Epidinium* and *Ophryoscolex,* though current reports from others indicate they may be cellulolytic. There is cellobiase activity in *Epidinium* (Bailey *et al.,* 1962), and various hemicelluloses are digested. These genera have been grown *in vitro* (Gutierrez and Davis, 1962a; Mah, 1964) with no cellulose other than that in the dried grass used as part of the substrate. They do not ingest cellulosic plant parts in the same fashion as do the cellulolytic entodiniomorphs. The green chlorophyllous materials often seen within *Epidinium ecaudatum* (Oxford, 1958b) and *Ophryoscolex* (Mah, 1964) rarely include plant cell wall material. Cellulose is not attacked in manometric experiments (Williams *et al.,* 1961). Starch is required in the culture medium for both genera. They avidly ingest starch, as do all entodiniomorphs, and it is probable that it serves as their chief source of carbohydrate.

Pectin is not digested by *Entodinium* (Abou Akkada and Howard 1961; Bailey and Clarke, 1963a), but is actively demethylated by *Ophryoscolex purkynei,* and the resulting polygalacturonate is split by a transeliminase enzyme (Mah, 1962). Added polygalacturonate and galacturonate are not fermented by this protozoon, perhaps because they do not get into the cell. Polygalacturonase and pectinesterase (whether of the transeliminase or hydrolytic type was not determined) occur in *Entodinium* (Abou Akkada and Howard, 1961). Galactosidases splitting melibiose, lactose, galactosyl glycerol, and digalactosyl glycerol occur in *Eudiplodinium maggii, Ophryoscolex, Polyplastron,* and *Epidinium* (Howard, 1963b).

Although the digestion of cellulose within the rumen protozoa occurs through the action of enzymes elaborated within the protozoa, it is possible, as in the case of starch, that intracellular bacteria are concerned. The hypothesis that intracellular symbioses have played a role in the evolution of the cell has many attractive aspects (Hungate, 1955, p. 194). If the abundance of symbioses displayed by so many modern forms of life indicate the extent of this phenomenon in the past, there would have been many opportunities for profitable partnerships which long evolutionary changes could so modify that the original microbial participants are no longer easily recognizable, much less cultivable. This attractive explanation is much in need of critical experimental tests. It seems best to refrain from speculation on intracellular cellulolytic symbionts and to hope that investigations will ultimately provide conclusive evidence as to the validity of the hypothesis.

2. The Fermentation by the Protozoa

Evidence of a fermentation by the rumen protozoa was obtained (Hungate, 1946a) when entodiniomorphs grown *in vitro* were tested manometrically. Carbon dioxide, hydrogen, and volatile acid were demonstrated as products. Hydrogen was a product also of *Isotricha* obtained directly from the rumen.

Holotrichs separated from rumen contents (Heald and Oxford, 1953; Gutierrez, 1955; Howard, 1959a) form acetic, butyric, and lactic acids as the chief fermentation acids, with traces of propionic acid. The average rate of endogenous fermentation is about 30% of the rate with sugar (Gutierrez, 1955). Single isotrichs produced 2.35 mμmoles of acid per hour, and single dasytrichs produced about 0.062 mμmoles. These values have been questioned because with the numbers of protozoa observed in other rumens fermentation products at this rate would exceed the food used (Clarke, 1964c). More reasonable values were found for *Ophryoscolex* (Mah and Hungate, 1965), i.e., 0.146 hydrogen, 0.071 carbon dioxide, 0.042 acetic, 0.004 propionic, and 0.032 butyric acid, all in mμmoles per cell per hour.

The fermentation by *Epidinium ecaudatum* (Gutierrez and Davis, 1962a) forms carbon dioxide, hydrogen, and acetic and butyric acids, with traces of formic, propionic, and lactic acids. *Entodinium caudatum* (Abou Akkada and Howard, 1960; Howard, 1963a) forms carbon dioxide, hydrogen, and acetic and butyric acids as the chief products, with traces of propionate, formate, and lactate. *Ophryoscolex purkynei* (=*Ophryoscolex caudatus* Williams *et al.*, 1961), obtained from the rumen, fermented starch to form acetic, butyric, and lactic acids, and carbon dioxide and hydrogen. *Ophryoscolex purkynei* grown in culture (Mah and Hungate, 1965) gave similar results. The proportions in which fermentation products are formed by the various protozoa are shown in Table III-6.

3. NITROGEN NUTRITION OF THE PROTOZOA

Early experiments (Ferber, 1928, 1929a; Mowry and Becker 1930) showed that the concentration of protozoa in the rumen increased when protein was added to the ration. This was thought to result from improved appetite and greater availability of food. The rumen protozoa are actively proteolytic (Warner, 1955; Schlottke, 1936), the mixed protozoan proteinase of the rumen being most active at pH's near neutrality (Sym, 1938). *Entodinium* has a trypsin-like proteinase (Abou Akkada and Howard, 1962).

By analogy with free-living ciliates it has long been assumed that the rumen protozoa utilize bacteria as a source of nitrogenous food (Margolin, 1930). Conclusive evidence was obtained when Gutierrez (Gutierrez and Hungate, 1957; Gutierrez, 1958) demonstrated ingestion of bacteria by *Isotricha prostoma,* with formation of vacuoles similar to those in *Paramecium.* Individual bacterial cells could be seen to stream from the mouth into the endoplasm. Previous starvation of the holotrichs facilitated detection of the ingestion. Bacteria have been shown to be ingested also by entodinia and diplodinia (Gutierrez and Davis, 1959).

Epidinium ecaudatum grows well in *in vitro* cultures supplied with rice starch, ground alfalfa hay, and cells of *Streptococcus bovis* and *Peptostreptococcus elsdenii* (Gutierrez, 1959). Addition of these bacteria is not essential since (Mah, unpublished experiments) flourishing cultures of *Epidinium ecaudatum* develop from rice and ground grass with no added bacteria. *Streptococcus bovis* and *Peptostreptococcus elsdenii* were probably growing in the cultures (Mah, 1964). The fluid in the epidinium culture remained remarkably clear, which suggests either that the bacteria were ingested or were not successful in competition with the protozoa for food.

Ingestion of bacteria by *Ophryoscolex purkynei* has been observed (Mah, 1964) when starved cells are provided with both wheat starch and mixed bacteria from the culture. The bacteria were not taken in unless starch was available. The bacteria resembled *Peptostreptococcus elsdenii* in appearance, but could not be cultured on lactate media. Lack of starch may explain the failure of *Entodinium* to assimilate casein (Abou Akkada and Howard, 1962).

Success in growing *Entodinium caudatum* has been contingent on inclusion of particulate material from an autoclaved suspension of rumen fluid containing both protozoa and bacteria (Coleman, 1964). By the use of C^{14}-labeled cells, it was shown that *Entodinium* ingested large numbers of *Escherichia coli* and other nonrumen bacteria, which suggests a lack

Table III-6

MOLAR RATIOS OF FERMENTATION PRODUCTS OF THE PROTOZOA

Protozoon (and substrate)	Carbon dioxide	Hydrogen	Formic acid	Acetic acid	Propionic acid	Butyric acid	Lactic acid	Reference
Eudiplodinium neglectum (endogenous)	13.5	12.4	(——————— 22.4 ———————)					Hungate (1946a)
Eudiplodinium maggii (endogenous)	2.4	2.3	(——————— 3.1 ———————)					Hungate (1946a)
Mixed holotrichs (endogenous)	—	—	—	0.33	—	0.16	0.2	Heald and Oxford (1953)
Mixed holotrichs (glucose)	2.7	3.6	—	0.96	—	1.1	2.0	Heald and Oxford (1953)
Mixed isotrichs (glucose)	5.9	5.4	—	0.78	—	1.2	4.2	Gutierrez (1955)
Mixed isotrichs (endogenous)	0.89	0.81	—	0.30	—	0.46	1.25	Gutierrez (1955)
Dasytricha (glucose)	3.44	4.02	—	0.89	—	0.63	2.2	Gutierrez (1955)
Dasytricha (endogenous)	0.89	0.58	—	0.12	—	0.08	0.29	Gutierrez (1955)
Dasytricha (galactose)	92.1	—	—	28.0	—	12.9	66.5	Howard (1959a)
Entodinium	30.0	26.0	1.8	15.4	4.4	22.4	1.0	Howard (1963a)
Epidinium	—	—	—	0.09	—	0.24	—	Gutierrez and Davis (1962)
Ophryoscolex (ground wheat)	14.6	7.1	—	4.2	0.4	3.2	—	Mah (1962)

of specificity in the relationship between the protozoon and the food bacterium.

The decreased bacterial population in faunated sheep (2×10^{10} per milliliter) (Eadie and Hobson, 1962) as compared to defaunated (3.3×10^{11} per milliliter) could be due either to competition for food or to utilization of the bacteria as food by the protozoa or to both mechanisms.

Direct evidence on possible interrelationships between protozoa and bacteria in the rumen is provided by the results of analyses comparing the bacterial and protozoal nitrogen at various times after feeding (Weller et al., 1962) in animals fed once each 24 hours. In sheep fed wheaten (650 gm) lucerne (150 gm) hay the protozoal nitrogen constituted only 5% of the total microbial nitrogen 3 hours after feeding, 6% at 6 hours, 7% at 10 hours, 17% at 16 hours, and 13% at 24 hours. On lucerne hay the values were 3% (probably too low) at 3 hours, 17% at 6 hours, 25% at 10 hours, 19% at 16 hours, and 39% at 24 hours. Each analysis was on a single sheep slaughtered at the indicated time after feeding.

Manometric experiments with feed concentrates (linseed, soybean, and cottonseed oil meals) suggest that nonbacterial protein may be a source of nitrogen for the rumen protozoa (Williams et al., 1961). This was anticipated earlier (Ferber, 1928; Mowry and Becker, 1930) and is consistent with the high counts (80,000 per milliliter) of Entodinium longinucleatum obtained in cultures with added protein (Einszporn, 1961). The material was rapidly ingested, and the rate of gas production increased significantly. The increase was possibly due to starch in the feed concentrates tested, since concentrates freed of starch do not increase the fermentation of Ophryoscolex (Mah, 1964). Protein fermentation would not be expected to cause manometric pressure increase since there is little net production of acid or gas. The Na^+ or other cation associated with the carboxyl anion of the amino acid balances any carbon dioxide (HCO_3^-) freed by decarboxylation, and the ammonia from deamination is a stronger base than the amino group. These will bind carbon dioxide and neutralize the new acidic groups formed in the fermentation.

Predation on other protozoa by the rumen ciliates has been mentioned frequently (Eberlein, 1895; Lubinsky, 1957d; Dogiel, 1927b). This undoubtedly augments the nitrogenous regimen of the predator, but the regularity and quantitative importance of predation has not been studied.

Several lines of evidence suggest that the protozoa may ferment protein to obtain energy. Ammonia is released when concentrates are fed to the protozoa in vitro (Williams et al., 1961) and, less direct, the protozoa remain active for a considerable period after disappearance of all iodophilic reserves. Their source of energy during this period is unknown, but it could be protein. Nitrogen disappears from starved entodinia (Abou

Akkada and Howard, 1962), most of it as ammonia (50–70%) and amino acids (20%). The C^{14} liberated from labeled holotrichs (Heald and Oxford, 1953) in the presence of unlabeled glucose may have come from protein.

Further information is needed to understand the nitrogen metabolism of the protozoa. For the present it seems highly probable that: (1) they digest bacteria and assimilate a considerable fraction of the bacterial nitrogen, (2) they ingest and assimilate other proteinaceous materials, and (3) proteins may be a source of fermentation energy. These activities of the protozoa may be important in the nitrogen economy of the host.

Ciliatine, a new amino acid (2-aminoethyl phosphonic acid), has been found in sheep rumen ciliates (Horiguchi and Kandatsu, 1960). This is the first report of C-P bonds in organisms.

4. CULTIVATION

The early experiments (Hungate, 1942, 1943) on cultivation of *Eudiplodinium neglectum, Eudiplodinium maggii, Diplodinium dentatum, Polyplastron multivesiculatum,* and *Entodinium caudatum* have been mentioned in connection with the experiments on cellulose digestion. *Eudiplodinium neglectum (=Eremoplastron)* and *Eudiplodinium maggii* (Clarke, 1963), *Epidinium ecaudatum* (Gutierrez and Davis, 1962a; Clarke, 1963), *Ophryoscolex purkynei* (Mah, 1964), and *Entodinium caudatum* (Coleman, 1958, 1960a,b, 1962; Kandatsu and Takahashi, 1959; Sayama, 1953) have been grown more recently, with similar methods employed and with various modifications of food and culture vessel.

The holotrichs were originally maintained without transfer for as long as 30 days (Sugden and Oxford, 1952; Purser and Weiser, 1963), and they divided several times (Gutierrez, 1955), but could not be grown continuously. Encouraging results in the cultivation of *Dasytricha* and *Isotricha* have recently been obtained (Clarke and Hungate, 1966). The former has been grown continuously for several months, with a minimum observed division time of 24 hours. An interesting correlation between methane production and the health of the dasytrichs was noted.

Rumen fluid is not essential for successful *in vitro* cultivation of the entodiniomorphs (Hungate, 1942, 1943; Mah, 1964; Gutierrez and Davis, 1959, 1962a). Undiluted rumen fluid is not a favorable medium for maintaining the protozoa (Willing, 1933), even with various modifications. Its content of bicarbonate explains its necessity in media employing other buffers. For the holotrichs, filtered rumen fluid is beneficial.

It has not been possible to duplicate *in vitro* the high concentrations of protozoa sometimes seen in the rumen. This is due to the problem of maintaining *in vitro* the pH, food concentration, and waste product removal of the rumen. Recently (Rufener *et al.,* 1963) *in vitro* fermentation rates essentially similar to those of the rumen have been observed, with fairly good survival of the protozoa. A continuous culture method for regulating various factors affecting growth of masses of the protozoa may provide means for studying aspects of protozoal physiology not easily approached in batch cultures. *Isotricha,* grown only for brief periods and in small number (Gutierrez, 1955), is reported to survive in continuous culture better than do the other protozoa (Quinn, 1962) and may be amenable to cultivation by the continuous technique, although growth was not measured. The turnover rate of the culture was slow enough that dilution would not necessarily have removed all the initial isotrichs during the period of the experiment.

Much work needs to be done with low concentrations of protozoa in batch cultures to elucidate the exact relationships between the protozoa and bacteria. Some of the questions to be answered are: (1) Can bacteria supply the total nitrogen requirement? (2) Is polysaccharide essential or can it be replaced by protein? (3) If bacteria are used, is there a specificity in the requirement, the protozoa ingesting selectively only certain types? (4) If specificity in relation to bacteria exists, is it the same for all the protozoa, or do they exhibit species differences? The work of Coleman (1964) is a definitive approach to answering some of these questions, although the use of *Escherichia* rather than rumen bacteria reduces the applicability of the results to rumen ecology. Rapid developments in this field may be expected.

5. SEXUAL REPRODUCTION

Conjugation has been observed frequently in protozoa obtained directly from the rumen. Conjugating genera include *Dasytricha* (Warner, 1962b) (Fig. III-3B), *Entodinium, Diplodinium, Epidinium,* and *Ostracodinium* (Dogiel, 1925b; Winogradowa, 1936). Conjugation has been seen also in cultures of *Ophryoscolex purkynei* started from a single cell (Mah, 1962).

In the holotrichs the conjugating cells attach along a considerable portion of the surface. In the entodiniomorphs attachment occurs only in the mouth region and the migrant gametic nuclei exchange through this area. The nuclear divisions appear similar in general to those of other ciliates.

6. GROWTH RATES OF THE PROTOZOA

Relatively little accurate information is available on the division rates of the protozoa. An average of 7% have been observed to be in a state of division (Ferber and Winogradowa-Fedorowa, 1929). *Entodinium* division lasts for about 15 minutes (Warner, 1962b). *Eudiplodinium neglectum, Eudiplodinium maggii,* and *Diplodinium dentatum,* when cultured *in vitro* (Hungate, 1942, 1943), maintain their numbers when the culture is divided each day, which indicates a division rate of at least once per day. This would be sufficient to maintain these protozoa in a continuous feed system with a turnover time of 35 hours. *Polyplastron multivesiculatum* maintained its members when the laboratory culture was divided every 48 hours, but not with daily subculture. This was also found for *Isotricha* (Gutierrez, 1955) and *Ophryoscolex* (Mah, 1964), though the latter in some cases showed division rates approaching once per day. The maximum observed *in vitro* rate for *Dasytricha* is one division per day (Clarke and Hungate, 1966). *Entodinium* increased in numbers when carried on a daily culture division schedule (Hungate, 1943), which indicates a capacity for division more than once per day. A division rate of twice per day has been reported for *Epidinium* by Gutierrez (1959).

The division rate of once each 5.5 hours, reported for *Entodinium* (Warner, 1962b), was calculated by a method valid if the time of division is random, whereas dividing forms were more common between 9:00 P.M. and 3:00 A.M. than between 9:00 A.M. and 4:00 P.M. The true division time can be obtained by averaging the calculated division times at evenly spaced intervals during the day. Application of this procedure to the data gives a division time for entodinia of 15 hours. This is the only reported measurement of protozoal division time in the rumen. The doubling in the concentration of protozoa 16 hours after feeding (B. C. Johnson *et al.,* 1944) suggests a minimum average division time of less than 16 hours.

The relationships between the turnover rate of the rumen and the rapidity of division of the protozoa is complicated by the differences in turnover rate of various components of rumen contents (see Chapter V). The liquid may pass through the rumen more rapidly than the solids. The protozoa, especially certain types with slow division rates, may settle or collect around components moving slowly from the rumen and maintain themselves without being washed out. Irregularities in the distribution of *Dasytricha* in the rumen contents of sheep have been noted (Purser, 1961b), explained in part by a periodicity in division of these forms (Warner, 1962b) and possibly by some bursting after the host feeds (Warner, 1964a; Clarke, 1965b). The increase in the specific gravity of the holotrich protozoa when sugar or starch is available could cause partial settling in the rumen.

F. Influence of Various Factors on the Numbers of Protozoa in the Rumen

The holotrichs appear in largest numbers in grazing ruminants and those on a diet containing a considerable fraction of hay. This presumably stems from their requirement for dissolved sugars as a source of energy. *Isotricha* species are sometimes entirely absent from certain sheep (Purser, 1961b), and individuals of *Dasytricha* may constitute only 5% of the total protozoa. These animals contained predominantly *Entodinium,* 88% of the total protozoa. The explanation may be found in the fact that the sheep were kept in pens and fed only once per day, with most of the feed consumed in 4 hours. In addition to alfalfa and chopped oat hay the ration contained a good deal of starch. The sugars in these forages probably provided the dasytrichs with their energy, but its availability only once per day may have limited their population. The starch probably lasted in the rumen (or at least in the entodinia) for a longer time and provided food for *Entodinium* over a longer period than was the case for *Dasytricha.*

The type of forage has an influence on the kind of protozoa that can live in the rumen. *Ophryoscolex purkynei* grows better on fresh alfalfa than on grass hay (Mah, 1964). *In vitro* populations of *Diplodinium dentatum* fed on various forages were (Gradzka-Majewska, 1961): on *Glyceria fluitans* 5333 per milliliter, on *Poa pratensis* 4166 per milliliter, on *Alopecurus pratensis* 3655 per milliliter, on *Aira cespitosa* 400 per milliliter, and on *Trifolium hybridum* no growth. The size of the particles of the various forages was not an important factor influencing the result. The effect of the forage may have been direct or mediated through the bacteria.

High levels of salt in the ration cause a decrease in the size and number of protozoa (Rambe, 1938).

The author once observed large numbers of *Isotricha* in animals on a ration consisting of 80% mixed grains and 20% alfalfa hay. They did not seem to be ingesting the starch to any extent, though the numerous accompanying entodinia were filled with ingested starch. Production of more sugar than the bacteria and protozoa could absorb might explain the ability of the isotrichs to develop in numbers, but their inability to use maltose makes it necessary to assume that also some glucose was formed.

Ophryoscolex and *Epidinium* have been found in high numbers in animals on an alfalfa hay ration supplemented with some grains, and *Epidinium* is abundant in animals on high starch rations (Gutierrez and Davis, 1962a) and in cattle consuming large quantities of fresh red clover (Oxford, 1958b). In this latter case, the protozoa ingested quantities of chlorophyllous material in which starch was probably included. They contained enzymes capable of digesting clover starch grains, but not intact potato starch (Bailey, 1958c). Chlorophyll is ingested by *Ophryoscolex* (Mah, 1964), and stimulation of protozoal development by green plants in the ruminant ration was early noted by Trier (1926). Protozoa actively hydrogenate the lipids in chloroplasts (Wright, 1959) and in oleic acid (Gutierrez *et al.,* 1962), and fat droplets are reported to be ingested and to disappear (Ferber, 1928).

Diplodinium and related genera never seem to occur in large concentrations in the rumen. They are present both in animals on high grain and in those on pasture and hay rations. The cellulose-digesting species ingest starch and presumably digest it. Diplodinia may be limited by the types of bacteria usable as food or by bacterial competition for food. Until more definitive experiments on the food habits of these and the other protozoa are completed, no final conclusions on the factors affecting their numbers can be drawn.

It has been generally noted that the numbers of protozoa are higher in animals on a good ration and with considerable quantities of it than when the ration is poor or scanty. The counts of protozoa per milliliter increase during pregnancy and lactation (Ferber, 1928). Grain and protein-rich foods increase the numbers of protozoa (Mowry and Becker, 1930). The explanation can be twofold: the better ration contains more food usable by the protozoa and is supplied more frequently and continuously, which causes the turnover to be faster. Even on a ration in which 88% of the total nitrogen was in form of amide nitrogen, some protozoa maintained themselves in the rumen (Schmid, 1939).

As might be expected, frequent feeding of the host increases the concentration of the protozoa (Table III-3; Moir and Somers, 1956). Five sheep fed the total daily ration at 9:00 A.M. showed an average of 1.15×10^6 protozoa per milliliter, whereas the same sheep fed the ration in four equal parts at 9:00 A.M., 11:00 A.M., 1:00 P.M., and 3:00 P.M., respectively, showed an average count of 3.14×10^6 protozoa per milliliter. The differences were significant at the 5% level. Feeding in two separate daily allotments gave concentrations of protozoa intermediate between those of the above treatments.

Grinding and pelleting the feed may eliminate the protozoa (Christiansen, 1963). The faster rumen turnover of ground feeds increases the fermentation rate, with an increased acidity which can inhibit protozoal development. Any heat treatment in the pelleting process also tends to favor bacterial rather than protozoal utilization of starch. The protozoa can survive if the amount of ground feed is reduced. With long hay the protozoa can survive at all levels of feeding.

One of the most interesting aspects of the protozoa is the apparent incompatibility of certain species (Eadie, 1962b). *Polyplastron multivesiculatum* did not occur in lambs containing *Epidinium* and *Eudiplodinium maggii,* and if introduced into such animals the latter two genera disappeared. A tendency for *Diplodinium affine* to be associated with

Polyplastron was observed. The peculiar distribution was noted also in calves but was not universal. In thirty sheep *Polyplastron multivesiculatum* and *Diplodinium affine* occurred together in some, whereas *Epidinium, Diplodinium,* and *Eudiplodinium* were found in others, and the two groups were mutually exclusive (Warner, 1962b).

Diurnal variations in the numbers of rumen protozoa have been examined (Purser, 1961a; Warner, 1962b) for animals fed once daily at 8:00 A.M. Division occurred chiefly during the 8-hour period prior to feeding. Since the protozoa are protected from rhythmic diurnal changes such as light, temperature, and humidity, the daily rhythm in division is most likely correlated with feeding time. With animals fed frequently the divisions of the protozoa would be expected to occur more randomly through the day.

G. Digestion of the Protozoa

Gruby and Delafond (1843) observed that the protozoa were digested in the abomasum, and all subsequent observations have substantiated this early discovery (Pounden *et al.,* 1950). The protozoa are digested by pepsin and trypsin (Baker, 1943). The remains of the cuticles of the protozoa are seen in the abomasum (Fiorentini, 1890), but not in the intestinal contents.

In the omasum, the protozoa become inactive and start to disintegrate (observed in Japan by Kubo at Kodaira-shi and by Hirose at Hokkaido, personal communication). The factors concerned have not been elucidated, but acidity may be concerned.

The proposal (Baker, 1942) that the intracellular stored polysaccharide of the protozoa was quantitatively important in supplying carbohydrate to the host when the cells are digested in the abomasum and small intestine has been tested (Heald, 1951b). The results indicate that an insignificant quantity of sugar becomes available to the host by this route.

H. Relationships between the Protozoa and Bacteria

This is one of the least understood areas in rumen microbiology, particularly insofar as quantitative relationships are concerned. As previously mentioned, the numbers of bacteria in nonfaunated animals are greater than in normal faunated ones (Eadie, 1962a; Bryant and Small, 1960). The difference can be due to competition for food or to consumption of the bacteria by the protozoa. The fact that the protozoa ingest and presumably digest rumen bacteria indicates that part of the effect is due to this ingestion, but the extent to which the protozoa utilize nitrogenous constituents of the feed and whether these are used in preference to the bacteria has not been established. An importance of the source of feed nitrogen for the activity and numbers of the protozoa has been reported (Purser and Cline, 1962), with possible interrelationships between the protozoa and the bacteria.

The bacteria in the rumen of a sheep containing also *Entodinium* are less able to digest starch than if the protozoon is absent (Walker and Hope, 1964). Ingestion of starch by *Entodinium* would explain this, at least in part, but an ingestion of amylolytic bacteria could also be concerned.

Dasytricha has been freed of all but one or two strains of bacteria (Clarke and Hungate, 1965), but under these conditions could not be maintained *in vitro*. Restoration of bacteria from an unpurified dasytrich culture restored the capacity for growth. The interpretation of the result is difficult, since the effect of the bacteria could be indirect (e.g., a conditioning of the medium) rather than direct (i.e., used as food).

The finding that faunated lambs gain more weight than nonfaunated controls (Abou Akkada and el-Shazly, 1964) increases the importance of analysis of the actual nitrogenous and other foods of the protozoa, analysis of the protozoan bodies, and the relationship of these features to the host nutritional requirements.

CHAPTER IV

Ruminant Functions Related to Rumen Microbial Activity

The intimate functional mutualism between the ruminant and its microbiota has evolved by selection of adaptations in many elements of the association. The microorganisms depend on the ruminant for the intake of food, its mixing and propulsion, secretion of saliva, and removal and supply of substances through the rumen wall.

One of the most important host adaptations is the subdivision of the stomach into four compartments: the capacious rumen, adjacent smaller reticulum, omasum, and abomasum. The rumen is the largest of these compartments and accommodates a mass of moist fermenting digesta. The reticulum is at the anterior end of the rumen, with openings also from the esophagus and into the omasum. It accommodates some fermenting material, but its chief function is to regulate movements of digesta. The omasum is a special compartment for screening and absorption, and it connects with the secretory subdivision, the abomasum.

A. Fistulation

Just as the gunshot wound in Alexis provided Dr. Beaumont an opportunity to study the human stomach, so artificially introduced openings (fistulas) into the various compartments of the rumen and intestine have tremendously increased the diversity of experiments possible with the ruminant stomach. Most studies on rumen physiology have been with fistulated animals.

Numerous operative devices for fistulating the rumen have been recommended, starting with the studies of Fluorens in 1833, used also by Colin (1854). Successful methods have been described by many recent investigators, among which the descriptions by Phillipson and Innes (1939) and the well-illustrated account of Dougherty (1955) may serve as examples. The principle of the techniques for fistulating the rumen of cattle is to make an initial opening through the body wall into the peritoneal cavity but not into the rumen. The usual site of a rumen fistula is on the left side, behind the ribs, though it can be introduced also into the anterior rumen (Dussardier *et al.,* 1961). It should be as high as possible to reduce loss of fluid, but should be below the thicker muscles of the back. In the earlier procedures the rumen wall was sutured to the edges of an incision into the body wall, and the rumen wall was opened. The sutures often tore loose, and microorganisms infected the peritoneal cavity.

A preliminary operation to cause adhesion of the rumen wall to the body wall greatly diminishes the incidence of peritonitis. The rumen and the body wall in juxtaposition to it are scraped and roughened with a sterile instrument to set up an inflammatory reaction. The rumen is lightly sutured to the margin of the incision to hold the tissues together, care being taken not to insert the needle clear through the rumen wall. The body wall is then sewn up. After a week or 10 days the rumen will have adhered to the body wall, and an opening through this portion into the interior of the rumen can be made with little danger that rumen contents will enter and infect the peritoneal cavity. The opening can be enlarged to the size desired.

It is well to have ready a plug of some impervious material such as rubber or plastic to insert into the fistula as soon as it is opened. The wound then heals with the tissue closely apposed to the plug. A fistula left open tends to shrink in size; for this reason a plug once fitted should not be left out for more than a few hours.

Numerous types of plugs have been described for cattle (Yarns and Putnam, 1962; Bowen, 1962). Some are flexible and can be inflated to provide a gas-tight seal (Dougherty, 1955; Komarek and Leffel, 1961; Johnson and Prescott, 1959; Mendel, 1961). This is essential for studies involving production of gases. Other plugs allow some leakage of gas, but retain most of the rumen liquid. Plugs can contain an opening through which materials can be removed or added to the rumen. With fistula plugs for cattle, solid digesta may accumulate between the plug and the margin of the fistula. This causes necrosis by increasing the pressure to a level which cuts off circulation in the tissues around the opening.

For sheep a short rubber tube, a cannula, with an internal flange (Jarrett, 1948a) is almost universally employed. The plug opening is closed with a stopper. Stiff flanges or improperly positioned cannulas often cut into the rumen epithelium or cause ulceration, injuries which may be accompanied by decreased appetite.

Fistulas (Dougherty, 1955) can also be introduced into the esophagus (Haubner, 1837), reticulum (Dussardier *et al.,* 1961), omasum (Bouchaert and Oyaert, 1954; Oyaert and Bouchaert, 1960), abomasum

(Stewart and Nicolai, 1964; Phillipson, 1952b), duodenum (Phaneuf, 1961), and intestine and into combinations of these. Rumen pouches can be prepared (Ganimedov and Zajac, 1961; Trautmann, 1933b; Tsuda, 1956a) with the circulation intact but opening to the exterior. Multiple fistulas introduced into several parts of the alimentary tract permit a wide variety of experiments.

The gut can be transected at various points, e.g., between omasum and abomasum (Kameoka and Morimoto, 1959), each exposed and exteriorized, and the ends connected by an artificial duct to complete the alimentary tract. Samples of material leaving the omasum can be obtained in this way.

B. Ingestion and Deglutition of Food

A fairly continuous gathering of forage is characteristic of wild ruminants. Habits vary with climate and forage conditions, but grazing generally occurs most frequently during morning and evening periods (Sheppard et al., 1957), especially in hot climates. It is more common during daylight hours than at night, but in warm climates diminishes during hot periods of the day when the animals seek shade if it is available. Under field conditions a ruminant spends about one-third of the time grazing, one-third ruminating (Bergmann and Dukes 1926; Harker et al., 1954; P. N. Wilson, 1961; Kick et al., 1937a) and one-third resting.

Stall-fed or paddocked animals are often fed frequently, e.g., high-producing dairy cows, and the time intervals of feeding resemble those of grazing animals. In other cases most of the daily ration is provided at one or two daily feedings and is rapidly consumed. The rapidity of food ingestion after feeding varies with its nature, concentrates usually being ingested more rapidly than forage and ground and pelleted feeds more rapidly than hays. When the food consists entirely of concentrates with very little roughage, or when it is pelleted, the time devoted to feeding may be relatively brief in comparison with grazing or hay-fed individuals. Intake in these cases cannot be regarded as continuous.

The ruminant mouth is admirably adapted for rapid grazing. A tough pad on the upper jaw replaces the incisor teeth and provides a firm base against which the lower teeth press the plant stems and leaves and hold them while a movement of the head breaks the attacked plants or pulls them out of the ground. The tongue effectively shifts the bite to the back of the mouth. After several bites have been taken they are swallowed *en masse* as a *bolus* which is carried to the rumen by a peristaltic wave in the esophagus. The average weight of a bovine bolus is about 100 gm, including the saliva.

During feeding there is relatively little chewing of the food, just enough to mix it with saliva and form the bolus. Food consumption can thus be rapid, enabling the ruminant to gather a great deal of food in a short time when forage is plentiful. Camper (1803), also Perrault (1680, cited by Brugnone, 1809), stated, "Most of these animals, fearful of their enemies have little time to devote to feeding; in consequence they consume suitable forage as quickly as possible; they hide or rest like domestic animals, and ruminate at their ease the food which has already undergone a little alteration or digestion," (translation from French). Whether fear or appetite causes the avid feeding, extensive grazing meets ruminant requirements.

Nonnutritious material, such as sand, and hazardous materials, such as pieces of metal, may be ingested, the latter occasionally irritating or perforating the stomach wall, particularly the reticulum, in which they often come to rest. Although the chewing action of the ruminant during feeding does not thoroughly grind coarse feeds, it frees a surprising amount (65%) of cell contents from fresh forages (Mangan, 1959). Most forage feeds are quite bulky; the specific gravity of dry forages is low, and the dry matter of green material is low. In consequence the volume of ingested forage is large.

A bolus of food passes rapidly down the esophagus and enters the stomach through the cardia. It is ejected into the rumen with considerable force (Ash and Kay, 1959). (See Figs. IV-1 to 7 for various parts

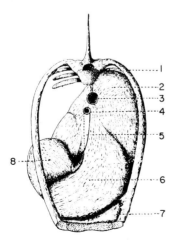

Fig. IV-1. Anterior view of the bovine stomach *in situ,* from Florentin (1952, p. 534). Key: (1) sixth rib; (2) dorsal sac of the rumen; (3) posterior aorta; (4) esophagus; (5) esophageal groove as if it could be seen through the other organs; (6) reticulum; (7) abomasum; (8) omasum.

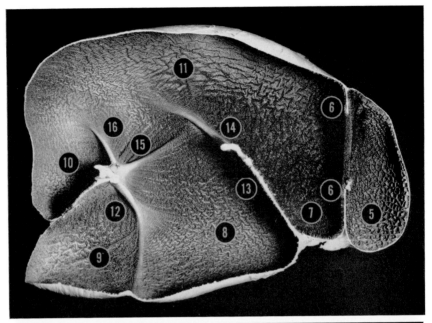

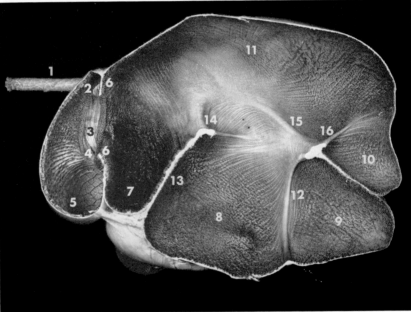

Figs. IV-2 and 3. The left (Fig. 2, *above*) and right (Fig. 3, *below*) halves of the rumen as seen from the inside. Photograph of a preparation made by N. J. Benevenga. Key: (1) esophagus; (2) cardia; (3) esophageal groove; (4) reticulo-omasal opening; (5) reticulum; (6) rumino-reticular fold; (7) anterior sac of the rumen; (8) ventral sac of the rumen; (9) ventral blind sac; (10) dorsal blind sac; (11) dorsal sac; (12) ventral coronary pillar; (13) anterior transverse fold; (14) anterior transverse pillar; (15) longitudinal pillar; (16) dorsal coronary pillar.

of the ruminant stomach.) The cardia is located at the dorsal end of the
esophageal groove, which is a special structure in ruminants, on the
anterior right side of the inner surface of the reticulum (Fig. IV-6).
The cardia is ordinarily tightly closed by contraction of a ring of muscle
tissue, the *cardiac sphincter,* but this muscle relaxes and the cardia opens
wide when boli are passing into or out of the rumen and when gas escapes.
There is no separation of reticulum from rumen on the right side. The
esophageal groove is slightly anterior to the level of the fold separating
rumen and reticulum on the left side. At the lower end of the esophageal
groove is the *reticulo-omasal* orifice through which digesta pass into the
omasum (Figs. IV-3, 5, and 6). Note in Figs. IV-4 and 5 that the eso-

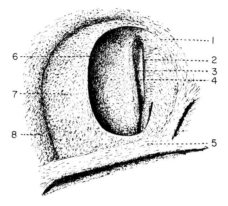

Fig. IV-4. View of the reticulum and rumino-reticular fold as seen from the
rumen. From Florentin (1952, p. 536). Key: (1) cardia; (2) right longitudinal lip
of the esophageal groove; (3) esophageal groove; (4) left longitudinal lip of the
esophageal groove; (5) anterior transverse pillar; (6) pillar at edge of rumino-
reticular fold; (7) rumino-reticular fold; (8) left wall of the rumen.

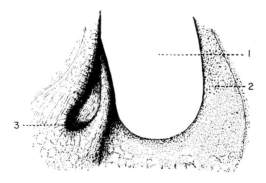

Fig. IV-5. View of the rumino-reticular fold and the lower end of the esophageal
groove as seen from the reticulum. From Florentin (1952, p. 537). Key: (1) rumino-
reticular orifice; (2) rumino-reticular fold; (3) orifice from reticulum into omasum.

phageal groove is twisted through almost a 100° angle; the opening into the omasum is toward the right and somewhat posterior, whereas the cardia opens from the anterior aspect. During swallowing (deglutition) the bolus, moving rapidly, is propelled into the rumen instead of falling into the reticulum. Muscular contractions in the rumen tend to mix it with previously ingested material.

The esophagus of ruminants differs from that of most mammals in containing contractile tissues of skeletal (striated) muscle. These muscle fibers act more quickly (Czepa and Stigler, 1926) and powerfully than smooth or nonstriated muscle, and boli traverse the esophagus faster than in nonruminants. The occurrence of striated muscle in the ruminant esophagus suggests a significance in connection with rumination, a unique activity in ruminants which will be discussed later, but no evidence for such significance has been reported.

In addition to the sphincter at the cardia there is the *pharyngoesophageal sphincter* (Dougherty and Habel, 1955) in the esophagus near the pharynx and the *diaphragmatic sphincter* where the esophagus passes through the diaphragm.

C. Structure of the Ruminant Stomach

1. THE RUMEN AND RETICULUM

The rumen and reticulum of forage-fed animals together hold an amount of digesta roughly equal to one-seventh the total weight of the animal (Table IV-1). On more easily digested rations this may be smaller—one-tenth of the weight (Yadawa and Bartley, 1964). The reticulum is small, but the opening into the rumen is large, which makes the reticulum resemble an anterior pouch of the rumen. The two compartments together form one large functionally integrated sac, the *rumen-reticulum,* filled with digesta and microorganisms. The reticulum lies against the diaphragm, with the esophagus on the right, and curves down and to the right, as seen in Figs. IV-1, 3, 6, and 7. On the inner wall are thin ridges arranged in a reticular (netlike) fashion from which the name of this organ is derived (Fig. IV-6). The ridge separating the reticulum from the rumen is called the *rumino-reticular* fold (Figs. IV-2–6). It extends from the ventral right side transversely toward the left, up the left side, and partially across the top from left to right. As mentioned previously it does not extend to the right lateral wall (Figs. IV-3, 4, and 6), and there is no demarcation of reticulum from rumen on the right side. The rumino-reticular fold contains muscle tissue which on contraction elevates the fold between the two compartments.

The rumen is divided internally into more or less distinct portions by

Table IV-1

WEIGHTS OF CONTENTS OF DIGESTIVE ORGANS

Animal	Live weight (kg)	Daily ad lib. intake, dry wt. hay (kg)	Wet weight of contents						Dry weight of contents				
			Rumen-reticulum (kg)	Rumen-reticulum (% live wt.)	Om-asum (kg)	Abom-asum (kg)	Small intes-tine (kg)	Large intes-tine (kg)	Rumen-retic-ulum (kg)	Om-asum (kg)	Abom-asum (kg)	Small intes-tine (kg)	Large intes-tine (kg)
Milk cow[a]	485	5.36	67.8	14.0	5.0	2.1	3.8	3.7	8.56	0.70	0.27	0.27	0.38
Milk cow	533	13.20	81.8	15.4	11.0	3.6	9.4	8.6	11.43	2.07	0.33	0.79	0.88
Milk cow	500	7.32	56.3	11.3	5.4	1.2	4.6	4.2	9.50	0.90	0.12	0.35	0.43
Milk cow	590	10.90	77.3	13.1	8.3	3.1	7.6	8.3	10.49	1.59	0.32	0.58	1.02
Milk cow	480	7.60	44.0	9.2	8.0	1.4	6.0	2.2	6.18	1.46	0.15	0.52	0.19
7-month bull	202	3.49	39.4	19.6	2.3	1.4	2.3	2.8	4.78	0.39	0.13	0.17	0.27
7-month bull	200	4.12	39.4	19.7	2.5	0.8	2.8	2.5	5.30	0.36	0.09	0.23	0.26
7-month bull	207	3.53	39.4	19.1	3.1	1.2	3.9	2.5	4.65	0.51	0.11	0.26	0.21
7-month bull	216	3.77	45.3	21.0	2.6	0.8	3.9	3.4	5.23	0.42	0.08	0.29	0.36
7-month bull	193	4.26	36.3	18.8	2.2	1.4	4.6	3.2	4.40	0.32	0.12	0.34	0.31
Zebu steer[b]	241	Grass pasture	31.4	15.0	—	—	—	—	—	—	—	—	0.91
Camel	339	—	56.4	17.0	—	—	—	—	—	—	—	—	3.41
Camel	523	Starved 24 hr	52.3	10.0	—	—	—	—	—	—	—	—	11.4
Feral ruminants													
Grant's gazelle	49.1	—	4.56	9.0	—	—	—	—	—	—	—	—	0.286
Thompson's gazelle	24.1	—	3.18	13.0	—	—	—	—	—	—	—	—	0.242
Eland	520	—	55.4	11.0	—	—	—	—	—	—	—	—	3.62
Suni	3.7	—	0.29	8.0	—	—	—	—	—	—	—	—	0.017

[a] From Makela (1956).
[b] From Hungate et al. (1959).

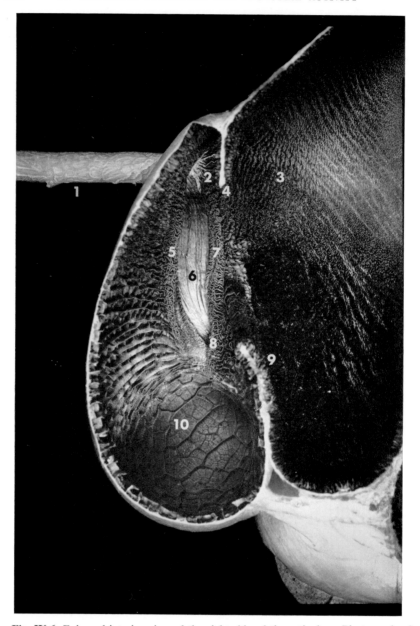

Fig. IV-6. Enlarged interior view of the right side of the reticulum. Photograph of the preparation by N. J. Benevenga. Key: (1) esophagus; (2) cardia; (3) right wall of the anterior rumen; (4) dorsal portion of rumino-reticular fold; (5) left longitudinal lip of the esophageal groove; (6) esophageal groove; (7) right longitudinal lip of the esophageal groove; (8) orifice between reticulum and omasum; (9) ventral portion of rumino-reticular fold; (10) reticulum.

an *anterior transverse fold* parallel and essentially similar to the fold separating the rumen from the reticulum, as well as by muscular *pillars*. The pillars assist in the mixing movements of the rumen and in the regulation of flow of digesta and gases. On the external surface of the rumen the pillars appear as shallow grooves. When contracted, the pillars become well-defined, fairly rigid structures, but in the relaxed state they are not prominent. The inner edge of the anterior transverse fold is well supplied with muscular tissue which makes up the *anterior transverse pillar*. The *anterior sac* of the rumen is separated from the reticulum by the rumino-reticular fold and from the ventral sac by the anterior transverse fold. These folds assist in retaining digesta and liquid in the rumen, while allowing small particles and some liquid to spill over into the reticulum.

From the ends of the anterior transverse fold, the *right* and *left longitudinal pillars* extend back on each side toward the *posterior transverse pillar* (Figs. IV-2 and 3). The anterior and posterior transverse pillars, with the connecting longitudinal pillars, divide the rumen into a *dorsal* and a *ventral sac*. The posterior part of the dorsal sac extends some distance caudally as the *dorsal blind sac,* and the ventral part similarly extends caudally as the *ventral blind sac*. The ventral blind sac is displaced somewhat toward the right and the dorsal toward the left. Extending dorsally on each side from the posterior junction of the longitudinal and transverse pillars are the two *dorsal coronary pillars* which demarcate the anterior limits of the dorsal blind sac. The *ventral coronary pillars* lie similarly at the junction of the ventral blind sac and the ventral sac.

The dorsal sac normally contains a greater proportion of dry matter and coarse material than does the ventral sac. The specific gravity of digesta in the dorsal sac is lower, due in part to the air in dry forages. As fermentation gases rise in the rumen, they carry up solid particles; this lowers the specific gravity of the upper material and increases the proportion of dry matter in the dorsal rumen.

Materials with high specific gravity, such as sand and metals, collect in the bottom of the rumen and reticulum and may occasionally be fairly abundant. The reducing conditions in the rumen make the metals bright and shiny.

The inner surface of the wall of the bovine rumen is covered with papillary protuberances of various shapes and sizes (Figs. IV-2, 3, and 6), from 1 to 10 mm in diameter and 3 to 10 mm long, chiefly semiconical, with a somewhat greater diameter at the base than at the tip. As Brugnone (1809) described it (translation): "Almost all of the internal face of the rumen wall is rough and uneven because of the innumerable papillae covering the surface. These papillae, variable in size, elevation, and number on the various areas of the wall, are also of various shapes. Some are small, short, and filiform, differing little from velvet. Some are almost as large as a myrtle leaf; others are conical, hooked, and pointed, almost resembling grains of rye with ergot." Many other shapes also occur.

The papillae are particularly well developed in the ventral sac and in the floor of the anterior sac, as well as on the anterior transverse pillar and adjacent surfaces. In many of these areas the papillae are so numerous that the wall actually does have a coarse velvety appearance and feel. The great absorptive surface provided by the elevated ridges of the reticulum and the papillae of the rumen permits ready movement in either direction of materials capable of diffusing through the epithelial cell membranes separating the stomach contents from the blood supply. The tissue on the entire inner surface of the rumen and reticulum is composed of squamous epithelial cells, the basal cells containing a columnar arrangement of mitochondria (M. J. Dobson, 1955; Schulz, 1963). Secretory cells have not been found in this compartment.

2. THE OMASUM

The omasum lies to the right of the rumen-reticulum and connects with the reticulum via the reticulo-omasal orifice at the bottom of the esophageal groove. The omasum is also called the manyplies because of the numerous flat parallel sheets of tissue within it.

In understanding the structure of the omasum it may assist to compare it with a loosely closed book, ovoid in shape and with the binding extending to cover also all the exposed edges of the pages, attached to them at the binding and top and bottom of the book, but not attached on the side. Because of this analogy the sheets of tissue are called *leaves.* The two ends of the channel on the side open into the reticulum and abomasum, respectively (Fig. IV-7).

The omasal leaves are of various sizes; all leaves attach to the wall of the omasum at the "binding" and at the ends, but only part of them extend to the channel at the opposite edge (Fig. IV-7). The free edges of these leaves extending to the channel lie against the muscular *pillar* of the *omasal groove,* an extension of the esophageal groove along the lesser curvature of the omasum from the reticulo-omasal orifice to the opening into the abomasum. All the edges of the omasal leaves are directed toward the side of the lesser curvature, as shown in Fig. IV-7. The omasal groove is opposite the free edges of the omasal leaves. The channel is open when the free edges of the leaves do not reach the groove. The free edges of the leaves are supplied with a number of conical papillae [according to Brugnone (1809) Aristotle called this part the *echinos*] arranged in a fashion which suggests that they strain out coarse particles

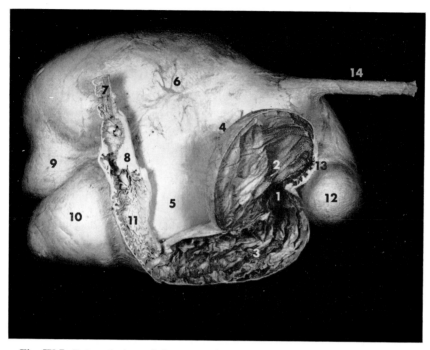

Fig. IV-7. External view of the rumen from the right side, showing also the left side of the omasum and abomasum, internal view. Photograph of the preparation by N. J. Benevenga. Key: (1) omasal-abomasal orifice; (2) leaves of the omasum; (3) longitudinal spiral folds of the abomasum; (4) greater curvature of the omasum; (5) region of the ventral sac of the rumen; (6) region of the dorsal sac of the rumen; (7) duodenum; (8) pylorus; (9) dorsal blind sac; (10) ventral blind sac; (11) pyloric portion of the abomasum; (12) reticulum; (13) vestibule of the omasum; (14) esophagus.

and prevent them from entering the spaces between the leaves. However, the spaces between the leaves are usually packed with *small* particles of solids, which makes it probable that both fluid and small particles enter between the omasal leaves.

The small space between the papillated free edges of the leaves near the reticular orifice and the floor of the omasal groove constitutes the omasal *vestibule*. Digesta from the reticulum enter the vestibule through the reticulo-omasal orifice. The omasal groove extends along only the anterior portions of the free edges of the omasal leaves. The posterior edges are attached to the wall, and the middle portions lie directly over the cavity of the abomasum (Fig. IV-7).

Examination of the omasum of slaughtered animals on a hay ration discloses rather dry digesta, containing no large particles, pressed between the omasal leaves. On lush green feeds omasal contents do not show this solidly packed state. The function of the omasum is the screening of large particles and absorption of water, acids, and other substances.

3. THE ABOMASUM

The tubular abomasum lies on the right side of the rumen and connects the omasum with the small intestine. The internal surface of the wall is much folded into somewhat spiralled ridges which are almost flaps of tissue (Fig. IV-7). At the opening from the omasum the longitudinal flaps are so large that they could conceivably serve as some sort of valve to prevent back flow of abomasal contents into the omasum. The acidic contents of the abomasum do not ordinarily gain access to the omasum, though in young milk-fed animals a fluid resembling milk serum has been reported (Trautmann, 1933b). The exact mechanism ordinarily preventing back flow from the abomasum into the omasum is not known.

The abomasum corresponds in function to the pyloric portion of the stomach of nonruminant animals. The epithelium is well supplied with specialized secretory cells which produce mucus, pepsin, and hydrochloric acid, respectively. At the posterior end of the abomasum the duodenal sphincter controls passage of digesta into the small intestine.

The secretion of somewhat alkaline digestive juices into the small intestine and colon and the formation and elimination of feces occur in ruminants about as in nonruminants. It is in the structure and functioning of the stomach compartments, particularly the first three, that the ruminant diverges most from other mammalian orders.

4. THE ESOPHAGEAL GROOVE

In the newborn ruminant the abomasum is relatively larger than the other compartments. It receives milk directly from the esophagus via the esophageal and omasal grooves (Figs. IV-6 and 7), which function as a bypass of the rumen-reticulum and omasum. The act of suckling induces reflex contractions of the muscles in the lips or ridges extending along each side of the esophageal groove and in the groove itself. The contractions narrow the gap between the two longitudinal lips and shorten the groove with a twisting motion which closes the gap to form a tube joining the esophagus with the omasum. Suckled milk passes through this tube, through the omasum, and into the abomasum.

If the young animal drinks milk from a bucket, the amount swallowed at one time may force open the closed esophageal groove and allow milk to enter the rumen or reticulum, though with goats this apparently occurs only to a small extent (Akssenowa, 1932). With modern starter feeds fed by bucket, a sizeable fraction of the food enters the rumen. It undergoes a lactic acid fermentation and gives an acid reaction to the rumen contents of the young animal.

Swallowed water also tends to go directly into the abomasum in the young animal, but in the adult it enters the rumen-reticulum. The rate of passage of dyed straw particles has been used to determine whether food bypassed the rumen (Lenkeit and Columbus, 1934).

The function of the esophageal groove was apparently first appreciated by Hernandez in Mexico in the first half of the seventeenth century, as cited by Brugnone (1809, p. 54) from Faber (1651). Brugnone wrote (translation), "All the milk which the young animal or adult swallows in suckling or in drinking, even that which they are forced to take, passes immediately from the esophagus into the groove, which carries it directly to the abomasum without its stopping in the reticulum, the omasum, or the rumen; that is perhaps why Jean Faber has given the term *via lactea* to the groove."

The closure of the esophageal groove is effected through a reflex originating from receptors in the mouth which are sensitive to a flow of liquid (Watson and Jarrett, 1941; Trautmann and Schmitt, 1933a; Comline and Titchen, 1951). The afferent impulse passes from the mouth to the central nervous system over the superior laryngeal nerve, with the efferent path in the dorsal abdominal vagus. (See Figs. IV-9 and 10 for the innervation of the rumen.)

It is dubious that the esophageal groove plays an important role in the normal adult animal. The muscles are relaxed during rumination and swallowing of food.

Although the esophageal groove does not ordinarily function in the adult ruminant, closure can be induced by administration of certain salts. Solutions (1 to 2%) of copper sulfate and concentrated sodium chloride, sulfate, or acetate solutions stimulate reflex closure in sheep (Watson and Jarrett, 1941). The older the animal the higher is the salt concentration required for closure. In cattle sodium bicarbonate and sodium chloride are more effective than copper sulfate (Riek, 1954). Induction of the reflex in adults can be of value for oral administration of drugs to the alimentary tract posterior to the rumen (Nanda and Singh, 1941).

D. Development of the Rumen-Reticulum

Prenatal development of the rumen has been described by Duncan and Phillipson (1951). Embryologically, the rumen is derived from the fundus, i.e., nonsecretory portion of the stomach; the abomasum is derived from this and from the secretory or pyloric part (F. T. Lewis, 1915). The omasum is early distinguishable as a swelling on the lower curvature of the embryonic stomach. The reticulum is the central portion. All parts are discernible in a 5-cm bovine embryo.

In the newborn ruminant the abomasum is about as large as the rumen and remains relatively large as long as only milk is consumed (Brugnone, 1809). As the young ruminant consumes forages, the reticulum, and particularly the rumen, develop rapidly. Rumen development is hastened by providing forage at an early age (Trautmann, 1932). By the time the animal can subsist on forage alone, the relative sizes of the various stomach compartments are the same as in the mature animal. There is no methane production in the rumen of a milk-fed calf (Reiset, 1868a), but it is produced when forage is consumed. The digesta of the adult rumen-reticulum compose roughly one-seventh the weight of the animal, and the combined contents of the omasum and abomasum are usually about one-seventh the weight of the contents of the reticulo-rumen. The contents of the rumen in calves 16 weeks of age constitutes a smaller fraction of the body weight (Kesler et al., 1951). In view of their relatively large food needs for growth, the rumen contents in these animals turn over more rapidly than in nonproducing adult ruminants.

Experiments (Flatt et al., 1958; Buffon, quoted by Colin, 1886), have shown that the rumen of a young animal can be distended to a large size by inserting sponges composed of indigestible material such as nylon. The wall of such an artificially enlarged organ is very thin and weighs no more than the wall of an undistended control. Similar results with feeding of hay were noted in calves (Blaxter et al., 1952) during initial stages. After continued feeding on roughage the weight of the rumen itself becomes significantly greater than in controls on a nonroughage diet. Products of the rumen fermentation, chiefly the volatile fatty acids (Sander et al., 1959), are necessary for this increased weight of the rumen wall and for the development of the papillae and muscle tissue. Microbial activity thus exerts an effect on the developmental physiology of the host.

As roughage is consumed the pH of the rumen gradually rises (Godfrey, 1961) and the concentration of ammonia diminishes. Both acid and ammonia are formed by microbial fermentation of the concentrates which enter the rumen. As forage is consumed the substrate concentration diminishes as fiber replaces concentrates. The rumen of roughage-fed calves showed stronger contractions and was more sensitive to stimuli than in milk-fed calves (Dzuick and Sellers, 1955b). Movements of the rumen-reticulum of suckling kids have been described in detail (Trautmann, 1933a).

Increased saliva production during forage consumption is also a factor in the decreased rumen acidity, though saliva may be secreted and accumulate in the rumen even before roughage is consumed (Akssenowa,

1932). Rumen tissues characteristic of the adult tend to develop even on a milk ration (Trautmann, 1932), due in part to fermentation of milk or milk serum which comes back into the rumen from the abomasum (Trautmann and Schmitt, 1933b), but their full development depends on fermentation of roughage-containing feeds, and a complete development of the villi takes place only if such feeds are not delayed too long during development of the young ruminant. Prolonged milk feeding retards the development of the omasum even more than the rumen-reticulum (Trautmann, 1932). A flexibility in the development of the rumen and reticulum is suggested by the observations (Trautmann and Schmitt, 1932; Piana, 1953a) that extirpation of these organs is followed by almost complete regeneration, provided roughage feeds are administered fairly soon after the operation. The omasum similarly regenerates after removal of most of that organ (Trautmann, 1933c).

Smith and Trexler (1961) have compared the structure of the rumen wall of normal and axenic (free of microorganisms) lambs up to the time that roughages are consumed. The axenic animals were obtained by aseptic Caesarian operation, kept free or almost free of bacteria in special cages, and fed sterile foods. In the normal animals exposed to numerous bacteria the propria was thicker and the squamous epithelium thicker and rougher.

The lactose tolerance of the calf decreases as forage is consumed (Atkinson *et al.,* 1957), and the blood glucose concentration drops. This latter phenomenon occurs independently of the change in type of feed (Lupien *et al.,* 1962; Nicolai and Stewart, 1965). The composition of plasma lipids changes as young ruminants change to a forage ration (Masters, 1964a,b,c).

E. Salivary Secretion

An importance of saliva in the digestive activities of ruminants was recognized very early. Sprengel (1832) wrote that ruminant saliva contained large quantities of an alkali combined with carbonic acid and thought this might explain the value of the Roman practice, reported by Pliny, of soaking straw in sea water. Sprengel believed the alkali was important in breaking down resistant plant parts. In very accurate quantitative experiments Colin (1852, 1886) cannulated the salivary duct and showed the continuous and copious secretion from the parotid glands, as well as the submaxillary secretion during feeding. Haubner (1837) considered saliva important as a source of water in the rumen, and Markoff (1913) recognized the importance of saliva in neutralizing fermentation acids.

1. The Nature of the Saliva

Different glands of various sizes secrete saliva. Insofar as composition is concerned the following salivary glands (Kay, 1960a) are alike in secreting a fluid containing a relatively high concentration of bicarbonate: the paired parotid, inferior molar (Scheunert and Kryzwanek, 1930) and buccal, and the unpaired palatine. These will be termed collectively the alkaligenic glands. They secrete little mucoprotein, in contrast to the mucogenic glands, which include the paired submaxillary, sublingual, and labial, the unpaired pharyngeal, and the numerous miscellaneous glands in the epithelium lining the mouth. The secretion of these latter glands has a bicarbonate concentration one-sixth that of the saliva from the alkaligenic type.

All the salivary glands secrete during feeding. The parotid and inferior molar glands secrete also in the periods between feeding, but the mucogenic glands do not (Ellenberger and Hofmeister, 1887b). Diversion of the saliva away from the rumen (McManus, 1962) of an animal with an active fermentation causes increased volatile fatty acid concentration and a more acid pH. Rumen motility is decreased. Sectioning of both parotid ducts diminishes the food intake, an effect that is partially but not entirely counteracted by providing artificial saliva (A. D. Wilson, 1964).

Because of its large size and continuous secretion the parotid gland is a particularly important producer of alkaline saliva and contributes about half the total (Kay, 1960a). Most studies of alkaline secretion have been made with the parotid because of its quantitative importance and also because the parotid duct can be diverted operatively to empty externally rather than into the mouth. This permits easy measurement of salivary flow and composition under a variety of conditions with little disturbance to the animal. This approach was used in 1831 (Tiedemann and Gmelin) and has been extensively employed since, with numerous improvements. The operation is performed on one side only. The activity on the unoperated side is sufficient to maintain digestive functions, provided the feed is supplemented by sodium equal to that lost through the exteriorized duct (Scheunert et al., 1929a; Trautmann and Albrecht, 1931; R. H. Watson, 1933). Failure to supplement with Na^+ leads to death (Scheunert and Trautmann, 1921a). The water lost must also be replenished. Supplementation of the remaining elements in the exteriorized secretion is not usually necessary.

Studies of parotid secretion have disclosed a rapidly variable activity controlled by extrinsic factors, superimposed on a continuous basal secretion. Chewing induces a rapid increase in secretion of both parotid glands, more on the side on which mastication occurs (Colin, 1852;

Denton, 1957), i.e., secretion from the ipsilateral is greater than from the contralateral gland. Estimates of saliva production based on exteriorized flow may be low, since the animal tends to chew less on the operated than on the normal side.

The basal secretion of the parotid and inferior molar glands does not appear to be under neural control (Ellenberger and Hofmeister, 1887b; Kay, 1958a) though it adjusts slowly to differences in the feed as the animal matures (Trautmann and Albrecht, 1931) and as the proportion of dietary roughage changes (Denton, 1957; Emery *et al.*, 1960).

The Na^+ in parotid saliva (ca. 180 mEq per liter) normally is about eighteen times as abundant as the K^+, but in animals deprived of sodium the level of potassium may rise as high as the norm for sodium (Denton, 1957). The sodium level drops reciprocally with potassium rise, the combined concentrations staying at 190 mEq per liter. After Na^+ deprivation, restoration by infusing the parotid gland with NaCl does not immediately increase the Na^+ concentration in the saliva. There is a period of 100–150 minutes in which the salt balance of the entire animal readjusts (I. R. McDonald and Denton, 1956). The infused sodium must be sufficient to satisfy the deficiency of the entire animal before the sodium content of the saliva returns to normal. Starvation and sodium depletion decrease the amount of parotid saliva, but the cation concentration stays constant (Dobson *et al.*, 1960).

The secretion of the alkaligenic glands is isotonic with blood, though differing in the concentrations of the individual ions (Kay, 1960a). Blood normally contains only 140–145 mEq Na^+ per liter. The anions in parotid saliva are Cl^- (13 mEq per liter), HCO_3^- (100–140 mEq per liter), and HPO_4^{--} (10–50 mEq per liter). The combined bicarbonate and phosphate concentrations are always about 150 mEq per liter, varying reciprocally.

The minimal phosphate concentration in mixed saliva is remarkably constant during phosphate deprivation, remaining at 11.6 mEq per liter (R. H. Watson, 1933; Clark, 1953). The inorganic phosphorus in the blood of the deficient animals is 1–3 mEq per liter, secretory work increasing the phosphorus concentration in the saliva. On a ration containing only 0.5 gm phosphorus, the salivary phosphate concentration remained constant until shortly before death of the animal from acute aphosphorosis. The saliva in these phosphorus-deficiency experiments was representative of the total secretion since it was obtained by placing a sponge in the mouth. This served to stimulate the mucogenic glands and also to absorb the total saliva, which then could be separated by squeezing the sponge.

The concentration of bicarbonate in parotid saliva is four times as great as in blood serum (McDougall, 1948), but Cl^- is only one-fourth as concentrated. Although the ionic composition of saliva and blood differ, the isotonicity of saliva with blood diminishes the work involved in secretion of water from the gland.

The secretion of the mucogenic glands is hypotonic to blood. It contains less bicarbonate and phosphate than the alkaline secretion (Ellenberger and Hofmeister, 1887b; Phillipson and Mangan, 1959), and the total ash content is lower, only 0.05–0.2% for submaxillary saliva as compared to 0.88% for parotid (Scheunert and Trautmann, 1921a). The secretory ducts of the submaxillary gland are lined with columnar epithelial cells presumably concerned with the reabsorption of salts.

The composition of the total saliva secreted in the resting animal is fairly constant and differs from the saliva secreted during rumination and feeding. The changes during these latter periods are due in part to the addition of the secretion from the glands which are quiescent during rest and also to changes with time in the composition of the secretion from each gland. The variations with time are not great unless there are deficiencies in the available ions (Coats and Wright, 1957; Coats et al., 1958; Kay 1960a,b) and the total osmolality remains the same (Coats and Wright, 1957).

The inorganic ion compositions of salivas studied by various investigators are shown in Table IV-2. The relative importance of saliva and food in supplying Na, K, P, and Cl has been evaluated (C. B. Bailey, 1961b). Saliva is the chief source of sodium, but other constituents are derived directly from the food to about the same extent as from the saliva.

The nitrogenous components in the two types of saliva differ. Sheep parotid saliva contains an average of 14 mEq of nitrogen per liter (Denton, 1957), of which 69% is in the form of urea (Somers, 1957, 1961a). Bovine parotid secretion is similar in the distribution of nitrogenous materials. Total saliva collected in a small metal capsule under the tongue while the animal is feeding (Lyttleton, 1960) shows 0.1–0.2% protein (an average of 17 mEq of nitrogen per liter), a value similar to the 0.18% protein noted by Markoff (1913). Most of the protein in saliva comes from the mucogenic glands. The submaxillary saliva of cattle contains 0.57% (w/v) protein (Phillipson and Mangan, 1959).

It has been reported (McManus, 1961) that the rate of secretion of nitrogen in saliva is fairly constant, not increasing proportionally with increased saliva production. The buffering capacity of rumen contents, and presumably of saliva, cannot be increased significantly by feeding forages which themselves possess a large buffering capacity (Ammerman and Thomas, 1953, 1955).

The total salivary protein in sheep is composed almost exclusively of

Table IV-2

COMPOSITION OF SALIVA

Gland	Type of stimulation inducing secretion	Nitrogenous constituents			Inorganic constituents					
		Urea N mEq/ liter	Protein N mEq/ liter	Total N mEq/ liter	HCO₃⁻ mEq/ liter	HPO₄⁻⁻ mEq/ liter	Cl⁻ mEq/ liter	Na⁺ mEq/ liter	K⁺ mEq/ liter	Ca⁺⁺ mEq/ liter
Bovine[a]										
Parotid	None or carbachol	9.9	1.3	12.4	107.5	20.6	15.4	136.6	13.5	3.3
	Inflation of rumen	10.3	1.6	12.5	110.8	22.4	10.6	123.0	14.7	3.4
Submaxillary	None or carbachol	8.1	10.9	19.4	15.7	0.5	30.9	13.6	13.9	7.1
	Inflation of rumen	13.8	66.0	98.0	18.6	2.6	49.2	—	15.0	7.7
Sublingual, cheek, and buccal	Carbachol	7.1	7.7	15.3	98.4	20.2	48.7	110.0	17.1	4.3
	Inflation of rumen	6.7	8.6	16.3	104.3	17.0	17.6	142.0	19.4	3.2
Total secretion	Feeding on red clover pasture	6.3	20.8	25.6	—	—	—	—	—	—
Bovine[b]										
Total secretion	Collection at cardia	—	—	11.3	—	9.8	—	134.	13.6	0.6
Goat[c]										
Parotid	None or feeding	—	—	—	40–130	10–100	10–40	175–195	5	—
Sheep[d]										
Parotid	None or reflex	—	—	—	95	75	13	186	5	—
Submaxillary		—	—	—	6	54	6	15	26	—
Sublingual		—	—	—	12	0.9	28	30	11	—
Labial		—	—	—	3	5	34	39	6	—
Inferior molar		—	—	—	134	48	10	175	9	—
Palatine and buccal		—	—	—	109	25	25	179	4	—

[a] From Phillipson and Mangan (1959).
[b] From Emery et al. (1960).
[c] From Kay (1960b).
[d] From Kay (1960a).

three fractions (Lyttleton, 1960). One of these, constituting 54% of the total, contains sialic acid (N-acetylneuraminic acid) and is the mucoprotein chiefly responsible for the viscous nature of the total salivary secretion. It is derived almost entirely from the submaxillary gland. Sialic acid constitutes 22.4% of the bovine salivary mucoprotein (Gottschalk and Graham, 1958) and N-acetyl galactosamine 15.7%. Other preparations (Nisigawa and Pigman, 1960) have contained 38% protein, 32.3% sialic acid, and 27.3% hexosamine. Sialic acid has a pK of 2.7 which makes the mucoprotein negatively charged, a characteristic which may explain the highly viscous nature of the mucoprotein solutions.

Although there is little mucoprotein in the parotid secretion, the gland itself contains mucoproteins in which sialic acid and hexosamines have been identified (Nisigawa and Pigman, 1960).

2. External Factors Influencing Salivary Secretion

The basal alkaline secretion varies in amount according to the nature of the feed, as does the additional secretion stimulated by feeding. Coarse dry feeds elicit more secretion than do lush green forages (Weiss, 1953a; Denton, 1957). Exposure to a new dietary regime must continue for several days before the salivary flow becomes relatively constant at the new level. In some experiments (Wilson and Tribe, 1963) drying of forage did not affect the quantity of saliva secreted.

The basal alkaline secretion from the parotid and inferior molar glands differs in composition from the total secretion during feeding, due to augmentation by secretions from the other glands. The secretion during feeding contains more mucoproteins and less bicarbonate. The variations in quantity and composition make it difficult to estimate average salivary secretion, as is evident from the abundance of investigations extending over a long period of years (Colin, 1852; Ellenberger and Hofmeister, 1887b; Scheunert et al., 1929b; McDougall, 1948; Denton, 1957; Balch, 1958).

The act of rumination stimulates salivary secretion, but not to the same extent as feeding. In rumination the mucogenic glands are not active. This may be due to failure of the moist regurgitated food to stimulate secretion or to inhibition from a rumination center in the brain.

3. Reflexes Concerned in Salivary Secretion

The increased parotid secretion during feeding (Clark and Weiss, 1952a; Comline and Kay, 1955; Ash and Kay, 1957; Kay, 1958a,b,c) results from a reflex initiated by stimulation of the walls of the rumen by coarse food particles in the vicinity of the cardia, stimulation of the rumino-reticular fold being most effective in eliciting increased parotid secretion (Comline and Titchen, 1957; Hill, 1957). Stimulation of the lower part of the esophagus also increases salivary secretion. Mild distension (up to 10 mm Hg pressure) in the lower part of the esophagus augments secretion (Kay and Phillipson, 1957), chiefly via the parotid gland, though other glands also contribute (Clark and Weiss, 1952a; Phillipson and Mangan, 1959). Salivary secretion is increased by rumen pressures of 5–20 mm Hg, but higher pressures inhibit (Phillipson and Reid, 1958).

Marked distension of the rumen can inhibit the augmentation of salivary secretion caused by the distension of the thoracic esophagus. The afferent path is over the vagus, but the site in the rumen wall for initiation of these impulses has not been identified (Kay and Phillipson, 1959). The submaxillary, sublingual, labial, and buccal secretions are increased by stimuli resulting from food in the mouth (Scheunert and Krzywanek, 1930; McDougall, 1948). The reflex path to the brain is over the glossopharyngeal nerve, the efferent path also lying along this route (Comline and Kay, 1955).

The afferent pathways of salivary reflexes are over the vagus for stimuli of abdominal origin and over the glossopharyngeal and lingual nerves for thoracic and buccal origins. The efferent pathway is over the ninth cranial (glossopharyngeal) nerve and to a lesser extent the seventh (facial) nerve and parotid branches from the postganglionic buccal nerve (Moussu, 1888; Coats *et al.,* 1956). Parotid secretory pressure can increase to the point that it exceeds the blood pressure and stops flow in the periacinar vessels (Coats *et al.,* 1956). This could occur only briefly since continued secretion depends on the blood supply.

The parasympathetic reflexes increase parotid secretion by increasing the blood flow through the gland (Coats *et al.,* 1956; Kay, 1958a). Brief efferent stimulation of the cervical sympathetic trunk (Kay, 1954, 1958b) causes a momentary increase in parotid flow due to contraction of muscle elements (Silver, 1954) within the gland, followed by a compensatory decrease, but has no effect on the true rate of secretion. It also causes a change in the electrolyte balance of the saliva. Prolonged sympathetic stimulation inhibits secretion. Adrenaline causes a quick increase in saliva due to contraction of myoepithelial cells (Silver, 1954; Kay, 1955) and then a more sustained secretion due to afferent impulses from the rumen and reticulum resulting from the action of adrenaline on these organs (Kay, 1955).

4. Amount of Salivary Secretion

The unusually copious salivary secretion of ruminants as compared with other mammals is due chiefly to the continuous activity of the parotid and inferior molar glands and the augmentation during feeding and rumination. There is much variation between individual animals of the same species. Experimental treatments such as anesthesia and induced recumbency can diminish salivary flow (Peel and Wilson, 1964). An idea of the magnitude of salivary secretion can be gained from inspection of Table IV-3. The total quantity when the ration consists of hay and other dried feeds appears to be at least equal to the volume of the rumen of the animal. It is evident that such a volume of saliva is an important source of liquid in the rumen. In sheep grazing semidesert forage, the saliva amounted to approximately two times the dry weight of the consumed forage (Edlefson *et al.,* 1960). During consumption of alfalfa hay by sheep the secreted saliva was 1491 gm during periods when 464.2 gm of feed was consumed (Bath *et al.,* 1956).

From the levels of bicarbonate, phosphate, and nitrogen in saliva and the estimate that an average of 8 liters is secreted each day in the sheep, calculation shows that a total of 1500 mEq of bicarbonate, 240 mEq of phosphorus, and 114 mEq of nitrogen enter the rumen in the saliva each day.

F. Stomach Movements

1. Mixing in the Rumen-Reticulum

Mixing of the rumen-reticulum contents aids in inoculating fresh ingesta with the mass of microorganisms in the fermenting digesta, spreads the saliva through the rumen, enhances absorption by replenishing the fermentation acids absorbed by the rumen epithelium, counteracts the flotation of solids during fermentation, and assists the comminution and passage of digesta to other organs of the alimentary tract. The mixing is accomplished by contractions of the wall of the rumen and reticulum, contractions coordinated with the movements of other digestive organs.

Rumen and reticulum movements were observed by Fluorens (1833) in a fistulated sheep. He also reported cessation of movement when both vagus nerves were cut. In 1872 Furstenberg and Rohde noted that rumen movements were more vigorous during rumination. In 1875 Toussaint devised the method of detecting rumen movements by pressure changes in balloons inserted through a fistula and connected to pressure recording devices. Ellenberger (1883) confirmed Fluorens' observation that paralysis of the nerves to the rumen interfered with its normal functioning. In these early observations no regularities in the sequence of movements in the rumen were reported.

In 1921 Schalk and Amadon reported that rumination in the goat usually preceded contractions of the rumen and reticulum. During the period following their first observations these investigators initiated a thorough study

Table IV-3

SOURCES AND AMOUNTS OF SALIVARY SECRETION IN SEVERAL RUMINANTS[a]

Animal	Parotid secretion	Sub-maxillary secretion	Other secretions	Total secretion	Reference
Cattle	9.6–33.6	2.9–5.8	—	—	Ellenberger and Hofmeister (1887b)
Cattle	76	24	—	—	Colin (1852)
Cattle	19–58	—	0.86–0.96 sublingual	—	Colin (1886)
Cattle (4 animals, 5 diets)	—	—	—	98–190	C. B. Bailey (1961a)
Cattle	—	—	—	100	Mendel and Boda (1961)
Sheep	3.2	0.5	—	—	Scheunert and Trautmann (1921a)
Sheep	2.6	—	—	—	McDougall (1948)
Sheep	2–8	—	—	—	Denton (1957)
Sheep	2.4	—	—	—	Somers (1961a)
Sheep	—	—	—	5	Somers (1957)
Sheep	—	—	—	9.6	Sperber *et al.* (1956)
Sheep	—	—	—	4–5	Watson (1933)
Sheep	4.1	—	—	—	Stewart and Dougherty (1958)
Sheep	3.4–7.2 during feeding	—	—	—	Luick *et al.* (1959)
Sheep (11)	3–8, 4.8 av.	0.4–0.8, 0.62 av.	0.1 sublingual; 0.7–2 inf. molar; 2–6 palatine, pharyngeal, and buccal	6–16 9.6 av.	Kay (1960a)
Goats	—	—	—	5.6	Trautmann and Albrecht (1931)
Goats	—	—	—	8.5	Kay (1960b)
Sheep during feeding	10 ml/min	2–4ml/min	—	—	Ash and Kay (1959)
Cattle	—	—	—	75[b]	Balch (1958)
Sheep	—	—	—	7–8	Sasaki and Umezu (1962)
Sheep	6.8–15.8	—	—	—	A. D. Wilson (1963b)

[a] All values in liters per day, unless otherwise indicated.
[b] Estimated.

of the motility of the various parts of the stomach of cattle and in 1928 published from North Dakota their classic contribution on the physiology of the ruminant stomach. A similar study with essentially the same results was completed somewhat earlier in Germany by Wester (1926). The results of both these investigations laid the groundwork for understanding the sequence of rumen movements; the principal features described by them have been corroborated by subsequent authors (Reid and Cornwall, 1959; Quin and van der Wath, 1938; Radeff and Stojanoff, 1955). The movements are important for an understanding of many aspects of rumen physiology and will be discussed in detail.

a. The Primary Mixing Cycle

The basic or primary mixing cycle consists of a fairly regular sequence of events occurring about once each minute, more often in small ruminants, and at an increased rate during feeding or rumination (Morello and de Alba, 1960). The cycle starts with two successive contractions of the reticulum. The first of these occupies 2 to 3 seconds and is incomplete, i.e., less than maximal. The second is complete and occurs before the muscle has entirely relaxed from the first contraction. It forces much of the reticulum contents over the rumino-reticular fold into the anterior sac of the rumen. After the second contraction the reticulum relaxes. It is reported (E. I. Williams, 1955b) that animals suffering damage due to metals in the reticulum often emit a grunt at the time of this contraction, this serving as one means of diagnosis of "hardware disease."

At the first reticulum contraction the rumino-reticular fold contracts, and with the second contraction both the fold and anterior sac of the rumen contract, as well as the anterior transverse pillar. While the anterior pillar is still contracted, the longitudinal pillars, the dorsal rumen sac, the posterior pillar, and dorsal coronary pillars contract, closely followed by the dorsal blind sac. These contractions take about 15 seconds (Reid and Cornwall, 1959). They diminish the volume of the dorsal rumen, and at their termination a greater proportion of the digesta is contained in the relaxed ventral sac. During the latter part of these contractions the anterior pillar and the anterior rumen sac relax.

The last contractions in the mixing cycle occur in the ventral sac, the ventral coronary pillars, the posterior pillar, and the ventral posterior blind sac. This causes the rumen liquid in the ventral portion to rise around the main mass of digesta, percolate into it, and mix with the more solid materials. The rising liquid also spills over into the reticulum and carries with it some of the smaller solids. As a result of these mixing movements the reticulum usually contains more liquid and less solid material than does the rumen, and the particles tend to be finer.

The efficiency of the rumen-reticulum movements in mixing the digesta is difficult to assess and varies with the physical state of the contents. The liquid and its suspended microorganisms and small particles are mixed more rapidly than are the solids, though not necessarily well equilibrated with the liquid within the solid mass. With soft rations or finely chopped feeds mixing is more rapid and complete than with hay. The speed of mixing decreases with increased particle size.

Soon after an animal feeds on long hay, much of the solid digesta forms a compact mass in the dorsal sac. Visual observation of this mass through a fistula discloses little mixing during a contraction cycle. When the ventral sac contracts, liquid wells up around the mass and may spill out of the fistula opening. No other marked result of rumen-reticulum contractions is apparent, except a very slow rotation of the mass of digesta; a sampling tube inserted into the digesta gradually moves across the fistula opening. The massing of solid digesta diminishes as soaking, rumination, and digestion break up the hay.

Addition of marker substances provides a means of assessing the degree of mixing in the rumen-reticulum (Corbett *et al.,* 1959). The contents in the posterior dorsal blind sac are the slowest to mix in cattle, an hour or two being required for equilibration.

b. *Stratification and the Problem of Sampling*

The dry matter in rumen contents may vary between 6 and 18%. In animals feeding on coarse hay many particles of ingesta float in the rumen because of air caught within the hollow stem. This gas ultimately dissolves and the particles become heavier than water. The specific gravity of the total contents exclusive of the gas phase does not become much greater than that of water, values of approximately 1.03 being reported (Balch and Kelly, 1950). In bloating animals when much gas is retained, the specific gravity of the digesta plus retained gas bubbles is much lower (Mendel and Boda, 1961).

If specific gravity of the particles themselves were the only factor influencing the distribution of solids digesta in the rumen, they would tend to settle, except as the mixing movements dispersed them. In actuality, the stratification in the rumen is in the opposite direction. The copious carbon dioxide and methane gases from the rumen fermentation tend to stratify the contents. The gases come out of solution as small bubbles which rise in the rumen and carry solid particles with them (Balch and Kelly, 1950).

There are many nuclei for gas bubble formation in rumen contents. Gas dissolved in the liquid is in equilibrium with the gas phase. Evidence for rapid equilibrium is disclosed in Warburg manometric experiments with vessels containing whole rumen contents, i.e., both liquid and solids. The rate of pressure increase when shaking is started is not significantly greater than the rate during the period with no shaking.

The mixing due to rumen contractions is insufficient to counteract the factors causing layering of digesta, and measurable differences in the rumen contents from different locations in the rumen have been demonstrated (Balch and Johnson, 1950; P. H. Smith *et al.*, 1956).

Obtaining a representative sample is difficult. Plant parts in various fragmentation stages make the rumen contents of pasture- and hay-fed animals extremely heterogeneous. The contents in animals fed ground pelleted rations are more uniform, though still not homogeneous. Entrances of food, saliva, and water introduce differences in concentrations of particular elements in various parts of the rumen, and fermentation gases float solids particles toward the top. Rumen contractions tend to mix the contents, but mixing is not complete.

Since the composition of rumen contents is influenced by ingestion of food, entrance of saliva, and passage of materials to subsequent stomach segments, the composition varies with time. A sample representative at a particular time is not necessarily representative of rumen contents at other times. Furthermore, since the influence of one host differs from that of others, even within the same species or breed, results obtained from one individual are not necessarily representative of the entire group. Though one can speak of an average or typical rumen, and this will be invoked repeatedly in this volume, it should be borne in mind that this is a concept rather than a real entity.

It is fairly easy to obtain samples of rumen contents. They are usually removed either (1) from a slaughtered animal (2) from a normal animal with a stomach tube, or (3) through a fistula. Each has its advantages and disadvantages.

(1) *From a Slaughtered Animal.* The rumen contents of a slaughtered animal can be mixed and yield a representative sample of the total material. It is possible also to measure the total volume of rumen digesta and thus extrapolate accurately from the sample to the whole rumen. The slaughter method is valuable for studies on game animals, provided the study can be made shortly after the animal is killed (Hungate *et al.*, 1959). Handicaps are the limitation to a single sampling, unavailability of facilities for slaughter adjacent to the experimental laboratory, and expense in sacrificing the animals for a single experiment.

In the commercial abbatoir, feed is usually withheld for 24 hours before slaughter. For this reason many data on slaughtered animals are not applicable to usual feeding conditions. Another disadvantage is that the previous feeding schedule of commercially slaughtered animals is not readily available. An unexpected feature in the small intestine of animals killed by shooting and bleeding is a shedding of epithelial cells into the lumen of the small intestine, plus a loss of material from Brunner's glands (Badawy *et al.*, 1957). Animals anesthetized with pentobarbitone do not show this phenomenon.

(2) *From a Normal Animal.* Where it is desired to obtain information on a large number of animals on certain rations or to obtain a rumen sample for diagnosis, rumen digesta can be withdrawn with a stomach tube.

It is difficult to obtain by stomach tube a single sample representative of the entire rumen contents, since the composition varies in the different parts. A stomach tube cannot be accurately directed to the various regions to obtain a composite sample. Samples drawn by stomach tube tend to be more liquid than is the average for the entire rumen contents. When unchopped hay is fed, the stomach tube may become plugged with coarse digesta. Sometimes the inner end of the tube plugs with solids which can be recovered by withdrawing the tube. When the rations are finely chopped or ground, excellent samples of rumen contents can be obtained by stomach tube. The same is often true of animals on a ration high in grain.

Stomach tubes should not be so hard as to injure the epithelium of the mouth and esophagus, yet rigid enough to retain shape. The larger the inner diameter of the tube the easier it is to remove the rumen solids, but there is a limit to the thinness of wall which can be tolerated. Insertion demands a certain degree of pliability, and a pliable tube collapses under vacuum unless the walls are fairly thick.

With large cattle a stomach tube with a lumen of 1¼ inches can be used, and samples can be obtained without a suction pump (Hungate, 1957). Pushing the tube back and forth in the rumen helps fill it with digesta. Sometimes the inner end of the stomach tube is near the top of the digesta and it is necessary to wait for a rumen contraction to force liquid up to the tube opening. For smaller cattle a garden hose or the thin-walled hose for automatic washing machines has proven satisfactory. Suction can be applied to the outer end of the tube by taping it into the opening of a glass filter flask with side arm attached to a suction line or vacuum pump (Markoff, 1913). An automobile tire pump with a reversed valve has been helpful in the field.

With smaller animals smaller tubes must be used—about 15-mm outside diameter for sheep—and with these it is usually necessary to employ suction. It is more difficult to obtain a large sample of rumen contents from sheep, and since these animals are less expensive it is more common to fistulate them. However, stomach tube samples can be taken from sheep, and Mangold and Schmitt-Krahmer (1927) found good agreement in nitrogen content between different stomach tube samples. The fine grinding to which the sheep rumen contents are subjected aids in making it feasible to obtain a sample even with the small size tubes that must be employed. The author has on one occasion obtained rumen contents from an unfistulated sheep by using an inoculating syringe fitted to a trocar. A quantity sufficient to inoculate a culture series was obtained.

It is possible to stomach tube an animal every day for an extended period and even to sample several times a day if the tube is inserted carefully. For repeated sampling it is well to use a tube with an outside diameter somewhat less than the maximum that can be inserted. As the tube is introduced, it should be directed toward the upper part of the throat in order to avoid entrance into the trachea. As the tube penetrates the esophagus there will be some resistance when the cardia is encountered, but this is easily overcome, and the inner end of the stomach tube can be pushed well into the rumen.

In using the stomach tube it is important to obtain or make a speculum to prop open the mouth and prevent chewing on the tube. It is often helpful to restrain the animal in a squeeze chute. Many less fractious individuals can be snubbed up against a post.

Samples withdrawn through a stomach tube contain variable amounts of saliva. When this affects the rumen characteristic to be measured or studied, e.g., bicarbonate or pH, removal through a fistula is preferred, or a device closing the inner end of the tube until it is in the rumen can be used. Except for the dilution which it causes, the saliva has relatively little effect on the rumen bacteria.

(3) *Through a Fistula.* The fistulated animal is extremely useful in rumen studies, and almost indispensable for an intensive program of investigation. The fistulation operation itself is relatively easy to perform, and the animals heal readily with little disturbance in their physiology. The rumen fistula, as usually inserted, causes no significant change in rumen function (Hayes *et al.,* 1964a), provided it is closed with a plug which does not allow leakage of much liquid or gas and which is well fitted to avoid necrosis of the tissues around the fistula margin. Some animals live remarkably well for long periods with an open fistula. Others go off feed. Frequent sampling of the rumen contents of fistulated animals does not introduce undue disturbance of function. Each 4 hours for 1 day total rumen contents were removed through the fistula and sampled, the rest being returned to the rumen without noticeable disturbance of rumen function (Burroughs *et al.,* 1946a). The period out of the rumen should not be too long, nor should any sample be too large.

Plugging of the fistula is more essential in cold than in warm climates, presumably because the heat loss through evaporation from the moist inner wall of the rumen is more serious. Maintenance of the fistula plug is more difficult than the creation of the fistula itself, and much time is spent repairing, redesigning, and manufacturing suitable plugs. Many successful ones have been reported in the literature. Fistulas and plugs have been adapted to permit sampling all parts of the alimentary tract and can be modified to suit purposes of many diverse experiments. The fistula makes the rumen more accessible, but does not eliminate the heterogeneity in rumen contents which makes sampling so difficult.

More accurate methods for sampling rumen contents in fistulated sheep fed on ground ration have recently been developed (Ellis *et al.,* 1962; Sutherland *et al.,* 1962). Two fistulas into the rumen are connected by a large bore tube containing a pump. Rapid circulation of the rumen contents with the pump mixes the contents thoroughly and reduces sampling error. Except in the case of ammonia, conventional sampling methods gave values quite close to those obtained with the pumped material, presumably because the ground ration facilitated sampling by both methods.

Drawbacks to the fistula method are that (1) it is more expensive to maintain fistulated animals, (2) clinical samples cannot be obtained since the method requires a previous operation, and (3) it is usually not feasible to fistulate a large number of animals for short-term experiments.

(4) *Sampling of Feed.* Sampling of the feed of stall-fed animals is ordinarily easy as compared to the rumen sampling problem. With grazing animals the situation is reversed. Recently methods have been developed for sampling the ingesta of grazing animals (Weir and Torell, 1959). A fistula is introduced into the esophagus, and a three-way tube is inserted. Two ends are attached to the proximal and distal ends of the cut esophagus. The other end is brought to the surface of the neck. It is stoppered ordinarily, but for collection of the grazed feed a bag or sack is attached, and the ingesta collects there rather than in the rumen.

(5) *Validity of Samples.* In selecting rumen samples it is important to consider the type of information to be derived. For qualitative studies of the kinds of microorganisms or acids many types of samples obtained directly from the rumen can be used. For quantitative work, in which the rate of fermentation, the total count, the content of fermentation products, and other characteristics are to be measured, the validity of the sample is critical.

With fistulated animals material can be removed from several different parts of the rumen and mixed before subsampling, or the entire contents can be removed, mixed, a sample obtained, and the rest returned to the rumen through the fistula. The problem arising with samples removed by stomach tube—that a sample contains relatively more liquid and less solids than is characteristic of the contents as a whole—can be alleviated in part by straining the sample through muslin or other coarse sieve to reconstitute the solids to a proportion resembling the rumen contents. Observation of the consistency of the contents in slaughtered animals or in animals with a fistula gives an idea of the consistency to be sought in the reconstituted sample (Hungate *et al.,* 1960).

Because of the difficulties in obtaining representative samples, in most quantitative studies of rumen contents it is necessary to make many measurements and subject them to statistical analysis in order to derive an average value indicative of the state in the rumen.

When mixed rumen bacteria are needed they are usually obtained by removing only the rumen liquid. It contains all the kinds of rumen microbes, but they are not necessarily in the same proportions as in the total contents (see *Lachnospira* Chapter II). If it is desired to study the nutrition of the mixed rumen bacteria (Bentley *et al.,* 1954a), a washed-cell suspension (Elsden and Sijpesteijn, 1950) is most useful.

The extent to which samples represent the rumen diminishes with time. Storage at refrigerator temperature diminishes some changes, but introduces others by killing some of the protozoa and bacteria. Maintenance *in vitro* without food, but with other conditions favorable, results in the death of some cells (Hungate, 1956a), and, if substrate is supplied, differences in growth rates of various species cause the composition of the population to change. The incubation time of batch experiments thus becomes an important factor influencing the applicability of the results to the rumen. This will be considered further in later discussions of batch culture experiments (Chapter VI).

2. RUMINATION

This process of rechewing the food or chewing the cud, is one of the most distinctive features of ruminants. It is intimately connected with the use of herbage as food.

Galen (reported by Tiedemann and Gmelin, 1831) observed that young ruminants did not ruminate until they consumed solid food (forage). Diminution of rumination in animals on pasture as compared to hay was reported by Aristotle (cited by Brugnone, 1809). Haubner (1837) believed that comminution of large particles occurred not only through rumination, but also through the influence of the alkaline saliva and acid fermentation products, together with the abundant moisture and fairly high temperature. Subsequently, enzymatic digestion has been postulated as a mechanism causing fragmentation of ingested forage (Mangold, 1943).

Several hours after pasturing steers on lush ladino clover, there are few intact plant parts in the rumen, aside from the long strings of undigested vascular bundles. Relatively little rumination is observed in such animals (Gordon, 1958c; Aristotle, from Brugnone, 1809). Mixing and enzymatic digestion in the rumen are the chief mechanism for comminution of the soft plant parts. With more resistant plant materials, lignified to a greater extent, enzymatic activity is insufficient to fragment it. This was proven by Columbus (1934) when he fed carefully cut fragments of oat straw, 0.5 by 2.0 mm, to sheep and goats and, except for a small fraction of broken pieces, recovered them unchanged in the feces. It also is shown by a greatly increased retention time when rumination is prevented (Pearce and Moir, 1964).

Rumen contraction and enzymatic activity can comminute rumen digesta to an extent permitting slow passage through the alimentary tract (Pearce and Moir, 1964), but the slow rate of comminution greatly limits food consumption in comparison to animals which can ruminate.

Although comminution of resistant plant parts in rumination has long been postulated as essential for complete action of the microbial digestive enzymes, this factor is probably minor compared to the importance of rumination in reducing particles to a size that can proceed through the alimentary tract (Ellenberger and Scheunert, 1925). However, the demonstration that extremely fine grinding increases the digestibility of cellulose in forage (Dehority and Johnson, 1961) indicates that rumination may increase digestibility to some extent.

As the first step in rumination, digesta in the reticulum are regurgitated into the mouth. Regurgitation is initiated by contact of coarse particles with the wall of the rumen and reticulum. Ellenberger and Scheunert (1925) state that rumination can be induced artificially by rubbing coarse plant material against the rumen wall, and Schalk and Amadon (1928) stimulated rumination by rubbing the mucosa of the reticulum, an observation confirmed by numerous rumen physiologists (Ash and Kay, 1957). The ease with which this reflex can be induced varies with the animal. One of the fistulated cows studied by Mendel at Davis, California, could regularly be induced to ruminate by gently stroking the fingers on the wall adjacent to the cardia. During rumination the electroencephalogram of goats resembles that obtained during somnolence (Bell, 1958b).

A certain degree of comminution of feed is required before rumination occurs. Long hay added through the fistula caused a diminution in rumination and a loss of appetite which lasted for several days (Glawischnig, 1962). Chopped hay added through the fistula was less disruptive.

Regurgitation is accomplished by a complicated set of muscular contractions preceding the two reticular contractions of a mixing cycle. Contraction of the reticulum elevates the digesta in this compartment well above the level of the cardia. The cardia opens, and the animal contracts the diaphragm and inspiratory muscles (Tiedemann and Gmelin, 1831), and at the same time closes the glottis (Bergmann and Dukes, 1925). This creates a lower than atmospheric pressure in the pleural cavity (Colvin et al., 1957), and since the wall of the esophagus is quite pliable the decreased pressure is transmitted to the interior of the esophagus. In part as a result of this lowered pressure, but chiefly due to the hydrostatic pressure from the liquid in the reticulum, digesta flow into the esophagus. The cardia closes, and the bolus is carried to the mouth by reverse peristalsis (Downie, 1954). Ash and Kay (1959) noticed a slight forward extension of the head at the instant of regurgitation in sheep, and Wester (1926) has considered this to be important in bringing digesta into the esophagus.

Relatively little force per unit area is required to move digesta into the esophagus, since with relaxation of the cardia the opening into the esophagus becomes very large. The pressures in the rumen (Bell, 1958c) appear quite adequate to move the material. A lowering of intrapleural pressure may not be essential, as rumination can occur also in animals with a tracheal fistula (Colin, 1875, quoted by Wester, 1926).

The liquid accompanying the solids is necessary for transporting them (Colin, 1873). Most of this liquid is immediately swallowed in one or two portions (Toussaint, 1875), and the solids are chewed (Downie, 1954). The rechewing during rumination is more thorough than the initial mastication during feeding. The jaws operate with a lateral grinding motion, and each bolus is chewed between 30 and 85 times. Ellenberger and Hofmeister (1887a), found 50–70 mastications per bolus; Downie (1954) observed 56.4 jaw movements per bolus; Gordon (1958c) observed an average of 78 chews per bolus, the number depending on the kind of animal and the nature of the feed. Digesta from coarse forage is rechewed more than that from less resistant feeds. The total time for chewing one bolus varies between 25 and 60 seconds. As mentioned earlier, during rumination there is an increase in secretion of parotid and inferior molar saliva, but very little secretion from the mucous glands. The submaxillary gland is almost completely inactive.

The rechewed bolus is swallowed, and another is regurgitated almost immediately. Since each regurgitation precedes the initiation of a primary mixing cycle, rumination is synchronized with the contraction cycles, one bolus per cycle. The swallowed ruminated bolus enters and mixes in the rumen in the same fashion as boli swallowed during feeding. There is no means for shunting the rechewed solids directly to the omasum, but, as will be discussed, the finer state of division of the particles in the ruminated bolus favors their passage from the rumen.

In cattle the ruminated bolus as it is reswallowed contains about 100 gm of material including 15 gm of dry matter. Direct observations on grazing cattle have shown that the maximum time spent in rumination is about 8 hours per day (Harker *et al.,* 1954; Gordon, 1958a). In such animals, if 50 seconds is the duration of each rumination cycle, 57.6 kg of digesta would be rechewed each day.

Since there is not complete separation of ruminated from unruminated digesta some material enters more than one rumination cycle. Schalk and Amadon (1928) noted that the ruminated bolus contained very little of the recently ingested food, probably because the latter collected in the dorsal sac and was slow to mix.

3. ERUCTATION

Methane and carbon dioxide are metabolic products of the rumen fermentation, and carbon dioxide is liberated from bicarbonate by the other fermentation products, the volatile acids. This copious production of gas necessitates a mechanism for its escape. Much of the methane and carbon dioxide liberated in the rumen is voided by eructation.

The carbon dioxide, if alone, might conceivably escape by diffusion into the blood and out into the cavity of the lungs, but the solubility of methane in blood is so low that this pathway is insufficient for escape of all the methane formed. Escape of methane by diffusion into the bovine lungs amounts to 25–96% (mean 61%) of the total formed before feeding and 8–53% (mean 21%) of the total methane formed after feeding according to Hoernicke *et al.* (1964), though others (Cresswell, 1960; Zuntz, 1913) found relatively little eliminated via the blood.

Absorption of some carbon dioxide through the rumen wall is indicated by the fact that the proportion of methane in the gas in the rumen (25–30%) is greater than the proportion in which it is produced in fermentation (15–20%). These values suggest that approximately one-fourth to one-third of the carbon dioxide liberated in the rumen escapes via the bloodstream. The amount would be greater than this figure to the extent that methane also diffuses into the blood and out through the lungs. Hoernicke *et al.* (1964) estimated that 11–99% of the carbon dioxide released in the rumen was voided through the lungs.

In addition to the sequence of muscular movements described above as the rumen mixing cycle there are often intercalated one or more rumen contractions in addition to those directly following the two reticulum contractions. Each of these additional contractions immediately precedes an eructation. The contraction is chiefly in the dorsal sac. It causes the large gas volume above the rumen contents to be displaced forward. At the same time the rumino-reticular fold contracts, and lifts up and holds back the forward surge of solid digesta (Dougherty and Meredith, 1955).

Contraction of the anterior pillar of the rumen is also important in holding back the digesta (Johns *et al.*, 1958). These movements cause the gas to move into the dorsal part of the reticulum. The cardia then opens, as do the diaphragmatic and pharyngeal sphincters (Dougherty and Habel, 1955), and the gas escapes through the esophagus.

The glottis remains open briefly while the external nares are closed, and considerable quantities of the eructating gas are forced into the lungs (Dougherty *et al.*, 1962a,b, 1965). This was detected by the appearance in the carotid artery of brief spurts of dark colored blood similar in appearance to that in the pulmonary artery.

In a normal eructation the gas is emitted with little apparent evidence of its passage. Rarely, in an animal tending to bloat, a noise of eructation may be heard. If gas production in fermentation is very rapid, or when large quantities of gas are artificially insufflated into the rumen, numerous eructations occur. As long as the animal's eructation mechanism functions properly the disposal of the large quantity of gas produced in the rumen, provided it separates from the solid ingesta as a pocket at the top of the rumen, poses little difficulty. When the gas fails to separate from the solids, as in certain types of bloat, the situation is quite different, as will be discussed in Chapter XII.

The eructation reflex is stimulated by distension of the rumen. Receptors capable of inhibiting eructation are located in the vicinity of the cardia (Dougherty *et al.*, 1958). These receptors are stimulated by contact with solids or liquid, and they transmit an excitation to the central nervous system which inhibits the reflex for sphincter opening initiated by rumen distension. This inhibition acts on all esophageal sphincters, but is most effective on the pharyngeal sphincter, the latter blocking digesta forced past the cardia and the diaphragmatic sphincter (Dougherty *et al.*, 1958). Normally, the holding back of solid and liquid digesta by the rumino-reticular fold assures that only gas is adjacent to the cardia area containing the receptors for the inhibitory reflex. Under these conditions the stimulus by distension initiates eructation.

Eructation can occur even after rumen motility is abolished by acetylcholine-blocking agents. Epinephrine completely inhibits both the rumen and reticulum and eructation does not occur, even though gas is insufflated in an amount normally causing the eructation reflex. Histamine also inhibits eructation, but does not affect deglutition.

Attempts to block the eructation reflex by introducing water to cover the cardia are usually unsuccessful in the normal animal. The abdominal cavity swells to accommodate the additional volume, and the relaxation of the reticulum and contraction of the rumino-reticular fold are sufficient to clear the cardia for eructation. If the reticulo-rumen is completely filled with fluid the eructation reflex can be inhibited. To accomplish complete filling it is necessary to raise the animal's hind legs.

By isolating a small pouch of the rumen-reticulum adjacent to the cardia it was shown (Dougherty *et al.*, 1958) that gaseous insufflation of this small compartment initiated the eructation reflex. Liquid pressures failed to induce a complete eructation reflex. At a pressure of 40 mm Hg the cardia and diaphragmatic sphincter allowed the liquid to move into the esophagus as far as the pharyngeal sphincter, but the latter prevented further movement. The eructation-inhibiting reflex is abolished by butyn sulfate (butacaine).

The existence of neuromuscular mechanisms by which the ruminant under one set of circumstances regurgitates digesta for rumination, yet under other circumstances retains solid digesta while allowing gas to escape, emphasizes the complex nature of the controls necessary for successful rumen function.

One digestive activity rarely observed in ruminants is vomiting. There is a vomiting center in the brain (Andersson *et al.*, 1959), and vomiting can be induced with certain drugs (Dougherty *et al.*, 1965), but it almost never occurs naturally. It is not a mechanism by which the ruminant can normally rid itself of undesirable rumen digesta. It appears that in the process of evolution of the complex neuromuscular mechanisms coordinating the ruminant stomach the capacity for disposing of digesta by vomiting, so commonly operating in some mammals, has become nonfunctional (Arduini and Dagnino, 1953) under circumstances in which the animal could employ it to advantage, as in bloat.

4. Separation of Coarse from Fine Particles

The correlation between the extent of rumination and the coarseness of the feed suggests that rumination is necessary in order that materials may pass through the alimentary tract. This was recognized early by Haubner (1837). It implies a mechanism by which coarse particles are retained in the rumen.

The material leaving the rumen is not representative of the total rumen contents. Comparison of the size of particles in rumen digesta with those in the feces of hay-fed animals shows only fine particles in the latter, as compared to a great many large fragments of plant material in the rumen. The sharp dividing line between coarse and fine particles lies between reticulum and omasum. There is a mechanism for partial retention of coarse particles in the rumen and a very effective mechanism for their complete retention in the reticulum.

The partial separation of coarse from fine particles between the reticulum and the rumen depends on the spatial configuration of these compartments and on the mixing contractions. The reticulum contents are more liquid than those of the rumen, due in part to the saliva which is secreted into this compartment during nonfeeding periods and in part to the rapid reticulum contractions which expel into the rumen the coarse materials which float to the top. During the slow contractions of the ventral sac of the rumen, liquid wells up and around the mass of solids digesta and spills over into the anterior sac and the reticulum; some solids particles are carried with the liquid, but the main mass of solids is left in the rumen, particularly the incompletely soaked large particles of forage in the upper layers.

The reticulo-omasal orifice is the site of the complete separation of fine from coarse particles (Ellenberger and Scheunert, 1925). This orifice is adjacent to the cardia, where coarse particles are regurgitated. Mixing cycles cause coarse and fine particles to spill over into the reticulum. Some of the fine ones pass into the omasum; the coarse ones accumulate due to the screening action at the entrance to the omasum and lie in a favorable spot for regurgitation. They will have accumulated to a maximum just before the next mixing cycle starts, and this is the time when rumination occurs.

The mechanism for retaining coarse particles at the reticulo-omasal orifice is not established. As has been mentioned, the edges of the omasal leaves which form the vestibule, just within the reticulo-omasal orifice, are well supplied with rough papillae. The open end of the small, roughly conical vestibule is the reticulo-omasal opening; the tapering portions are composed of the omasal groove and the papillated edges of the omasal leaves. This arrangement seems fitted to direct liquid and fine particles between the leaves and to screen out coarse particles (Becker et al., 1963). The coarse retained particles stimulate the wall of the reticulum and elicit the rumination reflex which leads to their comminution.

Absorption of water has been thought to be the chief function of the omasum. In cattle the amount absorbed was 100 liters per day (Sperber et al., 1953). In one case 48–55% of the water entering the omasum was absorbed (Badawy et al., 1958a). In other cases 33–64% (Gray et al., 1954) or one-half of the water in the omasum was absorbed in 3 to 4 hours (Aggazzotti, 1910). The material entering the omasum contains 90–95% water (Balch et al., 1951), the percentage dry matter in the reticulum being somewhat less than in the rumen. Absorption reduces the liquid content of the material passing into the abomasum. The liquid then increases in the abomasum as a result of secretion of gastric juice.

In addition to water, the omasal leaves absorb volatile fatty acids (VFA's) and other dissolved constituents (Gray et al., 1954; Johnston et al., 1961). Their close juxtaposition decreases the average distance between dissolved substances and an absorptive surface and renders the omasum more efficient than the rumen in lowering the concentration of the fermentation acids, which occur in the effluent from the abomasum in one-seventh to one-twentieth the rumen concentration (Masson and Phillipson, 1952). The removal of VFA salts before the digesta enters the abomasum may prevent a buffering against the acidity of the gastric juice.

Extrusion of omasal contents into the abomasum by omasal contractions has been noted in goats (Kameoka and Morimoto, 1959) with a fistula between these two organs. The results of these investigations throw some doubt on the water-absorbing function of the omasum. The material leaving the omasum contained an average of only 6.9% solids, a value in agreement with other results (Oyaert and Bouchaert, 1961), which is approximately the dry matter percentage of the reticulum contents. The high flow rates may be due to failure to return the digesta to the abomasum (Kay, 1963). Flow of digesta from the omasum to the abomasum when the omasum contracts (Kameoka and Morimoto, 1959) may involve both water and solids from between the omasal leaves. The omasal fluid entering the abomasum exceeds the quantity of abomasal secretion (Masson and Phillipson, 1952).

The high percentage of dry matter in omasal contents, as determined on slaughtered animals, could be explained as well by a squeezing out of fluid as by absorption, and both may occur. The fact that most of the VFA's entering the omasum are absorbed suggests that conditions for absorption are favorable, as would be the case if the fluid passed between the leaves.

Absorption of water in excess of solvents could occur only with performance of work. Since absorption of the VFA's in the omasum would decrease the osmotic pressure of the omasal contents, water would also tend to be absorbed, but the extent of this may not be nearly as great as the usual analyses of omasal contents indicate. In experiments with an anesthetized goat the omasum was tied off from the rumen and abomasum, and 40–80 ml per hour of added water was absorbed (Raynaud and Bost, 1957). This could have been due in part to the lowered osmotic pressure resulting from the added water.

The balance of evidence favors the hypothesis that the omasum absorbs some water and salts, most of the VFA's and also screens out and prevents further passage of coarse particles.

5. Movements of the Omasum

The reticulo-omasal orifice and the omasal pillar contract simultaneously with the first reticular contraction in the normal mixing cycle and slightly with the second reticulum contraction (Stevens *et al.,* 1960). During the first and second contractions the reticulo-omasal orifice is open, but it is blocked by the ventral ridges of the esophageal groove which form a valve allowing backflow into the reticulum. The contraction of the omasal pillar may force digesta back into the reticulum at this point in the cycle.

At the height of the second reticulum contraction in the primary mixing cycle, the lips open and a gush of digesta passes through the reticulo-omasal orifice into the omasum (Balch *et al.,* 1951; Stevens *et al.,* 1960). The orifice then closes as the wall around the vestibule contracts and forces digesta into the body of the omasum between the leaves and along the edges adjacent to the pillar. This occurs as the reticulum is contracting. Coarse particles are strained out by the edges of the leaves and collect in the vestibule. As the reticulum relaxes the reticulo-omasal orifice closes. The omasal canal then contracts as the rumen contracts. At the second or eructation contraction of the rumen the omasal-abomasal orifice may open, and contents are extruded into the abomasum. If the orifice is closed (this is the case when the pillar on the groove of the omasum contracts) the material in the vestibule is forced out through the reticulo-omasal orifice which remains loosely open after the vestibule has contracted and during the contraction of the omasal body. The reverse flow from the omasum presumably returns to the reticulum any coarse particles strained out in the vestibule. Reverse flow also occurs when the vestibule contracts at the time of the reticulum contraction associated with rumination, as well as with the first reticulum contraction of the regular mixing cycle. Some of the coarse particles forced out with the rumination contraction of the reticulum would conceivably be rechewed in that rumination cycle. This possibility has not been examined experimentally.

In addition to contracting at the time of the rumination contraction of the reticulum, and also at the time of the first regular reticulum contraction, the omasal vestibule contracts at the time of each rumen contraction. Then the vestibule relaxes, whether the rumen contraction is a regular one or an eructation contraction, and the reticulo-omasal orifice remains loosely open. During this period there is a slow even flow of material from the reticulum into the omasum.

A grinding action of the omasal leaves has been postulated by many investigators, particularly the earlier ones, as an important factor in the comminution of the food particles. It is extremely doubtful that this is of quantitative significance as compared with the comminution due to rumination. A rubbing compression of wet hay particles between omasal leaves lacking hard and abrasive surfaces seems ill adapted for grinding.

X-ray photographs to show omasal movements have been published (Benzie and Phillipson, 1957).

6. MOVEMENTS OF THE ABOMASUM

Abomasal contractions mix the digesta with the abomasal fluids. Both hydrochloric acid and pepsin are secreted, as well as mucus. The microorganisms which enter from the omasum are killed by the acid (Gruby and Delafond, 1843; Pounden *et al.*, 1950) and begin to digest. Abomasal contractions are not coordinated with those of the other compartments of the stomach.

The functioning of the ruminant abomasum differs from that of other mammals chiefly in the continuous nature of its activity (Hill, 1955; Phillipson, 1952b). Digesta enter more or less continuously from the omasum, even when the ruminant is not feeding, and the abomasum secretes continuously. If the rumen is emptied artificially or the animal is fasted, the abomasum gradually ceases to secrete acid (Fig. IV-8),

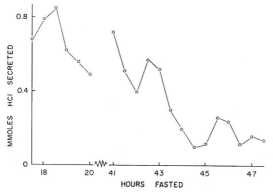

Fig. IV-8. Millimoles of HCl secreted by the abomasum. Each value represents the amount secreted since the previous measurement. From Hill (1955, p. 35).

which indicates that rumen contents stimulate secretion. Polypeptides appear in the abomasal contents, but they are not further digested in that organ (Raynaud, 1955).

The rumen fluid entering the abomasum stimulates abomasal secretion (Ash, 1959a), but in addition a movement of digesta to the duodenum is important (Ash, 1961a). The entrance of neutral digesta from the omasum and the exit to the duodenum of the more acidic abomasal contents keeps the pH in the abomasum above the level at which acidity of the abomasal contents inhibits further secretion of acid. This level is at about 0.01 N hydrogen ion concentration (pH 2.0) (Ash, 1961a). Distension of the abomasum also increases the acid secretion.

In sheep on a maintenance ration of poor-quality grass hay, the quantity of fluid leaving the abomasum was 10 liters per day. Similar values were obtained with a ration of hay-linseed meal-oats (Hogan and Phillipson, 1960). An increased rate of passage was noted during rumination (Harris and Phillipson, 1962), and a decreased rate during feeding. In other experiments 0.9 ml per hour per gram of organic matter in the ration entered the duodenum of sheep on a high straw ration, and 0.6 ml per hour per gram with a low straw ration (Phillips and Dyck, 1964).

7. Neural Coordination of Stomach Activity

The contractions of the ruminant stomach and esophagus are coordinated almost exclusively by reflexes through the parasympathetic system (Titchen, 1958, 1960). The afferent and efferent impulses are conducted chiefly via the tenth (vagus) nerves. In the neck region the right and left cervical vagus nerves are separate. Either one may be cut without destroying the functions of the stomach, but section of both cervical vagi causes atony, i.e., cessation of muscular activity (Ellenberger, 1883; Mangold and Klein, 1927). Food no longer moves through the stomach, food ingestion stops, and the animal dies unless it is fed directly into the abomasum (Duncan, 1953). Cutting of the nerve to the omasum causes that organ to diminish in size and become empty of digesta (Hoflund, 1940, after Stevens et al., 1960).

The right and left cervical nerves of the vagus join in the region of the esophagus; their fibers intermingle and then proceed posteriorly to the stomach as dorsal and ventral abdominal branches of the vagus. (Figs. IV-9 and 10). Each of these branches carries fibers from both the right and left cervical vagi. The dorsal branch innervates the left side of the rumen and reticulum. The ventral branch innervates the rest of the rumen and reticulum, all of the omasum, and part of the abomasum. The esophagus is also innervated (Dougherty et al., 1958). The dorsal branch can be cut without permanent disruption of digestive activity, but cutting of the ventral branch is fatal (Mangold and Klein, 1927).

Rumen reflexes are mediated through centers in the brain (Iggo, 1951, 1956; Bell and Lawn, 1955). These centers are concerned with the initiation and coordination of the rhythmic mixing cycles, rumination, and eructation. The brain centers are probably affected by chemical substances. Hyperglycemia inhibits rumen motility (Simmonet and LeBars, 1953b; Vallenas, 1956), and hypoglycemia increases it (LeBars et al., 1953b). It is also inhibited by acetate, propionate, and butyrate

(LeBars *et al.*, 1954a). Normal rumen motility can occasionally be initiated by stimulating the vagus in an intact conscious animal (Popow *et al.*, 1933).

The fact that destruction of the nerves to the stomach causes complete

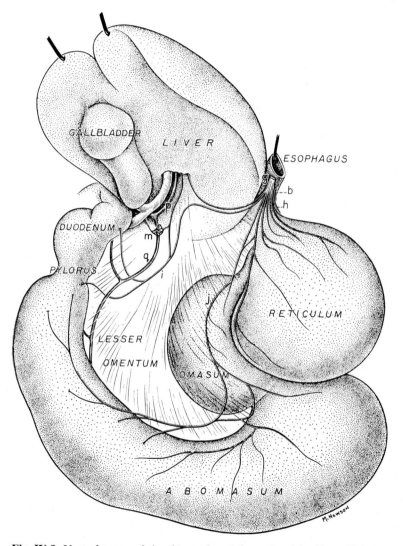

Fig. IV-9. Ventral vagus of the sheep, viewed from the right. From Habel (1956, p. 562). Key: (b) anastomosis to dorsal vagus; (g) ventral vagus; (h) branch to left side of cephalad portion of dorsal sac of the rumen; (i) long pyloric nerve; (j) continuation of ventral vagus; (m) sympathetic plexus; (p) hepatic artery; (q) right gastric artery.

loss of function shows that the organs themselves have little capacity for nonreflex initiation and coordination of muscular activity. Small localized spontaneous contractions which may show some spreading have been noted on the surface of the denervated rumen, but no rhythmic contractions of rumen or reticulum tissues have been observed. Strips of omasal and abomasal wall may contract periodically when transferred to physiological media, but this characteristic does not enable a denervated omasum to perform successfully its various complicated functions.

In contrast, the abomasum functions after nerves to the stomach have been severed, and a denervated animal lives if fed concentrate through an abomasal fistula. This effectively shunts the first three compartments and makes the ruminant functionally similar to nonruminants.

The centers controlling rumen motility are in the medulla. They do not

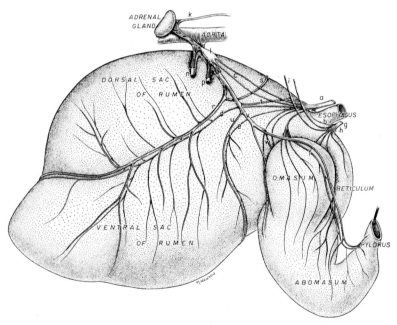

Fig. IV-10. Dorsal vagus of the sheep, viewed from the right. From Habel (1956, p. 563). Key: (a) dorsal vagus; (b) anastomosis to ventral vagus; (c) branches to coeliad plexus; (d) right ruminal nerve; (e) left ruminal nerve; (f) continuation of dorsal vagus; (g) ventral vagus; (h) branch to left side of cephalad portion of dorsal sac of the rumen; (i) long pyloric nerve and branch to the liver; (k) splanchnic nerve; (l) coeliacomesenteric ganglion; (n) cephalad portion of mesenteric artery; (o) coeliac artery; (p) hepatic artery; (r) right ruminal artery; (s) splenic artery; (t) reticular artery; (u) left ruminal artery; (v) left gastric artery; (w) left gastroepiploic artery.

depend on connections from the spinal cord or the cerebral hemispheres; transection between these parts and the medulla does not affect the reflexes. The center for rumen-reticulum motility is in the same area of the medulla as the center through which movements of the esophagus are controlled. Stimuli in the mouth attendant on ingestion and rumination initiate reflexes via afferent fibers in the ventral abdominal vagus (Stevens and Sellers, 1959) through the medulla, which increase the frequency of stomach mixing cycles. Rumination can be induced by stimulation of appropriate medullary areas (Andersson, 1951; Andersson *et al.,* 1959). The afferent fibers for the eructation reflex lie in the dorsal abdominal vagus (Stevens and Sellers, 1959).

Receptors sensitive to distension and muscular contractions are postulated (Iggo, 1955). These can affect salivary secretion (Kay and Phillipson, 1957). Histological examination of the wall of the rumen and reticulum has disclosed (Hill, 1957, 1958) numerous beaded neural fibrils which extend between the epithelial cells and terminate in bulbs in the layers near the inner surface. They are found in the esophageal groove, parts of the dorsal sac of the rumen, along the right longitudinal and posterior pillars of the rumen, and in the rumino-reticular fold and adjacent areas. Single nerve fibers from receptors near the esophageal groove have been identified and shown to increase their rate of firing when the reticulum is inflated. These receptors are excited by distension of the wall and by contraction of wall muscle fibers. A continuous contraction due to reexcitation of the receptors by the contracting muscle is apparently avoided by a refractile period in the synapses of the central nervous system. A reflex inhibition of reticulum contractions occurs with distension of the abomasum, the afferent impulses in this case being transmitted to the central nervous system over the splanchnic nerve, a branch of the sympathetic rather than the parasympathetic system. The efferent fibers of the reflex are all contained in the vagus. The sympathetic system exerts relatively little control over salivary secretion and the contractions of the ruminant stomach.

The rumen function centers in the brain are susceptible to influence from other central nervous system centers and can be affected by external environmental factors. For example, rumination ceases when the animals are confronted by danger. As in other mammals many digestive activities are suspended during periods of extreme stimulus to the sympathetic system. Conversely, milking of goats stimulates rumination (Andersson *et al.,* 1958a).

The factors governing passage of food through the alimentary tract are not without import from a practical standpoint. As will be discussed later, one of the important parameters of food utilization is the retention time in the alimentary canal. This is a function of appetite which in turn is influenced in part by the afferent neural impulses from the stomach.

8. ACTION OF DRUGS ON THE RUMINANT STOMACH

The ruminant stomach is affected by a number of drugs, some excitory, others inhibitory (Dougherty, 1942b). Atropine and adrenaline inhibit by blocking the nerve-muscle synapse (Dussardier, 1954a), and adrenaline may also decrease motility by causing hyperglycemia (LeBars *et al.,* 1953c). Nicotine inhibits nerve impulses at the ganglionic synapse. Histamine inhibits by acting on the medullary center (Brunaud, 1954). Prussic acid also is inhibitory on the center. Other factors such as pH below 5.0 in the rumen or too high an alkalinity can cause inhibition via effects on the central nervous system. Injected adrenaline has been reported to stimulate rumination (Kay, 1959), but the significance is not clear.

Excitory drugs are veratrine, eserine, neostigmine, acetylcholine, and other choline derivatives. Most of these either simulate the action of acetylcholine or inhibit the cholinesterase which normally decomposes the acetylcholine after its release at the synapses. Acetylcholine is the chemical mediator at the intraganglionic synapses as well as at the neuromuscular connection and in the centers in the central nervous system (Dussardier, 1954a,b). Administration of eserine hastened the passage of dyed oat husks through sheep (Aliev, 1958). Rumination could not be blocked with amphetamine, atropine, and small doses of apomorphine (Andersson *et al.,* 1958).

G. Absorption

1. VOLATILE FATTY ACIDS

Acetic and butyric acid were early known to occur in rumen contents (Tiedemann and Gmelin, 1831; Sprengel, 1832), and propionic and branched-chain lower acids were detected by von Tappeiner (1884a). Accurate quantitative separation of the individual volatile fatty acids (VFA's), especially in microamounts, was not possible until the liquid-liquid partition chromatographic method was adapted (Elsden, 1946) to the separation of the VFA's in rumen fluid. Subsequently a great number of reliable measurements have been made (Table IV-4). In the earlier analyses only acetic, propionic, and butyric acids were determined, higher acids than butyric being collected and titrated with the butyric acid. In others the higher acids were lumped as a separate category (Rook, 1964).

With the development of gas-liquid chromatography (James and

Martin, 1952), el-Shazly (1952a) demonstrated that small quantities of higher straight- and branched-chain fatty acids occur also in the rumen. Subsequently more refined methods have been applied, and numerous additional VFA's occurring in small concentrations have been identified (Table VII-6).

Formic acid has been found occasionally in rumen contents. The maximum reported proportion is 5% of the VFA's (Gray *et al.,* 1952). If rumen contents are sampled several hours after food is consumed only traces of formate can be found. Formic acid is a constituent of hay (0.0049% in hay according to Matsumoto, 1961), and its occasional occurrence may be explained by recent feeding. It is metabolized more rapidly than it is formed in the rumen (Carroll and Hungate, 1955), and as a result the concentration is usually less than 0.02 μmoles per milliliter (see Chapter VI).

The VFA's produced in the rumen are almost completely absorbed by the animal. This was first shown by Wilsing's (1885) demonstration that the total volatile acid in the urine and feces of a goat was only 4 gm, whereas more than 157 gm were estimated to have been produced in the rumen.

The first proof that fermentation acids were absorbed from the rumen was obtained by Barcroft *et al.* (1944; Marshall and Phillipson, 1945) when they showed that the portal blood leaving the rumen contained a greater concentration of VFA's than did the entering arterial blood (Table IV-5). Many subsequent investigations have confirmed these findings (Gray, 1947a, 1948; Kiddle *et al.,* 1951; Bensadoun *et al.,* 1962; Schambye and Phillipson, 1949). The acids are metabolized in the liver, acetic to a less extent than the other acids (McClymont, 1951a; Table IV-6).

The VFA's are metabolized to a certain extent during passage through the rumen wall (Pennington, 1952, 1954). This is particularly true of butyrate; very little of it reaches the blood at low planes of nutrition (Kiddle *et al.,*1951) (see Table IX-1). Tissue slices of rumen epithelium show greater oxygen consumption with butyrate as substrate than with the other acids or with glucose (Pennington, 1952, 1954; Pennington and Pfander, 1957). Excised epithelium shows a polarity in its action on butyrate (Hird and Weidemann, 1964), the latter appearing as β-hydroxybutyrate on the muscle side of the epithelium regardless of the surface, papillary or muscle, to which it was applied. The butyrate is converted chiefly to β-hydroxybutyrate also during passage through the omasal wall (Joyner *et al.,* 1963).

Propionate is also utilized to some extent in the rumen wall, though only in the presence of carbon dioxide. Presumably propionyl CoA is

Table IV-4

VOLATILE FATTY ACID CONCENTRATIONS IN THE RUMEN-RETICULUM

Animal	Ration	Total acid concentration (μmoles/ml)	Acid concentration in moles percent				Location and Reference
			Acetic	Propionic	Butyric	Higher	
Sheep	—	99	63	21	15	—	Scotland, Pfander and Phillipson (1953)
Sheep	Dried grass	65	68	19	13	—	Scotland, Armstrong et al. (1958)
Sheep	Grass pasture	81–178	65.5	19.5	15	—	England, Elsden (1945b)
Sheep	—	—	63	18	14	4.5	England, Kiddle et al. (1951)
Sheep	Wheaten hay	49–109	61–69	17–24	13–15	—	Australia, Gray and Pilgrim (1951)
Sheep	Alfalfa hay	52–193	68–69	15–20	10–16	—	Australia, Gray and Pilgrim (1951)
Sheep	16 different low protein rations	29–93	60–69	21–30	8–16	—	Africa, Topps and Elliott (1964)
Sheep	Rye grass and clover pasture	101–187	50–62	21–30	12–17	3–10	New Zealand, Johns (1955b)
Sheep	Rye grass and clover pasture	32–153	38–64.5	15–30	9.5–22	1.5–12	New Zealand, Jamieson (1959a)
Average	—	104.5	62.5	20.5	17.0 (butyric + higher)		—
Cattle	—	133	68	18	14	—	England, Elsden et al. (1946)
Cows	16–18 lb. hay, 20 lb. concentrate	110–180	62.1	21.0	11.5	5.4	England, Balch et al. (1955b)
Cows	2 lb hay, 24 lb concentrate	60–200	44.7	34.8	9.2	11.3	England, Balch et al. (1955b)
4 cows, 2–4 hr after feeding	Grass hay and concentrates	108–160	56–65	17–30	10–18	2.6–5.1	Australia, McClymont (1951a)

2 cows, 24 hr after feeding	Grass hay and concentrates	56–60	58–69	16–24	12–14	2.8–4.1	Australia, McClymont (1951a)
5 cows	Mixed hay	—	60	21	19	—	New York, Card and Schultz (1953)
6 cows	Mixed hay and grain	—	58	19	23	—	New York, Card and Schultz (1953)
4 cows	Pasture only	—	56	18	26	—	New York, Card and Schultz (1953)
10 cows	Pasture and grain	—	53	20	27	—	New York, Card and Schultz (1953)
3 cows	Grain only	—	47	23	30	—	New York, Card and Schultz (1953)
5 cows	Early silage	—	52	24	24	—	New York, Card and Schultz (1953)
3 cows	Barn-dried hay	—	57	20	23	—	New York, Card and Schultz (1953)
3 cows	Late silage	—	55	23	22	—	New York, Card and Schultz (1953)
5 cows	Field-cured hay	—	60	20	20	—	New York, Card and Schultz (1953)
Cattle	High protein	119	59	21	15	5.0	Maryland, Woodhouse et al. (1955)
	Medium protein	114	60	20	15	4.7	Maryland, Woodhouse et al. (1955)
	Low protein	105	61	19	14	4.7	Maryland, Woodhouse et al. (1955)
Cattle	61% barley, 16% soybean oil meal, 22% alfalfa meal	138–173	57	26	13	—	Maryland, Jacobson and Lindahl (1955)
Cows, not starved	—	89±10	65	21	14	—	Maryland, Brown and Shaw (1957)
Cows, starved 18–24 hr	—	51±12	70	16	14	—	Maryland, Brown and Shaw (1957)
Cattle	Alfalfa hay	120	63	23	14	—	Washington, Carroll and Hungate (1955)
	2/3 grain, 1/3 hay	81	58	25	17	—	Washington, Carroll and Hungate (1955)
	Green pasture	78	68	21	11	—	Washington, Carroll and Hungate (1955)
Milking cows	Hay and concentrate	108–224	67–80	9–19	7–17	—	California, Hungate et al. (1961)
Average		111	59.6	21.0	18.2	—	—

formed and carboxylated to methylmalonyl-CoA. This is rearranged to succinyl-CoA which in turn reacts with propionic acid to regenerate propionyl-CoA and releases succinic acid for use by the tissue (Pennington and Sutherland, 1956b).

Table IV-5

CONCENTRATION OF VOLATILE FATTY ACIDS IN BLOOD[a,b]

	Carotid artery	Jugular vein	Posterior rumen vein	Anterior rumen vein	Retic-ulum vein	Om-asum vein	Abom-asum vein
Range	0.2–1.3	0–1.2	1.7–7.5	5.1–10.9	2.8–4.0	0.5–2.7	0.1–09
Average	0.55	0.41	3.57	7.26	3.60	1.96	0.37

[a] From Marshall and Phillipson (1945).
[b] Values in microequivalents volatile acids per ml of blood.

Removal of fermentation acids from rumen fluid is important because accumulation of the acids tends to diminish the rate of fermentation. This has been shown precisely in the case of a pure culture of *Butyrivibrio* (Lee and Moore, 1959) and with crude cultures (Stranks, 1956). Addition of acetic, propionic, or butyric acid to a medium buffered at pH 6.5 caused a decrease in the production of the added acid during the fermentation.

Table IV-6

PROPORTIONS OF VOLATILE FATTY ACIDS IN BLOOD

	Percentage of total				
	Acetic	Propionic	Butyric	Higher	Reference
Peripheral blood, cattle	93.3	2.4	2.5	1.8	McClymont (1951a)
Ruminal vein, sheep	69.5	18.8	6.4	6.3	Kiddle *et al.* (1951)

Fermentation acids leave the rumen by absorption through the rumen epithelium and by passage to the omasum. In a well-fed animal, absorption of VFA's never reduces the concentration in the rumen below 40 μmoles per milliliter, i.e., about 0.3% (w/v) Table IV-4), and usually the concentration is between 0.6 and 0.9% (w/v). Fermentation is continuous in the large mass of digesta, and, even with mixing, the great distance acids must diffuse to reach the rumen wall prevents their removal to a low concentration, except during starvation. As fermentation increases after ingestion of food, the concentration of the acids builds up (Table IV-7) until the gradient across the epithelium into the blood causes an average rumen-to-blood diffusion rate equal to the average rate of production minus the rate of passage to the omasum. Pfander and Phillipson (1953) found, in agreement with earlier studies (Elsden *et al.,* 1946; Schambye, 1951a; McClymont, 1951a), that the average steady state distribution of the three principal VFA's in the rumen of sheep was about 63% acetic, 21% propionic, and 15% butyric acid. Concentrations approximating these are maintained under widely varying feeding conditions and in distant geographic localities, as may be seen from the values collected in Table IV-4.

Table IV-7

AMOUNTS AND PROPORTIONS OF LOWER VOLATILE FATTY ACIDS IN SHEEP RUMEN CONTENTS AT VARIOUS TIMES AFTER FEEDING[a]

Ration	Hours after feeding	Total VFA (μmoles/ml)	Percentage of total		
			Acetic	Propionic	Butyric
Wheaten hay	0	49.0	69.4	16.7	13.9
	½	58.1	67.3	19.6	13.1
	1	67.6	66.0	20.4	13.6
	4	93.4	61.3	24.4	14.3
	7	108	61.9	23.5	14.6
	12	109	66.8	19.6	13.6
	24	82.5	69.1	17.0	13.9
Alfalfa hay	0	51.4	69.6	14.8	15.6
	½	112	69.4	18.7	11.9
	1	138	68.1	20.2	11.7
	4	176	70.1	18.9	11.0
	7	193	71.5	18.7	9.8
	12	187	71.8	17.7	10.5
	24	75.7	68.3	16.1	15.6

[a] From Gray and Pilgrim (1951).

As production of VFA's diminishes, absorption and passage to the omasum exceed the rate of production, and the concentration of the acids in the rumen falls (Fig. IV-11). Though the concentration of the VFA's in the rumen is a function of production, it is not directly proportional and can thus serve only as a very rough index to the rate of production.

In the omasum, conditions for fermentation are less favorable and those for absorption are more favorable than in the rumen, with the consequence that the concentration of the fatty acids or their salts in the digesta leaving the omasum is very low. Any VFA's entering the abomasum are absorbed (Ash, 1961b; Johnston, *et al.,* 1961; Joyner *et al.,* 1963).

Most of the studies on absorption have measured the disappearance of acids from the rumen of sheep in which the reticulo-omasal orifice has been ligated to prevent exit of contents into the omasum. The rumen contents are washed out and replaced with a balanced solution of inorganic salts plus concentrations of acetate, propionate, and butyrate similar to those in the original contents. By suitable addition of the respective acids

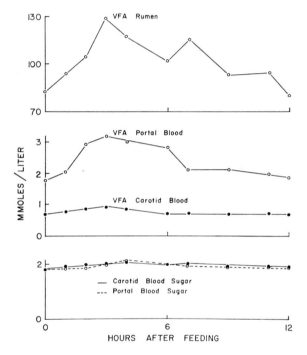

Fig. IV-11. Portal and carotid blood sugar and volatile fatty acids and rumen volatile fatty acids at various times after feeding a sheep. From Schambye (1951a, p. 562).

during the experiment their concentration is maintained near the rumen norm. Any acids added plus any decrease in the amount in the experimental rumen represent the quantity absorbed. The results obtained by most investigators (Gray, 1947a; Pfander and Phillipson, 1953; R. B. Johnson, 1951b; Tsuda, 1956b) show that per unit concentration the order of absorption is: butyric > propionic > acetic.

There is an appreciably faster absorption at pH 5.7 than at 7.5

(Danielli *et al.,* 1945; Masson and Phillipson, 1951; Parthasarathy and Phillipson, 1953; Tsuda, 1956b). The undissociated acid penetrates the membrane more readily than does the anion. This makes absorption of the VFA's more rapid whenever a high acid production rate causes the pH to drop. Since the dissociation constants of acetic, propionic, and butyric acids are greater than 10^{-5}, the VFA's are chiefly dissociated as the anions in normal rumen contents (pH range of 5.7–6.7). There is relatively little difference in the dissociation constants of the three chief VFA's, though acetic acid dissociates slightly more. This might in part account for the slower absorption as compared to the other acids. It may also be due to a lower solubility of the acetic acid in the lipoidal materials of the cell membranes. Increased absorption of VFA's across the rumen epithelium increased the rate of transport of ammonia into the blood (Hogan, 1961).

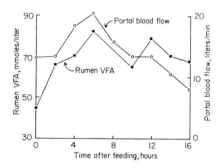

Fig. IV-12. Portal blood flow and rumen volatile fatty acids at various times after feeding a sheep. From Bensadoun and Reid (1962, p. 542).

Simple diffusion of the undissociated VFA's through the rumen wall accounts for their movement into the blood (Tsuda, 1956b). The conversion of butyrate and propionate in the rumen epithelial tissue makes the diffusion gradient for these acids steeper than for acetic, another factor which may account for their greater specific absorption rates. A greater portal blood flow during periods of maximum fermentation (Dobson and Phillipson, 1956; Bensadoun and Reid, 1962; Waldern *et al.,* 1963) also assists absorption (Fig. IV-12).

Insofar as the ruminant is concerned, the important factor in the absorption of the VFA's is the nature and total quantity absorbed, both in the rumen and omasum. This has been estimated by determining the rate at which the acids are produced. The amounts estimated to be supplied the host are discussed in detail in Chapter VI.

2. OTHER SUBSTANCES

Numerous organic acids (see Chapter VII for branched-chain acids) have been demonstrated in rumen contents, usually in very low concentrations, though occasionally in abundance, and they are absorbed. Lactic acid is usually present in very low concentrations (Jayasuriya and Hungate, 1959), but on certain feeds, particularly during changes from a high-forage to a high-concentrate ration, it may accumulate to a concentration of several grams per liter. This is discussed in Chapter XII. Lactic acid is absorbed only slowly under either acid or alkaline conditions, possibly due to the polar nature of the molecule (Ash, 1959b).

Other acids reported to occur in rumen contents include oxalic, quinic, glyceric, malic, glycolic, malonic, succinic, fumaric, and adipic (Bastié, 1957). Some of these nonvolatile fatty acids are constituents of forage, others may be intracellular cell metabolites, and some may be waste products. The higher fatty unsaturated acids are hydrogenated in the rumen (Chapter IX).

It has been generally concluded that no glucose is absorbed from the rumen, and on a quantitative basis any sugar is much less important than the VFA's. In the careful studies of Schambye (1951a,b) one experiment showed a slight excess of glucose in rumen portal blood as compared to carotid, but the results of all experiments indicate that glucose absorption is negligible under usual conditions of feeding. If glucose does accumulate in the rumen it is readily absorbed (Tsuda, 1956b).

3. INORGANIC IONS

The absorption of the anionic form of the acids, though at a slower rate than for the undissociated form, is aided by a potential difference across the rumen epithelium. Sodium ions can be actively absorbed from rumen contents into the blood, until the electrochemical gradient against the movement is 80 millivolts, i.e., until the blood is 80 millivolts positive to the rumen contents (Dobson, 1959). This absorption involves work, the energy presumably coming from the mitochondria in the basal epithelial cells.

Normally the blood is 30–40 millivolts positive to the rumen contents. This increases the tendency for the acid anions to enter the blood from the rumen, and it is increasingly important as the pH rises. Potassium and other cations are not actively absorbed, nor are the anionic inorganic radicals, under ordinary conditions (Dobson and Phillipson, 1958). A study of interrelationships between Na^+ and K^+ and the electric potential indicates that the latter is determined in part by factors in addition to these two cations (Sellers and Dobson, 1960). When the potassium ion concentration is very high in the rumen, chloride ion may be actively transported from the blood into the rumen and vice versa (Sperber and Hydén, 1952). Evidence of bicarbonate absorption from the omasum has been reported (Ekman and Sperber, 1953).

The polyvalent ions, Mg^{++}, Ca^{++}, and HPO_4^{--}, are absorbed chiefly from the abomasum. The acidity in that organ increases their solubility. Some may be absorbed through the rumen wall (Stewart and Moodie, 1956).

Although active absorption of Na^+ occurs across the rumen epithelium, the cells do not exhibit the histological structures usually present in absorptive and secretory tissue. This may explain why the important movements of materials through the rumen wall were not demonstrated earlier.

Sodium is extremely important in the economy of ruminants. In one study in sheep (Dobson, 1959) the amount of Na^+ in the ration was about 70 mEq per day. Each day 55 mEq were excreted in the urine and 15 mEq in the feces. The secreted saliva contained 1200 mEq per day, most of which was reabsorbed, approximately 600 mEq through the rumen wall and the remainder from the omasum (250 mEq per day) and the large intestine. Since the total amount of VFA's absorbed from the rumen and omasum of a sheep per day is about 4 moles, calculation shows that approximately one-fourth to one-fifth of the VFA's are absorbed as an anion in association with a sodium ion.

Ammonia is also absorbed readily from the rumen, probably as the ammonium ion at usual rumen pH's. The concentration in blood is extremely low, only 0.143 mEq per liter (McDonald, 1948b), whereas in rumen contents it is usually about 10 mEq per liter, though on nitrogen-poor rations much lower values may prevail. With the big difference in concentration favoring absorption of ammonia, it is highly probable that absorption occurs by simple diffusion.

The low blood ammonia values are the result of conversion of ammonia to urea in the liver. Some of this urea may diffuse from the blood back into the rumen to be again changed to ammonia. There is thus a recycling of nitrogen.

H. Additional Characteristics of the Rumen

1. OSMOTIC PRESSURE

The osmotic pressure of rumen contents is ordinarily similar to that of blood (Davey, 1936). The freezing point depression of rumen fluid is between 0.46 and 0.695°C in sheep (Parthasarathy and Phillipson, 1953), and values of 0.54 and 0.59°C have been reported for cattle (Hungate, 1942). A usual osmolality corresponding to a freezing point depression of 0.5°C has been found in sheep (Warner and Stacy, 1965). These latter authors noted a marked drop in rumen fluid osmolality after water was drunk and a steady rise back to normal over a period of as much as 10 hours. The rumen wall is permeable to water; therefore equilibration of osmotic pressure is possible after each fluctuation due to ingested water or feed. Since the osmotic pressure of saliva is only slightly less than that of blood, the continuous addition of saliva to the rumen has little effect on the osmolality.

Increases in rumen osmotic pressure caused by feeding excess salt (NaCl) diminish dry matter consumption (Galgan and Schneider, 1951) without apparently affecting the activity of rumen bacteria (Cardon, 1953). The protozoa are diminished in numbers and some kinds are eliminated when excess salt is fed (Koffman, 1938).

2. ACIDITY

The fact that the undissociated VFA's are absorbed more rapidly than the dissociated makes the specific absorption rate higher when the rate of acid production increases. Within the pH range 4.5–7, the lower the pH the more rapid the absorption. Coordination of salivary secretion with food ingestion also assists in acidity control by supplying sodium bicarbonate ($NaHCO_3$) and phosphate at the time of the most rapid fermentation. In starved ruminants the pH of the rumen moves toward the alkaline side. A ration very high in protein also gives an alkaline reaction, but one sufficiently rich in protein and poor in carbohydrate to produce this effect is rarely employed.

Because of the low buffering capacity, alkaline drinking water has little effect on the rumen pH (C. E. Johnson, et al., 1959). In general, more accurate estimates of rumen pH will be obtained by determining the concentration of bicarbonate in the rumen fluid and calculating the pH from the Henderson-Hasselbach equation (Turner and Hodgetts, 1955a), assuming a carbon dioxide concentration in the rumen fluid of about 0.017 M. With 0.5% sodium bicarbonate the pH is about 6.7. The pH can be calculated after measuring the carbon dioxide released when rumen contents are acidified. The pH of rumen contents of dairy cows on alfalfa hay was 6.27, and 6.00 for alfalfa hay plus beet pulp (V. R. Smith, 1941).

3. TEMPERATURE

A factor of great importance to the continued microbial fermentation is the relatively constant temperature of the rumen. The temperature tends to rise following ingestion of food, due to the evolution of heat in the fermentation process. This evolution of heat has been used as a measure of the fermentation rate (Walker and Forrest, 1964).

In cold climates the heat of fermentation assists in maintaining body temperature. The temperature in rumen contents is usually somewhat above that of the ruminant body, which is 38°C (Wrenn *et al.,* 1961), the difference in temperature decreasing as fermentation decreases with time after feeding (Dale *et al.,* 1954; Nangeroni, 1954). Very cold drinking water can lower the temperature of the rumen contents much below normal, but this exerts little affect on food utilization (M. D. Cunningham *et al.,* 1964). In hot climates decreased heat loss may cause a diminution in consumption of food.

The temperature in the rumen may rise as high as 41°C (Trautmann and Hill, 1949; Gilchrist and Clark, 1957) during active fermentation of alfalfa, but is usually between 39 and 40.5°C (Krzywaneck, 1929). In a fed heifer the rumen temperature was more than 2° higher than the rectal temperature (Dale *et al.,* 1954) and dropped to less than 1° difference after a 24-hour fast. Since 42°C is the upper temperature limit which cattle can survive (Findlay, 1958) there is little leeway in hot climates. The protozoa do not survive temperatures much above 40°C.

4. OXYGEN

It is sometimes concluded that rumen contents are not completely anaerobic because oxygen can be detected in the gas pocket in the dorsal sac (Kandatsu *et al.,* 1955). Because of the low oxygen tension in mammalian tissues and the fairly thick layer of metabolically active epithelial cells through which oxygen would have to diffuse in going from the blood into the rumen, it is doubtful that oxygen diffusion from the blood into the rumen can account for the oxygen in rumen gas. It is probably due to swallowed air.

If the oxygen in the rumen gas pocket were in equilibrium with the rumen contents, the latter would be aerobic, but the rate of diffusion into the great mass of rumen contents, even with the aid of the mixing cycle, is so slow that the amount of oxygen entering the mass of contents is insignificant. The small amounts entering are quickly reduced (Broberg, 1957d). Whether this is due to the oxidative metabolism of the relatively few aerobic and euryoxic microbes in the rumen or to a chemical reduction (not of biological significance) by intermediate metabolites of the anaerobes is not known. Regardless of the mechanism for removal of oxygen in the rumen contents, it is so effective that the oxidation-reduction potential (E'_0) is in the neighborhood of -0.35 volts (Broberg, 1957c; P. H. Smith and Hungate, 1958). In measuring the potential with the platinum electrode a lower and much more stable level is reached if a little benzyl viologen or other electrode-active dye is added to link the electrode with the oxidation-reduction systems in the contents. Admission of oxygen to incubating rumen contents increases the heat production (Walker and Forrest, 1964), which indicates reactivity of the oxygen.

Cellulose digestion in the rumen has been inhibited somewhat by bubbling oxygen into the rumen (Hoflund et al., 1948), and the loss of appetite in fistulated animals without a plug, more common in cold climates, might be due to oxygen, particularly under conditions of reduced food. An inhibition by oxygen has been noted also in in vitro cultures employing washed cell suspensions of rumen bacteria (R. R. Johnson et al., 1958).

5. SURFACE TENSION

Surface tension of rumen fluid is 45–59 dynes per centimeter (Blake et al., 1957a). Other values (Nuvole, 1961) are 46.52 ± 2.33 dynes per centimeter for bovines and 51.23 ± 2.80 for ovines.

6. COMPOSITION OF RUMEN GAS

In addition to carbon dioxide and methane, rumen gases have been reported to include carbon monoxide (Dougherty, 1940), nitrogen, oxygen, hydrogen sulfide, and hydrogen, but the methods of analysis are not always as precise as is desirable. Recently (McArthur and Miltimore, 1961), gas-solid chromatographic methods applied to the rumen have yielded the values shown in Table IV-8.

Table IV-8

COMPOSITION OF RUMEN GAS[a]

Gas	Mean percentage	Sample variance	Analyses variance
Hydrogen	0.18	0.0073	0.0027
Hydrogen sulfide	0.01	0.000007	0.000005
Oxygen	0.56	0.0151	0.0004
Nitrogen	7.00	0.6922	0.1277
Methane	26.76	0.7475	0.0620
Carbon dioxide	65.35	0.9293	0.1170

[a] From McArthur and Miltimore (1961).

I. Summary

In this chapter the physiological characteristics of the ruminants concerned with the ingestion, mixing, comminution, and passage of food have been described. Special activities such as eructation, rumination, salivary secretion, and straining of large particles are essential in providing favorable conditions for the rumen microbiota, and these activities distinguish this group of mammals from the nonruminants. Some of the quantitative aspects of the ruminant functions will be considered in the following chapters.

CHAPTER V

The Rumen as a Continuous Fermentation System

A. Introduction

As is evident from the discussions in Chapter IV, the ruminant alimentary tract is not a simple tube through which food and food residues are propelled, ingestion of fresh food at one end causing discharge of spent digesta at the other. Rather it is a regulated system in which the digesta are variously treated in the different parts.

The mechanics of the treatment in the rumen are essentially comparable to those in a vat in which fresh feed and saliva mix with the fermenting mass, and fluid and feed residues leave in quantities equivalent to those entering. This is a continuous-feed system. The rumen is not an exact duplicate of this model, but it is sufficiently similar to make continuous-feed theory and mathematics applicable to rumen phenomena (Hydén, 1961a). Movement of digesta through the rumen of pasture-fed animals approaches closely the continuous-feed model (Lampila, 1955). A small part of the fresh ingesta escapes into the omasum (Stalfors, 1926; Brandt and Thacker, 1958).

B. Characteristics of Continuous Fermentation Systems

A continuous-feed system is described by the volume, V, and the feed or dilution rate, D. In place of the volume, X can be used to represent the quantity of a component in the rumen. It can represent the liquid exclusive of solids, the dry matter, or any other material fed in, mixed, and passed into the omasum. The feed rate can be expressed as the amount of food entering the rumen per unit time, or, more usefully in a continuous system, as a fraction:

$$\frac{\text{intake of feed per unit time}}{\text{amount of food in the rumen}}$$

If the total amount in the rumen is taken as 1, the amount fed per unit time can be expressed as a fraction of this total. For example, if the digesta in the sheep rumen studied by Wildt (Table V-1) represent (i.e., are the remains from) 4.4 kg of dry feed, and the feed rate is 240 gm per hour, the rate can be expressed as

$$\frac{240 \text{ gm}}{4400 \text{ gm hr}} = \frac{1}{18.33 \text{ hr}}$$

The amount fed per hour is $1/18.33$ of that in the rumen. The reciprocal of this fraction, 18.33 hours, is the *cycle* or *turnover time*. It is the time required for entrance of an amount of feed equal to that represented in the rumen. Note that "represented" is used instead of "present" because particles in the rumen may have been partially digested.

If a single dose of an inert reference or marker material not present in the feed is added to a continuously mixing constant-volume rumen, the amount of the marker material leaving the rumen at any particular instant will be determined by the amount of marker in the rumen at that time. Because of mixing of marker with the unmarked contents, and with later ingested unmarked material, not all the marker leaves during one cycle or turnover time, even though during this period the total material leaving will be equivalent to the total entering.

If x is the fraction of the original substance in the rumen at time t (with the initial amount 1, i.e., all of it), the rate of change in the fractional amount of initial marker still present will be

$$dx/dt = -kt \tag{V-1}$$

The constant k is the feed rate, negative in value because the amount of x diminishes.

Integration of Eq. (V-1) gives

$$\ln x = -kt + C$$

At the instant the labeled material is added, $x = 1$, $t = 0$, and $C = 0$, and the integrated equation becomes

$$\ln x = -kt \tag{V-2}$$

or

$$x = e^{-kt} \tag{V-3}$$

Integration between limits

$$\int_{x_1}^{x_2} \frac{dx}{x} = \int_{t_1}^{t_2} -k \, dt$$

gives the equation

$$\ln x_2/x_1 = -k(t_2 - t_1)$$

or

$$2.3 \log x_1/x_2 = k(t_2 - t_1) \tag{V-4}$$

When one turnover has occurred, i.e., when 18.33 hours have elapsed in Wildt's sheep, kt will equal 1.0; $1/18.33$ of the feed represented by the total rumen digesta will be fed each hour, and the food consumed in 18.33 hours will equal the amount represented in the rumen, or 1. This steady state amount of a material represented in the rumen, equal to the amount entering or leaving during one turnover time, is the pool, designated as A.

The fraction, x, of the tracer in the rumen after one turnover will be 0.37 of the amount initially added [from Eq. (V-2), ln $x = -1$; log $x = -1/2.3 = -0.43478 = 9.56522 - 10$, for which the antilog is 0.37]. This means that, of the material in the rumen at any time, 0.37 will still be in the rumen one turnover cycle later. The time, T, for the feeding of an amount equal to that in the rumen, i.e., for one turnover, is the reciprocal of the feed rate, i.e., $k = 1/T$. As previously stated, for Wildt's feed rate of 1.31 per day, T would be 0.764 days or 18.33 hours. This value, T, is the turnover time, or cycle time. The value of k is the *turnover rate*, also called the feed rate or dilution rate. It can also be designated as the turnover rate constant.

The average time that particles of digesta remain in the rumen is equal to the turnover time. This can be illustrated mathematically and graphically in connection with the curve for Eq. (V-2), as shown in Fig. V-1. When T is the average time particles remain in the rumen, the area in the rectangle 0, 1.0, i, T equals the total area between the curve and the x and t axes. The area in the rectangle is T. The area under the curve is equal to $\int_{t=0}^{t=\infty} x\,dt$.

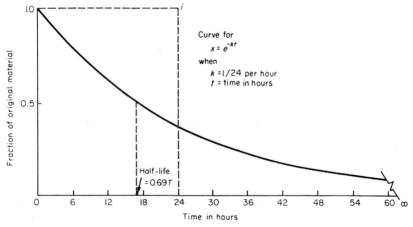

Fig. V-1. Curve showing the dilution of a pulse marker added to a continuous fermentation system with a turnover time (T) of 24 hours.

Since $x = e^{-kt}$ [Eq. (V-3)],

$$\int_{t=0}^{t=\infty} e^{-kt} \, dt = -\frac{1}{k} e^{-k\infty} + \frac{1}{k} e^0 = \frac{1}{k} \qquad (V-5)$$

Since $1/k = T$, the area under the curve is equal to the area of the rectangle.

It is obvious that over a long period the average time in the rumen must equal the time required for an equivalent amount of food to enter. Otherwise the amount in the rumen would change.

C. Passage of Digesta through the Rumen

The most direct method for estimating the extent to which the entire feed or any component of it is utilized in the ruminant is to measure it in the feed and feces and by the difference to calculate the quantity removed during passage through the alimentary tract. This method does not take account of the weight of microorganisms in the feces nor of the alimentary secretions which are not reabsorbed, but these errors are minor.

Nearly all components of forages are digested to some extent, though the exact utilization is variable, and some components of some feeds are almost completely indigestible. Digestibility values derived from analyses of feed and feces are provided by Schneider's data (1947).

The food-feces difference measures the total extent of digestion during passage through the entire animal, but does not show the extent of digestion at various points in the digestive tract. Analyses of digesta at these points will disclose the degree of digestion, provided an inert reference substance or marker is added to the feed to provide a standard against which the other components can be compared.

1. Use of Markers

Markers for the rate of passage and digestion of feed may be of two sorts: (1) internal, including indigestible components of the feed, or (2) external, i.e., added inert components. An intermediate method is the addition of material, such as a dye, which marks a component of the feed (Usuelli, 1956).

The following feed components have been employed as internal markers or reference substances: lignin (Hale *et al.*, 1947a), iron (Rathnow, 1938; Knott *et al.*, 1936), silica (Wildt, 1874; Van Dyne and Meyer, 1964), and chromogen (Squibb *et al.*, 1958; J. T. Reid *et al.*, 1950).

External markers include: iron oxide (Moore and Winter, 1934), chromic oxide (Edin, 1918, 1926, from Anderson, 1934), polyethylene glycol (Sperber *et al.*, 1953; Hydén, 1961a), rubber (Moore and Winter, 1934; Campling and Freer, 1962), plastics (Campling and Freer, 1962), and monastral blue (Lambourne, 1957a). There is a good correlation between digestibility as determined by chromic oxide and by food-feces weight difference of various components (Anderson, 1934). This marker has probably been used more extensively than others for determining total digestibility, and numerous refinements and limitations (Blaxter *et al.*, 1956, see p. 82 for curve) of the method have been recorded. Administration of chromic oxide (Cr_2O_3) after preliminary mixing with paper pulp gives a more representative marker (Corbett *et al.*, 1958a), and mixture with pelleted feed gives good daily recoveries, although irregularities in the concentration of chromic oxide recovered in the feces were observed (Elam *et al.*, 1959), presumably due to variations in salivary secretion or drinking water. In other experiments (Hardison *et al.*, 1959) a 12% error in recovery of chromic oxide was encountered. With goats, the fecal concentration of chromic oxide showed one 24-hour peak in animals fed at 9:00 A.M. and 4:00 P.M., but two daily peaks when fed at 12-hour intervals (Kameoka *et al.*, 1956). The peak concentration times for chromic oxide and lignin differed in cows fed chromic oxide (Kane *et al.*, 1952) and in sheep (D. E. Johnson *et al.*, 1964), as might be expected if the passage of lignin is more subject to retention until ruminated, with rumination occurring chiefly at night. Increased accuracy in analysis for chromic oxide can be obtained by colorimetric measurements after conversion to dichromate (Kimura and Miller 1957).

The percentage digestibility of a feed can be derived from the concentration of an inert marker from the relationship

$$\left(1 - \frac{\%\ \text{marker in feed}}{\%\ \text{marker in feces}}\right) 100 = \%\ \text{dry matter digested}$$

Percentage digestibility of a component, *x,* can be derived from the relationship

$$\frac{100\left[\%\ x\ \text{in feed} - \%\ x\ \text{in feces}\left(\dfrac{\%\ \text{marker in feed}}{\%\ \text{marker in feces}}\right)\right]}{\%\ x\ \text{in feed}} = \%\ \text{digestibility of } x$$

By substituting for the percentage in feces the percentages of the marker and of *x* in any part of the alimentary tract the extent of digestion of *x* at that point can be calculated.

a. *An Example with Silica as the Reference Substance*

One of the earliest uses of reference substances was that of Wildt (1874) with sheep. In fact, this was the first study to show conclusively that the rumen is the chief site of fiber digestion. The analyses were detailed and sufficiently well executed to warrant tabular portrayal of the results, even though subsequent studies have shown that feed silica is not a completely inert reference substance, some of it being absorbed during passage and excreted in the urine (Forman and Sauer, 1962). The principle of the method was well stated (translation), "In those parts of the alimentary tract where phosphorus and calcium are absorbed, the ratio of silica to these substances should be greater than in the feed. Thus the greater or lesser ratios of silica to the other minerals, as well as to the organic components of the feed, show their small or large absorption from the various parts of the alimentary tract." Wildt's results are shown in Table V-1.

Table V-1

RELATIONSHIP OF SILCA TO CRUDE FIBER IN VARIOUS PARTS
OF THE ALIMENTARY TRACT[a]

Organ	SiO_2 content (gm)	Expected fiber (gm)	Found fiber (gm)	Digested fiber (gm)	Crude fiber digested (%)	Organic matter digested (%)
Rumen-reticulum	29.700	366.32	336.41	29.91	8.16	32.7
Omasum	2.707	33.38	20.07	13.31	57.50	47.2
Abomasum	1.891	23.32	15.16	8.16	34.99	35.9
Small intestine	3.587	44.24	22.40	21.84	49.35	27.1
Cecum	11.498	141.81	62.10	79.71	56.19	61.4
Large intestine	2.113	26.06	12.61	13.45	51.61	64.5
Rectum	5.943	73.30	35.06	38.24	52.17	53.8

[a] After Wildt (1874).

It is uncertain how much time elapsed between the feeding and killing of the sheep of Table V-1, but it was probably fairly short since the crude fiber in the rumen had not been digested to a very great extent. However, the material in the abomasum showed digestion to the extent of 57.5%, the same as the percentage disappearing after passage of the digesta through the entire alimentary tract. The materials leaving the rumen had undergone approximately the crude fiber digestion characteristic of the total tract. The absorption of some of the silicate would make the digestibility estimates lower than the true values.

b. *Relation between the Amount Fed and the Amount Represented in the Rumen*

On the basis of the amount of silica (SiO_2) in the rumen (29.7 gm) and that in a day's feed (38.792 gm), Wildt concluded that the average time in the rumen was $29.7/38.782 \times 24$ hours, or 18.33 hours. If all the feed materials passed the alimentary tract in the same fashion as the silica it would be concluded that the quantity fed per day was 1.31 times that represented in the rumen.

c. *Lignin as a Reference Substance*

Lignin is relatively indigestible and has been used as a reference substance in a number of investigations. The results of several such studies are collected in Table V-2.

The calculated turnover times differ according to whether any of the lignin is digested. In these particular experiments, 6.8 to 28.8% of the "lignin" was digested during passage of the feed through the animal. With hay alone in the ration the extent of lignin digestion is usually slight, though with very young forages a good deal of "lignin" is digested. Note in Table V-2 that for one low-fiber feed the fecal "lignin" exceeded that in the feed.

On the assumption that any lignin digestion occurs largely in the rumen, the turnover times based on daily fecal lignin are more accurate than those based on feed lignin.

d. *Use of Dyed Hay to Determine Rate of Passage*

Dyed hay has been used (Lenkeit, 1930) to study the rate at which this forage passes through the rumen. The procedure is as follows: Several pounds of hay are boiled with a dye such as fuchsin (Lenkeit, 1930), magenta, rhodamine, brilliant green, crystal violet, or chrysoidene (Balch, 1950) until the color is fixed and does not leach out during passage through the animal. The dyed hay is dried and fed just prior to a regular feeding.

Feces are collected at frequent intervals over a period of 1 week to 10 days. A sample of each collection is mixed with water, and the mixture is spread on a dry weighed cheese cloth. With a fine spray of water the smallest particles and the liquid are washed through the cheese cloth.

The dyed particles left on the surface of the cloth are counted. The cloth and particles are weighed, and the weight of the total particles, dyed and plain, is calculated by difference. The count is reported as the number of dyed particles per unit weight of total particles.

The sum of all the daily counts, i.e., total particles collected, gives a measure of the total particles fed. The fraction of the particles retained in the animal at any time (total particles eliminated — total fraction eliminated up to the selected time) is plotted against time after consumption of the dyed hay. During an initial period in which no dyed particles appear in the feces, the digesta are passing from the rumen through the rest of the alimentary tract.

If the marked particles were instantly mixed with the rumen contents, when they left the rumen and passed straight through the rest of the tract, their concentration in the feces would be maximal initially and then would diminish as the concentration of those left in the rumen decreased. In practice, a few dyed particles appear initially, and the number per unit weight increases until approximately 5% of the total ultimately recovered has appeared (Balch, 1950), at which time the concentration begins to decrease. This will be discussed later. In practice, the time between feeding and 5% excretion of the marked particles is taken as the time for digesta to pass from the rumen to the feces (Balch, 1950).

If the rate at which dyed particles leave the rumen is proportional to the amount of retained dyed hay, the times at which certain percentages of the particles have appeared in the feces can be used to calculate the turnover. Dyed particles were recovered by Castle (1956a) from goats fed daily on 400 gm of calf nuts containing 20% crude protein and 7.5% crude fiber, plus medium-quality meadow hay *ad libitum*. The results are shown in Table V-3, together with turnover times calculated on the basis of the 5% and 95% values according to Eq. (V-4). Included also are Castle's estimates of mean retention time (turnover time) obtained by a graphical method.

Any arbitrarily selected points on the particle elimination curve can be used to calculate turnover time. Points near the beginning and end of the curve are to be avoided because of the diminished accuracy of analysis. As will be discussed, the latter portion of the elimination curve accords fairly well with Eq. (V-3), when plotted as the fraction remaining in the rumen.

The half-life is sometimes employed to designate turnover, as with isotopic tracers. The relationship between cycle time and any other measure can be seen from Fig. V-1. The half-life is approximately 0.69 of the cycle time. Of the various parameters, cycle time is most useful because it is easily related to the daily food intake, rate of food passage, and amount of digesta in the rumen.

2. Simultaneous Operation of Different Turnover Rates

Application of continuous fermentation theory to the rumen is complicated by the different rates of passage of different components of the feed. The soluble materials and bacteria proceed with the liquid component, but the solids may be caught in the mass of solids digesta, or, if free, retained by the straining action of the omasum, or settle in the bottom of the rumen due to high specific gravity, as with whole kernels of corn

Table V-2

CALCULATION OF TURNOVER TIME BASED ON LIGNIN IN THE RUMEN AND IN THE FEED

Expt.	Animals	Ration and time fed (gm)	Lignin in food (gm)	Lignin in rumen-reticulum (gm)	Lignin in feces (gm)	Turnover time based on food lignin (days)	Turnover time based on fecal lignin (days)	Reference
1	Sheep (2)	7:00 A.M. -450 gm hay 11:00 A.M. -150 gm linseed meal, 75 gm crushed oats 4:00 P.M. -450 gm hay	122.7	145	98.1	1.18	1.55	Badawy et al. (1958a,b)
2	Sheep (2)	7:00 A.M. -200 gm hay, 167 gm ground maize, 25 gm decorticated peanut meal 8 gm dried yeast 7:00 P.M. -same as at 7:00 A.M., plus 5 gm $Na_2HPO_4 \cdot 12\ H_2O$ and 5 gm NaCl	58.3	98	48.7	1.68	2.01	Badawy et al. (1958a,b)
3	Sheep (2)	7:00 A.M.-200 gm hay, 133 gm linseed meal, 67 gm crushed oats 7:00 P.M.-same as at 7:00 A.M.	88.4	109	65.9	1.23	1.65	Badawy et al. (1958a,b)
4	Sheep	8:00 A.M.-300 gm Rhodes grass hay (*Chloris gayana*) 4:00 P.M.-same as at 8:00 A.M.	59.0	74.35	58.7	1.26	1.27	Rogerson (1958)
5	Sheep	8:00 A.M.-250 gm hay and 150 gm cassava meal 4:00 P.M.-same as at 8:00 A.M.	50.4	52.38	50.9	1.04	1.03	Rogerson (1958)

6	Sheep	8:00 A.M.-300 gm kibbled maize 4:00 P.M.-same as at 8:00 A.M.	5.08	13.2	5.88	2.60	2.25	Rogerson (1958)
7	Cows	Timothy-clover hay at 80% of *ad lib.* consumption	842	1350	—	1.6	—	Makela (1956)
8	Cows	Timothy-clover hay at 73% of *ad lib.* consumption	534	1340	—	2.51	—	Makela (1956)
9	Cows	Timothy-clover hay at 65% of *ad lib.* consumption	487	988	—	2.03	—	Makela (1956)
10	Cows	Alfalfa hay	1544	3314	—	2.15	—	Hale et al. (1947a)

(Kick *et al.,* 1937b). In consequence, the turnover time for larger solids particles may differ from that for liquid and small particles.

a. *Estimation of Rumen Liquid Turnover Rates*

The rate of passage of rumen liquid digesta can be obtained by using polyethylene glycol as a marker. It passes through the alimentary tract without significant changes due to absorption or metabolism by rumen microorganisms or the host. The quantity in digesta can be determined easily and accurately (Hydén, 1961a). Difficulty in separating it from the feces may prevent total recovery (Christie and Lassiter, 1958), but recovery of polyethylene glycol is usually adequate. The method was introduced by Sperber *et al.* (1953) who found a cycle time of 13 hours in cattle. In some experiments (Corbett *et al.,* 1959) both polyethylene glycol and chromic oxide were used. Cattle were fed 400 gm of grass meal cubes containing 7.1% chromic oxide and 12.5% polyethylene glycol at 9:30 A.M. and 5:00 P.M. After 5 minutes, half (10 or 13 lb) of the daily ration of baled dried grass was fed. Water was available at all times. The results are shown in Table V-4.

Table V-3

CALCULATION OF THE RUMEN CYCLE TIME IN GOATS FED DYED HAY[a]

Animal no.	Expt.	Body wt. (kg)	95% retention time, t_1 (hr)	5% retention Time, t_2 (hr)	t_2-t_1 (hr)	Turnover time, t_2-t_1 /2.94[b] (hr)	Castle's mean retention time (hr)
1	A	34.9	16.0	65.0	49.0	16.7	19.8
	B	35.0	17.0	64.0	47.0	16.0	18.2
2	A	33.0	15.5	60.9	45.4	15.5	16.7
	B	33.3	15.5	69.0	53.5	18.2	18.3
3	A	34.2	13.5	78.0	64.5	21.9	22.7
	B	34.9	16.5	78.0	61.5	21.0	21.0
4	A	36.1	16.5	82.0	65.5	22.3	23.9
	B	36.8	17.5	85.0	67.5	22.9	27.3
5	A	27.1	15.5	84.0	68.5	23.3	24.5
	B	26.6	18.0	87.0	69.0	23.4	26.7
6	A	40.0	15.5	70.5	55.0	18.8	22.3
	B	42.8	17.0	67.0	50.0	17.0	18.2
7	A	43.4	15.0	78.0	63.0	21.4	25.3
	B	42.5	14.0	78.0	64.0	21.8	22.3
8	A	54.0	18.0	72.0	54.0	18.4	19.6
	B	57.3	16.5	77.0	60.5	20.5	24.8
Average		38.2	16.1	74.7	58.6	19.9	22.0

[a] From the data of Castle (1956a).
[b] $2.3 \times \log(95\%/5\%) = 2.94$.

Table V-4
RATES OF PASSAGE IN VARIOUS PARTS OF THE RETICULO-RUMEN, FROM POLYETHYLENE GLYCOL MARKER[a]

Time sampled	Liquid Digesta				Solids Digesta in Rumen							
	Reticulum		Rumen		Anterior dorsal		Posterior dorsal		Anterior ventral		Posterior ventral	
	$\frac{mg}{100\ ml}$	$\frac{k}{hr}$	$\frac{mg}{100\ ml}$	$\frac{k}{hr}$	$\frac{mg}{100\ ml}$	$\frac{k}{hr}$	$\frac{mg}{100\ ml}$	$\frac{k}{hr}$	$\frac{mg}{100\ ml}$	$\frac{k}{hr}$	$\frac{mg}{100\ ml}$	$\frac{k}{hr}$
9:30 A.M.	15.0	—	14.8	—	11.4	—	11.8	—	10.0	—	9.8	—
10:30 A.M.	67.6	—	83.5	—	74.3	—	49.3	—	50.3	—	83.6	—
1:30 P.M.	45.0	0.135	43.1	0.220	36.4	0.238	43.4	0.081	39.0	0.085	40.6	0.241
4:30 P.M.	30.0	0.135	29.6	0.123	24.4	0.138	26.8	0.159	24.9	0.149	25.4	0.153
5:30 P.M.	100.4	—	78.1	—	71.7	—	34.3	—	84.8	—	111.0	—
8:30 P.M.	51.4	0.222	53.0	0.128	46.6	0.146	48.1	—	50.8	0.170	49.1	0.271
11:30 P.M.	36.1	0.118	38.5	0.113	27.0	0.181	34.6	0.110	33.5	0.138	36.5	0.100
2:30 A.M.	21.4	0.177	23.0	0.171	19.2	0.121	23.4	0.130	17.3	0.219	21.4	0.173
5:30 A.M.	13.6	0.150	14.0	0.165	9.1	0.248	12.6	0.205	10.7	0.159	9.2	0.280
8:30 A.M.	8.8	0.146	8.4	0.170	4.8	0.213	3.8	0.400	4.9	0.261	4.1	0.268
Average	—	0.155	—	0.156	—	0.184	—	0.181	—	0.169	—	0.212
Turnover per day	—	3.7	—	3.7	—	4.4	—	4.3	—	4.1	—	5.1

[a] From Corbett *et al.* (1959).

In these experiments the average turnover times in the rumen and reticulum were almost identical, which indicates good equilibration between the liquid in the two organs. The faster passage of the polyethylene glycol in the rumen solids is difficult to understand. Perhaps saliva and water settled through the solids digesta and washed out some of the marker. Dried grass and grass meal cubes are readily digestible and also contain rather fine particles. The chromic oxide tended to be more concentrated in the solids digesta than was the polyethylene glycol. The results of the experiment as a whole suggest that the volume of the rumen contents of these animals was relatively small. Radioactive chromium (Cr^{51}) chelated with ethylenediaminetetraacetic acid (EDTA) behaves the same as polyethylene glycol and can be measured easily and accurately (Downes and McDonald, 1964; Hogan, 1964a).

The average value for k in Table V-4 is 0.176 per hour, which corresponds to 4.2 per day, a high value, possibly due to the nature of the feed. It is considerably greater than the turnover rate of 1.87 per day obtained by Sperber et $al.$ (1953) with cattle on usual rations.

Estimates of rumen liquid turnover rates can be obtained from rates of salivary secretion, since ingested water is absorbed from the rumen (Trautmann, 1933b) rather than contributing to turnover by flowing into the omasum. Reference to Table IV-3 indicates that the quantity of saliva in the sheep varies between 5 and 10 liters per day. It is difficult to estimate the liquid volume that turns over in the rumen because some of it is trapped in the interior of the larger solids particles and can move out only by diffusion (Hydén, 1961a). If 4.5 liters is taken as a rough estimate of the average volume of the rumen liquid pool turning over independently of the large solids, the turnover rate for sheep rumen liquid, based on salivary flow, is 1.1 to 2.2 per day.

If 120 liters per day is assumed to be an average total salivary secretion in cattle and 60 liters is assumed to be the average bovine rumen liquid pool, the liquid turnover rate is 2.0 per day. The turnover of 1.87 found by Sperber is of the same order of magnitude as this estimate from the amount of saliva.

A method for estimating the rumen liquid by use of antipyrene and N-acetyl-4-aminoantipyrene has been used (J. T. Reid et $al.,$ 1957). The antipyrene penetrates all body water, including that in the gut, whereas the N-acetyl-4-amino derivative penetrates only slowly into the gut.

Particles small enough to pass through the omasum do not readily settle out in the rumen. They are hydrated, and the specific gravity is not greatly different from that of the rumen liquid. The bacteria and some of the protozoa are in this size category and would be expected to pass from the rumen at approximately the same rate as the liquid. The large size of some of the protozoa, and the rather high specific gravity when filled with starch, might cause them to turn over at a slower rate.

The rate of passage of particles small enough to enter the omasum has been determined by Castle (1956a). She separated particles from goat feces, stained them, and mixed them with the hay ration. Being of a size which had already passed through the alimentary tract, the particles should not be retained by the omasal filter and should show a rate of passage similar to that of rumen liquid. The excretion curve for the stained fecal particles is shown in Fig. V-2, together with the curve for stained hay.

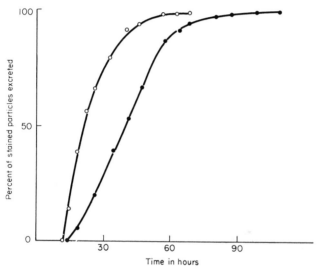

Fig. V-2. Curves to show the elimination of stained particles. From Castle (1956b, p. 115). Key: ○—○ stained fecal particles; ●—● stained hay.

From the amounts excreted at 13 and 40 hours, the turnover rate constant for the fecal particles is 0.086 per hour or 2.1 per day. The shape of the curve is that which would be expected if the elimination of the particles were proportional to their concentration in the rumen. Although liquid turnover rate was not estimated in the experiments of Castle, it seems quite probable that the values obtained would have been of the same magnitude as the value for the fecal particles, and this will be assumed in the following discussions.

Columbus (1934) cut straw into particles 2 × 1 mm in size and fed them to sheep. At periodic intervals rumen contents were withdrawn by stomach tube, and particle counts were made as with feces. At 2 days, 74% of the total particles had left the rumen; this value corresponds to a cycle time of 1.5 days, or a rate of 0.67 per day. These particles are of a size that would be expected to leave the rumen with the liquid, and their recovery in a relatively undamaged state in the feces supports this view. These were adult sheep fed 350 gm of oats and 1.5 kg of hay per day, whereas the turnover rate of 2.1 per day from Castle's data was for a 10-week-old kid. The slow elimination in the Columbus experiment could be caused in part by the small straw particles floating on top of the rumen fluid, which would slow their elimination (Campling and Freer, 1962).

b. *Estimation of Rumen Solids Turnover Rates*

At the same time the excretion pattern of the fecal particles was measured, Castle followed also the excretion of colored particles from a 10-week goat fed hay similar to that fed when the dyed fecal particles were prepared. This excretion curve is also given in Fig. V-2. It shows a slower elimination, and the curve is not shaped as would be expected from a pulse injection into a continuous-passage system. It is almost linear with time between 22 and 60 hours. The delay in elimination as compared to the fecal particles is caused by retention of coarse material until it is reduced in size by rumination and the enzymatic and mixing processes of the rumen.

Insofar as rate of passage is concerned, dyed coarse particles in the pool to be ruminated have not yet entered the pool of material leaving the rumen, but constitute a separate pool which feeds it. The average rate at which small dyed particles leave this rumination pool is proportional to the concentration of coarse dyed particles in the pool, and in consequence the passage from the coarse to the small particle pool follows the kinetics of a first-order reaction. This precedes the passage of the comminuted particle from the rumen liquid-small particle pool into the omasum. Total passage of initially large particles from the rumen is thus the resultant of two sequential first-order reactions. The plot of elimination of a single dose of marker against time will give a sigmoid curve, with initially a gradually increasing rate, and then a period of diminishing rate. Most curves for elimination of a single dose of dyed particles from animals fed long hay are of this shape.

The thorough mathematical analysis of Blaxter *et al.* (1956) established that the excretion curve for dyed particles fitted a model with two sequential first-order reactions. The constants for these reactions were calculated for various feeding levels of long dried grass, dried grass passing a $\frac{1}{4}$-inch sieve and a $\frac{1}{16}$-inch sieve, respectively. The constant for the first reaction increased from 0.031 per hour to 0.503 per hour as the feed changed from 600 gm of long grass to 1500 gm of $\frac{1}{4}$-inch-mesh grass daily, i.e., the long grass left the "rumination pool" more slowly than the ground material. This is the direction of change expected if the first constant represented the rate at which material left a coarse particle "rumination pool" to enter the rumen liquid-small particle pool. The second constant should be that for the rumen. It is small, .026 to .045 per hour, perhaps because the dried grass was quite digestible.

During the later stages of passage of dyed hay particles from the rumen, the elimination curve should approach that observed with fecal particles, since the longer the digesta remain in the rumen the greater the degree of comminution. It can be seen in Fig. V-2 that the portion of the curve above 85% resembles the comparable part of the curve for the fecal particles.

The difference in the rate of passage of the liquid-small particle fraction of rumen contents and the larger digesta particles is well illustrated by analyses (Weller *et al.,* 1962) for lignin and polyethylene glycol markers in the rumen of sheep fed once a day. The relative concentration curve for lignin and polyethylene glycol (Fig. 2 in Weller *et al.,* 1962) shows that the concentration of polyethylene glycol decreases much more rapidly during the first 8 hours than does the concentration of lignin, its passage rate being greater. The large plant particles are retained until they are comminuted by rumination. In the period of 8 to 24 hours after feeding, the difference is much less because the large particles are more completely comminuted.

It is possible that in hay-fed animals the liquid-small particle pool is smallest when the total rumen volume is largest. The rate of passage of digesta to the abomasum is greatest during and just after food is consumed. This would be material from the liquid-small particle pool. The rapid secretion of saliva and ready availability of digestible materials in the hay would support rapid microbial growth and counteract the more rapid passage during this period. The subsequent digestion and comminution of the rumination pool would feed the small particle pool during later periods. There is thus reason to postulate a fairly constant liquid-small particle pool approximating in size the rumen volume just before feeding.

The rumination pool includes the particles too large to pass through the omasum. It increases after a ration is consumed and is minimal just before feeding. The fraction of rumen space occupied by the rumination pool is greatest when the total volume is greatest; the liquid-small particle pool size does not increase in direct proportion with total rumen volume when coarse forages are consumed.

Under steady state conditions, or taking daily averages, the rate of ingestion of coarse plant parts will equal the rate at which they are comminuted to a size small enough to leave the rumen. They enter the pool of fine particles at the same rate as if they were ingested in the finely divided state, but the total rumen volume must be larger in order to accommodate the material in the rumination pool. The rumination pool occupies space which is not part of the liquid-small particle pool. This must be taken into account when calculating *turnover,* since turnover = turnover rate constant (k) $\times$ pool size (A). The rumination pool must not be included in the small particle-liquid pool when turnover is calculated from passage of small particles or liquid.

The relative size of the liquid-small particle pool and the rumination pool (in terms of dry matter) can be estimated by comparing the actual dry matter leaving the rumen (i.e., the amount recovered in the feces) with the amount calculated by multiplying the liquid turnover rate constant times the total rumen dry matter. This latter gives a higher quantity than the actual, since it is calculated from the sum of the rumination pool and the liquid-small particle pool. The actual amount divided by this high value, gives the fraction of the total dry matter in the small particle pool, the rest being in the rumination pool. So far as the author is aware, passage rate constants for the liquid-small particle fraction, fecal dry matter, and rumen dry matter (exclusive of volatile fatty acids) have not been simultaneously determined. Information on the dry matter in the feed and in the rumen is collected in Table V-5.

Although the quantity of lignin or silica can be used as an index, these substances are not always inert, and their estimation takes time. It would be useful if turnover rate could be derived from the dry material in the rumen, according to the relationship

$$\frac{\text{weight of feed per day}}{\text{weight represented in rumen}} = \text{average turnover rate per day}$$

To make this calculation it is necessary to know the extent to which digestion in the rumen has progressed:

$$\frac{\text{weight of rumen digesta}}{\text{percent digestion}} = \text{weight of food represented in the rumen}$$

With this value, the relative sizes of the rumen liquid and rumination pools can be calculated.

c. *Extent of Digestion in the Rumen*

The extent of digestion of dry matter in the rumen can be estimated from the relationship

$$1 - \frac{\%\text{ marker in feed}}{\%\text{ marker in rumen}} = \text{fraction of the feed digested in the rumen}$$

The percentages must be based on the dry matter. Some of the extents of digestion of dry matter in the rumen, obtained in this way, are shown in Table V-6. The data indicate that the average weight of dry material in the rumen amounts to about 55% of the weight of the food represented by that material. In calculating turnover on this basis

$$\frac{\text{dry matter in rumen}}{0.55 \times \text{daily intake}} = \text{average dry matter turnover time in days}$$

The exact degree to which digestion has occurred in rumen contents will increase with time after feeding, and the dry matter will decrease. Less error is involved if the dry matter and lignin content are determined just before feeding, since the rumination pool is minimal at this time. Estimates on turnover based on the dry matter content at 24 hours after feeding, and with estimates derived from Table V-6, have been calculated for the data of Table V-5 and are listed in the last column of that table. Inspection shows the considerable range of variation encountered with various animals on various feeding regimens. Note the relatively slow turnover in the sheep fed kibbled maize as compared to those on a hay or hay plus cassava ration; this will be explained later.

D. Expected Extent of Digestion in the Rumen, Based on *in Vitro* Fermentations and Continuous Fermentation Theory

The actual extent of digestion in the rumen can be compared with the amount expected on the basis of continuous fermentation theory.

The curve, ln $x = -kt,$ shows the rate at which rumen contents leave the rumen due to spillover, but does not portray the disappearance due to digestion and fermentation. If the relationship between degree of digestion of the food and time in the rumen is known, it can be used to correct the turnover curve and, from the area under the corrected curve, to estimate the extent to which digestion has progressed in a continuous fermentation. The time course of attack on alfalfa hay has been studied by subjecting a limited quantity of ground hay to an excess of rumen contents and measuring the rate of appearance of fermentation products (Fig. V-3).

It is important that the quantity of feed added be small in comparison with the rumen contents, to avoid inhibition of digestion by accumulation of fermentation products. Warburg manometry is well suited for measurements (McBee, 1953), and, since results are relative, error due to non-stoichiometry in the carbon dioxide release by fermentation acids is not important. With manometric vessels of 110-ml capacity, mercury in the

Table V-5

ESTIMATED DRY MATTER TURNOVER TIME BASED ON THE DRY MATTER IN THE RUMEN AND IN THE FEED

Animal	Weight (kg)	Feed	Time of sample in relation to feeding time	Dry Matter in rumen (kg)	Estimated % digestion of rumen contents	Turnover time (days)	Reference
6-month calf	218	3.175 kg dry wt. once daily, 60% brome timothy hay, 40% grain mixture	Just before	2.1	0.55	1.47	Agrawala et al. (1953a)
1-year steers, Holstein-Friesian	—	3.6 kg dry wt. once daily, Korean lespedeza hay	Just before	3.48	0.55	2.15	Cason et al. (1954)
			24 hr after	3.44	0.55	2.12	Cason et al. (1954)
		3.36 kg dry wt. once daily, Sericea lespedeza hay	Just before	4.55	0.50	2.7	Cason et al. (1954)
			24 hr after	3.98	0.50	2.37	Cason et al. (1954)
		2.70 kg dry wt. once daily, prairie hay	Just before	5.26	0.45	3.68	Cason et al. (1954)
			24 hr after	3.92	0.45	2.65	Cason et al. (1954)
5-year ewes	—	0.2 kg hay, 0.5 kg ground maize, 0.8 kg decorticated groundnut meal, 0.025 kg dried yeast, mineral salts, at 12-hr intervals	12 hr after	0.72	0.70	1.49	Boyne et al. (1956)
Sheep	—	0.8 kg lucerne hay, once daily	24 hr after	0.33	0.60	1.03	Gray et al. (1958a)
1-year sheep	40	0.91 kg meadow hay daily	—	0.46	0.50	1.1	Wildt (1874)
2 steers	273	4.55 kg timothy hay, 2.28 kg dairy average concentrate, once daily	Just before	3.6	0.60	1.32	Balch et al. (1957)

Sheep	—	600 gm Rhodes grass hay (*Chloris gayana*), 1/2 at 8:00 A.M., 1/2 at 4:00 P.M.	16 hr after	0.35	0.426	1.02	Rogerson (1958)
	—	500 gm hay, 300 gm cassava meal	16 hr after	0.32	0.52	0.77	Rogerson (1958)
	—	600 gm kibbled maize	16 hr after	0.35	0.74	2.25	Rogerson (1958)
Cows	510	6.4 kg hay, 2.3 kg beet pulp	—	6.8-9.6	0.55	1.89	Balch and Line (1957)
		8.2 kg hay	—	8.0 ± 0.9	0.50	1.95	Balch and Line (1957)

Table V-6

DEGREE OF DIGESTION OF TOTAL DRY MATTER IN THE RUMEN

Animal	Feed	Reference substance	Percentage reference in feed	Percentage reference in rumen	Percentage digestion of rumen contents	Reference
Sheep	Hay	SiO_2	2.25	3.2	30	Wildt (1874)
Sheep	500 gm wheaten hay + 250 gm wheat straw	Lignin	9.3	13.0	23, 5 hr after feeding, 34, 24 hr after feeding	Gray et al. (1958a)
	800 gm wheaten hay	Lignin	8.6	13.8	28, 2 hr after feeding, 45, 24 hr after feeding	Gray et al. (1958a)
	400 gm wheaten hay + 400 gm lucerne hay	Lignin	7.8	14.4	39, 3 hr after feeding, 53, 24 hr after feeding	Gray et al. (1958a)
	800 gm lucerne hay	Lignin	7.4	16.4	51, 2 hr after feeding, 60, 24 hr after feeding	Gray et al. (1958a)
Sheep	600 gm Rhodes grass (*Chloris gayana*) hay	Lignin	11.32	19.73	42.6, 52.4 at omasum 48.9 at feces	Rogerson (1958)
	500 gm hay, 300 gm cassava meal	Lignin	7.26	15.12	52.0, 58.1 at omasum, 62.4 at feces	Rogerson (1958)
	300 gm kibbled maize	Lignin	0.94	3.61	74.0, 79.6 at omasum, 89.8 at feces	Rogerson (1958)
Cows	Hay	Lignin	7.89–9.04 8.31 average	13.46–14.88 14.06 average	39–43, 41 average	Makela (1956)
Sheep	Chaffed wheat straw, lucerne hay	Lignin	—	—	41	Gray (1947b)
Goats	Hay and concentrate in various combinations	Lignin	—	—	36.0–61.8, 45.7 average	Kameoka and Morimoto (1959)
Sheep	800 gm clover hay plus 100–600 gm concentrate	Lignin	—	—	28.3–48.5, 38.0 average	Oyaert and Bouckaert (1960)

manometers, and 0.1–0.2 gm of hay in 20 ml of rumen fluid (with relatively little solids), the acidity of the milieu remains within the range for normal rumen contents. Conditions during the entire *in vitro* incubation, which lasts up to 40 hours, are not identical with those in the rumen, but it is doubtful that the changes seriously modify the supply of enzymes attacking the feed, even during the later periods of the incubation, and, since the rate of digestion is limiting, it can be measured. The effects of substrates other than added hay are corrected by a control rumen fluid sample without hay. The rate in the experimental vessel does not be-

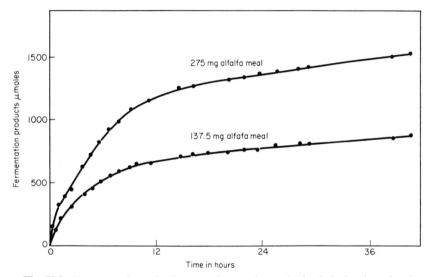

Fig. V-3. Curves to show the fermentation products obtained during long incubation of a small quantity of hay. Corrected for a control without hay. Rumen liquid (20 ml—no forage solids) incubated in 110-ml round-bottom Warburg flasks. Mercury in the manometers. Temperature: 39°C.

come identical with the control, even after several days, and the control maintains a slow production of fermentation gas over a very long period. Essentially similar results have been observed *in vivo* (Yadava *et al.,* 1964) when C^{14}-labeled alfalfa is introduced into the rumen.

1. Degree of Digestion of Soluble Carbohydrates in the Rumen

In a number of experiments in which the concentration of added alfalfa hay was 0.07 percent (w/v), fairly sharp breaks have been observed between an initial rapid fermentation lasting for less than an hour, a second which is slower and almost constant between 1 and 6 hours, and a third still slower fermentation which continues for a long time and decreases gradually until the slope of the curve is almost constant.

The initial rapid fermentation is interpreted as representing conversion of soluble carbohydrates (the water-soluble material in the hay studied amounted to 35% of the dry weight). A similar rapid initial fermentation followed by an almost linear rate has been found in sheep fed on ground pelleted rations (Sutherland et al., 1962).

The relationship between concentration of dissolved substrate, S, velocity, v, of its utilization, and maximum velocity, V, in enzymatic reactions is given by the Michaelis-Menten equation

$$v = \frac{VS}{K_m + S}$$

in which K_m is the concentration of substrate which gives half-maximum velocity. For most soluble substrates K_m is a very low concentration. When K_m is small in comparison with S, $v = V$, and utilization is maximum and constant until concentration of substrate becomes of the same order of magnitude as K_m. The concentration of water-soluble material in alfalfa hay is above the usual values for K_m, and, although dispersion in the rumen reduces the concentration of the sugars, mixing does not occur rapidly enough to dilute the soluble material in each ingested bolus with the total rumen liquid. It is postulated that the sugars remain in the vicinity of the ingested particles and raise the concentration to a point exceeding K_m, with a resulting period of constant rate of utilization and then a rapid fall to a very low level after K_m concentration is reached.

According to these assumptions, and from Fig. V-3, sugars ingested with the feed are utilized at a linear rate over a period not exceeding 1 hour, at which time all the sugar has been converted. According to this, if x represents the sugar, the fraction as yet unused at any time, t, in hours after ingestion would be $x = 1 - t$, $(t < 1)$. The fraction of x left in the rumen as calculated from the feed rate would be $x = e^{-kt}$. The fraction of x remaining and undigested at any time after its addition would be the product of these two,

$$x = e^{-kt}(1 - t) \tag{V-6}$$

between values of $t = 0$ and $t = 1$, after which any sugar not leaving the rumen would have been utilized. This curve is shown as B in Fig. V-4.

To calculate the concentration of sugar in the rumen it is necessary to turn again to the curve for a first order reaction, $x = e^{-kt}$. This was used to portray the change in the concentration of a marker with time after adding a pulse to the continuous fermentation system. It also represents the composition of the total rumen digesta with respect to the time each portion has been in the rumen. For example, in Fig. V-4, curve A, the quantity of material which had been in the rumen for 3.5 hours is twice as abundant as that which had been in the rumen for 20 hours.

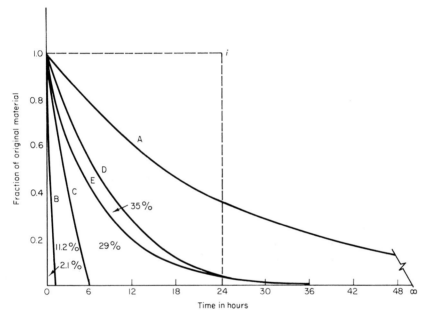

Fig. V-4. Curves to show the disappearance of feed components due to digestion and passage from the rumen. Curve A: $x = e^{-kt}$; curve B: $x = e^{-kt}(1 - t)$; curve C: $x = e^{-kt}[1 - (t/6)]$; curve D: $x = e^{-kt}[-uat + 1/2(ut^2) + 1]$; curve E: disappearance calculated on the basis of an harmonic series up to 36 hours, and on $x = e^{-kt}$. A turnover time of 24 hours has been assumed.

The area under curve A represents the total contents or the total quantity of any component, provided spillover is the only mechanism by which material leaves the rumen. Since the material leaves the rumen also by fermentation, digestion, and absorption, the product of the two equations, Eq. (V-6), represents the fraction of the original material which will be in the rumen at any time after its addition, and the area under this curve, divided by the area under the curve, $x = e^{-kt}$, gives the fraction in the rumen under continuous feed conditions. Integration of the curve $x = e^{-kt} (1 - t)$ gives

$$\int_{t=0}^{t=1} x \, dt = \int_{t=0}^{t=1} e^{-kt} (1 - t) \, dt = \int_{t=0}^{t=1} (e^{-kt} - te^{-kt}) \, dt =$$

$$\int_{t=0}^{t=1} e^{-kt} \, dt - \int_{t=0}^{t=1} te^{-kt} \, dt = \underbrace{\left[-\frac{e^{-kt}}{k^2} (-kt - 1) + \frac{e^{-kt}}{-k} \right]}_{t=1} +$$

$$\underbrace{\left[\frac{e^{-kt}}{k^2} (-kt - 1) - \frac{e^{-kt}}{-k} \right]}_{t=0} = \frac{1}{k} + \frac{1}{k^2} (e^{-k} - 1) \qquad \text{(V-7)}$$

When $k = 1/24$

$$\int_{t=0}^{t=1} e^{-kt} (1 - t) \, dt = 0.4992$$

$=$ the area under curve B in Fig. V-4.

The total sugar entering the rumen would be represented by the entire area of the curve $x = e^{-kt}$, which area is $1/k$, from Eq. (V-5), or 24 for the present example. It is the area 0,1.0, i, 24 in Fig. V-4. The fraction of the sugar in the rumen in the steady state would be 0.4992/24, or 0.021. This fraction of the ingested sugar would be lost to the rumen fermentation by spillover into the omasum. The fraction fermented would be 0.979.

With a cycle time of less than 24 hours, more of the sugar would escape to the omasum. If, with continuous ingestion of feed, the time for utilization is less than 1 hour, less of the sugar would escape fermentation. With frequent feeding it is probable that much less than an hour is required for sugar utilization.

These calculations show that continuous fermentation theory predicts almost complete utilization of soluble fermentable materials in the feed. This is in accordance with observations; the sugar concentration in the rumen is normally very low. Béranger (1961) has estimated that soluble sugars are 100% utilized in the rumen-reticulum.

2. DEGREE OF DIGESTION OF MATERIALS GIVING THE CONSTANT FERMENTATION UP TO SIX HOURS

This fairly steady fermentation period, lasting up to 6 hours, can be variously interpreted. It may represent soluble polymer components of the feed that are digested linearly with time, for example, long-chain molecules (xylans, pectins, glucans) digested from one end, or even the storage starch laid down from the soluble sugars. It may represent diminishing utilization of one set of components, coupled with beginning utilization of fiber materials. The second explanation is supported by the fermentation curve for the water-extracted alfalfa meal (Fig. V-5) and by the fermentation products of the water extract, the latter plotted as the difference between whole and water-extraced alfalfa meal. The water extract shows a diminishing rate during the 1–6-hour period, while the rate for the water-extracted meal increases. The sum gives the almost straight line for the whole alfalfa meal. During this period the degree of cellulose digestion gradually increases, as indicated by the curve for alfalfa holocellulose. The holocellulose was obtained by chlorite treatment of the

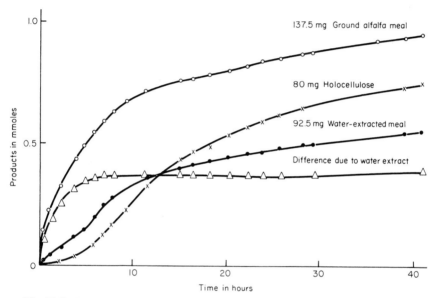

Fig. V-5. Curves to show the fermentation products from ground alfalfa meal, the water-soluble extract of ground alfalfa meal, and holocellulose from ground alfalfa meal. Experimental conditions the same as for Fig. V-3.

same amount of meal as that used for the other curves. The fermentation curve (Fig. V-6) for added cellulose in the form of pebble-milled filter paper also shows a delay before fermentation starts.

It is evident from Figs. V-5 and V-6 that soluble components contribute to the initial fermentation and that the effect of the soluble materials lasts for about 6 hours. It is also clear that added pure cellulose is digested only after a lag of several hours (McBee, 1953). This is interpreted as the time necessary for cellulase to adsorb, either in connection with adjacent bacteria or with adsorption from a low concentration of cellulase in solution. The same delay is shown by holocellulose from alfalfa hay (Fig. V-5), but the fermentation does not rise as rapidly and continues longer. These observations indicate that the almost linear rate of fermentation between 1 and 6 hours results from a trailing off of the effect of the soluble materials and the gradual increase in fiber utilization during this period. Part of the delayed effect of the soluble materials may reflect utilization of the starch stored from it.

Regardless of the explanation for the practically linear rate of utilization following the initial burst, it can be treated mathematically in the

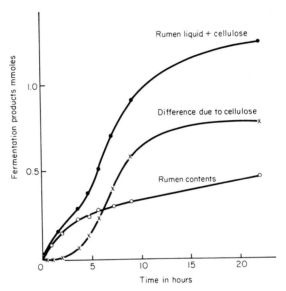

Fig. V-6. Fermentation product formation from cellulose added to indubated rumen liquid. Conditions similar to those used in the experiment of Fig. V-3, with 68 mg of pebble-milled cellulose used.

same fashion as the postulated linear rate of sugar utilization. The formula in this case becomes

$$x = e^{-kt} [1 - t/6] \qquad (V-8)$$

The values of x according to this equation are plotted in Fig. V-4, curve C, for values of t between 0 and 6 hours. Integration to obtain the area under the curve gives 2.688, which is 11.2% of 24, the total amount ingested. The material spilling out of a continuously fermenting rumen with a 24-hour cycle would carry with it 11.2% of this component in a still undigested state. The rumen-reticulum digestion of hemicellulose was found to be 88% complete (Béranger, 1961), an interesting but possibly coincidental agreement.

3. DIGESTION OF THE MORE RESISTANT FRACTIONS

The curves showing the fate of the most resistant portion of forage are not as easily handled mathematically because they are not linear with time. Two approaches have been attempted, both based on the assumption that the material undergoing digestion during this late period consists of cylindrical fibers containing long polymers arranged in concentric rings, a single chain in the center, a layer of polymer surrounding it, and with successive layers building the fiber up to the initial size; a further assumption is that the surface is saturated with the enzymes splitting the polymeric chains.

In both approaches it is assumed that the radius of the fiber diminishes at a linear rate with time and that digestion is complete in 36 hours. Digestion is probably not complete at 36 hours, but the rate is so low that no great error is involved by neglecting digestion after 36 hours. As evidence of prolonged fermentation, after 24 hours the rate of methane production is 25–50% of the initial level (Blaxter and Graham, 1955), and 5% of normal even after 64 hours of fasting (Blaxter, 1962). Part of this latter methane may be at the expense of microbial cells and cell products. The actual rate of digestion of residual food components after 64 hours is much less than the 5% indicates.

In the first treatment it is assumed that all the fibers are alike in initial effective diameter, i.e., it takes the same length of time for each fiber to be completely digested. In the second treatment it is assumed that the initial diameters of the fibers vary continuously and that the quantity of material making up each size is the same as the quantity making up any other size.

a. *Theoretical Equation Based on One Size of Fiber*

It is assumed that the diameter of all the fibers is such that digestion is complete at time a, in hours. The rate of digestion per unit of substrate is represented by the rate constant, u. Under these conditions digestion proceeds according to the equation

$$\frac{dx}{dt} = -u(a - t)$$

with $t \leqq a$

Integration gives

$$x = -uat + 1/2(ut^2) + C$$

When $t = 0$, $x = 1$ and $C = 1$, and

$$x = -uat + 1/2(ut) + 1 \tag{V-9}$$

The digestion constant, u, can be calculated for any assumed value of a by solving the equation when $t = a$, at which time $x = 0$. When $a = 36$, $u = 1/648$.

At any time, $t < a$, after ingestion of a particle, the value of x in Eq. (V-9) represents the undigested fraction of the slowly digestible component. The amount left according to the dilution rate of the rumen contents will be again the first-order reaction, Eq. (V-3), and the amount remaining from the action of both dilution and digestion will be

$$x = e^{-kt}(-uat + 1/2(ut^2) + 1) \tag{V-10}$$

The area under the curve for this equation will be

$$\int_{t=0}^{t=a} x\, dt = \int_{t=0}^{t=a} e^{-kt}(-uat + 1/2(ut^2) + 1)\, dt$$

Integration between these limits gives

$$\int_{t=0}^{t=a} x\, dt = \frac{ue^{-kt}}{k}\left(ta + \frac{a}{k} - \frac{t^2}{2} - \frac{t}{k} - \frac{1}{k^2} - \frac{1}{u}\right) - \frac{1}{k}\left(\frac{ua}{k} - \frac{u}{k^2} - 1\right)$$

The area under this curve, when $k = 1/24$, $a = 36$, and $u = 1/648$, is 8.57.

Thus, $8.57/24 = 0.357$, or the fraction spilling over; this makes the degree of digestion 64.3%. Note: this is the extent to which the *digestible* fiber is digested in the system.

The curve for Eq. (V-10) for the above values of a, k, and u is shown in Fig. V-4, curve D.

b. *Theoretical Equation Based on the Assumption of Different Sizes of Fibers*

The best fit has been given by an harmonic series in which it is assumed that digestion takes 36 hours and that the rate of digestion diminishes each hour by an amount proportional to the series, 1/1, 1/2, 1/3, 1/4, . . . , 1/36, respectively. This is the curve illustrated in Fig. V-7. In Fig. V-7 the points for kikuyu grass hay are taken from the start of the manometric experiment, whereas for alfalfa the time in excess of 6 hours is used. Alfalfa gives a much more rapid initial fermentation than is ordinarily observed with mature grass hays. On the assumption that the harmonic series model represents various-sized particles in rumen contents, more of them would have smaller diameters, but the total mass in each category would be the same.

If it is assumed that the third phase of digestion of alfalfa hay corresponds to this harmonic series, a curve showing the amount expected in the rumen can be calculated by adding the relative rates in each period (the sum of the relative rates is 4.1793 for 36 hours), determining the percentage which each constitutes of the total, and multiplying each value by the corresponding value in the curve $x = e^{-kt}$, Fig. V-4. The area under the curve has been estimated at approximately 7, which is 29% of the total area under curve A. Digestion would be 71% of the total possible if all the material resided in the rumen for 36 hours. An experimental value of 74% has been found (Béranger, 1961).

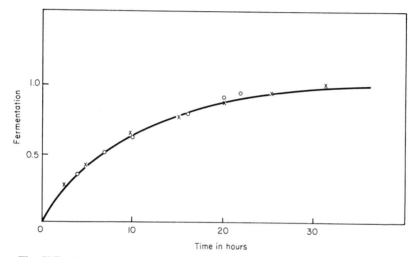

Fig. V-7. Curve for the harmonic series between 0 and 36 hours fitted to the amounts of fermentation products formed from kikuyu grass hay (O—O) after adding it to incubated rumen liquid, and from alfalfa hay (×—×) 6 hours after adding it to incubating rumen liquid.

4. THEORETICAL EXTENT OF DIGESTION OF TOTAL DRY MATTER IN THE RUMEN

The manometric pressure increases provide not only an estimate of relative rates of attack, they also represent the relative quantities of each of the fermented fractions (Fig. V-3). On the assumption that the proportions of products are similar for the different substrates, the quantity of each is proportional to the pressure increase during the period in which it is fermented. In Fig. V-3 the total pressure due to fermentation of 137.5 mg of alfalfa is 68. The pressure increase in the first hour is 14, in the next 6 hours 31, and in the remaining period 23. These are the values for the manometric vessel with no feed or spillover.

If the time course of fermentation of the three fractions in the rumen is the same, the amount fermented would be $14 \times 0.979 + 31 \times 0.888 + 23 \times 0.71$, or a total of 57.8. Digestion would be 57.5/68, i.e., 85% complete if all fractions moved through the rumen on a 24-hour cycle. Hay showing a digestibility of 65% in a 36-hour *in vitro* incubation would be expected to show 55% digestion in the rumen. The found values (Table V-6) are of this magnitude. The demonstration by McAnally (1942) that the material in sheep feces still contained material digestible in the rumen strikingly demonstrates the incompleteness of rumen digestion.

The formulas and curves developed in this chapter are not necessarily applicable to other forages under different conditions. They have been presented as a method whereby degree of digestion in a continuous feed system can be calculated. The mathematical treatment for the fiber part of the ration has been applied as at the start of digestion, whereas with alfalfa the harmonic series curve applied only to the period after 6 hours. If the fiber turns over at the same rate as the other components, the values for degree of digestion in the rumen are too high. If the fiber enters a rumination pool and is not released until a considerable time has elapsed, the degree of digestion in the rumen is greater than that calculated.

Because of retention in the rumination pool, it is necessary to distinguish between (a) the degree to which rumen contents has digested and (b) the extent of digestion accomplished in the rumen. The analyses of omasal contents by a number of investigators show that the percentage of lignin is higher in the omasum than in the rumen, which is interpreted as due to a greater degree of digestion of the escaping small particles than of the total rumen digesta. This is shown in Fig. V-8, for sheep fed once each 24 hours on alfalfa hay (Gray *et al.*, 1958a). Rogerson's figures (Table V-7) show that, even 16 hours after feeding, the lignin percentage in the rumen-reticulum was less than in the omasum.

The results of the rumen digestion studies of Kameoka and Morimoto (1959) fit the pattern expected on the basis of continuous fermentation theory. The gut was transected between the omasum and abomasum, and each cut end was brought to the surface separately. Omasal contents were collected, sampled for analysis, and the remainder introduced into the abomasum. Analyses of feces and omasal effluent for fiber disclosed that the percentage of total fiber in the two was approximately equal, which shows that the digestion of fiber (52–80%) occurred almost entirely in the rumen-reticulum. The digestion of nitrogen-free extract was also high (44–77%), but not as high as for the entire gut (52–92%). It would hardly be expected that the completeness of digestion of fiber in the rumen would be greater than that of the nitrogen-free extract, and it is probable that the extent of digestion of the fiber affords a more valid experimental index of the degree of rumen utilization than is available from the analyses for the other fractions. Production of microbial protein masks digestion of feed protein. Microbial components, fermentation acids, and other materials are included in the nitrogen-free extract, which makes it an unreliable index for digestion of any particular material.

The extent to which the rumen microbial fermentation exceeds that in the large intestine and cecum (Hungate *et al.*, 1959) is in agreement with a predominant digestion of fiber in the rumen-reticulum. The experiments of Trautmann and Asher (1939, 1941a), in which cellulose was introduced into the abomasum and recovered quantitatively in the feces,

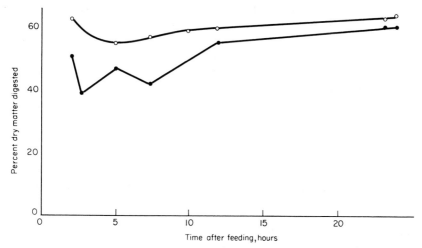

Fig. V-8. Curves to show the extent of digestion in the omasal digesta (○—○) as compared to the rumen digesta (●—●). From data of Gray *et al.* (1958a).

also indicate that the major site of fermentation is the rumen. The high direct counts of cecal bacteria as compared to the rumen (Brüggemann and Giesecke, 1963) is not contradictory to the hypothesis that the major digestion of fiber occurs in the rumen, since the cecal contents may have a much slower turnover and the volume is smaller. Direct measurements of fermentative activity in the cecum as compared to the rumen disclosed a much slower rate in the cecum (Hungate *et al.*, 1959).

With knowledge of the degree of digestion of organic matter in rumen contents, the dry weight of feed daily and the dry insoluble matter in the rumen become a basis for estimating average turnover of dry matter. In most cases this will be less accurate than the turnover based on lignin content of rumen contents as compared to daily fecal legnin, but will be quicker and easier.

Table V-7

EFFECT OF SCREENING ACTION OF OMASUM ON LIGNIN CONTENT OF RUMEN-RETICULUM AND OMASAL DIGESTA

Animal	Ration	Time after feeding (hr)	Lignin in rumen dry matter (%)	Lignin in omasal dry matter (%)	Rumen lignin turnover time (days)	Reference
Sheep	600 gm Grass hay	16	19.73	23.78	1.26	Rogerson (1958)
Sheep	500 gm grass hay 150 cassava	16	15.12	17.33	1.04	Rogerson (1958)
Sheep	600 gm kibbled maize	16	3.61	4.60	3.92	Rogerson (1958)
Average of 4 cows	2.4–2.48 kg timothy-clover hay	4	13.75	16.48	3.77	Makela (1956)
2 cows	3.38 kg timothy-clover hay	4	15.13	17.64	2.46	Makela (1956)
3 cows	4.08–4.14 kg timothy-clover hay	4	14.08	15.93	2.51	Makela (1956)
4 cows	6.0–6.67 kg timothy-clover hay	4	13.92	17.17	2.10	Makela (1956)
1 cow	10.67 kg timothy-clover hay	4	13.42	15.76	1.60	Makela (1956)
5 young bulls	3.49–4.26 kg timothy-clover hay	4	14.13	16.78	2.21	Makela (1956)
2 sheep	Ration 1 from Table V-2	17	32.8	38.3	1.37	Badawy et al. (1958c)

E. Significance of Rumen Turnover in the Ruminant Economy

In order to evaluate the importance of rumen turnover in ruminant nutrition, it is simplest initially to assume a constant rumen volume. Precise methods for measuring it are available for use in particular experiments (Hydén, 1961a). Two factors determine the quantity of fermentation products obtainable from this volume: (a) turnover and (b) the digestibility and fermentability of the ration in the rumen. When digestibility is influenced only by turnover rate, i.e., in comparisons of a single type of ration, increased turnover rate increases the yield to the animal. When more feed is consumed, more microbial fermentation products are formed; substrate is less limiting.

As turnover rate increases, the digestibility diminishes slightly because the average retention time is less, but the increased fermentation rate resulting from the increased supply of feed causes the same volume to turn out products at a faster rate, which supplies more fermentation products, including cells, to the host. The rate increases because the fermentation during the initial period after ingestion of feed is greater than at a later period. This can be illustrated by assuming that the turnover time for the animal consuming alfalfa hay is 12 hours instead of 24, i.e., feed consumption doubles and retention time halves. For the materials fermented in 1 hour, according to Eq. (V-6), the quantity escaping the rumen would be 4.52% as against 2.1% with a 24-hour cycle time. The quantity consumed is doubled, so the net to the animal is 97.9% for the 24-hour cycle and 190.96% for the 12-hour cycle—almost double.

The quantity of the slowly digestible materials escaping digestion on the 12-hour turnover is 43.5% (from curve E in Fig. V-4 transposed to the curve $x = e^{-(1/12)t}$), or 56.5% digested as compared to 71% of the slow component digested on a 24-hour cycle. However, with the doubled quantity of feed, the net to the animal is 71% on the 24-hour cycle and 113% on the 12-hour cycle.

If the turnover rate increased to the point that the organisms could not reproduce fast enough to maintain the rumen population, the microbiota would wash out. However, the turnover rates of the rumen are considerably below the growth potential of most bacteria. The average time for cell division must be 69% of the turnover time if an organism is to maintain its numbers in a continuous feed system (Hungate, 1942).

If the rumen is compared with a batch growth curve (Adams and Hungate, 1950), it represents a stage at which the population is high and the growth rate is decelerating. Faster feed rate at fixed volume reduces the population, but the specific growth rate increases to keep pace with the new feed rate.

When the rumen volume is held constant, the turnover rate *based on indigestible feed components,* as with lignin, will be inversely related to the digestibility, i.e., the higher the digestibility in the rumen the slower the turnover, provided the rate of feed consumption is constant. This follows because (1) there is less undigested daily food left in the rumen and (2) the total minimal dry matter in the rumen remains approximately constant (Makela, 1956). The rumen can contain the undigested residues from a greater amount of highly digestible feed, i.e., more daily rations, than when the digestibility is low. This explains the slower turnover of concentrates than of forage (Freer and Campling, 1963), when turnover is based on an indigestible component, and it also explains the effect of hay in increasing the rate of passage of dyed corn (Eng *et al.,* 1964). With such rations the liquid turnover rate is a better index of rumen function.

These considerations lead to the conclusion that, for a given ration and rumen size, the greater the turnover rate the greater the yield to the host. For different rations, the greater the digestibility the slower the turnover of an indigestible component. The total quantity of feed fermented in the rumen is the measure of the contribution of the microbiota to the host. It is increased by faster turnover.

A few records of animal gains provide information from which also turnover rate can be estimated. An example is shown in Table V-8. The kids with more rapid turnover (less retention time) tend to show the greatest weight increases.

Records in the literature from which turnover can be calculated reveal much variability in the values obtained (Table V-9). These variations derive from animal differences, kinds of feed, amounts of feed, experimental errors, and other factors which will be discussed in Chapter X.

F. Divergence of the Rumen from the Continuous Fermentation Model

Feed is not continuously consumed by ruminants and, in consequence, the rumen does not fit exactly the continuous fermentation model (Kay and Hobson, 1963). The volume is not constant, the rate of salivary secretion is not constant, and the rate at which material leaves the rumen is not constant, regardless of whether it leaves because of digestion, fermentation and absorption, or because of spillover into the omasum. Variations in the dry matter in the rumen with time after feeding (Burroughs *et al.,* 1946b) are shown in Table V-10.

The divergence from continuous fermentation is not as marked as the figures in Table V-10 suggest. The feeds all contained a high proportion of concentrate, which digests more rapidly than hay. Food *represented* in the rumen should be considered, instead of actual dry matter. Thus,

Table V-8

BODY WEIGHT AND FOOD RETENTION TIME IN KIDS[a]

Age of kid	Kid no. 1, female		Kid no. 2, female		Kid no. 3, male		Kid no. 4, male	
	Body wt. (kg)	Retention time[b]	Body wt. (kg)	Retention time[b]	Body wt. (kg)	Retention time[b]	Body wt. (kg)	Retention time[b]
4 weeks	6.7	52.4	8.3	61.0	8.5	42.6	9.3	71.3
7 weeks	8.7	42.9	11.2	43.0	11.3	44.1	12.0	49.7
10 weeks	11.1	43.4	13.1	41.0	14.5	39.0	15.2	42.7
13 weeks (weaned)	12.7	44.4	15.2	44.5	16.2	44.7	17.3	42.9
16 weeks	13.9	41.8	16.4	43.9	17.4	39.4	18.1	44.5
9 months	18.7	44.3	23.5	45.2	26.9	39.0	25.1	45.4
12 months	21.2	46.9	22.7	41.5	31.5	40.5	29.6	43.2
15 months	28.2	46.1	30.3	46.7	41.2	40.9	38.0	42.1

[a] From Castle (1956b).

[b] Time in the entire alimentary tract.

Table V-9

TURNOVER RATES CALCULATED FROM VARIOUS SOURCES

Animal	Daily ration	Information from which turnover was calculated	Turnover time (days)	Reference
Cattle	Long hay	Dyed particle elimination	1.54	Balch (1950)
Sheep	Hay	Half life of polyethylene glycol	0.54	Sperber et al. (1953)
Cattle	Dried grass	Polyethylene glycol, average of many determinations	0.23	Corbett et al. (1959)
Sheep and goats	350 gm oats, 1–2 kg hay	Dyed hay particles	2.32	Columbus (1934)
Sheep	—	Dyed oat hulls	1.45	Lenkeit (1930)
Sheep	Dried young grass	Chromic oxide and monastral blue markers	0.35–0.80	Lambourne (1957a)
Cows	Hay and concentrate	Stained hay	1.39	Balch (1950)
Cows	Hay and concentrate	Stained hulls in concentrate	0.61	Balch (1950)
Cows	Hay and concentrate	Dyed hay	0.51–0.61	Rodriquez and Allen (1960)
Sheep	—	Volume marker in the rumen measured at the duodenum	0.89	Bullen et al. (1953)
Zebu cows	Kikuyu grass hay	Dyed hay	1.0	Phillips et al. (1960)
Sheep	—	Cr_2O_3 administered, recovered 8 hours later	0.68–1.23	Purser (1961b)
Sheep	Chopped hay or chopped hay, beets, and concentrate	Dilution rate of polyethylene glycol	0.56–0.77	Hydén (1961a)
Sheep	350 gm wheaten chaff and 350 gm lucerne chaff	Cr^{51} chelated with EDTA	0.56	Hogan (1964a)
Sheep	700 gm chaff and 100 gm liver meal	Cr^{51} chelated with EDTA	0.63	Hogan (1964a)
Sheep	800 gm lucerne chaff	Cr^{51} chelated with EDTA	0.86	Hogan (1964a)
Sheep	1700 gm lucerne chaff	Cr^{51} chelated with EDTA	0.53	Hogan (1964a)
Sheep	5–6 kg fresh lucerne	Cr^{51} chelated with EDTA	0.40	Hogan (1964a)
Sheep	5 kg fresh grass	Cr^{51} chelated with EDTA	0.53	Hogan (1964a)
Sheep	3–5 kg fresh clover	Cr^{51} chelated with EDTA	0.45	Hogan (1964a)

the approximate doubling of dry matter content after feeding is due to ingested material which has not yet digested, as well as to indigestible materials, whereas the material in the rumen prior to ingestion of feed has undergone a considerable amount of digestion and represents more ingesta than is indicated by the dry weight.

Table V-10
DRY MATTER IN THE RUMEN AT VARIOUS TIMES

Time	Cattle receiving 3.25 lb feed at 8:00 A.M. and 4:00 P.M., water at 8:30 A.M.[a] (lb)	Cattle receiving 6.25 lb feed at 8:00 A.M. and 4:00 P.M., water at 4:30 P.M.[b] (lb)	Cattle receiving 8 lb feed at 8:00 A.M. and 4:00 P.M., water at 8:30 A.M. and 4:30 P.M.[c] (lb)	Sheep receiving 800 gm feed at 8:00 A.M. and 6:00 P.M., water *ad lib.*[d] (gm)
8:00 A.M.	2.75	4.37	7.26	
8:30 A.M.	5.73	10.15	14.35	
10:00 A.M.				1615
12:00 P.M.	4.36	7.73	10.48	1810
2:00 P.M.				1430
4:00 P.M.	3.63	6.55	8.66	1050
4:30 P.M.	6.61	12.32	15.74	
6:00 P.M.				930
8:00 P.M.	5.29	9.90	12.11	720
12:00 A.M.	4.11	7.97	9.61	
4:00 A.M.	3.77	6.23	8.27	
8:00 A.M.	2.89	5.19	7.20	

[a] From Burroughs *et al.*, 1946b. Ration composition: 2.5 lb maize, 3.5 lb chopped alfalfa hay, 0.5 lb protein supplement.

[b] From Burroughs *et al.*, 1946b. Ration composition: 7.5 lb maize, 3.5 lb chopped alfalfa hay, 1.5 lb protein supplement.

[c] From Burroughs *et al.*, 1946b. Ration composition: 10.5 lb maize, 3.5 lb chopped adfalfa hay, 2.0 lb protein supplement.

[d] From Boyne *et al.*, 1956. Ration composition: 400 gm hay, 990 gm ground maize, 160 gm decorticated groundnut meal, 50 gm dried yeast, 25 gm minerals.

In an animal with a 24-hour rumen turnover time, if lignin or other marker be taken as the measure for volume difference before and after feeding one-half the daily ration, assuming a 50% digestion of dry matter in the rumen, the feed contains half as much lignin as does the rumen before feeding, and the increase on feeding (insofar as turnover is concerned) is 50% instead of the 100% indicated by the dry weight increase. Added to this factor is the greater quantity of coarse fibers requiring rumination before they can leave the rumen and the relatively more rapid exit of liquid and fine particles from the rumen during and soon after feeding (Balch *et al.,* 1955b). In Table V-10 dry matter equal to approximately 10% of the feed had already left the rumen by the time the determinations of dry weight were made. In consequence of these factors, and the lower percent of dry matter before feeding, the volume of material turning over in the rumen does not change as much as is indicated by the dry matter values. The minimal rumen volume is rarely less than 50% of the maximum (Balch and Line, 1957). The expected slower rumen turnover at night, when the animal is not feeding, is offset in part in forage-fed animals by the fact that rumination occurs during the night (Gordon, 1958b; Harker et al., 1954), the comminution permitting more rapid flow of solids into the liquid pool and the secreted saliva increasing the liquid turnover.

If the volume of a continuous fermentation system does not change, changes in rate of turnover during a unit of time do not invalidate application of an average turnover rate. This follows from the fact that the quantity of an initial marker left in a continuous fermentation system will be the same if the turnover is once per day as if it is ¾ per day for half the time and 1¼ per day the rest of the time, and the order can be reversed without changing the degree of dilution of the final material.

For the one day at constant dilution, k,

$$ln\ x = kt$$

For the situation in which the dilution rate is first $¾ k$ for one-half the time, and $1¼ k$ for the other half,

$$ln\ x = ¾ k(t/2) + 1¼ k(t/2) = kt$$

Average dilution values can thus express the total passage rate and dilution even though they are not constant with time. When theoretical or actual extents of digestion with time are applied to the rumen, error is introduced by the departure from constant turnover, but unless rates differ in the extreme, the average value approaches the true value, and, when time intervals of a day or more are used, the error is small.

As Blaxter et al. (1956) showed, the analysis of the kinetics of passage of materials through the ruminant is greatly facilitated by subdividing the daily ration into portions which are fed at regular and frequent intervals over the 24-hour period. Under these circumstances the movement of materials during 1 day closely resembles the continuous fermentation model. By collecting and analyzing a marker in the total daily feces, an accurate average cycle time can be derived from animals fed only once per day, even though the rumen volume changes. The average volume and turnover will be the same each day, though varying within the day.

Finally, it should be realized that the value of the continuous fermentation model is not so much that it is exactly applicable, but that it provides a model for reference.

CHAPTER VI

Quantities of Carbohydrate Fermentation Products

A. The Metabolic Significance of Carbohydrate Fermentation

Although the rumen volatile fatty acids (VFA's) are of value to the host, they are waste products insofar as the microorganisms are concerned. The valuable product is microbial cells. Carbohydrate and other fermentable substrates, plus nitrogen and phosphate, other minerals, and growth factors, are converted simultaneously into cells and into wastes, with a concurrent evolution of heat (Walker and Forrest, 1964). In the initial conversions, high-energy molecules such as acetyl CoA and adenosine triphosphate (ATP), as well as amino acid precursors such as pyruvate, oxaloacetate, and α-ketoglutarate, are synthesized. These in turn react to form additional cell-building materials and to link them into the protoplasmic proteins, nucleic acids, lipids, and other components of the microbial cells.

This partitioning of food anaerobically into wastes with a relatively low energy content and into reactive molecules with high energy content was recognized early by A. J. Kluyver (see Kluyver and van Niel, 1956) as being characteristic of the substrate metabolism of all anaerobes. The high-energy molecules are reactants in the chemical reactions of synthesis. This is the mechanism of growth.

The chief energy supply for anaerobic microorganisms is carbohydrate. The carbon in carbohydrate is at an intermediate state of oxidation, CH_2O, with an oxidation-reduction state of zero. Carbon compounds in the completely reduced state, e.g., CH_4, or the oxidized, CO_2, have much less energy anaerobically than systems with carbon at an intermediate level of oxidation, as in carbohydrate (Hungate, 1955, p. 195). Carbohydrate fermentation consists in a rearrangement of the atoms of carbohydrate into molecules in which some carbon atoms are more oxidized and others are more reduced. The rearrangement occurs spontaneously if suitable catalysts are present, and the nature of the catalysts or enzymes of each anaerobe determines the pathways of rearrangement of food and the kind of products formed, including the microbial cell.

245

The waste products of the anaerobic rumen fermentation are relatively abundant in comparison with the quantity of cells formed. Many high-energy molecules participate in the various reactions (Gunsalus and Schuster, 1961) whereby a single carbohydrate molecule is converted into cell material. Since a single carbohydrate molecule converted anaerobically into waste products can yield only a few high-energy molecules (ATP), the quantity of carbohydrate converted anaerobically into wastes exceeds by a considerable amount the quantity synthesized into cells (Gunsalus and Schuster, 1961). When the acid wastes of the anaerobic rumen microbes are oxidized in host tissues they yield many more molecules of ATP than were available to the anaerobic microbes. This point will be stressed in Chapter VII in connection with the limitations imposed on the host by its dependence on an anaerobic fermentation process.

B. Measurements of the Rumen Carbohydrate Fermentation

1. *In Vivo* MEASUREMENTS

a. *Methane Production*

Methane production is a valuable index (Swift and French, 1954) to the extent of the rumen fermentation because it is a direct measure of activity in the entire rumen contents and does not involve rumen sampling errors or volume changes. It is a measure of the total fermentation, integrated over a period of time. It can be measured in an intact, non-fistulated animal. If the *ratio* of methane to total or single acid production is measured at several times after feeding, on samples removed by stomach tube, the total methane production can be used to calculate the production of total or individual acids in the entire animal (Swift *et al.,* 1948; Hungate, 1965b).

The concentration of methane in the rumen gas is usually between 15 and 30% (von Tappeiner, 1883; Olson, 1940; Reiset, 1868b; Markoff, 1911, 1913). Although conceivably the methane concentration in the rumen gas could be a function of its production rate in the contents, measurement of the amount given off is more accurate.

The quantities of methane produced in a ruminant can be measured by keeping the animal in a special metabolism chamber and analyzing the gas produced (Reiset, 1863a). There is no evidence that methane is further converted into other materials after it is formed in the rumen fermentation. The quantity given off from the intact ruminant represents the total quantity produced in the microbial fermentation. Most of it (95% or more) arises in the rumen, less than 5% arising in the large intestine and cecum (Hungate *et al.,* 1959; Markoff, 1913).

The amount and kind of digested food markedly influence the quantity of methane formed. In certain experiments, 3 to 12.5% of the digestible energy of the ration appeared as methane (Armsby and Fries, 1903; Blaxter and Graham, 1956). With sheep, 14.6% of the digestible energy was lost as methane when the intake was 508 kcal, and only 6.5% when it was 2475 kcal (Marston, 1948b). Some of the values for total methane production obtained by various investigators have been collected in Table VI-1.

Inspection shows that the ratios for moles of methane formed to moles hexose digested are between 0.4 and 0.5 for the carbohydrates and fiber of forages. For these calculations a molecular weight of 171 for the hexose in the forage, or 162 for the hexose in cellulose, is assumed. For sucrose and glucose the ratio is lower, 0.3 to 0.4, i.e., less methane per hexose is produced, and for starch the value is intermediate. Presumably the ration studied by Kellner (1911) contained a good deal of readily utilizable carbohydrate.

b. *Portal-Arterial Difference in VFA Concentration*

This was the method used initially (Barcroft *et al.,* 1944) to prove that the acids from the rumen fermentation are absorbed and utilized by the host. The vessels carrying blood from the rumen of anesthetized sheep are exposed, and blood samples are removed for analysis. The volatile acid content of the portal blood is greater than that of blood collected from the carotid artery (Tables IV-5 and IX-1, Fig. IV-11). The excess enters from the rumen and, except for acetate, is removed in the liver before the blood enters the arterial circulation. From the arterioportal difference in concentration, and the volume of blood flowing through the organ (Barcroft *et al.,* 1944; Schambye, 1951d), the acids can be shown to constitute an appreciable fraction of the food supplied to the ruminant (Table VI-2).

Measurements of portal blood flow disclose a good deal of variability (Table VI-3). The blood flow in the portal system of normal sheep (Bensadoun and Reid, 1962) shows an increase after a meal is consumed, due to an increased concentration of VFA's in the rumen (Dobson and Phillipson, 1956). The effect may be mediated through the metabolism of acids in the rumen wall. The portal circulation correlates with the level of VFA's in the rumen during the period up to 10 hours after feeding (Bensadoun and Reid, 1962), but not at later periods. During the later periods, salivary flow and absorption of the undissociated acid raise the pH and dissociate the acids. This could explain the lack of correlation.

The time for blood to flow between the portal vein and the heart is 15 seconds, 13 seconds from heart through lungs to aorta, and 12 seconds

Table VI-1

QUANTITIES OF METHANE PRODUCED PER DAY IN VARIOUS RUMINANTS

Animal	Ration	Moles of methane per day	Moles per mole of hexose fermented	Reference
Sheep	—	1.7	—	Blaxter and Graham (1955)
Sheep	2/3 wheat straw, 1/3 chaffed alfalfa, (estimated 410 gm digested)	0.8	0.35	Lugg (1938)
Ruminant	100 gm starch digested	0.2	0.32	Kellner (Markoff, 1911)
	100 gm sucrose digested	0.18	0.31	Kellner (Markoff, 1911)
	100 gm cellulose digested	0.28	0.45	Kellner (Markoff, 1911)
Ox	10 gm dry organic matter of rumen	7 ml/hr/10 gm	—	Markoff (1913)
Ruminant	100 gm carbohydrate, including fiber, digested	0.29	0.50	Armsby (1917)
Steer	3.014 kg digestible organic matter, timothy hay (early result so may be low)	6.45	0.37	Armsby and Fries (1903)
Ox	3.5 kg digestible timothy hay	8.9	0.44	Armsby and Moulton (1925)
Ox	1 kg carbohydrate	2.6–3.1	0.49	Bratzler and Forbes (1940)
Sheep	900 gm grass hay (Assuming 60% digestible = 540 gm)	1.27	0.40	Armstrong et al. (1958)
Ox	—	11.3	—	Swift and French (1954)
Ox	5.038 kg digested crude fiber and nitrogen-free extract	12.0	0.41	Zuntz (1913)
Ox	5.132 kg digested crude fiber and nitrogen-free extract	12.7	0.42	Zuntz (1913)
Milk cow	100 gm glucose	0.16	0.29	Kleiber et al. (1945)
Milk cow	100 gm digestible carbohydrate in Sudan hay	0.275	0.47	Kleiber et al. (1945)
Milk cow	100 gm carbohydrates in barley	0.24	0.39	Kleiber et al. (1945)

Milk cow A	25.3 kg (feed − feces)	51	0.35	Kellner (1911)
Milk cow B	27.6 kg (feed − feces)	60	0.37	Kellner (1911)
Ox	1/2 maintenance	—	0.57	Forbes et al. (1928)
	maintenance	—	0.52	Forbes et al. (1928)
	1.5 × maintenance	—	0.46	Forbes et al. (1928)
	2 × maintenance	—	0.46	Forbes et al. (1928)
	2.5 × maintenance	—	0.47	Forbes et al. (1928)
	3 × maintenance	—	0.45	Forbes et al. (1928)
Sheep, 66 kg	Straw and beet pulp	1.43		Reiset (1863a)
Sheep, 65 kg	Straw and beet pulp	1.20		Reiset (1863a)
Sheep, 70 kg	Straw and beet pulp	1.18		Reiset (1863a)

from aorta to portal vein, a total circulation time of 40 seconds. Between 26 and 40% of the total cardiac output goes through the portal system (Schambye, 1951d).

The rumen arteriovenous difference (LeBars and Simmonet, 1954b) in acetate in fed animals was 2.3–2.8 mg/100 ml, 1.3–1.5 after a 24-hour fast and 0.6–0.9 after a 46-hour fast. A venous-arterial difference of 2.5 mg/100 ml in a 40-kg sheep with a blood flow of 1480 ml per minute would yield 133 gm of VFA per day. The arterial acetate concentrations were 7.3–9.2, 3.9–5.1, and 2.7–3.3, respectively. Other concentration values (R. L. Reid, 1950c) for sheep are 6.1–11.1 in fed sheep, 3.3–6.5 after 24 hours without food, and 2.7–3.3 after a 46-hour fast. It has been estimated (Conrad et al., 1958a) that 48–63 gm of acetic acid

Table VI-2
RATE OF ABSORPTION OF VOLATILE ACIDS FROM THE RUMEN
INTO THE BLOOD STREAM OF THE SHEEP[a]

Sheep no.	Blood flow per minute from posterior ruminal vein (ml)	Arteriovenous difference in concentration of volatile acid[b]	Amount of volatile acid absorbed per hour (gm)
13	87	43	3.9
15	59	55	3.7
16	107	35	3.5
18	126	44	5.3
19	67	35	2.6
29	92	26	2.3

[a] From Barcroft (1945).
[b] Expressed as milliliters of 0.01 N NaOH per 100 ml of rumen blood.

Table VI-3
PORTAL BLOOD FLOW IN RUMINANTS

Animal	Method of measurement	Flow rate	Reference
Goats	Calibrated-orifice flow meter	185–296 ml/min	Ambo et al. (1963)
Calves	P[32]-labeled erythrocytes	764 ml/min/100 lb	Conrad et al. (1958a)
Sheep	Thermodilution	400–1400 ml/min	Bensadoum and Reid (1962)
Calves, 152 kg	P[32]-labeled erythrocytes	2100 ml/min	Schambye (1951d)
Sheep, 30 kg	P[32]-labeled erythrocytes	840 ml/min	Schambye (1951d)
Sheep	P[32]-labeled erythrocytes or added dye	31–43.9 ml/min/kg (37.1 average)	Schambye (1955a)

and 12–25 gm of propionic acid were absorbed into the blood per pound of feed dry matter.

Most of the butyric acid is metabolized to β-hydroxybutyrate during passage through the rumen wall (Pennington, 1952). Part of the propionic acid similarly does not reach the blood (Table IX-1). This utilization of acids in the rumen wall prevents accurate estimation of the total VFA supply by arteriovenous difference and blood flow measurements.

c. *Rate of Disappearance from the Rumen*

Difficulties due to acid utilization in the rumen wall have been avoided by measuring the rate at which the individual VFA's disappear from the rumen (Pfander and Phillipson, 1953). The openings to the esophagus and omasum of an anesthetized animal are tied off to prevent passage into those organs. The emptied rumen is filled with a mixture of VFA's resembling those in rumen fluid. Acids are added at about the rate they are normally produced, and, from the amount added and the change in the amount in the rumen, the amount absorbed is calculated. The absorbed acids per hour in a sheep (Pfander and Phillipson, 1953) were 32 mEq of acetic, 18 mEq of propionic, and 23 mEq of butyric acid. Per day this would amount to 0.77 moles of acetate, 0.43 moles of propionate, and 0.55 moles of butyrate.

Production of VFA's has been studied by excising the rumen, artificially pumping saline or blood through the organ, and measuring the increment of VFA's in the rumen contents and the quantity absorbed into the circulating fluid (Brown *et al.,* 1960). In this fashion 1.04 moles of acetic, 0.38 moles of propionic, and 0.24 moles of butyric acid were estimated to be produced daily in the rumen of a mature female goat. Perfusion of an excised rumen does not necessarily maintain rumen conditions. A shift from a VFA fermentation to one producing some lactate has been observed almost immediately after starting perfusion (McCarthy *et al.,* 1957).

d. *Rate of Disappearance from Containers Suspended in Rumen*

This technique appears to have been used by Spallanzani (1776) and Réaumur (1752) when they placed green leaves in brass capsules which were fed to sheep and recovered either from the feces or after killing the animal. The technique was used again by Colin (1886) when potato starch was placed in a bag in the rumen and found undigested after 20 hours. Modern use appears to have started with Quin *et al.* (1938), with silk bags containing test materials suspended in the rumen. McAnally (1942) used this technique to study digestion of wheat straw hemicellulose in the rumen; Umezu *et al.* (1951) also used it to study the digestion of cellulose. The most recent technique uses indigestible synthetic fiber bags (nylon or Dacron) to suspend feed in the rumen (Hopson *et al.,* 1963; Lusk *et al.,* 1962; van Keuren and Heinemann, 1962). These are removed after various intervals, and the materials digested are determined by analysis.

The method is reliable for qualitative comparisons of cellulose-digesting power in rumen contents of animals on different feeds, but the results are not easily adapted to the quantitative evaluation of the rate of disappearance of feed cellulose. A method using membrane filters has also been developed (Fina *et al.,* 1962).

e. *Amounts Required to Raise the Steady State Level*

The quantities of an acid produced in the rumen have been estimated by infusing a solution of the acid into the rumen at a steady rate and measuring the increase in the concentration of the acid relative to the others. This has been done for acetic, propionic, and butyric acids (Bath *et al.,* 1962). The results indicated that the acids are produced in about the same proportions that occur in the steady state in the rumen (Hungate *et al.,* 1961).

f. *Radioactive Tracers*

Volatile fatty acids containing radioactive carbon can be used to measure turnover of the acids in the rumen. This has been done for butyrate-2-C^{14} in cattle (Hungate *et al.,* 1961), and a rate of 0.44 moles per hour has been found. This is slightly higher than the rate of 0.358 moles per hour estimated in the same animal by the zero-time rate method. An average rate of production of the VFA's in a 40-kg, 24-hour-fasted sheep, with radioactive acids as markers to determine turnover, has been found to be 2.4 moles of acetic, 1.05 moles of propionic, and 0.5 moles of butyric per day (Gray *et al.,* 1960).

In cattle (Brown and Davis, 1962), dilution rates of acetate-1-C^{14} showed that 21.4 mmoles of acetate were produced per hour per kilogram of rumen contents. The rate was almost constant, though slightly greater 1 hour after feeding. In a roughage-fed calf the rate was 12 mmoles per kilogram hour. In other experiments (Cook and Ross, 1964) a value of 45 mmoles per liter per hour was found. The rates for propionate (Brown and Davis, 1962) were 16 mmole per kilogram hour 0–4 hours after feeding and 4 mmoles per kilogram hour 4–7 hours after feeding. With sheep on chaffed wheat hay, 0.099 moles of acetate, 0.044 moles of propionate, and 0.021 moles of butyrate were formed per hour (Gray *et al.,* 1960). In other experiments 1.52 moles of acetate per hour and 0.7

moles of propionate per hour were formed in a cow (Brown and Davis, 1962).

Four hours after administration of labeled acetate the specific activity of rumen butyrate was equal to that of the acetate (Brown and Davis, 1962). This rapid equilibration between rumen acetate and butyrate has been noted by others (Gray *et al.,* 1951a; Hungate *et al.,* 1961; van Campen and Matrone, 1960). Decreased specific activity is not due entirely to turnover of the acid through production by the microbes and utilization by the host, but in part to the exchange reaction. There is a little exchange of C^{14} between acetate and propionate, but this is not as rapid as the interconversion between acetate and butyrate.

The acetate-butyrate interconversion is surprising. On the basis of the intermediary metabolism of the clostridia, a conversion of acetate to butyrate via acetyl CoA, acetoacetyl CoA, β-hydroxybutyryl CoA, crotonyl CoA, and butyryl CoA might be expected. However, the step from crotonyl CoA to butyryl CoA involves a decrement in free energy and should not be easily reversible. The finding that a considerable conversion of butyrate to acetate occurs suggests active enzyme systems. It would be of interest to analyze the rumen biochemical mechanism for the interconversion of butyrate and acetate.

The amount of propionate supplied the ruminant has been estimated by infusing C^{14}-propionate into the jugular vein, together with sufficient unlabeled acid to raise the concentration to the point that the total quantity can be measured accurately (Annison and Lindsay, 1962). The method does not take into account any propionate used by rumen tissues during diffusion from the rumen into the blood. In the case of propionate this is less important than with butyrate, which is more readily used by rumen tissue. In the above experiments on propionate utilization (Annison and Lindsay, 1962), no entry of butyrate from the rumen into the blood of an animal starved 24 hours could be detected when propionate entry could still be measured (Table IX-1).

For acetic acid a rate of entry from the rumen of 3 mg per minute per kilogram during periods of maximum acetate concentration in blood was found (Annison and White, 1962a). This value is corrected for endogenous acetate production in the animal.

2. *In Vitro* EXPERIMENTS

One aim of *in vitro* experiments is to maintain continuously a microbial population essentially similar to that of the rumen. The term "artificial rumen" has been applied to such cultures, though this adjective is more applicable to a successful *in vitro* culture reconstituted from the individual microbial components grown in pure culture. In practice, the term "artificial rumen" has been used for *in vitro* cultures inoculated with rumen material and incubated for periods of one to several days. In the present discussion these will be termed batch cultures, in contrast to long-term cultures involving transfers. A third type of experimentation makes no attempt to culture the rumen microbial population, but maintains it for a few seconds to a few hours *in vitro* and examines a selected parameter.

a. *Long-Term Culture Experiments*

The most successful continuous culture experiments, from the standpoint of the length of time a rumen population has been maintained *in vitro*, are those reported in Chapter III in connection with the cultivation of the rumen protozoa (Hungate, 1942, 1943; Mah, 1964; Coleman 1962). The daily addition of substrate and fresh fluid medium and the removal of half the culture resemble to some extent the feeding regimen and rumen turnover of sheep fed once daily. These cultures produced methane throughout the culture period and supported growth of both protozoa and cellulolytic bacteria. No rumen fluid was employed in the culture medium (Hungate, 1942; Mah, 1964). The concentration of substrates (combinations of dried grass, ground cellulose, and ground wheat) was only 0.08% (w/v), much less than in the rumen (8–18%, w/v). Subsequently, the concentration of substrate has been increased (Gradzka-Majewska, 1961; Einszporn, 1961), with resulting higher populations, though still significantly less than the concentrations characteristic of the rumen. *Streptococcus bovis* and an organism resembling *Peptostreptococcus elsdenii* survived in long-term *in vitro* cultures of *Ophryoscolex* (Mah, 1962).

Numerous *in vitro* continuous cultures have attempted to use substrate concentrations characteristic of the rumen and to substitute artificially for the supply and removal functions of the ruminant (Louw *et al.*, 1949; Marston, 1948a; Gray *et al.*, 1962; Davey *et al.*, 1960; Quinn, 1962; Bowie, 1962; Adler *et al.*, 1958, 1960a; Adler and Cohen, 1960; Stewart *et al.*, 1961; Harbers and Tillman, 1962; Rufener *et al.*, 1963; Dawson *et al.*, 1964). Very few have duplicated rumen conditions sufficiently well to support the fermentation rate characteristic of the rumen and maintain the microbial population.

Survival of the protozoa in numbers and kinds comparable to the rumen is the easiest means to ascertain whether the rumen population is maintained. Rates of production of fermentation acids and gases are criteria for comparing *in vitro* activity with that in the rumen. According to these two measures, a few experiments (Rufener *et al.*, 1963; Slyter *et al.*, 1964) have approached rumen conditions. Use of resins to absorb the fermentation acids has been helpful, coupled with an appropriate balance between feed rate and concentration of substrate. The feed rate was 1.88 turnover per day for the experiment with concentrate and 2.21 per day for the run with hay. The method avoids the Na^+ inhibition noted by Walker and Forrest (1964).

Radioactive cellulose and hemicellulose have been fermented (Bath and Head, 1961) in the continuous culture system of Davey *et al.* (1960), together with the total ration constituents. The proportions of the VFA's were determined. A continuous culture method for cellulose digestion by rumen bacteria has been developed (Harbers and Tillman, 1962). *Ruminococcus albus* (Hungate, 1963a), *Selenomonas, Streptococcus bovis,* and a lipolytic bacterium (Hobson and Smith, 1963) have been grown in continuous culture.

Further experiments along these lines may disclose mechanisms for maintaining a more constant fermentation than can be achieved in the rumen; experiments also will permit easier tests of model systems. Interest in this approach should not distract attention from the basic problem, however, the functioning of the rumen itself.

b. *Batch-Culture Experiments*

This method as commonly employed consists in inoculating rumen preparations of various sorts into flask cultures, incubating for periods of one to several days, and measuring the changes which occur. It has been extensively employed in empirical studies on the effects of various sources of carbohydrates and protein (Nikitin, 1939), organic growth factors (Belasco, 1954a,b; Bentley *et al.,* 1954a; Burroughs *et al.,* 1950a–e, 1951a,b; Cline *et al.,* 1958; Dehority *et al.,* 1957), digestibility of natural and artificial feeds (Burroughs *et al.,* 1950a–e; R. R. Johnson *et al.,* 1962a,b; Henderson *et al.,* 1954; Hofmeister, 1881; Huhtanen *et al.,* 1954a; Louw *et al.,* 1949; Marston, 1948a; Woodman and Evans, 1938; McNaught *et al.,* 1954b; Cheng *et al.,* 1955; Claydon and Teresa, 1959; Hershberger *et al.,* 1959b; Barnett and Reid, 1957a,b,c; el-Shazly *et al.,* 1961a,b; Hinders and Ward, 1961; Hanold *et al.,* 1957; Quicke *et al.,* 1959a,b; Bowden and Church, 1962a,b), ammonia and urea utilization (Belasco, 1954c, 1956; Arias *et al.,* 1951; Burroughs *et al.,* 1951a,b; Henderickx, 1960a,b; McNaught, 1951; Hershberger *et al.,* 1959a; Felinski and Baranow-Baranowski, 1959a,b; Rojahn, 1960; Hudman and Kunkel, 1953; Müller and Krampitz, 1955a), proportions of fermentation products (Elsden, 1945b; Marston, 1948a), minerals (Cline, 1952; Swift *et al.,* 1951; Burroughs, 1950; Burroughs *et al.,* 1950e, 1951b; McNaught *et al.,* 1950b), protein digestion (Sym, 1938), storage product formation (Thomas, 1960), interaction of various nutrients (Eusebio *et al.,* 1959; el-Shazly *et al.,* 1961a), proportions of fatty acids (Gray *et al.,* 1951b;

Elsden, 1945b; Gray and Pilgrim, 1950, 1952b; Barnett and Reid, 1957a–d), lipolysis (Hill *et al.,* 1960), and cellulose digestion (R. R. Johnson *et al.,* 1962a,b; Quicke *et al.,* 1959a,b; Salsbury *et al.,* 1958a).

Various criteria for judging the applicability to the rumen of *in vitro* batch tests have been proposed (Warner, 1956a; Gray *et al.,* 1962). In essence, such criteria require that the effects detected in the culture be shown to occur in the rumen (Gray *et al.,* 1962).

As was evident in Chapter V, quantitative application of batch-culture results to the rumen is complex, because the turnover of materials maintains a flux equilibrium rather than the final static condition of the batch. A further difficulty is that incubation for more than 2 or 3 hours may change conditions from those in the rumen. Distinct differences in the

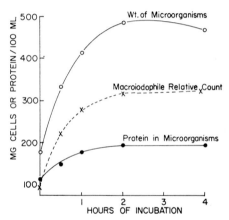

Fig. VI-1. Changes in microbial protein, count, and weight when 100 ml of rumen fluid was incubated with 840 mg of maltose and 100 mg of urea. Data from J. A. B. Smith and Baker (1944).

proportions of VFA's produced *in vitro* and *in vivo* have been encountered (Gray and Pilgrim, 1950). The rapidity with which rumen processes can come to a stop *in vitro* is illustrated (Fig. VI-1) by experiments in which 1% maltose and 0.1% urea were added to incubating rumen liquid.

Some of the criticism of the batch technique can be avoided if a small concentration of substrate is used. The fermentation products do not reach an inhibitory level, and special devices to remove them are unnecessary. If the inoculum is small (Gray *et al.,* 1951b), particular attention must be paid to culture factors such as the composition of the salt solution used as a culture medium, complete anaerobiosis, and a control to show the extent of digestion without the inoculum. With a large inoculum, culture conditions are not as critical because the larger volume of rumen contents supplies a favorable salt concentration, a low redox potential, and an abundance of bacteria.

The batch method with low concentration of substrate is especially useful for rapid *in vitro* determinations of digestibility of feeds. As mentioned (Chapter II), it has been used to determine the ability of various pure cultures of rumen bacteria to digest the components of hay (Hungate, 1957) (Fig. V-3) and to estimate the metabolic products arising from weighed portions of substrates (Phillips *et al.,* 1960).

In batch cultures, considerable accuracy in determination of the amounts of the VFA's or other parameters is achieved, since the difference between the initial and final state is large, but the conditions during the

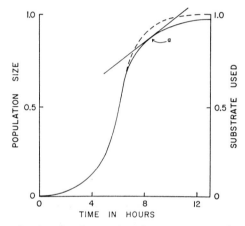

Fig. VI-2. Curve showing the changes in substrate concentration (broken line) and population size (solid line) during growth of a batch culture. A tangent has been drawn at "a" to indicate the growth rate at that point.

in vitro incubation may not resemble those in the rumen closely enough to permit direct application of the values obtained to the rumen itself. In many cases the protozoa are omitted from the inoculum. To the extent that the bacteria in the liquid and solid fractions of rumen contents represent different bacterial types, the washed cell suspensions used as inoculum for the *in vitro* cultures may not be representative for all purposes.

On the other hand, a comparison of parameters in a batch culture with those in a continuous system (Adams and Hungate, 1950) throws light on the conditions in the rumen. In Figure VI-2 is shown a typical growth and substrate utilization curve for a batch culture. The relative or specific growth rate, i.e., rate of cell production divided by cells present, is greatest during the early logarithmic stage during which substrate is still in excess and inhibitory fermentation products have not accumulated. Bacteria in this stage of the batch growth curve usually show a generation time of 20 minutes to 3 hours. Later, as substrate concentration diminishes and products accumulate, the specific growth rate, i.e., $dx/dt/x$ decreases.

The rumen resembles a continuous fermentation system with a growth rate similar to a late stage of a batch culture, e.g., as at the point "a" in Figure VI-2, where the specific growth rate is 0.066 populations per hour, or one turnover in 15.2 hours in a continuous feed system. Substrate is limiting, and products are inhibitory. The fact that substrate is limiting in the rumen is proven by the increased fermentation when substrate is added. The bacteria, at least, can grow more rapidly if more substrate is provided (Thaysen, 1945; el-Shazly and Hungate, 1965).

c. *Short-Term in Vitro Methods*

The utility of this approach to rumen microbial function appears to have been appreciated first by Zuntz (1913) and a student, Markoff (1913). Markoff states in his introduction, "We will see in the following sections of this work that the most exact results are obtained, when the rumen contents are removed as quickly as possible and the fermentation allowed to proceed for only a few hours." Markoff's experiments are particularly noteworthy for his efforts to conduct the *in vitro* fermentation in such fashion that quantitative results applicable to the rumen were obtained. *In vitro* experiments as usually conducted do not provide reliable estimates of the rates at which the phenomena under study occur in the rumen. Very short-term *in vitro* methods can provide such information.

The theory behind this technique is that a sample of rumen contents removed from the rumen continues to function as in the rumen until accumulation of fermentation products, exhaustion of substrate, availability of new foods, or other factors cause it to change. This method is theoretically always applicable and in many cases practicable for evaluating microbial activity in the rumen. In a short-term *in vitro* incubation of a few seconds to an hour, changes in the composition of the rumen microbiota in the sample lie within the range characteristic of the rumen, particularly if anaerobiosis, proper temperature, and suitable acidity are maintained. Results are subject to sampling errors, but since the procedure is not time-consuming, replication and repetition of experiments is feasible, and statistical methods can be used to evaluate the results.

In practice, it cannot be assumed that incubation, even for short periods, has not changed the nature or magnitude of microbial activities in a sample of rumen contents. This is evident from the data in Figure VI-1. If it were possible to determine the kind of activity and its magnitude at the instant the sample is removed, conditions would be the same as in the rumen, and the activity could be considered quantitatively representative of the activity in the rumen. However, changes in the kind and amount of a material can only be determined by a difference between two analyses made at different times. Incubation must be long enough to give a change in concentration of the analyzed material greater than the error in analysis. When it is possible to analyze accurately enough to determine the difference in the amount of material before and after an incubation interval of a few seconds, as in isotope-turnover measurements, the increment or decrement rate is very close to the actual rate in the rumen. Errors in most biological analyses are so large that they exceed changes in the concentration of a microbial fermentation product during a few-second incubation period.

Various rumen preparations have been used in short-term experiments. Washed cell suspensions of bacteria obtained directly from the rumen have been commonly employed, as have washed suspensions of protozoa. For quantitative studies, samples of the total contents have been used. The applicability to the rumen of results of these methods depends in part on the degree to which the material studied *in vitro* represents the natural rumen. In studying microbial activity on a particular ration, the inoculum should be obtained from an animal on that ration (Rice *et al.,* 1962). The importance of sampling and rumen volume measurement must again be stressed, when results are to be applied quantitatively to the rumen.

(1) *Washed Cell Suspensions of Bacteria.* A technique with washed bacteria obtained directly from rumen contents was used by Johns (1951a; Sijpesteijn and Elsden, 1952) to assay the magnitude of the succinic decarboxylation reaction. It has been used to test fermentation of organic acids (Doetsch *et al.,* 1953) and purines (Jurtshuk *et al.,* 1958) and in nutritional studies (Bentley *et al.,* 1954a). Manometric techniques may be employed, and numerous determinations may be completed within a relatively short time. Care should be taken to insure that conditions for the experiments simulate those of the rumen. Factors such as the use of bicarbonate-carbon dioxide buffer, complete anaerobiosis, growth factors, amount of inoculum and substrate, and temperature are important. The time subsequent to removal from the rumen should be kept short (Bladen and Doetsch, 1959). Even with these precautions the *in vitro* results may differ in some cases from the *in vivo* (Hueter *et al.,* 1958). The rapid deposition of polysaccharide storage materials *in vitro* (Hoover *et al.,* 1963), to the extent that the average protein content of the bacterial cells is reduced from 61% of the dry weight to only 30%, illustrates a possible change during incubation.

As mentioned in the discussion of sampling problems, it has not been established with certainty that the bacterial population floating loose in rumen liquid, and which composes the large part of the washed bacterial suspension, is identical with the total rumen flora (Doetsch *et al.,* 1955). It is certain for some bacteria, e.g., *Lachnospira* and *Bacteroides succinogenes,* and possibly for others, that the number associated with plant particles differs from the number floating loose in the rumen. A close spatial relationship to the substrate undergoing decomposition characterizes many cellulolytic bacteria. The washed cell suspension may differ from the total rumen in the proportions of different kinds of rumen bacteria, even though all kinds are present.

The washing of rumen bacteria can introduce conditions injurious to some components of the population, as, for example, by creating aerobic conditions (Hungate, 1957; Ayers, 1958). The washing should be performed under conditions simulating the rumen.

Similarly, the activity of the washed cell suspensions should be tested under conditions permitting participation by all cells capable of the activity under study. Again, influence of oxygen provides an example. Washed cell suspensions will not produce methane if the medium is not sufficiently reduced.

For determining the *potential rate* at which the rumen bacteria can attack a particular substrate, the washed bacteria technique is useful, provided that, as mentioned before, the washed bacteria are representative of the total bacterial population of the rumen, and if conditions permitting their maximum expression of activity are provided. Washed cell suspensions are extremely useful for qualitative examination of possible biochemical pathways and for studies on nutritional requirements.

The washed bacteria technique is ill suited for determining the actual magnitude of the activities of bacteria in the rumen, chiefly because in the washed state it is impossible to provide the relationship to substrate that characterizes the rumen. Although hay or grain of the same batch as that used to feed the animal can be added to the washed bacteria suspension, the degree of digestion will differ, and the extent to which the bacteria have attached themselves to the insoluble constituents will not be the same as in the rumen, where fresh substrate mixes with partially attacked digesta.

(2) *Washed Cell Suspensions of Protozoa.* The protozoa are easier to separate from rumen contents than are the bacteria, and washed suspensions are a preferred material for studying protozoan activities, in part because they avoid the problem of cultivation. The various methods for separating and washing the protozoa have been discussed (Chapter III).

Addition of sugar to incubated rumen contents changes the nutritional state of the protozoa, which makes them unsuitable for assays of activity to be referred to the rumen. It is possible to free the protozoa from bacteria and digesta by differential centrifugation, by electrophoretic migration, and by using various mesh bolting silks for preliminary separation (Hungate, 1942, 1943) with further purification by gentle agitation in a shallow dish. The plant debris localizes in the dish, and the protozoa can be removed relatively free of other materials. Such techniques have been used (in the author's laboratory) to free protozoa for measurments of their nitrogen content and for replicates subjected to various treatments (Christiansen *et al.,* 1962).

(3) *Use of Whole Rumen Contents.* Whole rumen contents is the material best suited for estimating rumen activity, since it contains both the microorganisms and their food. The food is chiefly in the solids fraction, as shown in Fig. II-5. Unless representative solids are included in samples, low and similar values are obtained (Stewart *et al.,* 1958; Grieve *et al.,* 1963), which prevents measurement of differences. Once collected, rumen contents change least if held at 20–23°C under anaerobic conditions rather than at higher or lower temperatures (Dirkson and Wolf, 1963). Techniques employed with whole rumen contents include manometry, the zero-time rate method, and calorimetry.

(a) *Manometry.* Gas production can serve as an index to rumen activity (Markoff, 1913; Stone, 1949), and, in consequence, manometric methods can be applied. In a first attempt (Zuntz, 1913), the rate of vacuum decrease in an evacuated Torricelli tube was used to measure the rate of gas production. This produced conditions more alkaline than normal due to the decreased carbon dioxide in solution at the reduced pressure, which may account for the cessation of gas accumulation after 3–4 hours, although it is possible that the acidity inhibited fermentation.

Conventional manometric methods can be used for studies of samples of rumen contents (McBee, 1953), but have the limitation that methane is not separately determined, and, as a consequence, carbon dioxide and methane production cannot be estimated. A modified method uses mercury in the manometers and 10–20 gm of rumen contents in large vessels (Hungate *et al.,* 1955a). Use of a pair of experimental and control vessels permits determination of the total methane, carbon dioxide, and acid. The total fermentation is measured continuously, but the separate products are measured only at the end of the run. To follow the time course of the fermentation it is necessary to use a number of replicate vessels and controls and to terminate the experiment in each pair at different times.

Mercury is employed in the manometers to allow measurement of greater pressures, as well as to accommodate rather high concentrations of carbon dioxide-bicarbonate buffer. Experience has shown that, in manometric measurements of the rumen fermentation with 0.5% bicarbonate and an atmosphere of carbon dioxide, the release of carbon dioxide from bicarbonate by the fermentation acids is about 85% of the expected stoichiometric value. It is necessary to calibrate the vessels for the conditions of the particular experiment by measuring the pressure increase from a known quantity of acid products.

If it is desired to assay the magnitude of the fermentation as it occurs in the rumen, no substrate should be added. The digesta solids in a representative sample maintain the fermentation at the rate it was occurring in the rumen. It may be necessary to add some bicarbonate if the rumen contents are fairly acid, or the gas released by a known quantity of the fermentation acids added to a control rumen sample can be used to calibrate for the particular experiment. Another procedure is to adjust to a constant pH (about 6.8) with the bicarbonate.

As has been emphasized already, since most of the substrate is insoluble plant cells and fibers, any differences in the proportions of liquids and solids in the sample as compared with rumen content may change the intensity of the *in vitro* fermentation as compared to the *in vivo*. Removal of solids by straining through cheese cloth or other filters, by removing the substrate as well as part of the organisms, prevents valid comparisons if the liquid draining off is used; however, separation of solids on a filter can also increase the validity of the experiment if the retained solids are used at a concentration characteristic of the rumen (Hungate *et al.,* 1960).

Since it is not possible to obtain completely representative samples of rumen contents, numerous determinations of microbial activity in removed samples must be made in order to obtain reliable values for the magnitude of the average fermentation. Some fermentation rates of whole rumen contents as measured by the Warburg manometric method (Hungate *et al.,* 1955a) are shown in Table VI-4.

If manometric methods permitted initial readings to be taken at the instant of removal of contents from the rumen, the methods would be ideal for accurate determination of rates of rumen metabolic reactions producing or using gas. Unfortunately, the time required to fill the vessels with the sample, remove all oxygen by passing oxygen-free gas through the vessels, and obtain temperature equilibrium in a water bath prevents use of the manometric method for determining the rate of fermentation at the instant of removal of contents from the rumen. By having the respirometric apparatus adjacent to the animal supplying the rumen contents, by having vessels at the proper temperature and already gassed, and by rapid weighing out of the sample, the interval can be much reduced. Without these special precautions, 30 minutes to 1 hour usually intervene between collection of the sample and the first manometric readings.

Table VI-4

RELATIVE AMOUNTS OF FERMENTATION PRODUCTS, AS MEASURED MANOMETRICALLY

| Animal | Ration | Products[a] | | | Ratio of methane to carbon dioxide | Reference |
		Acid[b]	Carbon dioxide	Methane		
Steers	Ladino clover pasture	4.74	6.02	1.71	0.285	Hungate et al. (1955a)
Steers	2/3 grain, 1/3 hay	3.5	2.11	0.93	0.44	Carroll and Hungate (1954)
Zebu	Grazing fresh grass pasture	2.83	2.63	1.36	0.517	Hungate et al. (1959)
Grant's gazelle *Gazella granti*	Grazing	3.02	3.34	0.75	0.225	Hungate et al. (1959)
Thomson's gazelle *Gazella thomsoni*	Grazing	1.98	3.15	0.84	0.266	Hungate et al. (1959)
Eland *Taurotragus oryx*	Grazing	2.83	2.95	1.47	0.5	Hungate et al. (1959)
Suni *Nesotragus moschatus*	Browsing	10.9	6.5	2.0	0.308	Hungate et al. (1959)
Zebu	Alfalfa hay	2.27	2.22	1.02	0.46	Hungate et al. (1960)
Zebu	Grass hay	2.50	2.37	1.08	0.455	Hungate et al. (1960)
Milking cows	Hay-grain	3.93	3.32	1.25	0.377	Hungate et al. (1961)
Average value					0.38	

[a] Values in μmoles per minute per 10 gm.

[b] These values were obtained by dividing the found value by 0.85, because the VFA's liberate only 85% of the theoretical CO_2.

Constant pressure manometry is well suited to measurements of the rate of fermentation in a large number of animals and in field experiments (Clarke, 1964c; el-Shazly and Hungate, 1965). The only requisites are a means of incubating at a constant temperature between 37° and 40°C, a wide-mouth vessel in which the rumen contents can be placed, addition of dry sodium bicarbonate, and a means of measuring the increase in gas with time. A half-pint milk bottle or a pint jar with an air-tight rubber stopper through which a syringe needle can be inserted has proven useful and precise (el-Shazly and Hungate, 1965). The syringe plunger, well lubricated with water, is forced out of the barrel as gas is produced. The displacement with time is recorded. When the syringe is filled with gas, it can be withdrawn and the gas expelled, and then reinserted for the next measurement. Use of the method to measure growth is described in Chapter VII. Others have placed the rumen contents within the syringe (Shibata, 1961; Ogimoto *et al.,* 1962) and followed gas production by the displacement of the barrel.

(b) *Zero-time rate method.* A short-incubation method for measuring the rate of fermentation by the rumen microbiota has been termed the zero-time rate method (Carroll and Hungate, 1954). It consists in removing a sample of rumen contents and incubating it *in vitro* under conditions resembling those of the rumen as closely as possible except that the sample is divorced from all activities related to the host. Carbon dioxide is used to assist in maintaining anaerobic conditions during the *in vitro* incubation, and special care must be used to prevent evaporative cooling of the warm digesta. As with other methods, the sample must have a solids content similar to that of the rumen (Zuntz, 1913; Carroll and Hungate, 1954). Utilization of rumen liquid obtained with a pipette gives low values (Stewart *et al.,* 1958).

Immediately upon removal of the sample from the rumen a subsample is removed and treated with a reagent to kill the microbes and stop their activities. This is the zero-time sample. After intervals of incubation, additional samples are removed and similarly killed. Sulfuric acid has been used to kill the samples because it also acidifies them preparatory to analysis of the fermentation acids. These killed samples must be stored in glass-stoppered containers to avoid loss of the acids through cork or rubber. Killing with alkali avoids this loss, but necessitates later acidification before analysis.

The numerical value of the zero-time rate is determined by plotting the concentration of fermentation products in the sample at zero time and after various periods of incubation (Carroll and Hungate, 1954; Hungate *et al.,* 1961). If the analyses have been sufficiently precise and accurate, a smooth curve results in which the rate of production of acid, the slope of the curve, diminishes with time. By graphically estimating the slope at zero time, a measure of the rate of production of the acid in the rumen is obtained.

Removal of the sample of rumen contents isolates it immediately from the absorptive activities of the host. The fermentative activities continue, but gradually slow down as the fermentation products accumulate and the food supply diminishes. At the instant of removal of the rumen sample the fermentative activity was the same as in the rumen. The zero-time rate is the rate in the rumen. The validity of this method depends on uninterrupted maintenance of rumen conditions during transfer and *in vitro* incubation. If an initial rate which gradually changes is obtained, it is concluded that the extropolated zero-time rate is correct.

The theory of the zero-time rate was fully appreciated by Zuntz (1913), "If, for example, the fermentation during 5 hours of incubation was the same as in the first hour, it is justified to conclude that the observed fermentation is the same as that in the rumen." As a modern version, this needs only to be revised to read, ". . . during 5 hours of incubation fell continuously from that in the first hour, it is justified to conclude that extrapolation back to zero time would give the rate in the rumen."

The intervals for subsampling are kept short, since accumulation of fermentation products and exhaustion of substrate decrease the rate rapidly after contents are removed from the rumen. This is particularly true for acetic acid (Hungate *et al.*, 1961). Figure VI-3 shows the type of curve obtained by the zero-time rate analysis.

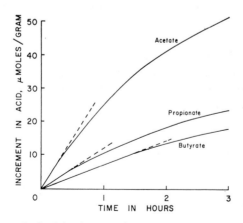

Fig. VI-3. Curves obtained in the zero-time rate method for measuring acid production. The dotted lines show the estimated rates at zero time, assumed to be the same as in the rumen.

It is necessary to perform the analyses very accurately and to reduce sampling and other errors to a minimum, since the rate of production is determined as a difference between two values, and such a difference includes the errors in two analyses. By refining analytical methods and replicating analyses, a satisfactory degree of accuracy can be attained. Gas-chromatographic measurement of the individual VFA's (Fenner and Elliot, 1963; Packett and McCune, 1965) can materially decrease the time required for the numerous VFA analyses involved in zero-time rate determinations. The values obtained by the above method and others are shown in Table VI-5.

(c) *Calorimetry.* In the rumen anaerobic fermentation of carbohydrates to wastes, and in the processes of biochemical work involved in the synthesis of high-energy molecules and building materials of the cell, there is an inevitable conversion of a certain amount of the chemical energy of the substrate into heat (Marston, 1948b). The amount of heat evolved bears a relatively fixed quantitative relationship to the quantity of substrate used and products formed. Heat production can therefore serve to measure metabolic rate. *In vitro* methods have been developed (Walker and Forrest, 1964) to measure the heat evolved in rumen contents.

Two identical containers are placed in a water bath, one with water and the other with an equal amount of rumen contents. A thermistor in each container records the temperature, and heat is added electrically to the control to balance that produced in the fermentation.

The rate of fermentation can be related quantitatively to the amount of heat produced. Gas evolution by the fermenting mass correlates closely with heat production. Krogh and Schmidt-Jensen (1920) estimated that 50 cal of heat appeared for each liter of methane produced in the rumen fermentation.

The method has the advantage that a short period of time (20 minutes) elapses between sample collection and the first measurement in the calorimeter. This delay is relatively unimportant except for periods immediately after feeding.

C. Stoichiometry of the Rumen Fermentation of Carbohydrate

By the application of one of the various previously described methods, a number of estimates of the proportions in which the VFA's are produced in the rumen have been obtained. Some of these have been collected in Table VI-6.

If the quantity of carbohydrate represented in the cells formed is subtracted from the total carbohydrate fermented, the remainder repre-

Table VI-5

QUANTITIES OF VOLATILE FATTY ACIDS FORMED IN THE RUMEN FERMENTATION

Method of measurement	Animal	Ration	Daily production[a]			Reference
			Acetate	Propionate	Butyrate	
Isotope dilution	Sheep	Chaffed wheat hay	2.4	1.06	0.504	Gray *et al.* (1960)
Zero-time rate	Sheep	—	0.64	0.24	0.23	Halse and Velle (1956)
Isotope dilution	Sheep, fasted 24 hours	—	0.21	0.048	0.032	Mukherjee (1960)
Zero-time rate	Cattle, based on 70-kg rumen contents	Hay	14.4	5.05	4.44	Carroll and Hungate (1954)
		2/3 grain, 1/3 hay	25.0	8.95	5.55	Carroll and Hungate (1954)
		Grass pasture	9.6	3.72	2.64	Carroll and Hungate (1954)
Zero-time rate	Lactating cows	Hay-concentrate	40.1	12.8	10.5	Hungate *et al.* (1961)
Steady infusion	Dry cows	Hay, 15 lb	28.5 ± 3.5	7.2 ± 0.5	4.1 ± 0.3	Bath *et al.* (1962)
Excised stomach	Goat	24-hour starvation, then 2 lb alfalfa hay and 1 lb concentrate	1.04	0.38	0.24	Brown *et al.* (1960)

[a] All values in moles.

Table VI-6

RATIOS IN WHICH VOLATILE FATTY ACIDS ARE PRODUCED IN THE RUMEN

Animal	Ration	Time after feeding	Molar percentages of the acids			Reference
			Acetic	Propionic	Butyric	
Sheep	—	0–8 hours	57.5	21.9	20.6	Halse and Velle (1956)
Sheep	—	24 hours	72.6	16.4	11.0	Mukherjee (1960)
Cattle	Alfalfa hay	Various	60.3	21.1	18.6	Carroll and Hungate (1954)
	Grain	Various	63.4	22.6	14.0	Carroll and Hungate (1954)
	Pasture	Various	60.2	23.3	16.5	Carroll and Hungate (1954)
Lactating cows	Hay and concentrate	0–7 hours	63.3	20.1	16.6	Hungate et al. (1961)
Sheep	Chaffed wheat hay	—	60.4	26.8	12.8	Gray et al. (1960)
Cow	—	—	71.8	18.0	10.2	Bath et al. (1962)
Cow	—	—	56.1	24.8	19.1	Brown et al. (1960)
Two 1500-lb Holstein steers	Alfalfa hay, some grain	—	59.5	19.1	21.4	Stewart et al. (1958)
Average			62.5	21.4	16.1	

sents the carbohydrate converted into wastes. This is a stoichiometric conversion, and if the ratios of the products are known it can be written as a chemical reaction by making a few assumptions regarding the pathways by which the VFA's are formed (Hungate, 1960).

1. STOICHIOMETRY OF BIOCHEMICAL PATHWAYS OF WASTE PRODUCT FORMATION

In nearly all microbial fermentations of carbohydrate, acetic acid arises by the oxidative decarboxylation of pyruvate.

$$CH_3COCOOH + H_2O \rightarrow CH_3COOH + CO_2 + 2 H$$

The pyruvate arises in turn from triose by a rearrangement and removal of 2 hydrogen atoms.

$$CH_2OHCHOHCHO \rightarrow CH_3COCOOH + 2 H$$

In the formation of each acetic acid from a triose there would arise one molecule of carbon dioxide and two molecules of hydrogen, and one molecule of water would be used.

$$1 \text{ hexose} + 2 H_2O \rightarrow 2 CH_3COOH + 2 CO_2 + 4 H_2$$

but if the hydrogen is used to reduce carbon dioxide to methane, i.e.,

$$CO_2 + 4 H_2 \rightarrow CH_4 + 2 H_2O \tag{VI-1}$$

then

$$1 \text{ hexose} \rightarrow 2 CH_3COOH + CO_2 + CH_4. \tag{VI-2}$$

Note that 420 kcal of the 672 kcal per mole available through oxidation of glucose are stored in acetate.

Propionic acid is derived by a circuitous pathway from elements equivalent to triose plus two hydrogen atoms (Fig. VI-4), and water is formed.

$$1 \text{ hexose} + 2 H_2 \rightarrow 2 CH_3CH_2COOH + 2 H_2O \tag{VI-3}$$

The energy conserved in the propionate, 734 kcal per mole, is slightly greater than in the glucose, provided hydrogen is available.

Butyric acid is formed by the condensation of two molecules of acetic acid to form acetoacetic acid, which is then reduced to butyric acid with four atoms of hydrogen.

$$1 \text{ hexose} + 2 H_2O \rightarrow 2 CH_3COOH + 2 CO_2 + 8 H$$
$$2 CH_3COOH \rightarrow CH_3COCH_2COOH + H_2O$$
$$CH_3COCH_2COOH + 4 H \rightarrow CH_3CH_2CH_2COOH + H_2O$$

$$1 \text{ hexose} \rightarrow 1 CH_3CH_2CH_2COOH + 2 CO_2 + 2 H_2 \tag{VI-4}$$

The energy in the butyrate is 524 kcal. Each butyric acid molecule will be accompanied by two carbon dioxide molecules and two hydrogen molecules.

If the amounts of acetic, propionic, and butyric acids formed are known, the theoretical amounts of carbon dioxide and methane expected from the rumen fermentation can be calculated. It is evident that the production of acetic acid will lead to the greatest relative production of methane, since from a single sugar molecule converted to acetic acid, four molecules of hydrogen could arise. With butyric acid, only two molecules of hydrogen could be formed, and with propionic acid the equivalent of a molecule of hydrogen could be taken up and appear in the propionic acid. Relatively high propionic acid production would thus be associated with low methane production, whereas relatively high acetic and butyric acid formation would be accompanied by increased methane.

The theoretical stoichiometric relationship and overall equation to represent the fermentation of carbohydrate to the average ratios in which acetic, propionic, and butyric acids are produced, as well as carbon dioxide and methane, are given below.

$$31 \text{ hexose} + 62 \text{ H}_2\text{O} \rightarrow 62 \text{ CH}_3\text{COOH} + 62 \text{ CO}_2 + 124 \text{ H}_2$$
$$11 \text{ hexose} + 22 \text{ H}_2 \rightarrow 22 \text{ CH}_3\text{CH}_2\text{COOH} + 22 \text{ H}_2\text{O}$$
$$16 \text{ hexose} \rightarrow 16 \text{ CH}_3\text{CH}_2\text{CH}_2\text{COOH} + 32 \text{ CO}_2 + 32 \text{ H}_2$$

$$58 \text{ hexose} + 40 \text{ H}_2\text{O} \rightarrow 62 \text{ HAc} + 22 \text{ HPr} + 16 \text{ HBut} + 94 \text{ CO}_2 + 134 \text{ H}_2$$
$$134 \text{ H}_2 + 33\frac{1}{2} \text{ CO}_2 \rightarrow 33\frac{1}{2} \text{ CH}_4 + 67 \text{ H}_2\text{O}$$

$$58 \text{ hexose} \rightarrow 62 \text{ HAc} + 22 \text{ HPr} + 16 \text{ HBut} + 60\frac{1}{2} \text{ CO}_2 + 33\frac{1}{2} \text{ CH}_4 + 27 \text{ H}_2\text{O}$$

38,976 kcal 13,000 kcal   8070 kcal 8340 kcal   7030 kcal

$$\text{(VI-5)}$$

$$58 \text{ moles hexose} \rightarrow 100 \text{ moles acid} + 60\frac{1}{2} \text{ moles CO}_2 + 33\frac{1}{2} \text{ moles CH}_4 \quad \text{(VI-6)}$$

The moles of water are approximately equal to one-half the moles of hexose fermented, and the quantity of methane is roughly half the quantity of carbon dioxide. The exact theoretical proportions of products are, on a moles percent basis: 51.6% acid, 31.2% carbon dioxide and 17.2% methane.

The total heat of combustion of the products is 36,440 kcal, compared to 38,976 kcal in the hexose. The energy dissipated in the reaction, chiefly as heat, would be 2536 kcal, or 6.5% of the initial energy.

2. ACTUAL RATIOS OF PRODUCTS

Results of experiments measuring the proportions in which acid, carbon dioxide, and methane are actually formed are shown in Table VI-7.

In many of the experiments the proportions of products approximate the theoretical expected from the ratios of VFA's produced (Table VI-6). The results for other substrates differ markedly from the theoretical. The ratios of fermentation products expected will depend also on the extent of

MANOMETRICALLY DETERMINED PROPORTIONS OF FERMENTATION PRODUCTS FROM VARIOUS SUBSTRATES

Substrate	Acid		Carbon dioxide		Methane		Moles products per mole hexose
	μmoles	%	μmoles	%	μmoles	%	
40 mg glucose (225 μmoles)	228	59	119	36	18	5	1.62
40 mg glucose	229	67	79	27	18	6	1.45
100 mg glucose (852 μmoles)	723	64	357	31	61	6	2.29
38.8 mg cellulose	208	52	118	30	74	18	1.67
38.8 mg cellulose	240	54	136	30	74	16	2.0
62.5 mg cellulose	419	54	270	34	94	12	2.03
Rumen contents from Zebu and European cattle fed on Kikuyu grass hay[a]	—	37.8	—	42.6	—	19.6	—
Rumen contents from steers pastured on ladino clover[b]	—	34.1	—	51.1	—	14.8	—
Alfalfa meal	—	32.9	—	60.4	—	6.7	—
Water-extracted alfalfa meal	—	40.1	—	52.9	—	7.2	—
Water extract, by difference	—	23.7	—	71.0	—	6.3	—
Alfalfa holocellulose	—	40.5	—	47.3	—	12.2	—
Hay-grain ration[c]	—	55.3	—	31.5	—	13.2	—
Cellulose, dextrin, and glucose, 61 μmoles, as hexose[d]	63	59.5	33	31	10	9.5	1.74

[a] From Hungate et al. (1960).
[b] From Hungate et al. (1955).
[c] From Hungate et al. (1961).
[d] From Bowie. (1962).

microbial growth during fermentation and on the composition of the cells produced. These variations will be discussed in Chapter X.

The total quantity of fermentation products should bear a stoichiometric relationship to the substrate fermented, if a constant fraction of the substrate is converted into cell material. For the ratios of acids in Table VI-6, 1.7 moles of acid are formed per mole of hexose fermented. The actual ratio between total fermentation products and the weight of substrate fermented are shown in the last column of Table VI-7. D. A. Balch (1958) found 4.0 mmoles of acetic acid, 1.8 mmoles of propionic acid, and 1.65 mmoles of butyric and higher acids, a total of 7.45 mmoles produced per gram of feed fermented, i.e., about 1.27 mmoles per mole feed of an average molecular weight of 171, in experiments *in vitro* with rumen fluid from cattle. According to the theoretical stoichiometric relationships such as those shown above, these acids could have been formed as waste products from 4.55 mmoles of hexose, or 778 mg of carbohydrate with a molecular weight of 171. Twenty-two percent of the feed disappearing is not accounted for by the products. Much of this presumably became cell material. Louw *et al.* (1949) found approximately 100 mmoles of acid from 69 mmoles of carbohydrate cellulose, or 1.45 moles of acid per mole of carbohydrate fermented.

According to Eq. (VI-5), 0.57 moles of methane should appear for each mole of hexose fermented, or 0.52 moles of methane per mole of hexose, if 10% of the substrate is assimilated into cells. This compares with the 0.45 mmoles actually found (Table VI-1). The amount of methane is less than theoretical because the carbon in cells is, on the average, more reduced than in carbohydrate, and some of the substrates, e.g., the uronic acids, are more oxidized than carbohydrate. The decreased methane production on starch and sugar substrates (Table VI-1) may be due to a greater relative production of propionate, which correlates with decreased methane production (Hungate *et al.,* 1961), or lactate may have been formed.

D. Conversions of Certain Intermediates in the Rumen Fermentation

In Table II-9 the fermentation products of pure cultures of rumen bacteria grown *in vitro* were listed. These included carbon dioxide, methane, hydrogen, formate, acetate, propionate, butyrate, lactate, succinate, and ethanol. Of these materials, carbon dioxide, methane, acetate, propionate, and butyrate are final products. Hydrogen, ethanol, formate, lactate, and succinate do not ordinarily accumulate in the rumen to an appreciable extent, except that occasionally large quantities of lactate are found. If the bacteria produce in the rumen the same products that are formed in pure culture, the absence of hydrogen, ethanol, formate, lactate, and succinate can be accounted for by postulating that they are intermediates converted into other final products. The carbohydrate conversions assumed to occur in the rumen are shown in Figure VI-4.

1. HYDROGEN

Hydrogen is rapidly utilized by rumen contents incubated *in vitro* (Carroll and Hungate, 1955; Lewis, 1954; McNeill and Jacobson, 1955). There are two possible products of anaerobic hydrogen and carbon di-

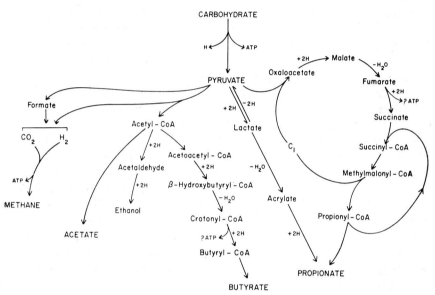

Fig. VI-4. Chart to show the pathways leading to the principal products of the carbohydrate fermentation in the rumen.

oxide conversion, namely, acetic acid and methane. Both are formed in the rumen fermentation. If they arise from hydrogen the reactions forming them exhibit the following stoichiometric relationships:

$$CO_2 + 4\ H_2 \rightarrow CH_4 + 2\ H_2O$$

and

$$2\ CO_2 + 4\ H_2 \rightarrow CH_3COOH + 2\ H_2O$$

The ratio of carbon dioxide to hydrogen consumed in the first reaction is 1:4, whereas in the second it is 1:2. Studies of the carbon dioxide-hydrogen utilization ratio in rumen contents incubated *in vitro* show that it approaches 1:4 (Carroll and Hungate, 1955). This makes it probable that the conversion of hydrogen into methane is the chief fate of hydrogen in the rumen. Bacteria accomplishing the conversion of hydrogen and carbon dioxide into acetic acid have not been found in significant numbers in the rumen.

If hydrogen is an intermediate in the rumen fermentation, it would be expected to be a constituent of rumen gas. Until recently, the literature on the occurrence of hydrogen in rumen gas has been quite contradictory. Methane and hydrogen were determined earlier by combustion of the two gases with oxygen and calculation of methane from the amount of carbon dioxide produced. Oxygen not accounted for on the basis of the carbon dioxide recovered was assumed to be used in the oxidation of hydrogen. The several measurements concerned involved enough error to make it uncertain whether the hydrogen calculated to be present was, in fact, there. Lugg (1938), in an extensive series of experiments over a long period of time, came to the conclusion that no hydrogen was contained in the respiratory (and eructated) gases given off by the ruminant.

The recent sensitive and specific techniques for chromatographic analysis of gases have permitted unequivocal demonstrations of hydrogen in the gas in the normal rumen. It is present in very small concentrations, of the order of magnitude of 0.05% or less, and would have been missed by former techniques unless the hydrogen occurred in abnormally large concentration, which is occasionally the case soon after feeding or after feeding starved animals (Pilgrim, 1948). Hydrogen concentrations up to 21% have been reported for the goat rumen (Kandatsu et al., 1955).

The usually low concentration of hydrogen indicates either that hydrogen is not a quantitatively important product of fermentation in the rumen or that it is rapidly converted to other materials. The fact that added hydrogen is rapidly converted to methane shows that the small pool of hydrogen turns over rapidly. The large production of hydrogen in fed, previously starved animals prior to reestablishment of methane bacteria is additional evidence that hydrogen is an important intermediate. When rumen methanogenesis is inhibited with metabolic analogs, hydrogen gas accumulates. Its rate of accumulation is not sufficient to account for the quantity of methane formed, which suggests either (1) that substances other than hydrogen gas supply hydrogen atoms to form methane or (2) that the analog partially inhibits hydrogen production.

The occurrence of hydrogen as a rumen fermentation product can explain the low oxidation-reduction potential of rumen contents, although other systems undoubtedly are also effective. Failure to observe a measurable increase in methane production when hydrogen was sparged into the rumen (Nelson et al., 1960) can be explained by the difficulty in dispersing the gas through the rumen contents (Bauchop and Hungate, unpublished experiments).

The restriction of *Methanobacterium ruminantium* to formate, hydrogen, and carbon dioxide as substrates for methane production (P. H. Smith and Hungate, 1958) points toward hydrogen as an important precursor of methane in the rumen. Hydrogen turnover cannot be measured by adding deuterium or tritium gas to *in vitro* incubated rumen contents because of rapid exchange with water (Bauchop and Hungate, unpublished experiments). The methane formed has the same specific activity as the water.

2. FORMATE

Some investigators have found formate in the rumen contents (Gray *et al.,* 1952), whereas others did not detect it (Matsumoto, 1961). This can be explained by the time of sampling in relation to feeding and the sensitivity of the analytical method. Alfalfa hay, and presumably grass hays, often contain formate (Claren, 1942; Matsumoto, 1961), and, if a large quantity of forage has been recently consumed, the formate in it will not have been completely converted.

Formate produced in the rumen is rapidly converted. The pool size is only 0.01–0.02 μmoles per milliliter or less, except after feeding. Formate added to rumen contents rapidly disappears (Claren, 1942; Jacobson *et al.,* 1942; Beijer, 1952; Carroll and Hungate, 1955) and can be recovered as methane and carbon dioxide. The rate of added formate decomposition in the bovine rumen was 0.628 ± 0.33 μmoles per gram minute (Carroll and Hungate, 1955) and 0.547 ± 0.103 μmoles per gram minute in a goat (Matsumoto, 1961). This, plus the normal turnover without added formate, is the maximum *capacity* for formate metabolism.

Conversion of formate to hydrogen and carbon dioxide has been demonstrated by preparing a washed suspension of rumen bacteria in a Warburg vessel and adding formate (Doetsch *et al.,* 1953). Equal quantities of hydrogen and carbon dioxide were produced. A further conversion of hydrogen and carbon dioxide to methane did not occur in these experiments because the bacterial suspension was not sufficiently anaerobic. In other experiments (Beijer, 1952; Carroll and Hungate, 1955), when formate was added to incubated rumen contents, the methane expected on the basis of the stoichiometric relationships

$$4 \text{ HCOOH} \rightarrow 4 \text{ CO}_2 + 4 \text{ H}_2$$

and

$$\text{CO}_2 + 4 \text{ H}_2 \rightarrow \text{CH}_4 + 2 \text{ H}_2\text{O}$$

was formed. When formate-C^{14} is added to incubated rumen contents, the resulting methane is not labeled, which shows that there is not a direct reduction of the formate to methane (Carroll and Hungate, 1955). These lines of evidence point toward the conversion of formate to hydrogen and carbon dioxide and their subsequent conversion to methane, although the possibility that the hydrogen atoms in formate are in part transferred directly to form methane has not been experimentally refuted.

The labeling of methane ultimately at least equals the labeling of eructated carbon dioxide when carbonate-C^{14} is infused into the rumen (W. F. Williams *et al.*, 1963) of fed animals. This equivalence was reached only after 2 hours and was never reached in fasted animals.

The first-order turnover rate constant, $k,$ for disappearance of formate-C^{14} added to *in vitro* incubated rumen contents is about 15 to 25 per minute in the initial 8 seconds. Coupled with a pool size of 0.03–0.05 μmoles per milliliter, this gives a formate turnover of 0.45–1.25 μmoles per gram minute, compared to a methane production of 0.13 μmoles per gram minute. The hydrogen in formate thus accounts for more hydrogen than is found in methane. It is also more than would be expected on the basis of the acetic and butyric acid formed. These anomalous results may be due to exchange reactions of formate with carbon dioxide or other substances, which would cause the turnover rate constant to include more than a net conversion of formate to carbon dioxide and hydrogen.

Experiments to compare methane production rate with the dissolved hydrogen pool in rumen contents, using infused formate as a source of hydrogen, show a linearity between hydrogen concentration at low methane production rates, but a falling off at high rates, with formate presumably being in part directly dehydrogenated by carriers linked to methane production rather than hydrogen formation (Bauchop and Hungate, unpublished experiments).

The reported occurrence of carbon monoxide in rumen gas (Olson, 1940) is of interest because it could be an intermediate in methane production. A carbon compound two hydrogens more reduced than carbon dioxide could be expected as an intermediate, and could be either formate or carbon monoxide, or, more likely, a single carbon bound to an enzymatic complex.

3. SUCCINATE

The discovery by Johns (1951a) that *Veillonella gazogenes* occurred in large numbers in the rumen of sheep, and its effectiveness in decarboxylating succinate to propionic acid and carbon dioxide (Johns, 1948, 1951b), focused attention on the possible importance of succinate in propionate formation in the rumen. Succinate is formed by a number of important rumen bacteria, *Succinivibrio, Bacteroides succinogenes,* and other *Bacteroides* species, but does not appear as an important end product of the rumen fermentation. Succinate added to rumen contents is converted to propionate and carbon dioxide (Sijpesteijn and Elsden, 1952), and the succinate arising as an intermediate in the normal fermentation presumably also undergoes this change. Washed suspensions of rumen bacteria produce carbon dioxide from succinate (Doetsch *et al.,* 1953) on a mole for mole basis, with propionate presumably the other product. The bacteria effecting the decarboxylation have not been identified. *Veillonella* is not sufficiently abundant to account for the rate of decarboxylation of succinate in the rumen (Ogimoto and Suto, 1963).

The turnover of succinate in the rumen has been measured (Blackburn and Hungate, 1963). The turnover rate constant is 10 per minute, and the pool size 0.004 μmole per milliliter, which gives a turnover of 0.04 μmoles per minute, compared to a propionate production rate of 0.12 μmoles per milliliter per minute. This indicates that at least one-third of the rumen propionate arises from succinate produced by one cell and decarboxylated to propionate and carbon dioxide by another.

4. LACTATE

For many years the chief known microbial conversion of lactate was to acetic and propionic acids and carbon dioxide. Lactate has often been postulated as the precursor of most of the propionic acid in the rumen. Occasionally propionibacteria are found in large numbers in rumen contents (Gutierrez, 1953), but these appear, at least occasionally, to have entered the rumen from the feed on which they can occur in unbelievably large numbers. Added lactate is not fermented rapidly by rumen contents (Jayasuriya and Hungate, 1959; Baldwin *et al.,* 1962).

Determinations of lactate turnover in rumen contents freshly removed from hay-fed animals fail to disclose that it is an important intermediate. The pool size of L-(+)-lactate in hay-fed cattle is small (Jayasuriya and Hungate, 1959), usually below the level (0.12 μmoles per milliliter) easily detectable by the available analytical methods, and the turnover rate is not large, 0.03 per minute. Pool sizes of 0.17–0.44 μmoles of lactate per milliliter have been reported (Baaij, 1959) in cattle. The turnover of DL-lactate-2-C^{14} is inadequate to account for the amount of propionic acid formed (Jayasuriya and Hungate, 1959). It is highly probable that succinate rather than lactate is the chief precursor of the propionic acid.

The major part of added lactate-C^{14} is converted to acetate (Jayasuriya and Hungate, 1959; Baldwin *et al.,* 1962; Bruno and Moore, 1962), although infusion of large quantities of lactate into the rumen of dairy cows (Montgomery *et al.,* 1963) caused an increase in the relative proportion of butyrate, and feeding of lactate increased the proportion of propionate (Ekern and Reid, 1963).

Acrylate is intermediate in the rumen conversion of lactate to propionate (Baldwin *et al.,* 1963), and relatively little rumen propionate usually arises by this route in hay-fed cattle (Satter *et al.,* 1964; Jayasuriya and Hungate, 1959). The 4-carbon dicarboxylic acids are intermediates in the rumen formation of propionate from glucose (Baldwin *et al.,* 1963). In hay-fed animals this accounts for 92% of the propionate formed (Satter *et al.,* 1964), but in animals receiving large amounts of grain, as much as 23% of the porpionate arises via the acrylate pathway.

Certain rations containing large proportions of flaked maize or barley favor the production of unusually large proportions of propionic acid (Phillipson and McAnally, 1942; Phillipson, 1952a), and lactate sometimes accumulates in large quantities (D. A. Balch and Rowland, 1957), particularly during the initial period of adaptation to the feed. The factors concerned in this distinctly different balance of products have not been distinguished. Continued feeding of the ration may result after 5 or 6 weeks in a return to the ratios found before the high starch ration is fed (Jacobson and Lindahl, 1955).

There are numerous indications in the literature that lactic acid accumulates when sugars and other readily fermented feeds are added to rumen contents *in vivo* (Phillipson and McAnally, 1942; Waldo and Schultz, 1955, 1956; Hungate *et al.,* 1952) and *in vitro* (Bruno and Moore, 1962; Walker and Forrest, 1964). Accumulation of lactate is evidence against the normal occurrence of lactate as an important intermediate since, if it were important, the microbial population would have a high capacity for metabolizing it. Accumulation would not be expected.

Additional work is needed before the importance of lactate as an intermediate can be fully evaluated. The relatively high culture counts (Gilchrist and Kistner, 1962) on lactate substrates incubated for 5 days might indicate a relative abundance of intermediate lactate in the rumen, but the seeming prevalance of lactate utilizers in these studies may not be real. The low counts in glucose and starch may have been due to the short incubation time of 16 hours.

5. ETHANOL

Of the various products of the pure cultures of rumen bacteria, the most difficult to explain, insofar as its putative production and further conversion are concerned, is ethanol. It is formed in pure cultures of a number of species of rumen bacteria, but does not occur in the normal rumen in a significant concentration, and the turnover rate (0.003 per minute) is not sufficient to make it important (Moomaw and Hungate, 1963). Ethanol added to the rumen exerts little effect on the oxygen consumption and methane production (Emery *et al.,* 1959). If sufficient quantities are fed it may increase the concentration of acetate in the rumen and could serve as a minor source of energy (Chalupa *et al.,* 1964; Leroy, 1961). Most of the ethanol is absorbed and metabolized in the ruminant, as in nonruminants (Leroy, 1961).

Of 888 gm per day administered to cattle, about 3% was recovered as ethanol in the respired gas (Zuntz, 1913). Ethanol was not rapidly converted in the rumen of sheep fed rations containing ethanol, nor was it rapidly attacked *in vitro* (Leroy, 1958; Emery *et al.,* 1959). Alcohol slowly disappeared from the rumen, presumably by absorption and passage to the omasum. These results show that ethanol is not an important intermediate in the usual rumen fermentation.

An alcoholic fermentation occasionally occurs in lambs prior to roughage feeding (Cunningham and Brisson, 1955) if large quantities of glucose are included in the diet. The acid condition in the rumen of the young animal favors alcoholic fermentation. Alcohol also accumulates in the acid rumen contents of animals overfeeding on grain (Allison *et al.,* 1964a).

6. Explanation for Relative Unimportance of Ethanol and Lactate

The explanation for the discrepancy between the appearance of ethanol as an important product of pure cultures and its usual absence in the rumen is uncertain. The greater acidity which commonly develops during the later stages of batch pure cultures may cause the increased production of alcohol. Excess soluble sugar may favor the formation of lactate. Another possibility is that, in the competition to achieve maximum growth, selection is against lactic acid and ethanol production. Formation of each of these products entails loss of available ATP. Conversion of pyruvate to lactate sacrifices the ATP available in the conversion of pyruvate to acetyl CoA, and the reduction of acetyl CoA to ethanol similarly entails loss of a potential ATP. The acetyl CoA can yield an ATP unless it must be used for hydrogen disposal. Methane formation is a means of hydrogen disposal in the rumen which increases rather than decreases the ATP available. The significance of methanogenesis may lie in the removal of hydrogen which would otherwise reduce molecules available to yield ATP. This is a possible mechanism by which the mixed rumen fermentation yields a greater crop of cells than would the separate pure cultures.

Propionibacterium shows a greater yield of cells from glucose than is expected on the basis of the Embden-Meyerhof-Parnas scheme (Bauchop and Elsden, 1960). The cytochrome-linked fumaric reductase in *Bacteroides ruminicola* (White *et al.,* 1962a,b) is a possible source of additional ATP in succinate formation; if so, the importance of succinate as a rumen intermediate might be explained. According to this view, the propionate formed in the rumen represents an additional synthesis of ATP. This is also true for acetate on the hypothesis that pyruvate is split to acetyl CoA. A production of ATP in butyrate formation has been surmised, but efforts to demonstrate it have been unsuccessful. If there is a selection for maximum biochemical work, butyrate should also represent an end product accompanying additional energy conservation by the cell.

According to these views, the carbon dioxide, methane, acetate, propionate, and butyrate, final products in the rumen fermentation, are formed because pathways leading to them also provide the most efficient conversion of fermentable substrate into microbial cells.

CHAPTER VII

Conversions of Nitrogenous Materials

Thus far we have dealt almost exclusively with rumen conversions of carbohydrate and other nitrogen-free derivatives of ruminant feeds. The carbon- and hydrogen-containing waste products of the microbial fermentation, energy-yielding materials for the host, were described, and the quantities produced were estimated.

We are accustomed to considering as fermentation products the various acids, gases, and other substances given off into the medium, and they do constitute waste fermentation products. However, cell material is the most interesting product, biologically, of the rearrangements of food during fermentation. The success of a microbe is measured by the quantity of food material and energy conserved in the form of cells. The amounts of acids, gases, and other excreted wastes are an inevitable accompaniment of cell production under anaerobic conditions. The microbial cells are usually more complex chemically than is the average for ruminant food, and on an equal weight basis usually contain as much or more energy than carbohydrate.

As discussed in Chapter VI, the high-energy phosphate intermediates in metabolism conserve some of the substrate energy during the substrate-level conversions. Their participation in the further spontaneous reactions leading to cell material makes them extremely important metabolic intermediates in growth. In addition, the pyruvate, α-ketoglutarate, oxalacetate, and other intermediates in the conversions of carbohydrates are used in cell synthesis.

The bodies of the microorganisms and the host contain other materials than the carbon, hydrogen, and oxygen of carbohydrate. The most abundant of these is nitrogen.

A. Proteins

Proteins are the most common nitrogenous materials in forages; in nature proteins provide the major part of the nitrogen used by rumen microbes and the host. Plant proteins are partly soluble, but there are also

281

relatively insoluble proteins (Synge, 1957), some associated with the plant fiber (Ter-Karapetjan and Ogandzanjan, 1960a).

The first step in protein utilization is digestion. In theory, it would seem simplest and most profitable for the rumen microorganisms to digest the food proteins and assimilate them directly into microbial protein, using the high-energy phosphate available from fermentation. The amino acids of the food would become the amino acids of the microbe. This undoubtedly occurs, but amino acids may be used also in other ways. Some anaerobic microorganisms use amino acids to yield high-energy phosphate. The nitrogen is removed, and the carbon- and hydrogen-containing residues are rearranged in much the same fashion that carbohydrates are fermented. Some amino acids, e.g., arginine, can yield high-energy phosphate in the process of splitting off nitrogen.

When protein is fermented to provide energy, ammonia is formed. The ammonia may in turn be assimilated again into amino acids, but it cannot all be assimilated unless carbohydrate is available. The relative importance of proteins as energy or as monomers for cell synthesis depends on the ratio of carbohydrate to nitrogen in the rumen. Of the numerous pathways of protein utilization possible in the rumen, the simplest are shown diagrammatically in Fig. VII-1.

1. DIGESTION

Regardless of the ultimate use of the protein, digestion is the first step in utilization. The problem of following the digestion of feed proteins is

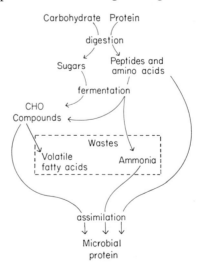

Fig. VII-1. Diagram to show some of the carbohydrate and nitrogen interrelationships in the rumen.

complicated by the fact that, as feed protein is digested, microbial protein is synthesized, and the rates of the two processes are not easy to measure separately.

Purified proteins have been added to feeds or inserted directly into the rumen, and the rates of their disappearance have been followed. The first studies along this line were those of McDonald (1948a,b,c, 1952), who found that casein and gelatin disappeared rapidly. Zein was more resistant. Direct analysis for zein in the abomasum indicated that 40% of it had been digested in the rumen. A value of 50% for the degree of digestion of zein was calculated from the percentage of lysine (3.9–5.7%) in the abomasal protein as compared to that in the rumen microorganisms (7% lysine) and in zein (no lysine).

In later experiments (McDonald and Hall, 1957) the phosphorus split from casein was used as a measure of the amount of casein in abomasal contents. The highest percentage of casein in abomasal contents (47% of the initial rumen concentration) was found 2 hours after feeding. At other times the amount of casein in the abomasal contents was negligible. On a 24-hour basis it was estimated that 97% of the casein was digested before it reached the abomasum. In these experiments, casein nitrogen constituted 87% of the total nitrogen of the food.

Rumen microorganisms from sheep fed on casein, hay, and groundnut meal attack casein, arachin (groundnut protein), and soya protein rapidly, but bovine albumen, wheat gluten, and zein only slowly (Annison, 1956). Similarly, proteins differ in the speed with which they give rise to ammonia in the rumen (Lewis, 1962), presumably due to differences in their rates of digestion. Casein, gelatin, and arachin were digested most rapidly, soya protein and wheat gluten were intermediate, and bovine albumen and zein were least digested.

Relatively few studies have determined directly the rate at which forage protein disappears from the rumen, chiefly because of the previously mentioned difficulty in distinguishing between the plant, bacterial, and protozoal protein. Bacterial protein has been distinguished from plant and protozoal protein (Weller *et al.,* 1958) by measuring the amount of 2,6-diaminopimelic acid (Work, 1950, 1951). This component of the cell wall of most bacteria and blue-green algae does not occur in animals and higher plants (Synge, 1952b, 1953). Two hours after feeding chopped wheaten hay, 27% of the rumen nitrogen was in the form of plant nitrogen, 20% after 5 and 7 hours, 16% after 10 hours, and 11% at 16 and 24 hours (Weller *et al.,* 1958). Presumably most of the initial plant nitrogen was in the form of protein. On the assumption that the 11% of plant nitrogen undigested after 24 hours was not protein, digestion of the protein was complete within 16 hours after feeding. In a later set of experiments (Weller *et al.*, 1962), 80% of the plant nitrogen in the rumen was converted to microbial nitrogen, and it was shown that only plant materials with nitrogen digested to this extent escaped into the omasum. The conversion of plant to microbial nitrogen was very rapid. With a ration of wheaten hay-lucerne hay (650 gm–150 gm) containing 1.4% total nitrogen, the plant nitrogen decrease 6 hours after feeding was 7 gm, and the microbial nitrogen increase was 6.5 gm. With lucerne hay (2.9% nitrogen), the decrease in plant nitrogen was 17 gm, and the microbial nitrogen increase was 7 gm, a difference reflecting the wastage of the excess nitrogen in the legume.

Most of the enzymes responsible for the digestion of protein in the rumen have not been identified and purified. The proteolytic action of the rumen microorganisms was first studied (Sym, 1938) by following the disappearance of casein from rumen contents incubated in the presence of toluene. In 3 hours, 44% of the casein (0.5% w/v) was no longer acid precipitable. Measurements of amino nitrogen indicated splitting of one-third of the peptide bonds. Little proteinase was in the rumen fluid, but considerable activity was within the cells and in an extract of an acetone powder of the cells.

The disappearance of casein from *in vitro* cultures has been used to estimate proteolytic activity in the rumen (Blackburn and Hobson, 1960a). Casein was incubated with 2 ml of rumen contents from a sheep receiving casein as part of the ration. With 100 mg of casein in a total of 10 ml of culture medium, 48 to 99% of the casein was hydrolyzed in 48 hours. A sheep receiving no casein, but equivalent quantities of protein in other forms, showed equally rapid casein dissimilation *in vitro*. This also oc-

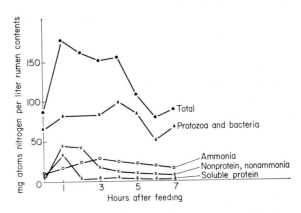

Fig. VII-2. Distribution of nitrogen in the rumen of a sheep at various times after feeding. From Blackburn and Hobson (1960c, p. 293).

curs *in vivo* (Fig. VII-2) and has been the general experience (Annison and Lewis, 1959; Warner, 1956b; Blackburn and Hobson, 1960a); the rate of digestion of a particular protein by rumen contents is about the same, regardless of the ration. In other experiments, the soluble protein and soluble nonprotein nitrogen in the rumen were maximal 1 hour after feeding and diminished as the ammonia nitrogen increased, the ammonia nitrogen becoming maximal after 2 hours (W. E. C. Moore and King, 1958).

The rate of digestion by rumen bacteria of various proteins has been correlated (el-Shazly, 1958; Blackburn and Hobson, 1960d; Henderickx and Martin, 1963) with the solubility of the protein in salt solutions. These results are collected in Table VII-1 and Fig. VII-3.

From a 10^5 dilution of rumen contents, a proteolytic euryoxic bacterium was isolated (W. G. Hunt and Moore, 1958) and its proteinase studied. The purified proteinase differed from trypsin and chymotrypsin in its susceptibility to various inhibitors. The optimum pH was 7.5, more alkaline than the optimum obtained in experiments with whole rumen contents (Sym, 1938; Annison, 1956). The significance of the enzyme in the rumen was not established. There is a peptidase in the omasal contents (Blaizot and Raynaud, 1958).

Relatively few rumen bacteria are unable to use carbohydrate and have to depend solely on protein as a source of energy (Blackburn and Hobson, 1960b, 1962). Instead, many saccharoclastic isolates produce a little proteinase, which, although not formed in large quantities per bacterium, is significantly large because of the abundance of the bacteria. Digestion of casein after 4 days of incubation (Abou Akkada and Blackburn, 1963) was 30–94% complete for *Bacteroides amylophilus,* 28–97% for *Bacteroides ruminicola,* 26–87% for *Butyrivibrio,* and 12–96% for *Selenomonas ruminantium.* Caseinolytic strains constituted only a small part of the total colonies developing. In other experiments, (Fulghum and Moore, 1963) with a skim milk medium, there were 10^9 caseinolytic bacteria per milliliter of rumen contents from cattle fed hay and grain at a maintenance level. The proteolytic colonies constituted 38% of the total colony count and included strains of *Butyrivibrio, Succinivibrio, Selenomonas lactilytica, Borrelia, Bacteriodes* sp. and selemononad-like bacteria similar to the B385 of Bryant (1956). Amino acids and polypeptides were demonstrated as products of casein digestion by pure cultures of rumen bacteria (Abou Akkada and Blackburn, 1963).

The most actively proteolytic rumen bacterium encountered by the author was *Clostridium lochheadii.* In spite of this capacity to digest protein, it can only occasionally be significant in the rumen since its occurrence in large numbers is rare.

Table VII-1

SOLUBILITY AND DIGESTION OF PROTEINS[a]

Proteins	Digestion (%)	Solubility[b]		Proteins	Digestion (%)	Solubility[b]	
		Mineral solution	Water			Mineral solution	Water
Gelatin F[c]	84	160	2.674	Fibrin F	22	44	32
Meat peptone F	82	160	3.169	Soya gluten	19	36	53
Casein F	80	140	6	Ovolivetin fraction	19	24	10
Casein	72	120	6	Collagen F	19	25	9
Hemoglobin	67	112	56	Keratin F	18	36	26
Casein peptone F	64	106	5.991	Glutenin F	17	48	57
Serum albumin F	60	98	414	Ovomucin	9	14	47
Soya albumin 1	58	96	8	Rumen protein	4	7	3
Ovo albumin F	48	83	19	Serum-β-globulin	4	4	6
Gliadin F	34	58	232	Alfalfa protein	4	5	16
Soya albumin 2	33	50	6	Soya globulin	4	6	2
Gluten F	33	48	26	Edestin F	4	4	3
				Elastin F	3	2	5

[a] From Henderickx and Martin (1963).

[b] Solubility in mg N/100 ml.

[c] The F proteins were obtained from commercial sources, the others isolated in the laboratory.

2. ASSIMILATION OF AMINO ACIDS

There is relatively little information on the extent to which amino acids in the rumen are directly assimilated by rumen microorganisms, although a good deal of information on asssimilation by pure cultures has been obtained (Table II-10). Bacteria capable of assimilating at least some amino acids (Bryant and Robinson, 1962a,b) include *Streptococcus bovis, Bacteroides succinogenes, Bacteroides ruminicola, Eubacterium, Selenomonas, Butyrivibrio fibrisolvens, Peptostreptococcus elsdenii, Succinivibrio dextrinosolvens,* and *Lactobacillus.*

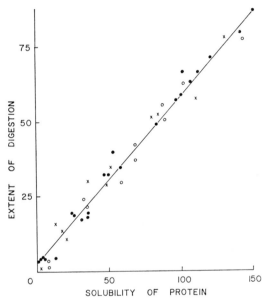

Fig. VII-3. The relationship between solubility of a protein and the extent of its digestion in the rumen. From Henderickx and Martin (1963).

For at least some strains of *Bacteroides ruminicola, Butyrivibrio fibrisolvens, Peptostreptococcus elsdenii,* and *Succinivibrio,* the amino acids of casein hydrolyzate were essential for growth. This suggests that some direct assimilation of amino acids occurs in the rumen. Experiments with C^{14}-labeled glutamate (Otagaki *et al.,* 1955) indicate that some amino acids were directly assimilated, and experiments of Portugal (1963) indicate that a small fraction of the C^{14} in trace quantities of added labeled glutamate or aspartate is assimilated as glutamate or aspartate, respectively.

Amino nitrogen in the rumen varies from 9 μmoles per 100 ml of rumen contents to more than 1000 μmoles (Lewis, 1955; Gutowski, 1960; Gutowski *et al.*, 1960). Prior to feeding, rumen contents contained 28–215 μmoles of α-amino nitrogen per 100 ml of rumen contents, much of it within the bodies of the bacteria, and 700–1070 μmoles after feeding dried grass (Annison, 1956). Amino acids have been demonstrated in the bodies of the protozoa (Spisni, 1952) by means of histochemical tests. Amounts of amino acids at various times after introducing gelatin into the rumen (Lewis, 1962) are shown in Fig. VII-4.

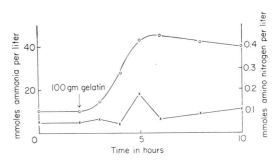

Fig. VII-4. Concentration of ammonia nitrogen (O—O) and amino nitrogen (×—×) in the rumen of a sheep at various times after dosing with 100 gm of gelatin. From D. Lewis (1962, p. 78).

In these experiments, digestion of the protein to amino acids was evidently slower than the fermentation of the acids to ammonia, since amino acids did not accumulate, except briefly at 5 hours. In addition, in sheep on lush spring grass pasture, only transient increases in the α-amino nitrogen of the rumen were found (Annison *et al.*, 1959a). The possibility that amino acids can be absorbed through the rumen wall has been demonstrated (Demaux *et al.*, 1961), but in view of the low concentration in the rumen it is doubtful that rumen absorption of amino acids is quantitatively important.

Although the amino acids formed during the digestion of protein do not accumulate to high concentrations in the rumen contents, traces of free amino acids can be demonstrated by paper chromatography. The kinds found by various investigators are shown in Table VII-2. Glutamic acid, aspartic acid, alanine, glycine, proline, serine, threonine, arginine, tyrosine, valine, leucine, phenylalanine, isoleucine, methionine, and cystine occur at least occasionally in rumen contents, with glutamic acid, aspartic acid, alanine, and proline usually present. Rumen proteins contain no homocysteine, but it occurs in small concentrations in the total

Table VII-2

FREE AMINO ACIDS DETECTED IN RUMEN CONTENTS

Amino acid	Blaizot and Raynaud (1957)	Amison (1956)	Gutowski (1960); Gutowski et al. (1958a,b,c)	Ter-Karapetjan and Ogandzanjan (1959)	D. Lewis (1955)	Lacoste-Bastié (1963)[a]
Glutamic acid	+	+	+	++	+	+
Aspartic acid	+	+	+	+		+
Alanine	+	+	+	+	+++	++++
Glycine			+		++	++++
Proline		+	+	+		
Phenylalanine			+			++
Isoleucine		+				
Methionine			+	++	+	+
Cystine			+			
Serine	++	+	+		+	+++
Threonine	++	+	+	+		+
Arginine		+		+	+	+
Tyrosine	++		+	+		+
Valine	++	+	+	+	+	+
Leucine	+	+	+	++	+	+
Histidine					+	+
Lysine				+	+	+
α-Aminobutyric				+		+
Tryptophan				+		
γ-Aminobutyric						+

[a] These analyses were on acetone–dried rumen bacterial cells.

contents (Henderickx 1961a), probably as an intracellular intermediate in synthesis of methionine.

Rumen amino acid determinations include both the amino acids free within the cells and in the fluid around the cells. Lacoste-Bastié (1963) has demonstated in the bacterial cells most of the amino acids reported for the total contents. Since many bacteria concentrate amino acids, the extracellular amino acid concentration may be much smaller than the analyses of total contents suggest.

In cell-free rumen fluid Portugal (1963) found, in μmoles per milliliter: 0.028 glutamic acid, 0.022 aspartic acid, 0.021 valine, 0.025 alanine, 0.008 glycine, 0.007 proline, 0.006 threonine, and 0.011 serine. The turnover rate of glutamic acid was 0.53 per minute, which, with the 0.0275 pool size, gave a total turnover of 84 mmoles of glutamic acid per day in the 4-liter sheep rumen. The quantity of glutamate entering with the feed was 80 mmoles per day, with possibly 3 mmoles entering with the saliva—values in good agreement with the measured turnover. The turnover rate of aspartate was 0.76 per minute, which gives a daily turnover of 98 mmoles of aspartate in the rumen as compared to an aspartate content of 88 mmoles in the feed.

Approximately 10% of the glutamate was recovered in rumen cell material and about 6% of the aspartate. This represents the extent to which the pulse-administered traces of C^{14}-labeled amino acids were assimilated. These results of Portugal (1963) are among the most precise which have been obtained for rumen phenomena, due partly to the analytical skill and imaginative experiments of the investigator, partly to the use of sheep which were continuously fed, and partly to the greater precision resulting from the method of Sutherland et al. (1962) for thorough mixing of the rumen contents prior to sampling.

3. FERMENTATION OF AMINO ACIDS

Bacteria from many habitats can live anaerobically on proteins and their split products; the bacteria ferment mixtures of amino acids to form cell material, ammonia, carbon dioxide, and acids. Such fermentations are usually characterized by offensive odors due to the various reduced compounds of sulfur and nitrogen liberated from amino acids when the latter are used for energy. The odor of rumen contents suggests that some protein fermentation occurs, but the odors are offset in part by the odor of the volatile fatty acids (VFA's).

a. Single Amino Acids

A few rumen bacterial species are able to ferment single amino acids. Elsden et al. (1956) found that LC (Peptostreptococcus elsdenii) fermented L-serine to carbon dioxide, ammonia, and acetate, and fermented L-threonine to carbon dioxide, ammonia, propionate, and small amounts of acetate and valerate. Pyruvate and α-ketobutyrate were intermediates in the latter fermentation (Lewis and Elsden, 1955). Acrylate was reduced to propionate and oxidized to carbon dioxide and acetate. These same results were found in a rumen fermentation of L-threonine and L-serine (van den Hende *et al.*, 1963b). *Peptostreptococcus* contains a threonine deaminase (D. J. Walker, 1958), and arginine is also metabolized by this organism (Hinkson *et al.*, 1964).

Two strains of *Escherichia coli,* from enrichments inoculated with rumen contents, together fermented lysine to ammonia, carbon dioxide, acetic acid, and butyric acid (Dohner and Cardon, 1954). Their significance in the rumen is not clear, since numbers were not determined. Strains of other euryoxic bacteria attacking amino acids have been found in the rumen (Lacoste-Bastié, 1963), but they were not abundant.

Experiments with washed cell suspensions of mixed rumen bacteria exposed to single amino acids (Sirotnak *et al.*, 1954; Lewis, 1955) showed that aspartic and glutamic acids are rapidly attacked. In other experiments, aspartate, cysteine, alanine, and threonine were rapidly dissimilated (Lewis, 1955), and slight production of ammonia was noted with each amino acid tested. In a later test (T. R. Lewis and Emery, 1962a) serine, cysteine, aspartic acid, threonine, and arginine were deaminated most completely, followed by glutamic acid, phenylalanine, lysine, and cysteine. Only slightly deaminated were tryptophane, δ-aminovaleric acid, methionine, alanine, valine, isoleucine, ornithine, histidine, glycine, proline, and hydroxyproline.

In Portugal's (1963) studies with glutamate-U-C^{14} almost 50% of the activity was recovered in VFA's and carbon dioxide; with aspartate the value was 55%. This is much greater than the 6–10% assimilated into cells, which indicates that these amino acids were extensively fermented. The sheep with which these experiments were performed were fed on a grass cube ration containing 2.62% nitrogen.

In vivo experiments have also been performed (Lewis, 1955), with single amino acids added directly to the rumen of fistulated animals. The increase in the ammonia level has been used as an index of attack. With 100 mg of added amino nitrogen per kilogram live weight of hay-fed sheep, the greatest increase in ammonia nitrogen was obtained with DL-glutamic acid. In decreasing order, glycine, DL-alanine, and DL-methionine were attacked. For a sheep fed hay and casein hydrolyzate, the order was DL-alanine > DL-methionine > DL-glutamic acid > glycine. With hay and grass, the order was DL-glutamic acid > DL- methionine > glycine > DL-alanine. When washed suspensions of rumen bacteria from these ani-

292   VII. CONVERSIONS OF NITROGENOUS MATERIALS

mals were tested, the observed rates were almost nil. A glutamic acid supplement to a ration containing chiefly urea gave no detectable improvement in ruminant nutrition (Oltjen *et al.*, 1964).

The effects of various amounts of ration protein on the ammonia in the rumen and the urea in the blood are illustrated in Fig. VII-5. A sheep on a hay ration with added protein concentrate showed more rapid production of ruminal ammonia when single amino acids were added than did an animal on a hay or hay-grass ration (Lewis, 1955).

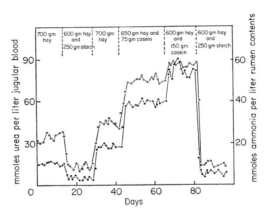

Fig. VII-5. Effects of various amounts of ration protein on the ammonia concentration in sheep rumen contents (●—●) and the urea concentration in blood (○—○). From D. Lewis (1957, p. 442).

The increased amino acid fermentation when protein concentrates are fed has been noted by several investigators (Butz *et al.*, 1958b) and contrasts with the failure of protein supplements to increase the concentration of rumen proteinase (see p. 285). It has been suggested (Synge, 1957) that there are always proteinases in the rumen, i.e., they are constitutive, needed for rapid hydrolysis of forage proteins. The amino acids formed are assimilated unless in excess. Accumulation of amino acids elicits adaptive enzyme formation, which accounts for the differences in deaminative activity. According to this view, most of the amino acids would be assimilated when carbohydrate is in excess.

Inclusion of additional amino acids in the substrate increases the rate of their fermentation, presumably by increasing the pertinent enzyme concentrations. Increased production of tryptophanase by rumen bacteria in media containing tryptophan has been observed (Lacoste-Bastié, 1963). The increase was due in these experiments to selection of strains rather than to induction of the enzyme. When digestible proteins are supplied, a large number of split products, including amino acids, are simultaneously available, and fermentation is quite rapid (Reis and Reid, 1959), as mentioned earlier in connection with estimates of protein digestion.

Aspartase is the enzyme concerned in the deamination of aspartic acid by washed cell suspensions, as evidenced by the pH optimum, inhibitors, and Michaelis constant (van den Hende *et al.,* 1959; Hoshino and Hirose, 1963). After deamination, aspartate (Sirotnak *et al.,* 1954) is converted initially to propionic acid, later acetic acid, and finally butyric acid. In addition, hydrogen and valeric acids are formed (Lewis, 1955).

Malate, oxalacetate, fumarate, succinate, and pyruvate are fermented to carbon dioxide, propionic acid, and acetic acid, the proportion of the acids depending on the oxidation level of the substrate (van den Hende *et al.,* 1959). Results of action of washed cell suspensions of rumen bacteria on aspartate and possible intermediates in its conversion are shown in Table VII-3 (Henderickx and Martin, 1963).

Table VII-3

THE FERMENTATION OF L-ASPARTIC ACID AND POSSIBLE INTERMEDIATE FOUR-CARBON DICARBOXYLIC ACIDS AND PYRUVIC ACID BY WASHED SUSPENSIONS OF RUMEN BACTERIA[a,b]

Substrate	Ammonia	Carbon dioxide	Total VFA	Butyric acid	Propionic acid	Acetic acid
Experiment 1						
L-Aspartate	100	100	100	0	100	0
D-Aspartate	100	100	95	0	95	0
Fumarate	—	97	100	0	95	0
Succinate	—	89	83	0	83	0
DL-Malate	—	138	99	0	62	37
Oxalacetate	—	187	72	0	14	60
Pyruvate	—	96	87	0	0	90
Experiment 2						
L-Aspartate	100	140	100	0	65	35
D-Aspartate	100	135	100	0	60	35
Fumarate	—	134	88	0	63	25
Succinate	—	116	100	0	80	20
DL-Malate	—	140	93	0	61	37
Oxalacetate	—	170	90	0	15	75
Pyruvate	—	95	90	0	0	90

[a] From Henderickx and Martin (1963).

[b] Incubation: volume 10 ml; 8 mg of bacterial nitrogen in phosphate buffer, pH 7, final concentration of 0.09 M; 100 μmoles L- or 200 μmoles DL-substrate; time, 20 hours. Results are expressed as μmoles per 100 μmoles of substrate and are corrected for blank values.

In a 48-hour incubation of a small quantity of glutamate-1-C^{14} with rumen fluid, 7% of the label was recovered in cells, and 30% in carbon dioxide and VFA's (Otagaki et al., 1955). Leucine-3-C^{14} was more completely attacked, but the rate was still very slow. Leucine is converted to isovaleric acid by *Bacteroides ruminicola* (Bladen et al., 1961b).

Histidine and glutamic acid are both fermented by rumen bacteria (van den Hende et al., 1963a) to hydrogen, carbon dioxide, ammonia, and acetic, propionic, and butyric acids. Similar results were observed by Lacoste-Bastié (1963), except that no propionate was formed. In the initial steps of the fermentation, histidine is converted to glutamic acid via urocanic acid and α-L-formamidinoglutaric acid.

Histidine can be converted to histamine (Dain et al., 1955), and tyrosine to tyramine, by rumen bacteria. Prolonged *in vitro* incubation of rumen contents (van der Horst, 1961) results in the formation of tryptamine, histamine, and tyramine. With feed added, cadaverine and putrescine were also found, the cadaverine derived from lysine and the putrescine from arginine.

Tryptophan, incubated 24 hours with washed rumen bacteria, was converted chiefly to indoleacetic acid (Scott et al., 1964; Lacoste, 1961). Tryptophanase was concerned in the production of indole and skatole (Lacoste-Bastié, 1963).

Labeled tyrosine added to rumen contents is converted to phenylpropionic acid (Scott et al., 1964; von Tappeiner, 1886), and phenylalanine is converted to phenylacetic acid (Scott et al., 1964; Lacoste-Bastié, 1963). The results of Lacoste-Bastié (1963) are shown in Table VII-4.

A steer on a diet of cottonseed hulls and molasses produced ammonia most rapidly when glutamic acid was added to the rumen, and at lesser rates from aspartate, alanine, and lysine. No ammonia was produced from glycine. In cows fed alfalfa hay concentrate, ammonia formed most rapidly when arginine was added to the rumen and ornithine was formed (T. R. Lewis and Emery, 1962b). Lysine gave rise to δ-aminovaleric acid, and tryptophan gave rise to indole and skatole. *In vitro* studies (T. R. Lewis and Emery, 1962c) gave the same results after 24 hours of incubation, and, in addition, putrescine was obtained from arginine and ornithine, and cadaverine from lysine.

b. *Mixed Amino Acids*

It was shown by Strickland (1934), though not for the rumen, that fermentations of mixtures of amino acids may proceed more rapidly than fermentations of single acids. Some amino acids are more readily reduced (hydrogen acceptors) and others more readily oxidized (hydrogen donors). If both types are present a fermentation may ensue, even though the acids singly are not fermented.

A mixture of the classic hydrogen-donating and hydrogen-accepting amino acids of the Strickland reaction, alanine and proline, supports a much more rapid fermentation in rumen contents (Lewis, 1955) than is observed with either amino acid alone (Table VII-5). The alanine is oxidized to acetic acid, ammonia, and carbon dioxide, and the proline is reduced to δ-aminovaleric acid (el-Shazly, 1952b).

Table VII-4

DECARBOXYLATION AND DEAMINATION OF SOME AMINO ACIDS
BY WASHED SUSPENSIONS OF RUMEN BACTERIA[a]

Substrate	Amount (μmoles)	Time of incubation (hours)	Carbon dioxide produced (μmoles)	Ammonia produced (μmoles)	Organic acid formed
Aspartic acid	5	4	4.9	5	Succinic, propionic
Glutamic acid	3	8	1.3	—	Acetic, butyric
Glutamic acid	3	24	3.1	3.2	—
Alanine	3	8	1.1	—	Acetic
Alanine	3	24	3.1	2.8	—
Histidine	3	8	1.4	3	Glutamic
Histidine	3	24	3.4	6.4	—
Tryptophan	3	24	3	2.9	Indoleacetic, indole, and indolepropionic
Phenylalanine	3	24	0.8	1	Phenylacetic
Tyrosine	3	24	0.4	1.6	Phenylpropionic and p-hydroxyphenyl-acetic
Proline	3	24	0	0.4	δ-aminovaleric
Hydroxypro-line	3	24	0	0.3	δ-aminohydroxyva-leric

[a] From Lacoste-Bastié (1963). The results have been corrected for the controls.

Amino acids in casein hydrolyzate are fermented by rumen liquid and washed suspensions of rumen bacteria (el-Shazly, 1952b). The amino acids attacked most rapidly included phenylalanine, tyrosine, proline, alanine, threonine, glycine, serine, aspartate, and glutamate. The fate of each amino acid in the mixture is much the same as when single amino acids are tested. This may result from the fact that in single amino acid experiments other amino acids are released by autolysis of the rather dense washed cell suspension. The extent of the conversion may be much less than if more acids are added, but qualitatively the same pathway is found.

El-Shazly (1952a) extended the analysis of the rumen VFA's to include those containing up to six carbon atoms. In addition to acetic, propionic, and butyric acids, he found isobutyric, isovaleric, and caproic acids, with possible traces of even longer-chain acids. One of the C_5 branched-chain acids was later shown (Annison, 1954a) to be 2-methylbutyric acid.

Table VII-5

AMMONIA PRODUCTION FROM MONOAMINOMONOCARBOXYLIC ACIDS AFTER INCUBATION WITH WASHED SUSPENSIONS OF RUMEN BACTERIA IN THE PRESENCE OF SEVERAL ADDITIVES[a,b]

Substrate	Glycine	DL-Alanine	DL-Valine	L-Leucine	DL-Isoleucine
Without additives	30	35	25	25	20
Sodium thioglycollate (1%)	15	10	10	15	10
Catalase (10 mg)	40	40	30	30	25
Glycine (200 μmoles)[c]	—	95	80	80	80
Proline (200 μmoles)[d]	50	50	55	40	45
Preincubation	5	10	0	0	0
Crude liver extract (0.5 ml)	80	90	75	65	60
Sterilized rumen liquor (2 ml)	50	55	50	55	50
Sterilized suspension of rumen bacteria (2 ml)	90	70	70	75	65

[a] From Henderickx and Martin (1963).

[b] Incubation: volume of 10 ml; 6 mg of bacterial nitrogen in phosphate buffer, pH 7, final concentration of 0.06 M; 100 μmoles L- or 200 μmoles DL-substrate; time, 15 hours. Additions as noted individually. Results, expressed as μmoles of ammonia from 100 μmoles L- or 200 μmoles DL-substrate, are corrected for blank values.

[c] Glycine incubation: 30 μmoles NH_3.

[d] Proline incubation: 25 μmoles NH_3.

El-Shazly found somewhat higher concentrations of the branched-chain acids when the ration included more protein, though the increase was by no means proportional to the increase in protein. The concentration of the branched-chain VFA's decreased when urea was substituted for proteinaceous concentrates as a source of nitrogen in the rumen (Brüggemann et al., 1962), but the concentration of the short straight-chain acids increased slightly. Acids found by several investigators are shown in Table VII-6.

The branched- and straight-chain VFA's (Bryant and Doetsch, 1954a; Bentley et al., 1954c), required nutrients for many of the rumen bacteria, arise through fermentation of the amino acids. Cellulose digestion by washed cell suspensions of rumen bacteria is stimulated by valine, leucine, and isoleucine (MacLeod and Murray, 1956). These amino acids are fermented to isobutyrate, isovalerate (Menahan and Schultz, 1964a), and 2-methylbutyrate, respectively, growth factors for many rumen bacteria (Table II-10). The amino acids are deaminated to the corresponding keto acid which is decarboxylated, and the resulting molecule with one less carbon is oxidized to the acid (Dehority *et al.,* 1958). Assimilation of the branched-chain amino acids appears to follow the reverse pathway (Allison and Bryant, 1963). Proline can be reduced to δ-aminovaleric acid, which can be converted to valeric acid, another growth factor for some of the cellulolytic bacteria of the rumen.

Pyruvate, hydroxypyruvate, glyoxylate, α-ketoisovalerate, α-ketoisocaproate, α-keto-β-methylvalerate, oxalacetate, and α-ketoglutarate have been detected in rumen fluid 30 minutes after the animal was fed (van der Horst, 1961). Some of these are probably intracellular intermediates in the metabolism of amino acids.

A number of pure cultures of rumen bacteria have been tested for production of ammonia from casein. *Bacteroides ruminicola, Selenomonas ruminantium,* and some strains of *Butyrivibrio fibrisolvens,* as well as *Peptostreptococcus* (Bladen *et al.,* 1961a), release ammonia, *Bacteroides ruminicola* being most active. In other experiments, significant amounts of ammonia were produced by 25% of proteolytic strains (Abou Akkada and Blackburn, 1963). Some of these might be expected to produce the VFA growth factors. None was able to grow with casein as the sole substrate. A carbohydrate or carbohydrate derivative was also necessary. Since little or no ammonia is produced from casein by pure cultures of *Lachnospira, Bacteroides succinogenes, Succinivibrio dextrinosolvens, Borrelia, Succinimonas amylolytica,* and ruminococci, they probably do not form the VFA growth factors.

Rumen bacteria able to digest casein and requiring no carbohydrate have recently been encountered by the author. Their characteristics fit them for conversions of amino acids to branched-chain VFA's, but further study is necessary before their significance in the rumen can be assessed. They digest other rumen bacteria to some extent. Activity of bacteria of this sort would be expected to remain relatively high, or even to increase proportionately with time after feeding. The maintenance of branched-chain acids in 48-hour starved animals at a concentration of 2.3 moles per liter, as compared to 2.8 moles per liter after feeding (Annison, 1954a), is in agreement with the postulate that rumen microbial cells are fermented. Autolysis of cells may also provide substrate for protein fermentation. The increase in ammonia concentration in rumen contents prior to feeding (D. Lewis and McDonald, 1958) suggests a continuing protein fermentation.

Table VII-6

QUANTITIES OF HIGHER FATTY ACIDS IN RUMEN CONTENTS OF SHEEP

Reference and ration	Time of sampling in relation to feeding	Total VFA's (μmoles/ml)	Molar percentage of total VFA's							
			Acetic	Propionic	Butyric	Iso-butyric	Valeric	Iso-valeric	2-methyl-butyric	C₆ and higher
Annison (1954a)										
1/3 groundnut meal	2 hr after	184	60	26	10.1	0.9	1.6	0.9	0.5	—
1/3 maize meal	5 hr after	142	44	32	17.4	1.9	2.6	1.4	0.7	—
1/3 hay	8 hr after	97	41	34	19.0	2.4	1.8	6.9	0.9	—
1/7 casein	2 hr after	77	60	22	7.9	2.5	3.1	2.8	1.7	—
3/7 ground maize	5 hr after	87	70	18	7.3	1.3	1.3	1.1	1.0	—
3/7 hay	8 hr after	92	68	18	9.0	1.4	1.2	1.4	1.0	—
6 parts hay	2 hr after	89	46	32	18.0	0.5	1.9	1.3	0.4	—
15 parts ground maize	5 hr after	89	56	28	13.3	0.4	1.0	1.0	0.3	—
	8 hr after	104	56	32	8.9	0.5	1.3	0.9	0.4	—
Hay	2 hr after	117	81	13	4.3	0.6	0.5	0.3	0.3	—
	5 hr after	113	77	19	2.4	0.5	0.5	0.3	0.3	—
	8 hr after	120	78	18	2.9	0.4	0.3	0.2	0.2	—
Jamieson (1959b)										
Rye grass-clover Pasture	Variable: 47 samples	32–153	38–64.5	15–30	9.5–22	1.0–4.5	0.5–3.0	0.5–4.0	0–3.5	
el-Shazly (1952a)										
Concentrate plus 50 gm of casein	Before	56	64.4	18.0	7.7	2.8	0.9	1.2		Trace
	After	118	56.9	23.7	8.5	1.5	1.4	3.9		Trace
Concentrate plus 50 gm of casein	Before	73	65.5	22.9	10.6	1.4	1.0	2.7		Trace
	After	141	45.5	23.5	10.0	2.0	2.3	3.5		Trace
Concentrate plus 100 gm of casein	Before	86	60.0	14.3	10.2	1.3	1.0	2.1		0.8
	After	106	67.0	18.5	11.7	2.5	2.4	5.0		1.9
Concentrate plus 100 gm of casein	Before	105	55.7	20.6	19.3	1.7	2.3	3.1		Trace
	After	160	53.1	21.7	16.6	2.7	3.3	3.6		Trace

Frozen grass	Before	86	65.7	17.1	9.3	2.1	1.3	3.1	0.4
	After	117	54.0	22.6	14.2	1.7	1.2	1.1	Trace
Frozen grass	Before	58	67.2	17.1	9.3	2.2	1.6	2.9	Trace
	After	138	55.2	24.3	12.5	1.2	1.6	1.3	Trace
Dried grass	Before	98	69.5	18.6	8.1	1.4	0.7	1.9	Trace
	After	140	57.7	22.2	8.4	0.8	0.8	1.4	Trace
Dried grass	Before	94	55.5	15.7	7.6	1.0	1.1	1.3	Trace
	After	135	63.5	26.0	11.9	0.5	1.5	0.8	Trace
Silage	Before	55	70.0	16.2	8.6	2.6	0.6	3.2	0.9
	After	110	53.2	25.4	7.4	1.3	1.7	2.5	0.8
Silage	Before	75	72.6	14.9	7.7	1.6	1.0	1.9	Trace
	After	145	66.6	23.6	10.7	1.5	2.3	2.7	Trace
Gray *et al.* (1952) Wheaten hay	—	—	62–70	16–27	6–9	0.3–0.6	1.6–3.2		.04–.05

B. Ammonia

The foregoing account of the fermentation of amino acids by rumen contents indicates that ammonia is always formed during the process. It is universally produced during the utilization of proteins as an energy source, and a parallelism between rumen ammonia and branched-chain VFA's has been reported (Jamieson, 1959b). Levels of ammonia in the rumen vary from 0 to 130 mg/100 ml (Johns, 1955b). Ammonia levels in the rumen contents at various times after feeding are shown in Fig. VII-6. Part of the material analyzed as ammonia may be methylamine

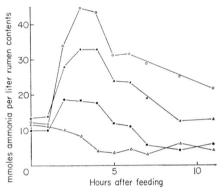

Fig. VII-6. Ammonia concentrations in the sheep rumen at various times after feeding different rations: daily ration of 600 gm of chopped meadow hay and 300 gm of groundnut cake (○—○), 600 gm of hay and 150 gm of groundnut cake (▲—▲), 700 gm of hay (●—●), 600 gm of hay and 200 gm of flaked maize (△—△). From D. Lewis (1957, p. 443).

(Hill and Mangan, 1964). The latter is produced in the rumen and further metabolized to some extent.

One of the most intriguing problems in rumen ecology is the extent to which ammonia serves as the nitrogenous material for synthesis of microbial cells. Many signs point in the direction of importance. It appears rapidly in rumen contents when the feed contains much nitrogen; many of the rumen bacteria assimilate it in preference to amino acids (Table II-10), and for some it is essential (Bryant and Robinson, 1963). Numerous studies indicate that considerable quantities of feed ammonia (usually given as urea) can be assimilated by the host animal, presumably via preliminary assimilation by the rumen microbes. Tracer studies with N^{15} (Boggs, 1959) have shown incorporation of ammonia into amino acids by rumen bacteria in sheep on a purified ration. Activity appeared first in glutamic, aspartic, and diaminopimelic acids, the first two being probable sources of amino groups for transamination to precursors of other amino acids. All these lines of evidence indicate a capacity for ammonia utilization by rumen microorganisms. The question is the extent to which this capacity is utilized.

There are limits to the amount of nitrogenous substrate which an organism can assimilate. Amino acids in excess of that limit will be available for supplying energy, and ammonia will appear to the extent that the available amino acids exceed their assimilation into microbial cells (Annison *et al.*, 1954). However, the fact that the nitrogen in excess of the assimilable amount appears as ammonia does not necessarily prove that ammonia is an intermediate in assimilation. Amino acids and peptides might also be the starting point for assimilation. There would seem to be an advantage to a rumen microorganism if it could capture and assimilate preformed peptides without having to synthesize them, but the actual importance of these intermediates in nitrogen assimilation can be assessed only after kinetic experiments on the turnover of ammonia and amino acids in the rumen have been completed. Exploration in this direction has been started (Warner, 1955; Boggs, 1959; Portugal, 1963), but additional studies are needed.

Ammonia is probably relatively more important for the nutrition of the fiber- and starch-digesting bacteria than for those utilizing soluble sugars. Immediately after forage is ingested, both soluble carbohydrates and soluble proteins are present. Digestion of the proteins is rapid, with release of at least small concentrations of amino acids which may be assimilated directly, and ammonia may also be assimilated with the available carbohydrate. Ammonia concentrations increase to a maximum at about 4 hours (Somers, 1961c) and then decrease.

With most forages, and particularly with lush spring grass and with legume forages, the quantity of protein is more than can be assimilated with the energy available in the soluble carbohydrates (Hogan, 1964b). Immediately after ingestion of the forage, the fiber-digesting bacteria and their enzymes are not yet established on the particles of new digesta and use relatively little of the amino acids available from the protein. By the time the bacteria are established, the amino acids in excess of those assimilable with the soluble carbohydrates will have been fermented as a source of energy.

By the time the fiber-digesting bacteria have started growth, and during the extended period in which they attack the more resistant components of the forage, amino acids will be scarce. It is thus not surprising that the fiber-digesting bacteria have the capacity to use ammonia as a source of nitrogen.

The starch digesters are not so clearly dependent on ammonia, although *Bacteroides amylophilus* can use it as a sole source of nitrogen. *Butyrivibrio* strains, *Succinivibrio dextrinosolvens,* and *Streptococcus bovis* can use it, as well as an amylolytic coccus (Provost and Doetsch, 1960), although the latter probably utilizes at least some amino acids, as judged from the growth on casein.

The increase in the rumen ammonia nitrogen concentration which often occurs shortly before the next feeding, is shown in the precise analyses of T. M. Sutherland *et al.* (1962) and is evident in the higher ammonia level before feeding than 11 hours after feeding (Fig. VII-6). This increase could result from a decrease in ammonia utilization and its continued production from feed protein. However fermentation of feed protein is essentially complete 6–8 hours after feeding, and the increases in ammonia occur after that. Endogenous metabolism of nongrowing microbes releases ammonia when soluble carbohydrate is scarce (Henderickx, 1960a), or cytolytic bacteria may digest and ferment other rumen organisms, with release of ammonia. This could explain also the high rate of ammonia production by rumen fluid *in vitro* in the absence of added substrate (Warner, 1956a). The average ammonia concentration in the rumen is less when sheep are continuously fed (Portugal 1963).

The effect of utilization of ammonia given as a salt is to increase the acidity in the rumen due to release of the anion. This can be deleterious (R. Clark and Quin, 1951) with poor rations and makes neutral forms of nitrogen such as urea a generally more desirable form for nitrogen administration. Nitrate can be used (R. Clark and Quin, 1951), but in too large quantities can cause methemoglobinuria (Chapter XII).

C. Utilization of Urea

Ever since the discovery by Weiske (1879) that asparagine could be used by sheep, amides have held a special interest for students of ruminant nutrition. In 1906 Müller performed *in vitro* experiments with asparagine and with ammonium salts, respectively, and reported that they could be synthesized into protein by rumen organisms. Warner (1964b) found that asparagine, glutamine, nicotinamide, and formamide were actively deaminated in the rumen, but acetamide and propionamide were only slowly attacked. Voltz (1919, 1920) showed that urea could replace feed protein. Subsequently, there have been innumerable experiments on the utilization of urea by ruminants (Hamilton *et al.,* 1948; Harris and Mitchell, 1941a,b; Hart *et al.,* 1938; Mills *et al.,* 1942, 1944; B. C. Johnson *et al.,* 1942, 1944; J. T. Reid, 1953; Repp *et al.,* 1955a; C. J. Watson *et al.,* 1949a,b; Wegner *et al.,* 1941a,b, 1940a; N. M. Williams and Tribe, 1957). One of these comparisons utilized identical twin cows (Schmidt and Kliesch, 1939), the urea-supplemented twin maintaining milk production when 37% of the protein was replaced by urea.

Urea is hydrolyzed in the rumen to ammonia and carbon dioxide (Huhtanen and Gall, 1955). An hydrolysis rate of 13 μmoles per hour per milliliter of rumen fluid was found in experiments *in vitro*. This indicates a rate of about 0.8 μmoles per milliliter per minute for fluid plus solids. In urea feeding, initial concentration of ammonia remains lower than if an equivalent amount of ammonium salts is fed directly. In some reports, urea appears not to be hydrolyzed rapidly, 50% being still present after 4.5 hours, and 13% after 12 hours in cows; in goats the rate was even less (Kaishio *et al.*, 1951). Since ammonia in too high concentrations is toxic to sheep, urea is more extensively used in the feed. Urea introduced directly into the abomasum was toxic to goats (Kaishio *et al.*, 1951).

There have been a number of rather thorough attempts to detect rumen bacteria hydrolyzing urea rapidly, to which a specific role in ammonia formation from urea might be assigned. Although a few investigators have reported actively ureolytic bacteria, they occur in small numbers (Gibbons and Doetsch, 1959; Sosnovskaja, 1959). Most investigations (E. J. Carroll, 1960; Gibbons and McCarthy, 1957; Gibbons and Doetsch, 1959; Sosnovskaja, 1959) have failed to disclose a special bacterial type particularly active in this respect and capable of accounting for the hydrolysis of urea added to rumen contents (Huhtanen and Gall, 1955). Presumably, many bacterial types contribute to the process (Muhrer and Carroll, 1964).

Since little energy is released, the splitting of urea to ammonia would be of value to an organism only to supply ammonia to be used in growth. A high concentration of urease is not necessary since this enzyme is extremely active.

Colin (1886) mentions that ruminant saliva contains urea, and this is a source of some rumen nitrogen. Simmonet *et al.* (1957) demonstrated that urea diffused from the blood back into the rumen and noted that this would make the nitrogen available for reutilization. Houpt (1959) and Somers (1961d) measured the quantity of urea entering the rumen from the blood when extra urea was infused into the jugular vein of sheep on a ration containing less than a maintenance level of nitrogen. Under these conditions it was estimated that 4.9 mmoles of urea entered the rumen each hour. Absorption of ammonia from the rumen was estimated at 2.4 mmoles per hour. The salivary secretion of urea amounted to 0.3 mmoles of urea nitrogen per hour. The total urea nitrogen entering the rumen from the sheep has been estimated at 1.5 gm per day (Dobson, 1961). About 700 mg has been estimated (Somers, 1961a) as the quantity entering with the saliva. The concentration of the salivary urea is related to the nitrogen level of nutrition (Somers, 1961b).

In the camel, reutilization of urea has been shown (Schmidt-Nielsen *et al.*, 1957) to be extremely efficient. Very small quantities were eliminated in the urine. Also in the sheep, urea is retained when the nitrogen in the food is low (Schmidt-Nielson *et al.*, 1958). Loss of nitrogen as urea diminishes markedly when the nitrogen in the ration is limiting and water intake is restricted (Livingston *et al.*, 1962) (Table VII-7). Nitrogen retention not only conserves this element, it diminishes the water lost in urine.

The possibility of surgically rerouting the ureters to empty into the rumen rather than the bladder, has been examined (G. H. Conner, unpublished experiments). Survival of unilaterally operated animals was obtained, but in bilaterally operated individuals uremia developed due

Table VII-7

ACTUAL AMOUNT UREA NITROGEN IN THE URINE AND PERCENTAGE UREA NITROGEN
IN THE TOTAL URINE NITROGEN OF *Bos taurus* AND *Bos indicus*,
IDENTICAL TWIN SETS DURING FIVE 10-DAY EXPERIMENTAL PERIODS[a]

Twin set	Species	Animal	Experimental period[b]	Crude protein of ration (%)	Urea Nitrogen per day (gm)	Urea nitrogen as percent of total urine nitrogen
1	*Bos taurus*	A_1	I	12	32.0 ± 2.19	67.7 ± 1.60
			II	12	28.6 ± 1.55	69.0 ± 1.44
			III	12	30.3 ± 2.52	62.6 ± 1.23
			IV	12	36.8 ± 1.46	67.7 ± 1.20
			V	12	36.7 ± 3.31	60.8 ± 2.37
		A_2	I	12	36.1 ± 2.28	70.2 ± 1.71
			II	8	29.0 ± 2.41	63.1 ± 1.76
			III	4	13.0 ± 5.28	55.4 ± 2.60
			IV	4	5.3 ± 1.03	32.5 ± 3.82
			V	4	2.3 ± 0.06	15.0 ± 2.95
2	*Bos indicus*	Z_1	I	12	29.8 ± 3.45	74.1 ± 1.70
			II	8	20.4 ± 2.07	64.9 ± 1.96
			III	4	16.6 ± 3.50	71.6 ± 2.54
			IV	4	9.0 ± 4.29	52.7 ± 6.61
			V	4	3.4 ± 0.06	35.9 ± 6.71
		Z_2	I	12	32.6 ± 3.53	72.0 ± 1.20
			II	12	28.4 ± 1.83	70.3 ± 1.35
			III	12	29.2 ± 3.76	68.5 ± 1.51
			IV	12	33.6 ± 0.10	71.0 ± 2.71
			V	12	39.6 ± 1.84	68.7 ± 2.23

[a] From Livingston *et al.* (1962, p. 1057).

[b] In periods IV and V, twins A_2 and Z_1 were deprived of water for 96-hour intervals.

to occlusion of the orifice into the rumen. The possibility that urine could be detoxified and reused holds theoretical interest.

Lewis (1957) has found that blood urea concentration varies according to the level of protein in the ration. With high levels of nitrogen the blood urea increased, as shown in Table VII-8. The urea would be most abundant in the bloodstream when it was least needed as a nitrogen supplement.

Because of the economic advantages in substituting simple forms of nitrogen for protein in the ration, there have been innumerable experiments in a great many countries testing the extent to which this substitution can be practiced (Brüggemann *et al.,* 1962). Most of these have been studies on urea supplementation.

Table VII-8

BLOOD UREA IN SHEEP ON VARIOUS RATIONS[a]

Experiment	Ration	Urea (mg/100 ml blood)	
		Range	Mean and standard deviation
1	700 gm of hay	30.0–41.8	36.9±2.5
	600 gm of hay + 200 gm flaked maize	12.2–15.6	14.1±0.9
	600 gm of hay + 150 gm of groundnut cake	50.1–63.0	57.5±3.3
2	700 gm of hay	32.8–44.6	37.4±2.7
	600 gm of hay + 200 gm flaked maize	13.7–16.0	15.0±0.7
	600 gm of hay + 150 gm of groundnut cake	48.9–64.3	55.8±4.7

[a] From D. Lewis (1957, p. 440).

Assimilation of simple nitrogenous compounds into microorganisms depends on two chief factors in the feed: (1) the amount of carbohydrate available (Pearson and Smith, 1943; Drori and Loosli, 1961) and (2) the lack of complex nitrogen in the feed. In those instances of ruminants receiving a ration containing all necessities except protein, and with plenty of digestible carbohydrate, there is a net assimilation of the nitrogen of urea. In practice, it is usually found that urea can supply about one-third of the required nitrogen. Even larger amounts can be used under suitable conditions. Instances in which urea supplementation does not significantly improve growth (Bartlett and Blaxter, 1947) may be due to lack of carbohydrate, minerals, or other growth factors, or to excess of other nitrogen sources.

In theory, it should be possible to devise a ration containing no protein, in which urea or ammonia would meet the total nitrogen requirements of the animal. This would involve supplying all of the nutrients usually supplied directly or indirectly by protein and would include the branched-chain fatty acids, the required vitamins, hemin and presumably numerous as yet unidentified nutrients required by the microorganisms. Selection could conceivably yield a population synthesizing even these factors. The importance of proteins as a source of inorganic ions and trace elements must also be kept in mind in devising a ration for maximal ammonia utilization. These questions will be discussed further in Chapter X. Also in that chapter will be discussed the utilization of various additional forms of nitrogen.

D. Rumen Synthesis of Amino Acids

Synthesis of ruminant-essential amino acids from simple nitrogen sources was assumed after these sources were demonstrated to replace protein. Thus, in lambs on a diet of 25% corn sugar, 42% corn starch, 20% cellophane, 5% minerals, 4% lard, and 4% urea it was shown that arginine, histidine, isoleucine, leucine, lysine, methionine, phenylalanime, threonine, tryptophan, and valine were formed (Loosli *et al.*, 1949).

The ammonia requirement of many rumen bacteria indicates a capacity for synthesis of amino acids. The extent to which the various rumen species can synthesize individual amino acids is incompletely known, but information is available for a few. Allison and Bryant (1963) and Allison *et al.* (1959, 1962a,b) showed that valine was synthesized by *Ruminococcus flavefaciens* from isobutyrate, and leucine from isovalerate, the branched-chain acids being required nutrients. Carbon from C^{14}-carbon dioxide was incorporated into the carboxyl group of valine and leucine and into the carboxyl of isoleucine, 2-methylbutyrate being the branched-chain precursor of isoleucine. Some strains of *Ruminococcus albus* require isobutyrate, isovalerate, and 2-methylbutyrate (Hungate, 1963a).

Isovalerate is incorporated *in toto* into leucine. The latter is not assimilated either in the presence or absence of isovalerate. The pathway of leucine synthesis, by addition of carbon dioxide to isovalerate to form the carboxyl, differs from the pathway in other bacteria, in which the methyl group of acetyl CoA provides the carboxyl carbon. In addition to serving as a precursor of leucine, isovalerate is synthesized into the terminal branched portion of some long-chain fatty acids in ruminococci (Allison *et al.*, 1962a).

The nutritional requirement of *Bacteroides succinogenes* for fatty acids suggests that this species also uses them to synthesize amino acids. Phenylalanine is synthesized in the rumen from phenylacetic acid (Allison, 1965).

A minimum of 2.3 gm of lysine was synthesized each day in a lactating goat supplied with a feed poor in lysine (Edwards and Darroch, 1956). On the basis of 6% lysine in the microbial protein of rumen contents, this indicates synthesis of at least 38 gm of microbial protein per day. Supplementation of alfalfa with cystine gave a 10% increased growth in lambs (Smuts *et al.,* 1941), but did not affect mature sheep.

Studies on individual amino acids, tested as the sole nitrogen source for mixed rumen bacteria (Raynaud, 1961), show that aspartic acid, asparagine, glutamine, serine, and histidine are good sources. Glycine, alanine, glutamic acid, cysteine, proline, and oxyproline were sometimes good, but poor with some bacterial preparations. Threonine, phenylalanine, tryptophan, valine, leucine, isoleucine, methionine, and lysine were poor.

E. Nature of the Microbial Protein

The rumen bacteria contain approximately 65% protein, based on total nitrogen content, and, as can be seen from Table VII-9, this value is not appreciably affected by various common feeds (Weller, 1957; Vrid-

Table VII-9

PERCENTAGE NITROGEN IN DRY RUMEN MICROBIAL CELLS FROM SHEEP ON VARIOUS RATIONS[a]

Microbial cells	Ration[b]			
	Diet 1	Diet 2	Diet 3	Diet 4
Bacteria	9.3	11.1	10.4	12.4
Protozoa	3.8	6.5	2.7	7.9

[a] After Weller (1957, p. 386).

[b] Diet 1: wheaten hay chaff (0.89% nitrogen) ad lib.
Diet 2: lucerne hay chaff (2.91% nitrogen), ad lib.
Diet 3: daily: 600 gm wheaten straw containing 2.3 gm of nitrogen, 200 gm of crushed oat grain containing 2.9 gm of nitrogen, 15 gm urea containing 2.3 gm of nitrogen.
Diet 4: mixed green pasture grazed by the sheep.

nik, 1961). The nitrogen content of the protozoa is low as compared to the bacteria, because of a higher polysaccharide content (McNaught *et al.,* 1954a).

Since microbial protein constitutes the major part of the ruminant's nitrogenous food, its value for the synthesis of mammalian tissue becomes important. The food value has been determined by feeding rumen microbes to mice or rats, because of the difficulty in obtaining sufficient microbial protein to feed a ruminant, and also because the use of nonruminants avoids the complications of the rumen attack on feed protein, even though the latter is microbial. Some of the results obtained are shown in Table VII-10. The studies on the synthesis of amino acids in cattle (A. L. Black *et al.*, 1952) indicate that the amino acids synthesized in the bovine tissue are the same as those synthesized in the rat.

The food value of nitrogenous materials is estimated in terms of true digestibility and biological value.

$$\text{net utilization} = \text{biological value} \times \text{true digestibility}.$$

$$\text{true digestibility} = \frac{\text{N in feed} - (\text{N in feces} - \text{metabolic N})}{\text{N in feed}} \times 100$$

Metabolic nitrogen is the fecal nitrogen on a nitrogen-free ration, i.e., it is tissue nitrogen lost in the feces (Ellis *et al.*, 1956).

$$\text{biological value} =$$

$$\frac{\text{feed N} - (\text{fecal N} - \text{metabolic N}) - (\text{urinary N} - \text{endogenous N})}{\text{feed N} - (\text{fecal N} - \text{metabolic N})} \times 100$$

Table VII-10
NET UTILIZATION VALUES OF RUMEN BACTERIA AND PROTOZOA AND SOME OTHER MATERIALS

Material	Test animal	True digest-ability	Biological value	Net utilization	Reference
Rumen bacteria	Rat	74	81	60	McNaught *et al.* (1954a)
Rumen protozoa	Rat	91	80	73	McNaught *et al.* (1954a)
Dried brewers yeast	Rat	84	72	60	McNaught *et al.* (1954a)
Rumen bacteria	Rat	73	88	64	McNaught *et al.* (1950a)
Rumen bacteria	Rat	55	66	36	B. C. Johnson *et al.* (1944)
Rumen protozoa	Rat	86	68	58	B. C. Johnson *et al.* (1944)
Rumen bacteria green feed	Rat	62	80	50	Reed *et al.* (1949)
Rumen bacteria dry feed	Rat	65	78	50	Reed *et al.* (1949)
Casein	Rat	101	80	81	Reed *et al.* (1949)
Urea	Sheep	80	54	43	Ellis *et al.* (1956)
Gelatin	Sheep	83	57	47	Ellis *et al.* (1956)
Casein	Sheep	86	73	63	Ellis *et al.* (1956)
Blood fibrin	Sheep	82	82	67	Ellis *et al.* (1956)
Soybean protein	Sheep	86	83	71	Ellis *et al.* (1956)

Endogenous nitrogen is the urinary nitrogen from an animal on a nitrogen-free ration. Biological value is the percentage of the digestible nitrogen assimilated and presumably used as protein in the body. The difference between urinary nitrogen and endogenous nitrogen represents the absorbed nitrogen that is excreted instead of built into cells. The net utilization values of rumen bacteria and protozoa and some other materials, determined in various laboratories, are shown in Table VII-10.

It is apparent that the rumen bacterial and protozoal nitrogenous materials are as good sources of protein for rats as are a number of proteins, with the protozoal slightly superior to the bacterial. The biological value of nitrogen sources has been studied (Ellis *et al.,* 1956) by adding them to the ration of a lamb fed a diet consisting essentially of carbohydrate and minerals plus the test nitrogen source. The feed without the nitrogen source contained only 0.004% nitrogen. The sheep received 80 gm of nitrogen supplement daily and were in approximate nitrogen balance. The metabolic nitrogen was 2.39 mg of nitrogen per gm of dry matter intake, or 7.17 mg per gm of feces. Endogenous nitrogen was 3.19 mg of nitrogen per kilogram of body weight.

Analyses of the rumen bacteria have yielded somewhat variable results, possibly associated with the time in relation to feeding, which would affect the concentration of polysaccharide reserves in the cells. J. A. B. Smith and Baker (1944) found 36.3% crude protein (nitrogen $\times$ 6.25), 46.8% carbohydrate, and 9.5% lipids in dried rumen bacteria which had received carbohydrate. McNaught *et al.* (1950a) found 44.4% protein, 40.3% carbohydrate, 3.1% lipids, 0.3% "crude fiber," 7.1% ash, and 4.8% unidentified material. Analyses of bacteria taken directly from the rumen show as much as 65% protein. An unexplained toxicity to the rat of rumen microorganisms from pasture-fed cattle has been reported (Ardo *et al.,* 1964).

The food value of the microorganisms can be estimated also on the basis of the content of amino acids essential for the ruminant. The amino acid content of the bacteria and protozoa from sheep fed the diets shown in Table VII-9 (Weller, 1957) is shown in Table VII-11, and the compositions found by a number of investigators are collected in Table VII-12 (C. W. Duncan *et al.,* 1953; Bouchaert and Oyaért, 1952; Holmes *et al.,* 1953; Chance *et al.,* 1953a; Ter-Karapetjan and Ogandzanjan, 1959; Johanson *et al.,* 1949).

Amino acids which the ruminant cannot synthesize include tryptophan, histidine, leucine, lysine, methionine, phenylalanine, tyrosine, and valine (A. L. Black *et al.,* 1952). Presumably arginine, isoleucine, and threonine, essential for the rat, are required also by ruminants (C. W. Duncan *et al.,* 1953). Methionine is believed to be an amino acid limiting rumi-

Table VII-11

AMINO ACID COMPOSITION OF HYDROLYZATES OF RUMEN MICROBIAL PREPARATIONS

(Amino acid nitrogen expressed as percentage of total nitrogen)

Amino acid	Bacteria[a]		Bacteria[b]				Protozoa[b]			
	Dry ration	Green ration	Diet 1	Diet 2	Diet 3	Diet 4	Diet 1	Diet 2	Diet 3	Diet 4
Aspartic acid	7.2	7.0	6.8	6.7	6.8	6.8	7.4	7.8	8.2	8.4
Threonine	2.9	2.6	3.5	3.6	3.6	3.8	3.1	3.4	3.5	3.7
Serine	—	—	2.8	2.5	2.8	3.0	2.6	3.0	3.2	3.1
Glutamic acid	11.5	12.1	6.6	7.0	7.5	6.9	7.9	8.3	8.7	8.4
Proline	—	—	2.8	2.1	2.4	2.2	1.9	2.9	2.1	2.2
Glycine	2.9	4.0	6.0	6.3	5.9	6.1	4.9	5.7	4.7	5.7
Alanine	2.9	4.0	6.5	6.5	6.4	6.5	4.1	4.6	4.1	4.1
Cystine	—	—	0.7	0.8	0.8	0.8	1.1	1.1	1.3	1.2
Valine	3.7	4.9	4.4	4.5	4.4	4.5	3.6	4.1	3.7	4.0
Methionine	1.1	1.2	1.5	1.5	1.5	1.5	1.0	1.4	1.4	1.4
Isoleucine	2.8	2.7	3.6	3.7	3.6	3.8	4.3	4.6	4.9	4.5
Leucine	4.5	4.0	4.5	4.6	4.7	4.6	5.0	5.7	5.3	5.2
Tyrosine	2.5	1.9	2.0	2.1	2.2	2.1	2.0	2.3	2.2	2.4
Phenylalanine	—	—	2.3	2.5	2.5	2.4	2.8	3.3	3.2	3.2
Histidine	4.8	6.2	3.0	2.6	2.6	2.8	2.6	3.4	2.9	3.1
Lysine	6.2	5.0	8.2	7.5	7.9	8.2	10.7	10.6	11.3	12.6
Arginine	16.3	11.8	9.1	8.6	9.3	9.3	8.1	10.0	8.4	9.3
Tryptophan	1.4	1.2	—	—	—	—	—	—	—	—

[a] From Holmes *et al.* (1953).

[b] From Weller (1957).

nant growth (W. E. Thomas *et al.*, 1951; T. K. Loosli and Harris, 1945; Fauconneau and Chevillard, 1953).

Of the amino acids required by the ruminant, lysine, leucine, isoleucine, and phenylalanine appear to occur in slightly greater concentration in protozoal protein. Methionine and valine are slightly greater in the bacterial protein, and tyrosine, threonine, and histidine compose about the same

Table VII-12

AMINO ACID COMPOSITION OF RUMEN MICROBIAL HYDROLYZATES
COMPARED WITH RUMEN DIGESTA AND HERBAGE PROTEINS
(Amino acid nitrogen expressed as percentage of total nitrogen)

Amino Acid	Bacteria	Protozoa	Rumen digesta[a]	Herbage proteins
Aspartic acid	6.7–6.8	7.4–8.4	—	4.7–5.4[b]
Threonine[c]	3.5–3.8	3.1–3.7	3.1–4.6	3.7–4.2
Serine	2.5–3.0	2.6–3.2	—	3.2–3.8
Glutamic Acid	6.6–7.5	7.9–8.7	—	6.4–7.8[b]
Proline	2.1–2.8	1.9–2.9	—	3.5–3.8
Glycine	5.9–6.3	4.7–5.7	—	6.4–7.0
Alanine	6.4–6.5	4.1–4.6	—	5.7–6.0
Cystine	0.7–0.8	1.1–1.3	—	1.1–1.6[b]
Valine[c]	4.4–4.5	3.6–4.1	3.6–4.5	4.5–4.8
Methionine[c]	1.5	1.0–1.4	0.5–0.9	1.2–1.6[b]
Isoleucine[c]	3.6–3.8	4.3–4.9	2.2–4.1	3.3–3.5
Leucine[c]	4.5–4.7	5.0–5.7	3.7–5.1	5.3–5.9
Tyrosine[c]	2.0–2.2	2.0–2.4	—	1.5–1.7
Phenylalanine[c]	2.3–2.5	2.8–3.3	1.7–2.1	3.3–3.6
Histidine[c]	2.6–3.0	2.6–3.4	2.7–4.0	3.6–4.0[b]
Lysine[c]	7.5–8.2	10.6–12.6	4.6–7.9	5.0–6.8[b]
Arginine[c]	8.6–9.3	8.1–10.6	9.5–12.0	12.0–14.0[b]

[a] Calculated from data of C. W. Duncan *et al.* (1953).

[b] Data given by Lugg (1949). The remaining values in this column are from the data of Kemble and Macpherson (1954), with threonine and serine values uncorrected for hydrolysis losses.

[c] Amino acids not synthesized by the ruminant (Black *et al.*, 1952) or assumed to be essential because they are required in the rat (C. W. Duncan *et al.*, 1953; Underwood and Moir, 1953).

proportion in both. The differences in amino acid content do not explain the greater biological value of protozoa for rats (Table VII-10). Possibly less bacterial nitrogen is available because a greater fraction of it occurs in nonamino acid cell wall components or in nucleic acids (Portugal, 1963).

Recent experiments of Abou Akkada and el-Shazly (1964) indicate a favorable effect of the rumen protozoa on the nutrition of the host, although the possibility that the results may have been due to lack of certain rumen bacteria has not been completely excluded. So little is known of the nitrogen interrelationships between the rumen bacteria and protozoa, particularly from a quantitative standpoint, that it is difficult to formulate explanations for the favorable effect of the protozoa, other than the greater biological value of their nitrogenous constituents. Failure to observe any influence of the protozoa (Preston, 1958; Pounden and Hibbs, 1950) in other experiments could be due to the high proportion of concentrates fed the calves studied or to use of mature ruminants.

The amino acid composition of whole rumen contents is about the same before and after feeding (C. W. Duncan *et al.*, 1953) (Table VII-13), presumably because most of the protein before and 6 hours after feeding is in the form of microbes. As shown in Table VII-14, the rumen microbial protein 6 hours after feeding is about 70% greater than the quantity just before feeding.

The similarities in the composition of the microorganisms from animals on various feeds (Table VII-9), and the similar biological values of widely different proteins fed to ruminants (Table VII-10), are due to the uniformity in the composition of the microbial cells. Any utilizable source of nitrogen is assimilated into the same cell material.

Relatively little attention has been paid to the nonprotein nitrogenous compounds of rumen microorganisms. The purines and pyrimidines are presumably digested and absorbed. The fate of the amino sugars in the microbial cell walls has not been studied.

F. Quantity of Microbial Nitrogen Available to the Host

There are several methods for estimating the amounts of microbial cell nitrogen synthesized in the rumen fermentation. Direct analyses indicate that more than half of the rumen nitrogen is in the form of microbial cells (Gray *et al.*, 1953; Weller *et al.*, 1958).

Estimates can be obtained from the quantity of energy-yielding materials fermented. This method assumes a close linkage between fermentation and growth. If paucity of nitrogen prevents growth, the breakdown of carbohydrate or other energy-yielding materials ceases; fermentation of carbohydrate is stopped by lack of nitrogen. The link between carbohydrate and nitrogen is assumed to be high-energy phosphate molecules such as ATP (Bauchop and Elsden, 1960). The substrate fermentation reactions yield ATP which participates in the various reactions of growth. When growth cannot occur because of lack of nitrogen, accumulation of ATP retards fermentation. The numerous observations (Klein *et al.*, 1939; Moir and Harris, 1962) on the loss of appetite in ruminants on

Table VII-13

PERCENTAGE OF AMINO ACIDS IN THE "TRUE" PROTEIN IN THE DRIED RUMEN DIGESTA BEFORE AND 6 HOURS AFTER FEEDING

Time	Arginine	Histidine	Isoleucine	Leucine	Lysine	Methionine	Phenylalanine	Threonine	Tryptophan	Valine	Reference
0 hours	5.1	2.15	5.25	6.8	4.78	1.15	3.83	5.14	0.58	5.46	C. W. Duncan et al. (1953)
6 hours	5.1	2.04	4.94	6.53	5.2	1.2	3.86	4.83	0.59	5.49	
0 hours	4.7	1.18	6.5	4.77	4.42	1.4	4.15	3.3	1.24	4.23	Bouckaert and Oyaért (1952)
3 hours	4.02	1.17	6.8	5.8	4.48	1.8	4.2	3.5	1.24	4.4	
16 hours	4.4	1.14	6.8	5.98	5.74	2.0	4.68	3.7	1.20	4.8	
Abbatoir cows	3.71	0.96	6.74	5.01	6.19	1.68	3.35	3.94	0.63	4.20	Bouckaert and Oyaért (1952)

Table VII-14

GRAMS OF AMINO ACIDS IN THE BOVINE RUMEN DRY MATTER BEFORE AND 6 HOURS AFTER FEEDING[a]

	Arginine	Histidine	Isoleucine	Leucine	Lysine	Methionine	Phenylalanine	Threonine	Tryptophan	Valine
0 hours	6.7	2.8	6.4	9.0	6.1	1.5	5.6	6.4	0.8	7.2
6 hours	11.9	4.7	11.3	15.3	11.0	2.7	9.7	10.3	1.3	12.3
% increase	77	68	77	70	80	80	73	61	63	71

[a] From C. W. Duncan et al. (1953, p. 46). Average of 8 trials.

nitrogen-poor rations may be due to a cessation of the production of VFA's.

The correlation between quantity of sugar fermented and cells formed is subject to variation according to the environmental conditions. The yield of cells per unit of carbohydrate fermented is greatest when carbohydrate is limiting and nitrogenous materials to be synthesized into cell materials are in excess. The yield per unit of nonenergy substrate is greatest when that substrate is limiting, but carbohydrate or other energy-yielding materials are in excess. Conditions in the rumen may fluctuate between limitation by carbohydrate and by utilizable nitrogen or even other factors. When the ammonia concentration in rumen contents is very low, it may limit growth. Other nitrogenous substances may limit growth when ammonia is sufficient, and on rations containing high percentages of nitrogen, carbohydrate may limit growth.

1. ESTIMATES BASED ON THE RATIO OF CARBOHYDRATE FERMENTED TO NITROGEN ASSIMILATED

It is difficult to measure directly the extent of conversion of carbohydrate substrate into microbial cells in the rumen itself. This can be measured *in vitro*. *In vitro* conditions do not duplicate exactly the conditions in the rumen, and it is necessary to check the *in vitro* results by direct estimates of rumen synthesis.

a. *Microbial Synthesis in Vitro*

Köhler (1940) added glucose to incubated rumen contents and recovered as much as 8% of this substrate in the form of microbial cells. A similar experiment in the author's laboratory with the soluble fraction of alfalfa hay gave a yield of 12% of cell material when the fraction of rumen contents containing only bacteria was used, and approximately 50% when the protozoa were included. The latter high value is due to conversion of the sugar into reserve starch by the holotrichs.

Block *et al.* (1951) found 10.44% nitrogen in rumen bacteria. In experiments *in vitro*, J. A. B. Smith and Baker (1944) found 10.3–10.8% nitrogen in the sedimented cells from rumen liquor. After 6 hours of incubation with 1% maltose, a cell yield of 152.2 mg, or 14.5% (based on the glucose equivalent weight of maltose), was obtained, with a nitrogen content of 6.7 mg or 4.4%. This would correspond to 64 mg of cells with a nitrogen content of 10.5%, the average for rumen bacteria. The cell yield on this basis was 6.4% of the maltose fermented (as glucose). In an experiment with 210 mg of maltose (0.21%) the cell nitrogen content increase was 3.7 mg, equivalent to 35 mg of cell material with a nitrogen content of 10.5%. This would be a cell yield of 17% of the substrate. In other similar experiments (Bloomfield et al., 1964) the cells constituted 17.4% of the carbohydrate substrate.

Bowie (1962) obtained 6–11% of fermented carbon as protein, i.e., cell yields of 9.5–17.5% on the basis of 10.5% nitrogen, in continuous culture experiments, with a substrate of the same synthetic diet fed to the animal from which the inoculum was obtained.

Henderickx (1960b) found that the protein formed by rumen microorganisms constituted 6.8% of the available carbohydrate (as glucose), i.e., the assimilated nitrogen was 1.09%. On the basis of a 10.5% nitrogen content, the microbial cells synthesized from the carbohydrate amounted to 10.3% of the substrate.

In a few cases, the efficiency of cell synthesis by pure cultures of rumen bacteria has been measured. *Ruminococcus albus* (Hungate, 1963a) assimilated 1.2–1.5% nitrogen per unit of cellobiose fermented in batch cultures, equivalent to 11.4–14.3% cells with a nitrogen content of 10.5%. In continuous cultures the cell nitrogen constituted 2.24% of the cellobiose utilized, or 21.3% cells with a nitrogen content of 10.5%. Similar efficiencies for conversion of substrate into cell material have been observed with *Selenomonas* in continuous cultures (Hobson and Smith, 1963).

b. *Estimates from in Vivo Measurements*

In a few cases, rations almost free of nitrogen except for one added component have been fed. In an experiment (Agrawala et al., 1953a) with cattle fed on 42% corn starch, 25% glucose, 20% cellophane, 4% lard, 5% mixed materials, and 4% urea, the rumen contents were removed just before feeding; the contents were weighed, sampled, and returned; then the animals were fed. The procedure was repeated 6 hours later. The protein in the rumen of seven calves increased by an average of 109 gm, and the dry matter decreased by 1.866 kg. The microbial cells synthesized, if assumed to contain 10.5% nitrogen, amounted to 8.9% of the dry matter disappearing. Since both dry matter and cells were leaving the rumen during the 6-hour period after feeding, the value is very approximate. On the assumption that cells left the rumen faster proportionately than the dry matter, the actual cell yield was higher than 8.9%.

In the experiments of Ellis et al. (1956) with sheep, 9.53 gm of urea nitrogen per day were fed with a nitrogen-free ration consisting of 40.3% Solka-Floc cellulose, 27.7% starch, 15% cerelose (glucose), 2% lard, and 5% complete mineral mixture. Daily consumption of the ration was 720 gm. Dry matter digestibility was 67%. The metabolic nitrogen was 1.91 gm per day, and the endogenous nitrogen was 1.04 gm per day. Nitrogen lost in the feces amounted to 1.68 gm, not including metabolic nitrogen. Of the 7.65 gm of nitrogen absorbed, 4.58 gm was eliminated as urinary nitrogen and, with the 1.91 gm metabolic fecal nitrogen, amounted to 6.49 gm not retained out of 7.65 gm absorbed. The difference, 1.16 gm of nitrogen, (the actual experimental value as weight gain was 0.78 gm of nitrogen) was retained within the animal. If the difference in biological value of the urea supplement and soybean protein supplement was due to loss of ammonia from urea which was absorbed and eliminated in the urine without assimilation, approximately 2.2 gm of the absorbed urea nitrogen was not assimilated. This leaves 5.45 gm of urea nitrogen as the amount synthesized per day into microbial nitrogen by the rumen microorganisms, equivalent to 34 gm of protein per day. Since 454 gm of dry matter of the 720 gm fed was digested and presumably fermented, the protein formed per 100 gm of carbohydrate fermented was 7.5 gm, a 12.6% yield of cells from the carbohydrate.

If it is assumed that the lambs in question (31.5-kg body weight) possessed a rumen with an average content of 4.5 kg, the average rate of production of microbial protein was 0.315 gm of protein per hour per kg rumen contents, or 38 gm of protein per day per 5 kg of rumen contents. In similar experiments, the digestibility of the ration was greater with 2.05% nitrogen than with 1.65 or 2.45% (Ellis and Pfander, 1957, 1958). M. J. Head (1953) concluded that 1% nitrogen in the ration was the quantity required for cellulose digestion by rumen microorganisms.

Microbial nitrogen synthesis can be estimated also from the results of experiments in which the available nitrogen in the feed approximates the amount required for fermentation of the digestible carbohydrates. In experiments with sheep (Gray *et al.,* 1958b), only 48% of the dietary nitrogen passed from the rumen into the omasum when the ration contained 2.9% nitrogen, 65% when the ration held 1.8% nitrogen, 100% with 1.1% nitrogen, and more than 100% with 0.7% nitrogen, the excess presumably coming from urea entering the rumen with the saliva. These results show that the rumen microorganisms assimilate an amount of nitrogen not exceeding 1.1% of the feed, a value very similar to the 1.2% found essential by Klein *et al.* (1939).

The true digestibility of the feed nitrogen was estimated at 82% (Gray *et al.,* 1958b), and the degree of digestion of the carbohydrates was estimated in the rumen at 50% (the 40% reported as the degree of digestion in the rumen apparently did not include soluble components of the hay). According to these estimates, approximately 5.5 gm of microbial protein were produced from 50 gm of fermentable carbohydrate, or 11%, equivalent to 18.5% cells with 10.5% nitrogen.

In another experiment (Oyaért and Bouckaert, 1960), little ammonia was lost from the rumen of sheep when 700 gm of dry matter in clover hay plus potato starch were digested. The clover hay contained 18.4 gm of total nitrogen. Of this nitrogen, 10.6 gm was recovered in the urine or retained in the animal. This amounts to 1.5% of the dry matter disappearing, equivalent to a yield of 14.4% cells with a nitrogen content of 10.5%.

Moir and Harris (1962) found that dry matter digestibility decreased when feed nitrogen fell below 1.4% of the total digestible nutrients. On the assumption of 80% true digestibility of the nitrogen, this would represent synthesis of 10.7% of the digestible material into microbial cells with a nitrogen content of 10.5%. In another study (Gallup *et al.,* 1952) the most efficient utilization of urea nitrogen was found when the ration contained 10% crude protein equivalent.

The extensive data, compiled by Schneider (1947), on the digestibility and composition of feeds has been analyzed (Glover and Dougall, 1960) to show that the variability in digestibility is determined in some cases by the crude fiber content and in others by the nitrogen content. When the nitrogen content of the feed falls below a protein equivalent of 5%, the digestibility diminishes. This is probably the lower limit for nitrogen assimilation (Glover and Duthie 1960). On the assumption of an average digestibility of dry matter of 55%, and a nitrogen digestibility of 80%, the percentage of nitrogen utilized, based on dry matter disappearing, would be 1.17, equivalent to a 11.1% yield of cells with a nitrogen content of 10.5%.

A formula (Garrett *et al.,* 1959),

$$\text{total digestible nutrients for maintenance} = 0.036 \times \text{weight}^{3/4}$$

gives a value of 2.9 kg of total digestible nutrients for a 455-kg cow. On the basis of the Haecker standard for maintenance nitrogen in this size animal, 320 gm of protein equivalent nitrogen is needed, or, according to Maynard and Loosli (1956), 273 gm are required. If 80% digestible, the protein nitrogen utilized would constitute 1.4% of the total digestible nutrients, which corresponds to a 13.4% yield of microbial cells. The estimate of 181 gm of digestible nitrogen (Kehar, 1956) corresponds to a cell yield of 9.5%. In the experiments of Harris and Phillipson (1962), the difference between feed and fecal nitrogen constituted 1.09% of the digested organic matter in sheep on a maintenance ration of poor hay. Henderickx (1960b) found that the microbial cells constituted 10.3% of the fermented substrate.

These various lines of evidence from *in vivo* experiments indicate that, on the average, the weight of microbial cells formed in the rumen fermentation is at least 10% of the substrate fermented. The microbial protein assimilated amounts to about 6.5% of the fermented substrate. For a 455-kg cow on a maintenance ration of 2.9 kg of total digestible nutrients, microbial nitrogen equivalent to 190 gm of protein would be formed. The estimate of Marston (1948a) that 7–16% of the energy of the cellulose fermented in the rumen was conserved in the form of microbial cells accords well with the above average.

In some instances, stimulation of fiber digestion is obtained by feeding up to 18% protein (Conrad and Hibbs, 1961). For example, in one trial (Burroughs and Gerlaugh, 1949) the digestion coefficient for corn cobs was 58.9% with 8% protein in the ration and 66.7% with 15% protein. On an average, each percent of protein up to 8% supported digestion of 8.25% carbohydrate, whereas the additional 7% supported an additional digestibility of only slightly more than 1% for each percent protein. The increased digestibility could be due to the protein itself. It would be necessary for protein to be very inexpensive for such increased digestibility to be economically advantageous. In later trials (Burroughs *et al.,* 1950a), supplementation of corn cobs with alfalfa ash gave increased digestibility. The added protein in these experiments may have increased digestion because of its mineral content. In other trials (Burroughs *et al.,* 1949b), 5% protein was reported adequate for digestion of corn cobs.

2. Estimates from Rumen Microbial Nitrogen and Turnover Rates

The rumen microbes have been estimated (Weller *et al.,* 1958) to contain between 63 and 81% of the nitrogen in the rumen. Blackburn and Hobson (1960c) found 47–77% of the rumen nitrogen in the protozoa and bacteria (Fig. VII-2). Values of 30–34%, obtained earlier (Schwarz and Steinlechner, 1925), were low because of failure to remove the bacteria from the solids. The rumen liquid contains 54–74% of the total rumen nitrogen (Weller *et al.,* 1958). This latter value was obtained by separating liquid from solids, washing the latter with two volumes of water, shaking, again separating liquid from solids, and combining this liquid with the first liquid fraction. Some of the values reported in the literature on nitrogen in rumen contents are collected in Table VII-15. Values at various times after feeding have been obtained by Fauconneau (1961).

On the basis of 1.25% (w/v) microbial protein in a 5-liter sheep rumen turning over 1.6 times per day, the animal would receive 100 gm of protein. On the assumption that only 80% of the protein was digestible, this would mean 80 gm of digestible protein. Maintenance protein is ap-

Table VII-15

NITROGEN IN THE RUMEN

Animal	Ration	Material analyzed	Nitrogen % (w/v)	Reference
Sheep	Hay and concentrate	Total dry matter	0.33 –0.54	Boyne et al. (1956)
Sheep	Chaff, wheat, starch, and molasses	Protein	0.226	Williams and Christian (1956)
Cattle	Abbatoir animals	Total	0.22	Köhler (1940)
Sheep	Pasture	Protein and NH_3	0.348	Jamieson (1959b)
Calves	Hay and grain	Protein	0.44	Agrawala et al. (1953a)
Sheep	Straw, molasses, and casein	Total	0.125	Blackburn and Hobson (1960c)
Heifers	Alfalfa pasture	Protein	0.23 –0.26	Gutowski et al. (1960)
		Total	0.40 –0.57	
Cattle		Bacterial protein	0.18	Thaysen (1945)
Cattle	Abbatoir animals starved for 24 hours	Total Protein	0.154	Raynaud (1955)
Cattle	Fistulatated cows	Total protein	0.232	Emery et al. (1960)
Sheep	Wheaten hay	Plant nitrogen	0.016–0.092	Weller et al. (1962)
		Bacterial nitrogen	0.023–0.126	Weller et al. (1962)
		Protozoal nitrogen	0.007–0.009	Weller et al. (1962)
		Soluble nitrogen	0.007–0.029	Weller et al. (1962)
		Total nitrogen	0.099–0.231	Weller et al. (1962)
Sheep	Alfalfa hay	Plant nitrogen	0.054–0.141	Weller et al. (1962)
		Bacterial nitrogen	0.069–0.185	Weller et al. (1962)
		Protozoal nitrogen	0.006–0.062	Weller et al. (1962)
		Soluble nitrogen	0.020–0.069	Weller et al. (1962)
		Total nitrogen	0.194–0.401	Weller et al. (1962)

proximately 45 gm. The estimate indicates that the rumen microorganisms constitute a large part of the nitrogenous foods needed by the host.

3. ESTIMATES FROM THE RATE OF RUMEN MICROBIAL GROWTH

The growth of the rumen microorganisms is another means for evaluating the quantity of microbial cells supplied to the host. Microbial growth in the rumen has been estimated by comparing fermentation rates when the microbial population is supplied *in vitro* with unlimited substrate (el-Shazly and Hungate, 1965; Hungate, 1956a); the zero-time rate method is used. The necessary precision in measuring the fermentation has been achieved by use of the constant-pressure syringe method which permits measurements almost immediately after removing the sample from the rumen.

Approximately 300 gm of rumen contents is removed from the rumen, care being taken to obtain portions from different regions and to keep the proper solids-liquid ratio. The sample is immediately incubated at 39°C. From this, a subsample of about 50 gm is removed to 200 ml of balanced inorganic salt solution at 39°C; the salt solution contains 20 gm of ground feed similar to that consumed by the animal. If the feed is acid, as most hays are, enough NaOH to bring the pH to 6.7 must also have been added to the mineral solution. The container, a pint jar, is immediately made anaerobic by displacing the air with carbon dioxide, stoppered, and the contents incubated at 39°C with occasional shaking. The gas production is determined by inserting a needle on a 10-ml syringe through the thin (5-mm) rubber in a recess in the stopper, and measuring the gas as it is produced. The barrel of the syringe must be kept well lubricated with water. The amount of gas produced is plotted against time.

After the main sample has incubated for an hour, a second subsample is removed, diluted with salts solution and hay as for the first subsample, and measured for its rate of gas production. Any growth or death of microorganisms in the 1-hour incubation of the main sample will be reflected in the difference between the fermentation in the two subsamples. The types of curves obtained are shown in Fig. VII-7 (el-Shazly and Hungate, 1965).

The rate of gas production by the main sample is also shown in the figure between the curves for the subsamples. The difference in the rate of gas production by the straight rumen sample, as compared to the diluted samples with substrate added, illustrates strikingly that in the rumen the microbes are not fermenting at their maximum capacity. Substrate is the chief limitation, but the concentration of fermentation products exerts some inhibiting influence (el-Shazly and Hungate, 1965). Sodium bicarbonate in the undiluted sample may be insufficient to measure acid production, but even with added bicarbonate the rate is far less than when additional substrate is provided.

The increase in the slope and the greater total gas production of the maximum-fermentation curves before and after the 1-hour incubation of the after-feeding sample indicate an increased capacity of the microbiota to ferment the substrate. This increase is interpreted as resulting from growth by the microbial population. In samples taken before feeding, when much less substrate is available in the rumen, the rate and total fermentation products after the 1-hour incubation are often slightly less than before incubation. This negative growth can be intrepreted as death of some of the microbial population, but could represent a depletion of metabolic machinery within still living cells.

The curves are interpreted as representing the net synthesis or growth of the microbial population. In Table VII-16 are shown the net microbial growths in rumen contents removed from hay- or grain-fed cattle at various times after feeding. The values for the two animals are quite similar, and, although some indications of differences in the rate of growth at different times after feeding are evident, the net growth rates per hour are never strikingly positive or negative. The average growth rate constant for

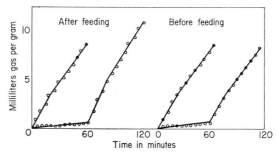

Fig. VII-7. Maximal fermentation rates as a measure of microbial growth during a 1-hour incubation. A sample of rumen contents is incubated for 1 hour under conditions resembling the rumen. A subsample is removed at zero time and incubated with salt solution and excess substrate to give the steep curve starting at zero time. After the 1-hour incubation a second subsample is removed and similarly incubated with excess substrate. An increase in the fermentation rate after 1 hour, as in the sample removed after feeding, indicates growth, the magnitude corresponding to the extent of the increased fermentation. In the sample obtained before feeding, no growth occurred; in fact death exceeded growth during the 1-hour incubation period. The relatively low gas liberation by the samples incubated under rumen conditions is due in part to limitation by the amount of bicarbonate, but chiefly by lack of substrate. The substrate added was the feed which the animal had been provided, ground to pass a 1-mm sieve.

all the experiments was about 0.08 per hour. In a continuous-fermentation system, this rate of growth would require a turnover of 1.92 times per day to keep the microbial population constant. This is about the turnover of the liquid-small particle fraction of the rumen contents. The agreement in the values indicates that the net growth equals the dilution rate, a requisite in a continuous-fermentation system.

Since the measurements are of net growth, the possibility of simultaneous growth and death of various fractions of the rumen population is not excluded, and undoubtedly occurs. However, the average net growth of any organism maintaining itself in the rumen must equal the dilution rate for that organism. In terms of division time, to maintain a constant concentration in the population, the average time between divisions of an organism must be 0.69 times the turnover time (Hungate, 1942). For a turnover rate of 1.92 per day, an average division time of once each 8.6 hours would be required. This is not a rapid division rate for most bacteria, but it seems probable that it exceeds the division rate of some of the protozoa. This would require that the latter turn over at a rate less than the 1.92 times per day for the liquid-small particle fraction of rumen contents.

Although net death of rumen bacteria may occur at various periods after feeding, particularly for some elements in the population, it is doubtful that simultaneous growth and death of particular species explain the relatively low net growth rates. Rapid growth of some cells and rapid death of others of the same species seem incompatible, unless the death is due to predation. The nutritional and inhibitory factors which would permit rapid growth of some of the cells should provide the same favorable milieu for others. Rapid growth and rapid death cannot simultaneously occur in the same species with all cells exposed to the same conditions. In a continuous-fermentation system all species grow at the same rate. To the extent that the rumen departs from a continuous-fermentation model, some species may grow at different rates at different times, but the average growth rate of each must equal the dilution rate, i.e., turnover rate.

It is obvious that the longer the interval between feedings the greater will be the differences in net growth at different times in relation to feeding. This emphasizes the advantage of keeping food continuously available to obtain maximum productivity or growth of the microbial population.

The results in Table VII-16, in which approximately the same growth rate was observed in the hay-fed and the hay- and grain-fed animal, emphasize that the net growth rate measures the turnover. Two animals at different levels of nutrition can show similar microbial growth rates, yet

Table VII-16

NET GROWTH OF RUMEN MICROORGANISMS AT VARIOUS TIMES AS MEASURED BY CHANGE IN MAXIMAL FERMENTATION RATE

Animal and feed	Time of sample	Rate of gas production of rumen contents (μl/gm/min)	Maximum rate of gas production before incubation (μl/gm/min)	Maximum rate of gas production after incubation (μl/gm/min)	Net growth (%)
Heifer, fed alfalfa hay	7:15 A.M.	8.45	89.2	82.6	−7.4
	9:00 A.M.	6.95	57.2	55.1	−3.5
	9:25 A.M.	5.88	61.4	64.8	6.0
	9:55 A.M.	10.70	37.2	38.5	4.0
	7:40 A.M.	18.30	73.4	84.4	15.5
	8:45 A.M.	10.01	50.0	63.5	27.0
	9:15 A.M.	8.05	50.4	61.5	22.0
	10:45 A.M.	9.47	54.5	60.6	12.0
	1:25 P.M.	8.06	56.1	63.2	13.0
	Average		58.8		9.8
Lactating cow, fed hay plus grain	Before feeding	10.50	123.0	124.0	1.0
	Before feeding	9.80	133.0	137.7	3.0
	45 min after feeding	17.25	119.0	125.5	5.5
	55 min after feeding	12.35	175.0	188.5	7.5
	75 min after feeding	13.80	118.5	143.0	20.0
	95 min after feeding	15.90	129.0	138.0	7.0
	3 hr after feeding	10.29	127.5	135.7	6.5
	6 hr, 10 min after feeding	9.10	131.0	150.0	14.0
	6 hr, 15 min after feeding	8.65	135.0	133.0	−1.3
	7 hr, after feeding	8.67	136.7	152.6	11.5
	7 hr, 45 min after feeding	9.60	114.6	135.5	19.0

1 hr, 35 min after feeding	11.55	128.5	127.8[a]	−0.5
2 hr, after feeding	—	119.0	113.5[a]	−4.6
2 hr, after feeding		119.0	121.0[a]	1.5
2 hr, after feeding	13.35	124.0	123.5[a]	0.0
6 hr, after feeding	9.00	126.0	149.0[a]	18.0
6.5 hr after feeding	9.84	135.0	152.0[a]	12.8
		135.0	157.0[a]	16.0
Average		129.4		7.6

[a] These samples were incubated for 1 hour in a plastic bag in the rumen, then removed, and the maximum fermentation measured *in vitro*.

receive from the rumen quite different quantities of microbial cells. This is because of the difference in the size of the microbial population. The net growth measurements are relative rates and show the percentage increase in the population. The average maximum fermentation rate with unlimited substrate is a measure of the average population size. If this is multiplied by the average growth rate, a value representing net growth rate is obtained. For the hay- and grain-fed cow this value is $129.6 \times 0.076 = 9.8$, and for the hay-fed heifer it is $58.8 \times 0.098 = 5.76$. The net cell production with hay-grain was 1.7 times the production in the hay-fed animal.

If turnovers of the microbial population in different animals are equal, i.e., if the net relative microbial growth rate is the same, the rumen volume and microbial concentration determine the quantity of cells supplied the host. The data in Table VII-17 (Moir and Williams, 1950) show

Table VII-17

EFFECT OF PLANE OF NUTRITION ON TOTAL DIRECT COUNT OF RUMEN BACTERIA[a]

Ration[b]			Percent of nitrogen	"True" digesti- bility of nitrogen (%)	Nitrogen balance (gm)	Biological value of nitrogen (%)	Bacterial Count $\times 10^{-9}$
Oaten hay	Starch	Casein					
499	173	0	0.595	72.8%	−23.48	83	29.1
499	148	24	1.157	93.2	−10.02	63.6	40.0
499	124	50	1.734	93.	+ 3.34	61.5	54.5
499	99	75	2.294	92.7	+13.33	62.7	55.3
499	73	102	2.893	91.2	− 1.61	39.7	61.2

[a] From Moir and Williams (1950, p. 387).
[b] All values in grams.

that the plane of nutrition influences the concentration of bacteria in the rumen. At the highest nitrogen intake much of the protein supplied must have been fermented. This is suggested by the low biological value, coupled with the high bacterial count.

The rate of disappearance of glucose added to the rumen is slow for sheep on poor hay as compared to better nourished animals (Elsden, 1945b), which reflects the difference in microbial population. Disappearance of glucose from *in vitro* cultures has been used to compare rumen microbial populations (Coop, 1949; McAnally, 1943).

G. Thermodynamic Limitations on Protoplasmic Synthesis Imposed by Anaerobiosis

The radiant energy converted to chemical energy during photosynthesis is usually stated as existing in the form of the synthesized carbohy-

drates. Strictly speaking, this is only half the picture. The equation showing energy storage in photosynthesis is usually formulated as

$$energy + CO_2 + H_2O \rightarrow CH_2O + O_2$$

The energy stored in carbohydrate is released only if oxygen is available
to combine with the carbon of the carbohydrate, or more accurately
speaking, with the hydrogen which can be removed. The energy is really
stored through the separation of the oxygen from the carbon and hydrogen and is only released when oxygen has again combined.

As previously emphasized, living organisms do not use food merely to
release energy as such. The energy is essential to perform biological
work, of which the synthesis of new protoplasm is of primary importance.
The amount of the food which can be stored in the form of new cells
(Marston, 1948a) is an important measure of the success of metabolism
in microbes, as well as in the metabolism of growing macroorganisms.

In the process of growth the food is rearranged into protoplasm, and
part of the energy in the food is stored in the chemical constituents of
protoplasm, the proteins, carbohydrates, lipids, and nucleic acids. Part
of the energy of the food is dissipated as heat (Marston, 1948b). This is
the unavoidable consequence of the fact that in any chemical reaction the
efficiency with which chemical energy in one form is converted into chemical energy in another form is never 100%. Usually the dissipated energy
appears as heat. The heat generated in metabolism represents the loss of
energy unavoidable during the conversion of substrate molecules into cell
constituents.

The amount of protoplasm which can be synthesized depends on two
things, the usable high-energy compounds that can be derived from the
substrate (usually expressed as ATP) and the amount and nature of the
food derivatives (intermediates easily synthesized into monomers) which
can be built into cells. In many cases the energy-yielding material and
the material transposed into cells are the same. This is the situation when
carbohydrate is the substrate (Hungate, 1955), insofar as the carbon and
hydrogen of the cell material are concerned, and is the case in the rumen
where most of the substrate is carbohydrate. The fermentable substrates
serve not only as a source of energy, but also as the source of the chemical compounds built into the microbial cells.

The utilization of feed constituents to form protoplasm poses much
the same problem for aerobes as for anaerobes, except that the availability
of oxygen makes the energy derivable from a carbohydrate substrate
much greater. Aerobic microorganisms can synthesize into protoplasm
as much as 60–70% of carbohydrate substrate, whereas for most anaerobic bacteria the quantity is usually of the order of 10%, rarely exceed-

ing 20%. *Anaerobiosis limits the extent to which food can be synthesized into cell material.*

Since the rumen is anaerobic, microbial synthesis is significantly less than that possible under aerobic conditions. Because the ruminant is aerobic, the greater energy supply gives it the potential for greater synthesis of protein, but its digestive arrangement makes it dependent on the protein synthesized in the anaerobic microbial metabolism. Since this anaerobic synthesis is small, the synthesis by the aerobic host must be small. Insofar as protein is concerned, the ruminant lacks the alimentary features necessary to give the conversion efficiencies characteristic of nonruminants. The rumen anaerobiosis imposes a thermodynamic limit on the extent of host protein synthesis.

The thermodynamic limitation on the possible extent of host protein synthesis has as its corollary an excess of energy-yielding materials available to the host. The VFA's from the rumen fermentation of carbohydrate contain 75% of the energy, disregarding the proteins, or about 65% if 10% of the substrate energy is assimilated into microbial cell bodies (Hungate, 1963a). This 65% is available to the host through oxidation in its tissues. Part of it is used in synthesizing host protein from the available microbial protein and part for the various other host processes requiring energy. The remaining energy, chiefly in the form of acetic acid, is stored as fat.

The limited storage of protein in the ruminant as compared to other animals, and the greater storage of fat are shown in Table VII-18 (from Meyer and Nelson, 1963).

It can be seen that the ruminants convert a smaller percentage of digested food into body protein, and a larger percentage into fat, except that swine are unusually high in fat storage. It is tempting to suggest an active fermentation in the swine cecum. Volatile fatty acids have been found in the colon (Friend *et al.,* 1962, 1963a,b), and the amounts in the portal blood have been estimated as sufficient to meet 15–28% of the porcine maintenance energy requirement (Friend *et al.,* 1964).

If nitrogen assimilation by the rumen microorganisms is the chief factor limiting the growth of the host, nitrogen assimilation by the host would appear to be a better measure of growth than is weight gain. Especially with ruminants nearing maturity, much of the weight gain is likely to be in the form of fat, which makes weight gain a poor index of nitrogen assimilation.

H. Summary

The microbial cells and the acids and gases formed during fermentation are concomitant products of the anaerobic rumen metabolism. Car-

Table VII-18

EFFICIENCIES OF FEED CONVERSION IN SOME DOMESTIC ANIMALS[a]

	High fiber ration[b]					Low fiber ration[c]			
	Rats	Chicks	Pigs	Sheep	Steers	Rats	Chicks	Pigs	Sheep
Daily gain (wet wt.), percent of intake (dry wt.)	26%	33%	27.5%	18.7%	16.3%	32.6%	35.6%	29.2%	16.5%
Protein stored, dry wt./dry wt. feed digested[d]	6.7%	12.6%	4.3%	2.9%	3.4%	7.4%	12.0%	3.6%	2.0%
Energy gain, kcal stored/dry wt. feed digested	690	894	1640	1120	1070	894	1055	2015	1100
Percent of fat in body wt. gain	9.2	3.7	40.0	41.3	38.7	10.4	5.9	49.6	48.5

[a] After Meyer and Nelson (1963, p. 345).

[b] 30% alfalfa hay, 46.5% barley, 15% soybean oil meal, 5% cottonseed oil meal, 2% tallow, 1% $Ca_3(PO_4)_2$, 0.5% NaCl.

[c] 5% alfalfa hay, 70.5% barley, 15% soybean oil meal, 5% cottonseed oil meal, 2% tallow, 2% $Ca_3(PO_4)_2$, 0.5% NaCl.

[d] In calculating the digestibilities of the feeds, the data of Schneider (1947) were used. The calculated digestibilities were 76% for the high fiber ration and 84% for the low fiber ration.

bohydrates are rearranged to yield high-energy molecules and building materials of the cell, and these are combined in further reactions with nitrogen and other elements to form the various materials composing the microbial bodies. The amount of cells which can be formed is a function both of the carbohydrate supply and the mineral elements and other required growth factors. Reciprocally, the quantity of carbohydrate which can be digested depends on the synthesis of new cells to replace those leaving the rumen. Fermentation and growth are inextricably linked.

This chapter has stressed the quantitative relationships between fermentation and growth, in an attempt to obtain an average value expressing the quantity of digested carbohydrate that can be assimilated into cells during the rumen fermentation. A rough estimate of 10% fits reasonably well with the available data. This value changes according to the quantity of storage carbohydrate in the cells. If the cell yield is based on nitrogen, about 1.1 gm of microbial nitrogen is assimilated for each 100 gm of carbohydrate fermented. These estimates are conservative. They do not take into account the fact that some of the digestible material is protein and other materials instead of carbohydrate, which would make the assimilation based on actual carbohydrate a little higher. However, this error is probably less than the total error involved in these rough average estimates. The values are reported more as a stimulus to the collection of accurate data than as a standard value.

The ratio between nitrogen assimilated and carbohydrate fermented undoubtedly varies with changes in the kinds and ratios of feed available. This modification will be discussed in Chapter X.

CHAPTER VIII

Vitamins and Minerals

In the last chapter it was pointed out that proteins and amino acids served in the rumen not only as a source of building materials for growth of microorganisms, but also were fermented to obtain energy if more were present than could be assimilated into microbial cells with the available fermentation energy. This holds for any particular time in the rumen. Even though the amount of feed protein is barely sufficient to form the maximum possible quantity of cells with the energy available from carbohydrate, it cannot all be assimilated immediately. Time is required for the digestion of the energy-containing insoluble carbohydrate components in order that nitrogen can be assimilated.

Immediately after the ration is ingested, protein is usually in excess over available carbohydrate, and some protein is fermented as a source of energy. This leaves a less proportion of protein or amino acids during the later stages of fermentation as fiber is digested. Microbial nutrients other than carbohydrate are present chiefly in relatively simple forms. The rumen bacteria active on fiber during later stages of the fermentation exhibit considerable powers of synthesis of cell material from simple substances. Water-soluble vitamins, inorganic minerals, and other growth factors are used.

A. Vitamins

The fat-soluble vitamins have been studied very little in relation to the rumen bacteria (Albaugh *et al.*, 1963). The destruction of vitamin A by rumen contents (Klatte *et al.*, 1964) is greater in animals on roughages than when concentrates are fed (Keating *et al.*, 1964). This destruction is more likely due to a reduction of carotene unsaturated bonds than to oxidation. Nitrate added to incubating rumen contents did not increase carotene destruction to a significant extent (Davison and Seo, 1963). A requirement for vitamin K has been found for a rumen strain of *Fusiformis* (Lev, 1959).

It was recognized early that ruminants were in a special category insofar as the water-soluble vitamins were concerned, since B vitamins were synthesized during *in vitro* incubation of rumen contents (Bechdel *et al.,* 1928). Studies on individual vitamins (McElroy and Goss, 1939, 1940a,b, 1941a,b; McElroy and Jukes, 1940; M. I. Wegner *et al.,* 1940b, 1941c) showed that vitamin K, pantothenic acid, vitamin B_6, biotin, and riboflavin were synthesized in the rumen. On some rations vitamin synthesis was augmented by urea (Teeri and Colovos, 1963), as would be expected for any limiting food.

The amount of B vitamins in rumen contents was several times that expected on the basis of the vitamin content of the feed. Thiamine was slightly more abundant in the rumen than in the feed, but the increase was not striking. Biotin was synthesized in the rumen (McElroy and Jukes, 1940), as was nicotinic acid (M. I. Wegner *et al.,* 1940b). Most of these early studies (reviewed by Kon, 1945; Kon and Porter, 1947, 1953a, 1954), highly suggestive of a rumen microbial synthesis of vitamins, have been substantiated by subsequent measurements. Even with poor rations the B vitamins are not limiting (Head, 1953).

Feeding of calves on purified rations containing abnormally low levels of vitamins (Agrawala *et al.,* 1953b) showed convincingly that riboflavin, pantothenic acid, and niacin were synthesized. Young ruminants on rations containing these vitamins show little rumen synthesis (Buziassy and Tribe 1960b). In general, the greater the quantity of vitamins in the ration, the less the synthesis in the rumen (Hollis *et al.,* 1954). In feeds containing minimal B vitamins it is important that all other nutrients for bacterial growth be present in adequate concentrations to support vitamin synthesis. The most conclusive evidence for mass synthesis of B vitamins has been provided in experiments with milking cows fed a vitamin-free ration (Virtanen, 1963). The rumen concentrations of thiamine, riboflavin, nicotinic acid, pyridoxine, folic acid, biotin, and pantothenic acid were at least as high as in control cows on conventional rations.

In *in vitro* experiments (Hunt *et al.,* 1943, 1952), added starch increased riboflavin synthesis 126% with rumen contents from alfalfa hay-fed fistulated steers; niacin synthesis was similarly increased by 225%, pantothenic acid by 52%, and vitamin B_{12} by 39%. With inoculum from timothy hay-fed steers, riboflavin synthesis was increased by 243%, niacin by 191%, pantothenic acid by 3135%, and B_{12} by 3%. The low synthesis of B_{12}, even though the medium contained added cobalt, suggests that the bacteria attacking the starch do not synthesize much cobalamine (see also R. B. Johnson *et al.,* 1956). In other experiments (Minato *et al.,* 1962), starch additions caused an increase also in thiamine.

The studies of Rérat and colleagues (Rérat and Jaquot, 1954; Rérat *et al.,* 1956, 1958a,b, 1959) have shown that, although the rumen wall is permeable to most of the B vitamins, these vitamins are absorbed very little in this organ, since they are bound within the bodies of the microorganisms. As digestion occurs the vitamins are released. Some release and absorption occurs in the omasum, more in the abomasum; release and absorption are maximal in the proximal part of the small intestine. Biotin, B_{12}, thiamine, riboflavin, niacin, pantothenic acid, and pyridixal phosphate were synthesized in the rumen, and B_{12} and biotin were synthesized to some extent also in the cecum. The synthesis of thiamine, riboflavin, and niacin was increased by supplementing an alfalfa hay ration with starch. Supplementation with gluten gave an increase in B_{12} synthesis. Ascorbic acid was destroyed in the rumen (Knight *et al.,* 1940), presumably by reduction. The rumen concentration of thiamine and riboflavin varied significantly according to the ration, whereas the biotin concentration did not vary (Hayes *et al.,* 1964b). The amounts of six B vitamins found in the various parts of the alimentary tract of sheep are shown in Table VIII-1.

Since the vitamins are chiefly in the rumen microbes, vitamin synthesis is related to the amount of growth. When nitrogen is limiting, vitamin synthesis is a function of the amount of nitrogen in the ration (Buziassy and Tribe, 1960a).

There is little information on the kinds and quantities of vitamins formed by the different rumen bacteria. The protozoa do not synthesize the B vitamins (Usuelli, 1956). The fact that the concentration of the B vitamins in the rumen is not influenced by the amount in the diet indicates not only that the rumen organisms have the capacity for their synthesis, but also that metabolic control prevents synthesis of the vitamins beyond the levels usually found.

1. THIAMINE

In a quantitative study (Phillipson and Reid, 1957) of thiamine in the feed and in the rumen, the daily intake of sheep on a ration of hay plus various supplements was 2.0 to 4.4 mg. Most of the thiamine was in solution in the fluid. The rumen contents of these animals contained an average of 0.55 mg thiamine per kilogram, i.e., a rumen containing 5 kg of contents would contain 2.75 mg of thiamine. With a turnover time of 18 hours, the thiamine leaving a 5-kg rumen would be 4.0 mg. This rough comparison suggests that the synthesis was not great, a conclusion in agreement with those of early investigators.

Little thiamine was found in the rumen contents of two fistulated cows on a low thiamine ration (McElroy and Goss, 1941a), but it could be demonstrated in two sheep and in the rumen contents of an unfistulated

Table VIII-1

VITAMIN IN MICROGRAMS PER GRAM DRY WEIGHT OF CONTENTS[a]

Ration	In ration	Rumen	Reticulum	Omasum	Abomasum	Small intestine	Cecum	Colon	Rectum
Thiamine									
A	2.4	3.1	7.	8.9	6.2	5.4	3.5	3.3	3.
B	18.4	20.3	29.4	25	20	4.8	6	5.6	6.1
C	5.7	6.8	16	14.9	18.9	13.3	2.2	2.2	1.9
Riboflavin									
A	4.9	7.1	12.8	19.6	10.2	12	8.5	6.8	5.3
B	26.9	23.7	36.7	24.8	20.9	27.7	12	11.4	9.9
C	11	9.7	21.7	13.3	14.4	26	6.8	6	3.9
Niacin									
A	10.8	59.6	76.9	140.6	119.1	96.2	35.6	33.7	24
B	91.2	180.6	232.9	159.2	219.4	234.7	66.6	65.9	60.1
C	50.4	46.1	82.3	42.3	127.6	196.3	24	19.7	18.6
Pantothenic acid									
A	23.1	17.5	20.2	33.6	28.5	58.6	11	9.4	5.3
B	29.7	37.9	56.5	60.7	54.8	138.3	25.7	17.9	15.7
C	42.2	16.8	41.9	25.6	45.9	130.8	20.5	22.2	16.6
Biotin									
A	0.04	0.13	0.18	0.32	0.18	0.45	0.72	0.76	0.46
B	0.47	1.16	0.52	0.36	0.88	1.43	2.25	1.99	2.35
B_{12}									
A	Traces	1.15	2.2	3.46	2.5	2.44	5.59	4.22	4
B	Traces	2.3	2.99	3.21	2.62	2.76	7.71	6.37	6.63
C	Traces	1.39	2.65	1.76	2.4	2.12	3.11	4.43	3.24

[a] From Rerat and Jacquot (1954).

cow. Four hours after feeding, the rumen contents contained a greater concentration of thiamine than did the ration (Hunt *et al.,* 1941), but 12–16 hours after feeding, the rumen thiamine concentration was less than in the ration. A daily ration of grass contained 8.1 mg thiamine, and the amount in the rumen was 1.32 mg per kilogram, or a total of 6.6 mg. This would amount to 8.8 mg daily leaving a rumen with an 18-hour turnover, which shows that, in this case also, little thiamine was synthesized by rumen microorganisms.

When rations contain relatively little thiamine, the vitamin is synthesized in the rumen (Buziassy and Tribe, 1960a; Teeri *et al.,* 1950; Blaxter and Rook, 1955). Administration of *p*-aminobenzenesulfonamidomethylthiazole, tested as an antibacterial chemical, greatly increased the quantity of thiamine in the goat rumen (Baldissera, 1951), presumably due to ready availability of the thiazole moiety. The thiamine content of rumen bacteria is greater than that of the protozoa (Manusardi, 1931: Conrad and Hibbs, 1955).

2. Riboflavin

Riboflavin may be synthesized in the rumen in relatively large quantities. The rumen contents of sheep feeding on a ration with 0.3 μg riboflavin per gram (McElroy and Goss, 1939) contained 25 to 33 μg of riboflavin per gram. When the ration consisted of yellow maize, lucerne hay, and a protein supplement, there was a net synthesis of riboflavin by the rumen microorganisms (Hunt *et al.,* 1941, 1943), but rumen contents from a steer fed only lucerne hay showed no synthesis of the vitamin.

The amount of riboflavin leaving the body of goats via the milk, urine, and feces was independent of the amount of riboflavin in the ration (Crossland *et al.,* 1958). The basal ration was practically free of riboflavin, and part of the animals received a supplement of 15 mg daily. On the low riboflavin diet only riboflavin was found in the urine, but additional lyochromes, inactive nutritionally for *Lactobacillus casei,* were eliminated by animals on riboflavin-sufficient rations. In two goats the quantity of riboflavin synthesized per day was estimated at 5.2 mg and 6.9 mg, respectively. Twenty milligrams of riboflavin were synthesized in 6 hours in a bovine rumen (Agrawala *et al.,* 1953b). In other experiments (P. B. Pearson *et al.,* 1953) with cattle, 30 mg riboflavin were excreted daily when the ration contained 60 mg. Synthesis on riboflavin-poor diets was not affected by changing the protein supply.

Nutritional studies on young calves (Wiese *et al.,* 1947) and lambs (Luecke *et al.,* 1950) indicate that riboflavin is required prior to development of the rumen microbiota, a requirement probably retained in the mature animal. Both bacteria and protozoa of the rumen are rich in riboflavin (Conrad and Hibbs, 1955).

3. NIACIN

The nicotinic acid content of lambs (P. B. Pearson *et al.*, 1940) and cattle (Agrawala *et al.*, 1953b) on a ration low in nicotinic acid was as high as in similar animals on a regular diet (P. B. Pearson *et al.*, 1940). Nicotinic acid is excreted by sheep fed a diet causing black tongue in dogs (Winegar *et al.*, 1940). The source of the vitamin was uncertain since the lambs were in a poor nutritional state. Synthesis of nicotinic acid in the sheep rumen increased when there was more protein nitrogen in the ration (Buziassy and Tribe, 1960a) and decreased when nicotinic acid was provided. Twenty-five milligrams of nicotinic acid were excreted in the urine of sheep on a semisynthetic ration containing 12.6% protein; approximately the same amount was excreted by sheep on a hay ration (P. B. Pearson *et al.*, 1953). When the protein of the semisynthetic ration was reduced to 3.6% the quantity of excreted nicotinic acid dropped to 7 mg daily. Nicotinic acid to the extent of 154 mg was synthesized in a bovine rumen during the 6 hours after feeding a ration relatively free of B vitamins (Agrawala *et al.*, 1953b).

4. PANTOTHENIC ACID

Reported values (Agrawala *et al.*, 1953b; McElroy and Goss, 1941b; Teeri *et al.*, 1950) show that pantothenic acid is actively synthesized in the rumen. Excretion of pantothenic acid by animals on a semisynthetic diet (P. B. Pearson *et al.*, 1953) was about the same as on regular rations, i.e., 50 mg per day, but when the protein content of the semisynthetic ration was decreased to 3.6%, the excretion was only 25 mg, presumably due to the reduction in the microbial population in these latter animals. Seventy micrograms of pantothenic acid per gram of dry matter were found in sheep rumen contents (McElroy and Goss, 1941b) and 60–90 μg per gram in the rumen contents of two cows. The amount of pantothenic acid in the milk of one of the cows exceeded the quantity in the ration.

Apparent inhibition of pantothenate synthesis by sulfur compounds was observed in experiments *in vitro* with rumen microorganisms (Hunt *et al.*, 1954). A stimulation of panthothenate synthesis by molasses in the ration has been reported (Teeri *et al.*, 1951).

5. VITAMIN B$_6$

Analyses (McElroy and Goss, 1940b) indicate that B$_6$ vitamins are synthesized in the rumen of cattle and sheep fed a ration containing

1–1.5 μg of B$_6$ per gram of dry weight. A sheep on the ration for 30 days showed 10 μg B$_6$ per gram dry weight of rumen contents. A composite sample of bovine rumen contents, removed after 57, 66, and 91 days on the low B$_6$ ration, contained 8 μg of B$_6$ per gram. Milk from one of the cows contained 0.17 μg per milliliter of B$_6$, a concentration equal to that in commercial skim milk, assayed by promotion of growth in rats.

Pyridoxamine was the vitamin that most effectively stimulated cellulose digestion by washed cell suspensions of rumen bacteria (MacLeod and Murray, 1956). It is required by *Ruminococcus albus* (Bryant and Robinson, 1961c; Hungate, 1963a).

6. BIOTIN

Rumen contents can relieve the symptoms of biotin deficiency in chicks fed egg white (McElroy and Jukes, 1940). Biotin is a required growth factor for many strains of *Ruminococcus, Bacteroides succinogenes, Bacteroides ruminicola,* and *Butyrivibrio.* For one strain of *Ruminococcus albus,* growth on cellulose was diminished when the concentration of biotin was less than .03 μg per 10 ml. Biotin stimulated cellulose digestion by washed rumen bacteria (Bentley *et al.,* 1954a).

7. COBALAMIN AND RELATED GROWTH FACTORS

The history of this vitamin is closely linked with that of cobalt as a nutritional factor, and both will be considered in this section.

The cobalt-deficiency disease characterized by poor appetite is variously named—in Australia as Coast disease, wasting disease, nutritional anemia, and enzootic marasmus; in New Zealand as bush sickness, Turanga disease, Morton Mains disease, and Mairoa dopiness; in Kenya as Nakuruitis; in America as salt sickness, Burton Ail, Grand Traverse disease, and weaner ill thrift; in South Africa as dune pining; in Brazil as peste de Secar and mal de Colete; in Eire as Galar Truagha; and in Great Britain as pining, vinquish, daising, moorcling, and moor sickness (A. I. Wilson, 1962).

The disease is relieved by oral administration of cobalt (Marston, 1935; Lines, 1935; Underwood and Filmer, 1935), which is incorporated into an organic form in the rumen (Abelson and Darby, 1949; Hine and Dawbarn, 1954). An effective preventive dose for sheep is 2 mg twice each week.

Oral administration of cobalt is by far the most effective route (Phillipson and Mitchell, 1952; Ray *et al.,* 1948; McCance and Widdowson, 1944), even though much of it is lost in the feces (Monroe *et al.,*

1952). If sufficiently large quantities of cobalt are injected into the animal, deficiency is relieved, the cobalt presumably diffusing from the blood into the rumen (Monroe *et al.*, 1952). Introduction of high doses of cobalt into the abomasum or duodenum is also effective (Phillipson and Mitchell, 1952). Cobalt oxide in the form of heavy pellets in the rumen is adequate for periods of at least a year (Dewey *et al.*, 1958). The cobalt is assimilated into the bodies of rumen microoorganisms (W. H. Hale *et al.*, 1950; Tosic and Mitchell, 1948) and is most concentrated in the small intestine (Rothery *et al.*, 1953). Radioactive cobalt added to the bovine rumen disappears at about the rate expected from turnover (Comar *et al.*, 1946).

Ruminants are unique in their requirement for cobalt. Horses grazing forages containing inadequate cobalt for cattle and sheep, i.e., 0.03–0.05 ppm on a dry weight basis, show no ill effects. Below 0.07–0.1 ppm, ruminants show signs of deficiency. It has been suggested (Underwood, 1956) that the use of volatile fatty acids by ruminants increases their need for cobalt-containing factors.

Plants do not require cobalt, but they contain a small concentration of the metal, usually 0.1–0.3 ppm on a dry weight basis, the amount varying according to the soil. Up to 160 mg of cobalt per 100 lb live weight is not toxic to sheep (D. E. Becker and Smith, 1951a), and 40 mg/100 lb is tolerated by cattle (Underwood, 1956).

With the discovery (Rickes *et al.*, 1948) that vitamin B_{12} contained cobalt, the effectiveness of B_{12} for relieving the cobalt-deficiency symptoms was tested. Cobalamin was ineffective when given orally, and in initial experiments it was found not to relieve cobalt-deficiency symptoms when given by intravenous injection or parenterally (D. E. Becker and Smith, 1951b; Becker *et al.*, 1949; McCance and Widdowson, 1944). Later studies using larger amounts of B_{12} showed that it relieved deficiency symptoms when injected parenterally (S. E. Smith *et al.*, 1951), intramuscularly (Marston and Smith, 1952; Anderson and Andrews, 1952), or intravenously (Hoekstra *et al.*, 1952a). Vitamin B_{12}-like compounds were more abundant in the rumen contents of cobalt-supplemented animals than in deficient animals (Hoekstra *et al.*, 1952b; Koval'skiĭ and Raetskaya, 1955). The B_{12} concentration in ruminant liver is higher than in nonruminants.

Lambs receiving 35 μg of B_{12} daily by injection showed growth as good as those given 500 μg of B_{12} daily by mouth (Hoekstra *et al.*, 1952a; Kercher and Smith, 1955). An adequate sheep dose is about 100 μg of B_{12} per week. Even though the ruminant stores more B_{12} than is characteristic of other animals, dosing must be frequent (Andrews and Anderson, 1954) since cobalt is continuously metabolized. Deficiency symptoms appear when the B_{12} concentration in the blood drops to 0.2 mμg per milliliter (Dawbarn *et al.,* 1957b).

In addition to B_{12}, cobalt occurs in cyanocobalamin (Holdsworth, 1953), pseudovitamin B_{12} (with adenine in place of 5,6-dimethylbenzimidazole) (Dion *et al.,* 1952; Holdsworth, 1953; F. B. Brown and Smith, 1954), factor B (cobalamin with no nucleotide) (Holdsworth, 1953; Gant *et al.,* 1954), vitamin B_{12}b (hydroxycobalamin), vitamin B_{12}c (nitrocobalamin), factor C (guanine cobalamin) (Holdsworth, 1953), factor A (2-methyladenine), and additional types. A number of these compounds have been identified in rumen contents, chiefly in the bacterial cells. The diversity in the form in which cobalt occurs suggests slight differences in function in different microbial species. Some microorganisms, e.g., *Ochromonas malhamensis,* can use only one form of B_{12} (Ford and Hutner, 1955), whereas others, e.g., some strains of *Escherichia coli,* can utilize a wide variety (Ford *et al.,* 1953).

Synthesis of B_{12} is restricted to microorganisms (Underwood, 1956). Bacteria and blue-green algae are the only groups in which its synthesis has been shown to occur, and not all bacteria and blue-green algae possess this ability. Animal tissues contain various quantities of B_{12}, most of it in the form of the vitamin itself (Ford *et al.,* 1953), obtained from the food or from intestinal bacterial synthesis.

The concentration of vitamin B_{12} in the rumen has been estimated by various workers (Dawbarn *et al.,* 1957a). Table VIII-2 shows the results (Kon and Porter, 1953b) for a number of vitamins. The amount per gram of rumen contents is 0.4 μg cobalamin and 1.4 μg total B_{12}-like factors (Ford *et al.,* 1953). Cobalamin-like factors are synthesized not only in the rumen (33 mg synthesized per day per bovine) (Cardoso and Almeida, 1958), but also in the posterior alimentary tract (7.6 mg per day).

Propionibacterium species form factor A, but are not numerous enough to account for the factor A in the rumen (Porter and Dollar, 1958). *Bacteroides amylophilus* does not form factor A, but an unidentified rumen species does. Vitamin B_{12}-like compounds are produced in significant quantities by *Selenomonas* (Dryden *et al.,* 1962) and to a less extent by strain B385, *Peptostreptococcus,* and *Butyrivibrio,* listed in the order of decreasing amount formed. Tested strains which do not synthesize appreciable quantities of B_{12}-like materials (Dryden *et al.,* 1962) include *Bacteroides succinogenes, Bacteroides ruminicola, Eubacterium, Lachnospira, Propionibacterium, Ruminococcus albus, Ruminococcus flavefaciens, Succinivibrio, Succinimonas,* and *Streptococcus bovis* (Hardie, 1952). There are no reports that cobalt stimulates growth of pure cultures of rumen bacteria, although high concentrations are inhibitory (Gibbons and

Table VIII-2

THE B VITAMIN CONTENT OF THE RUMEN OF STEERS GIVEN VARIOUS DIETS[a,b]

Vitamin	Hay		Hay and concentrates		NaOH-treated straw, casein		Hay and concentrates with penicillin	
	Diet	Rumen contents	Diet	Rumen contents	Diet	Rumen contents	Diet	Rumen contents
Thiamine	0.8	2.1	5.0	3.0	0	1.8	5.0	—
Riboflavin	13.0	11.0	9.0	13.0	1.0	12.0	9.0	20.0
Nicotinic acid	27.0	50.0	32.0	60.0	2.1	52.0	32.0	63.0
Pantothenic acid	11.0	10.0	19.0	28.0	1.2	18.0	19.0	—
B_6	2.7	2.8	2.5	2.5	0.25	2.4	2.5	3.0
Biotin	0.14	0.16	0.12	0.22	0.004	0.17	0.12	0.16
Folic acid	0.4	1.7	0.25	2.3	0.08	1.0	0.25	—
B_{12}, all kinds	0	5.0	0	6.5	0	8.3	0	5.0

[a] All values given as μg per gram of dry material; each value represents the mean of 40–50 samples.
[b] From Kon and Porter (1953b).

Doetsch, 1955; Salsbury *et al.,* 1956). Toxicity of silicon fluoride for ruminants, relieved by B_{12} injection, has been ascribed to inhibition of rumen microorganisms synthesizing cobalamin (Sahashi *et al.,* 1953).

The question arises, "Are the cobalt vitamins required by the microorganisms or the host?" An effect on the host is indicated by the finding that propionate metabolism is impaired in cobalt-deficient hosts, and propionate accumulates to higher steady state concentrations than are characteristic of normal animals (R. M. Smith and K. J. Monty, 1959; Annison, 1964). It is presumably concerned in the reaction converting succinyl CoA to methylmalonyl CoA. The fact that small amounts of B_{12} injected into the host can cure cobalt-deficiency symptoms and restore appetite, whereas much more must be administered orally, also indicates that the primary effect is on the host. If the animal receives B_{12}, appetence is maintained, and the rumen microbial fermentation is not affected (Hoekstra *et al.,* 1952a). When the ration of a cobalt-sufficient animal is reduced to the level imposed by the inappetence of the cobalt-deficient animal, the body weight changes are identical (J. Stewart, 1953). This suggests that the rumen fermentation in cobalt-deficient animals is limited by lack of food rather than specific lack of cobalt. Cobalt administration does not increase the digestibility of the ration (Bentley *et al.,* 1954b). In some experiments (D. E. Becker and Smith, 1951a,b) the digestibility of the total cobalt-supplemented ration was higher because the cobalt-sufficient animals consumed 50% more concentrate. The fiber digestibility was greater in the cobalt-deficient, probably due to the slower turnover accompanying the 25% decrease in consumption.

It is difficult to reconcile the above observations with the known occurrence of B_{12}-like substances in bacteria and with the report (L. E. Gall *et al.,* 1949) that the bacterial count in cobalt-supplemented sheep is higher (5.6×10^{10} per milliliter) than in deficient sheep (3×10^{10} per milliliter) consuming the same quantity of feed. Cobalt is reported to stimulate cellulose digestion by washed cell suspensions of rumen bacteria (Bentley *et al.,* 1954a), and a B_{12} requirement for some cellulolytic rumen bacteria has been reported (Dehority and Scott, 1963). Methylcobalamin is an intermediate in methanogenesis by *Methanobacterium omelianskii* and may be important also in *Methanobacterium ruminantium*. The carboxylation of phosphoenolpyruvate by *Succinivibrio* has been reported to be Co^{++}-dependent (Scardovi, 1963).

The facts that concentrate consumption decreased more rapidly during cobalt deficiency than did fiber consumption (D. E. Becker and Smith, 1951a,b), and that much B_{12} synthesis occurs in the cecum (Cardoso and Almeida, 1958) suggest that B_{12} synthesis is associated more with protein than with carbohydrate utilization in the rumen. The concentration of B_{12} in rumen bacteria was greater than in the protozoa (Kawashima et al., 1959) and less in the microorganisms from concentrate-fed cattle and goats than from those fed hay. The importance of B_{12}-like coenzymes in the bacterial fermentation of glutamate (Volcani et al., 1961) is well known. Fiber digestion may remain normal even during severe cobalt deficiency, because the fibrolytic bacteria do not require it. The evidence suggests that the ruminant is affected first, but it seems probable that deficiency would ultimately affect at least some of the bacteria. It is possible, however, that a flora requiring no cobalt might develop.

8. Folic Acid and Related Factors

The concentration of folic acid in the rumen is greater than in feeds selected for low vitamin concentrations (Kon and Porter, 1953b; Lardinois et al., 1944). Requirements of rumen bacteria for p-aminobenzoic acid (Bentley et al., 1954a; Bryant and Robinson, 1962b; Hungate, 1963a) indicate an importance of folic acid. A strain of Ruminococcus required the entire folic acid molecule (Ayers, 1958).

B. Other Organic Nutrients for Rumen Microorganisms

The importance of additional trace organic materials in the nutrition of rumen bacteria is recognized, but not all of them have been identified. The extensive investigations of Bryant and co-workers have demonstrated the requirements for the branched- and straight-chain fatty acids, acetic acid, and heme. The importance of the C_4 and C_5 acids in microbial nutrition had not previously been appreciated, but they are now being found essential for a number of nonrumen bacterial species. The stimulus provided by natural feeds and by rumen fluid to the growth of rumen bacteria in media containing known nutrients suggests that additional substances are important in the rumen. Their identification may materially assist in the design of ruminant rations, as well as in disclosing new biochemical reactions.

C. Inorganic Foods

The relative proportions and kinds of minerals needed by the ruminant are reflected in the composition of the mineral mixtures fed animals on largely synthetic diets (Tillman et al., 1954b; Rook and Campling, 1959; Virtanen, 1963; W. E. Thomas et al., 1951; Agrawala et al., 1953a; Ellis et al., 1956). The review of mineral nutrition by Underwood

(1957) gives a good account of several aspects of mineral metabolism, with references to others.

The inorganic requirements for the rumen can be divided into two categories: bulk minerals and trace minerals. Bulk minerals are sodium, potassium, calcium, magnesium, phosphorus, sulfur, carbon, and chloride. Some of these have been mentioned previously, e.g., sodium as an increased requirement when parotid saliva is lost.

Organic compounds of the elements carbon, hydrogen, and oxygen have been considered extensively in Chapter VI on carbohydrate fermentation, and nitrogen was discussed in Chapter VII. Hydrogen and oxygen are essential also in mineral form as H_2O, and carbon in the form of CO_2 or bicarbonate. A requirement for oxygen in the rumen as a whole has not been demonstrated, and, to the best of the author's knowledge no microorganisms requiring oxygen have been found to play an important role in the rumen.

1. CARBON DIOXIDE

Carbon dioxide is a common fermentation product, but it is not ordinarily a required nutrient. The requirement for carbon dioxide or bicarbonate as a nutrient in the rumen has been mentioned in connection with the metabolism of the succinogenic and methanogenic bacteria. In ruminants, the abnormally high alkalinity in the stomach of animals on a poor ration may in part be aggravated by inadequacy of the fermentative carbon dioxide production to keep the acidity within the normal range. The carbon dioxide pressure of the circulating blood (0.05 atm) is insufficient to maintain the 0.6–0.7 atm carbon dioxide pressure characteristic of the rumen, even if it equilibrated rapidly through the rumen wall.

Fixation of C^{14}-carbon dioxide in the rumen is almost exclusively into cells and into propionate and other acids with an odd number of carbon atoms (van Campen and Matrone, 1960). The chief avenue of fixation is through succinate, precursor of propionate. Very little C^{14}-carbon dioxide appears in acetate and butyrate. Some is fixed by the methane bacteria.

2. WATER

Water is important in the ruminant in the same fashions as in other mammals. In addition, a large quantity must be maintained in the stomach. Availability of water is an important factor influencing ruminant distribution, not only indirectly through plant growth, but also directly. Many of the great migrations of African antelope are a search for drinking water.

Water affects the rumen function, and various ruminants have evolved differing mechanisms for meeting the problems of water limitation. These will be discussed in Chapter X.

3. THE BULK MINERALS

These are provided in ruminant rations largely to meet the requirements of the ruminant itself, and, since they are given with the feed, their presence in the rumen permits them to meet the mineral requirements of the microorganisms as well. Calcium and phosphorus are important as elements used in bone formation. Their metabolism is reviewed by Underwood (1959). Chlorine is important as a constituent of the hydrochloric acid secreted in the abomasum; sodium and phosphorus are important as constituents of parotid saliva. The replacement of depleted sodium by potassium has been mentioned in connection with production of saliva. The dietary sodium requirement has been variously reported to be as low as 0.2 gm per head per day for cattle (Horrocks, 1964a,b) and as much as 1.01 gm for sheep (Devlin and Roberts, 1963).

Sulfur is an important constituent of many proteins and occurs in high concentrations in hair and wool. Magnesium is required in the proper functioning of many metabolic enzymes, and severe host damage can result when Mg^{++} is deficient.

The importance of the mineral elements in regulating the acid-base balance of the animal (Brouwer, 1961) will not be further discussed since it is only indirectly related to microbial activity. Success in influencing rumen acidity with minerals has been reported (Oltjen et al., 1962a). The quantities needed in the ration may depend on the magnitude of the microbial fermentation. Adjustments of ruminants to new feeds may include changes in the absorption and secretion of bulk minerals. A deficiency of calcium in Indian cattle fed on paddy straw has been ascribed (Talapatra et al., 1948b) to potassium oxalate in the straw in concentrations of up to 5% (Talapatra et al., 1948a). The oxalate is decomposed in the rumen to alkali bicarbonate which causes alkalosis. This would not be important with good rations, but on the rather poor paddy straw volatile fatty acid production is insufficient to counteract the alkaligenesis. The alkaline reaction could injure host nutrition by decreasing cellulolytic activity. Pure cultures of cellulolytic rumen bacteria do not digest cellulose at pH's above 7.0.

a. Absorption

The distribution of calcium and magnesium ions between alimentary fluid and particles of digesta is greatly influenced by the acidity of the contents (Storry, 1961a,b). Very little of these elements may be in solution in the rumen, in which case there is little absorption (Dobson,

1961). The acidity of the abomasum makes the calcium and magnesium compounds soluble, and most of the absorption of calcium is in the abomasum (Garton, 1951; Storry, 1961a). Some magnesium was absorbed from the rumen and abomasum, but more from the small intestine (J. Stewart and Moodie, 1956).

Since the reaction of the rumen is normally slightly on the acid side and occasionally quite acid, considerable quantities of ions should be in solution also in the rumen. Rumen contents of sheep on chopped meadow hay contained 7–11 mg of magnesium per 100 ml and 11–21 mg of calcium; the abomasal values were 11–16 mg of magnesium and 38–56 mg of calcium. Animals consuming abnormally high amounts of potassium show a movement of chloride into the rumen (Dobson, 1959). This leaves the blood more positive to the rumen contents and could impede absorption of Mg^{++}. Supplementation with calcium, magnesium, sodium, and potassium of a basal ration low in these elements does not significantly improve digestibility, but causes greater consumption of water.

b. *Requirements of the Rumen Microorganisms*

Relatively little has been done to determine the relationship of rumen microorganisms to the bulk inorganic elements. Precise studies of Bryant *et al.* (1959) show that *Bacteroides succinogenes* requires a minimum of 6 mg PO_4^{3-}, 0.2 mg Mg^{++} and 1 mg Ca^{++} per 100 ml culture medium. Sodium and potassium are both required, the amounts depending on the concentration of both. In the presence of 0.9 M sodium, the minimal concentration of potassium is 0.5 M. With 0.05 M sodium, the minimal concentration of potassium is 0.01 M.

In vitro cellulose digestion by a suspension of washed rumen bacteria (Hubbert *et al.,* 1958b; Martin *et al.,* 1964) is stimulated by sodium, potassium, sulfur, magnesium and calcium. Most bacteria profit from a balance of inorganic ions in the medium and cannot tolerate excessive deviations, particularly for single elements. The amounts of the macroelements required by the rumen microorganisms are probably less than the amounts usually present in the rumen, and the bacteria can function with concentrations, at least of certain salts, several times the usual levels in the rumen. Addition of large quantities of NaCl to the rations of cows on a timothy hay ration (Galgan and Schneider, 1951) did not significantly decrease the utilization of feed constituents. Large concentrations of NaCl eliminate some of the large protozoa (Koffman, 1938). Additional information is needed on the effects of concentrations of inorganic elements on the growth of bacteria in the rumen. This becomes particularly important with purified rations designed to foster maximum microbial growth.

Sheep and cattle require 0.15 and 0.2% calcium in the feed, on a dry matter basis, respectively, and 0.2 and 0.3% phosphorus (Blaxter, 1952). In India, 7–17 gm of calcium and 3.5–4.0 gm of phosphorus were needed daily for 500-lb buffalo and zebu cattle (Kehar, 1956). The cattle required 3.6–9 gm of magnesium per 500-lb body weight, and the buffalo required 5–9 gm. These amounts would provide in the rumen contents more than ten times the concentration of magnesium and calcium required for *Bacteroides succinogenes*. The amount of phosphorus is also in excess, but not to as great an extent. The concentrations of these elements required for *in vitro* culture of a single bacterial type may be considerably less than the total quantity needed in the rumen where much more substrate is metabolized.

c. *Phosphorus*

Some of the phosphorus in the bovine rumen enters via the saliva (A. H. Smith *et al.*, 1956; Garton, 1951), which contains approximately 0.03% phosphorus (w/v) based on the total secretion. Relatively little diffuses into the rumen from the blood, though in sheep some of the phosphorus is reported to enter by this route (A. H. Smith *et al.*, 1955). Very little leaves the rumen (Dobson, 1961) through the wall. Retention of soluble phosphorus in the rumen could involve work, since the concentration of phosphorus in the rumen is considerably higher than that in the blood (R. Clark, 1953). Levels of 10, 20, and 40 μg of phosphorus per milliliter of medium were used in studying phosphorus requirements of washed rumen bacteria (R. Anderson *et al.*, 1956). The 20-μg level is about the same as the concentration required by *Bacteroides succinogenes* (Bryant *et al.*, 1959). The soluble phosphorus in the sheep rumen has been found to vary between 290 and 370 μg per milliliter of rumen contents (Garton, 1951). From 32 to 100% of feed phosphorus is absorbed, chiefly in the abomasum (E. Wright, 1955a,b; Lofgreen and Kleiber, 1954).

The phosphorus content in grass hays in New York is 0.16–0.21% of the dry weight (Burke, 1950). Somewhat higher values are found for legumes. The phosphorus combined as phytate in forages is split off in the rumen (R. L. Reid *et al.*, 1947) and is completely available to rumen microorganisms (Tillman and Brethour, 1958; Raun *et al.*, 1956). Ortho-, meta-, and pyrophosphate were utilized equally well by ruminants in some experiments (P. G. Hall *et al.*, 1961), but in others the orthophosphate was the best (Ammerman *et al.*, 1957). An interrelationship between calcium and phosphorus has been shown (Barth and Hansard, 1962).

The phosphorus requirement for sheep has been estimated at ca. 3

gm of phosphorus per day for a sheep gaining 180 gm per day (R. L. Preston and Pfander, 1964), and at 0.15–0.2% (Blaxter, 1952) and 0.22% (M. B. Wise *et al.,* 1958) of the dry matter intake. Forage does not always contain this amount. Only 0.026–0.044% phosphorus was found in barley and wheat straws grown under severe drought (Adler *et al.,* 1960b), and a syndrome observed was attributed to a phosphorus deficiency. Although there is a fair amount of phosphorus in most forages, the needs of microorganisms are fairly high; the cells contain 2 to 6% phosphorus on a dry weight basis, and on occasion this element could be limiting. Availability of a phosphate lick has been found to increase the consumption of a poor forage (Brünnich and Winks, 1931).

d. *Sulfur*

Sulfur is one of the most interesting elements in ruminant nutrition because of the different types of conversions it can undergo. The sulfur-containing amino acids, cyst(e)ine and methionine, are important components of rumen bacteria (Johanson *et al.,* 1949).

A great many bacteria, though not all (Prescott, 1961), are able to derive their sulfur totally from sulfate, and numerous studies of assimilation from sulfur added to the rumen show that sulfate is as effective as any other form (Block *et al.,* 1951; Warth and Krishnan, 1935; W. H. Hale and Garrigus, 1953; Lofgreen *et al.,* 1953; W. E. Thomas *et al.,* 1951; Hunt *et al.,* 1954). Elemental sulfur is also assimilated (Starks *et al.,* 1953, 1954), but not as readily as sulfate (W. H. Hale and Garrigus, 1953), probably because of its lower solubility.

The sulfur from sulfate is rapidly incorporated into the cystine-cysteine and methionine portions of bacterial protein in the rumen (Henderickx, 1961a; Emery *et al.,* 1957a,b; Block *et al.,* 1951) and is soon found also in the proteins of blood serum and milk. The sulfate must be reduced prior to combination in amino acids. Homocystine occurs in traces in rumen contents, but not in rumen proteins (Henderickx, 1961a); this is expected from its role as an intermediate in methionine synthesis.

Radioactive sulfur in SO^{-}_4 was incorporated into the rumen bacteria, but not into the protozoa (Müller and Krampitz, 1955b). Ingested sulfate disappears from the rumen in 3–5 hours (Lofgreen *et al.,* 1953). Rumen strains of *Bacteriodes, Lachnospira,* and *Butyrivibrio* assimilate sulfur from sulfate into cells when grown *in vitro* (Emery *et al.,* 1957b), with ascorbic acid as a reducing agent rather than cysteine. Growth is more rapid with cysteine, and incorporation of S^{35} from sulfate is partially inhibited. With cysteine, most of the S^{35} incorporated from sulfate was in the form of glutathione.

Not only is the sulfate required for microbial cell synthesis reduced in the rumen (D. Lewis, 1954); an excess is reduced since sulfide accumulates. This was most striking when the sheep had been dosed with 40 gm of $Na_2SO_4 \cdot 10\ H_2O$ daily for 2 weeks prior to testing. Copious formation of sulfide did not invariably occur even in sheep receiving added sulfate, but it was more common in them (C. M. Anderson, 1956). The ration in these experiments was alfalfa hay, with a sulfur content of 0.28%. Hydrogen sulfide was released also from casein by the rumen microorganisms at an optimum pH of 6.0, as compared to 6.5 for optimum sulfate reduction. In this case the hydrogen sulfide came from the sulfur-containing amino acids. A concentration of 0.16% hydrogen sulfide in rumen gas (Kleiber *et al.,* 1943; Matsumoto and Shimoda, 1962) has been found. This concentration, if equilibrated with the rumen fluid, would give a hydrogen sulfide concentration of 0.1 μmole per milliliter, equivalent to a concentration of 0.002% $Na_2S \cdot 9$ H_2O.

From experiments on copper nutrition in sheep (Dick, 1954) it was suggested that sulfate reduction interferes with copper uptake by precipitating it as the sulfide. Sodium sulfide added to rumen contents decreased to one-half the initial concentration in 30 minutes, and a transient rise in sulfide level in the blood could be detected (C. M. Anderson, 1956). It was shown that hydrogen sulfide diffused through the rumen wall into the blood stream. In this connection the observation (Klein *et al.,* 1937) that the crevices and fissures in the interior surface of the rumen wall are black is of interest. It was proposed that *Piromonas* reduced sulfate and caused blackening in the crypts. Because of the readiness with which *Selenomonas* produces hydrogen sulfide from amino acids, one wonders whether it could be the *Piromonas*.

Because of the well known ability of *Desulfovibrio* to reduce sulfate to hydrogen sulfide, its presence in the rumen has been sought by a number of investigators. Coleman (1960c) found a bacterium which reduces sulfate to sulfide. It was present in only small numbers and was subsequently found (Postgate and Campbell, 1963; Buller and Akagi, 1964) to be a mesophilic strain of *Clostridium nigrificans,* recently renamed *Desulfotomaculum ruminis* (Campbell and Postgate, 1965). In most cases the numbers of *Desulfovibrio* are not large enough to indicate that this genus is important in the rumen (D. Lewis, 1954; Coleman, 1960c). One report of *Desulfovibrio* (Gutierrez, 1953) in the rumen was based on spiral shape and lactate fermentation with sulfide production, but *Selenomonas* could have been the causative organism and produced hydrogen sulfide from cysteine. Recently a fairly high count of sulfate reducers in the rumen has been reported (Matteuzzi, 1964).

One of the interesting features of sulfur metabolism in the rumen is the lack of toxicity to the host from the quantities of hydrogen sulfide formed in animals receiving added sulfate. Hydrogen sulfide introduced into the rumen may prostrate the animal if it eructates (Dougherty *et al.,* 1965), but causes little effect in sheep having a blocked trachea. In eructation, some of the hydrogen sulfide enters the lungs and is rapidly absorbed into the pulmonary veins and goes to the brain before it can be detoxified in the liver. Cattle fed metabisulfite silage for as long as 5 years show no ill effects (Luedke *et al.,* 1959). The author has rarely noted an odor of rumen contents suggesting hydrogen sulfide, even in animals on a silage ration to which 1% sulfite has been added as a preservative. Perhaps the odor was masked by other odoriferous rumen materials or sulfide was not produced under those circumstances. The exact nature of the conversions of sulfur in the rumen, and the organisms causing them, constitute one of the more interesting areas for further investigation. It is highly probable that hydrogen sulfide can be used as a source of sulfur by most rumen organisms. It can be assimilated by *Bacteroides succinogenes, Ruminococcus flavefaciens,* and *Streptococcus bovis.*

C. M. Anderson (1956) found that sheep rumen contents could produce as much as 12 gm of sulfide per day. Most of it diffused into the blood stream. A maximum concentration of 97 mg sulfide per liter was found in the rumen when 2.25 gm per liter of sulfur as sulfate was given. The sulfur fed in casein and cystine did not appear as sulfide in the rumen. Evidence that sulfide was removed from the blood by the liver and kidney was obtained (C. M. Anderson, 1956), the liver being more active.

The importance of chloride for rumen bacteria has not been investigated extensively, but it has been shown not to be required in a few instances (Bryant *et al.,* 1959). Bicarbonate is the chief monovalent anion in rumen contents and in the media giving excellent growth of many rumen bacteria. Conceivably it might be satisfactory as the only such ion, but it is possible that some chloride is also essential.

4. TRACE ELEMENTS

The part played by ruminant nutritionists in focusing attention on the role and importance of cobalt has already been discussed in connection with B_{12}-like vitamins. Other trace elements important in ruminant nutrition include iron, manganese, copper (Harvey, 1952; Havre *et al.,* 1960), molybdenum, zinc, (J. K. Miller and Miller, 1960), selenium, and iodine. In the rumen the microorganisms contain the major part of these trace elements (Mitchell and Tosic, 1949). Increased activity after addition of iron, cobalt, copper, manganese, and zinc to *in vitro* rumen fermentations containing metal chelating agents (O. Little *et al.,*

1958; McNaught *et al.*, 1950a) indicates that these trace minerals play a role in the microbial metabolism. Deficiencies lead to disturbances in the functioning of the microorganisms, the host, or both.

Molybdenum occurs in combination with some of the flavoproteins catalyzing essential metabolic reactions. If it is required by ruminants, the amount is very small. Growth of some sheep was normal with as little as 0.01 ppm molybdenum in the ration (Sheriha *et al.*, 1962), but in other experiments a stimulation by molybdenum has been reported (Ellis *et al.*, 1958). The chief significance of molybdenum is its adverse effect on copper utilization. High levels of molybdenum diminish the copper content of the sheep liver and can cause death (Scaife, 1956b). Levels as high as 50 ppm of molybdenum can be tolerated (Vanderveen and Keever, 1964) by cattle. Cattle are less susceptible to molybdenum poisoning than are sheep (Cunningham and Hogan, 1959), but are affected by higher concentrations (Lesperance and Bohman, 1963). The blood of cattle normally contains 1 μg of copper per milliliter, about equally distributed between cells and plasma (Adams and Haag, 1957). Molybdenum causes a reduction in the copper concentration. The effect of molybdenum is greater when the ration contains high levels of sulfate (Wynne and McClymont, 1956; Vanderveen and Keever, 1964), and absorption and elimination of molybdenum in ruminants is related to sulfate metabolism (Scaife, 1956b; Dick, 1952, 1954, 1956; Underwood, 1959; Wynne and McClymont, 1956).

These effects have not been demonstrated to be mediated through the rumen microbiota. In some experiments (McNaught *et al.*, 1950b) relatively high levels of molybdenum did not inhibit rumen bacteria *in vitro,* but in others (Uesaka *et al.*, 1962b) an inhibition was found. The importance of molybdenum in nitrogen fixation by some bacteria is well established, but no significant fixation of nitrogen has been found in the rumen. So far as the author is aware, nitrogen fixation has not been sought with critical methods, and, in view of its common occurrence among anaerobic bacteria, it could well occur also in the rumen. The quantities of nitrogen potentially available by this mechanism must be small, since no investigators have reported an increase in the total fixed nitrogen during passage of food through the ruminant.

Stimulation by molybdenum of *in vitro* cellulose digestion by rumen bacteria has been reported (Ellis *et al.*, 1958). The effect was not observed with other substrates. In one analysis dried rumen bacteria contained (in parts per million) 300 iron, 200 zinc, 100 copper and 1 molybdenum (Ellis *et al.*, 1958); in another analysis the bacteria contained (in parts per million) 535–990 iron, 242–400 manganese, 130–220 zinc, 2–5 molybdenum, 40–72 copper, 8–15 nickel, 0.9–2 vanadium, and 24–87 titanium. An inhibitory effect of molybdenum on two copper-containing enzymes obtained from the hide of black sheep was observed (Scaife, 1956a), but the molybdenum did not replace the copper in the enzyme.

Swine are much less susceptible to molybdenum poisoning than are cattle (M. C. Bell *et al.*, 1964).

Iron is assimilated by *Bacteroides ruminicola* as a constituent of the cytochromes, in the ferredoxin demonstrated in *Peptostreptococcus elsdenii* and *Veillonella alcalescens*, and undoubtedly in a number of other constituents of rumen microbes. The requirements of the microorganisms probably are less than those of the ruminant. A stimulation of cellulose digestion by iron has been shown *in vitro* with rumen bacteria (Burroughs *et al.*, 1951b; O. Little *et al.*, 1958), and in feeding trials with steers (Bentley *et al.*, 1954b), but in other cases no effect was observed (Hubbert *et al.*, 1958a).

The iron content of rumen fluid is rather low. The possibility for detecting iron objects in the rumen by analysis of the iron content has been explored (Meister, 1934). The iron content of the rumen dry matter was 13 mg per 100 ml (Rathnow, 1938). Because the iron concentration in the rumen liquid remained relatively constant at 0.7 mg/100 ml, it was concluded that iron was not going into solution from the solids. This agrees with the observed reduction of rusty iron objects to shiny metal in the rumen. High intake of iron has been associated with deleterious effects on milk production (Coup and Campbell, 1964), in contrast to earlier findings of other investigators.

Selenium in very small amounts, 0.1 ppm of the feed (Muth *et al.*, 1959, 1961), increases weight gain in deficient sheep (Hartley and Grant, 1961; Butler and Johns, 1961; Blaxter, 1963; McLean *et al.*, 1959), but in larger concentrations is poisonous. Administration of trace quantities of selenium can cure white muscle disease of sheep (Lagace, 1961). No evidence that it affects rumen microorganisms has been presented, and one hypothesis is that it protects vitamin E in the metabolism of the host and prevents dietary liver necrosis (Schwarz *et al.*, 1961). The effect may be to prevent oxidation of the vitamin. On the other hand, evidence for differences in the effects of vitamin E and selenium has been observed (Hopkins *et al.*, 1964). A role of selenium in microbial metabolism cannot be a priori eliminated, and the observation that selenium affected the formic hydrogenlyase system of *Escherichia coli* (Pinsent, 1954) suggests that functions of selenium in microorganisms may ultimately be found. In ruminants, much of fed selenium is eliminated in the feces rather than in the urine (Butler and Peterson, 1961), about one-half of it in organic form. This also suggests a microbial metabolic activity involving selenium.

In nonruminants selenium is excreted chiefly in the urine (Butler and Peterson, 1961).

Iodine as a trace element similarly exerts its influence on the metabolism of the host, and only secondarily on the rumen organisms through changes in the activities of the host.

Copper is essential for the host (D. N. Sutherland, 1952) and, as with the other trace elements, is probably concerned in the metabolism of some of the microorganisms, but no specific information is available.

Skin lesions in cattle in British Guiana were cured by oral or injected zinc (Legg and Sears, 1960). Zinc is important as a component of carbonic anhydrase and a number of other enzymes in the ruminant (Underwood, 1959). It assists in the catalysis of various metabolic reactions by fungi, but no role in the rumen microorganisms has been demonstrated.

Manganese is a cofactor for a number of enzymatic reactions and may be important in both the microbes and the host. Availability of manganese is reported to limit *in vitro* cellulose digestion by rumen bacteria (Chamberlain and Burroughs, 1962), and manganese deficiency has been induced in cattle (Rojas and Dyer, 1964).

The finding that alfalfa ash stimulated fiber digestion even when known mineral elements had been added (Chappel, 1952) could suggest an additional undetected trace element, but more specific experiments are needed.

CHAPTER IX

Host Metabolism in Relation to Rumen Processes

Dependence on the fermentative and synthetic activities of the rumen microorganisms has led to various modifications in the host metabolic patterns. The carbohydrate metabolism differs markedly from that of nonruminants. The protein metabolism is affected, as discussed in Chapter VII, and the fat metabolism is modified.

A. Carbohydrate Metabolism

Chief among the features of carbohydrate metabolism is the diminished blood sugar concentration. Newborn ruminants exhibit a blood sugar level of approximately 100 mg/100 ml of blood (McCandless and Dye, 1950), a value characteristic of most eutherian mammals; this changes gradually to lower adult levels after 6 to 9 weeks (R. L. Reid, 1953). The "pseudoruminants" such as the camel and llama exhibit a blood sugar level similar to that of young ruminants. After forage consumption starts and the rumen fermentation products are the chief source of energy for the host, the blood sugar concentration drops to 25 to 65 mg% (R. L. Reid, 1950a; McCandless and Dye, 1950) and ordinarily does not rise above 75 mg% during the remaining life of the animal.

The adult shows a low sensitivity to insulin (Jasper, 1953). Insulin administration lowers the blood glucose, though less effectively than in nonruminants. The volatile fatty acid (VFA) concentration increases as the glucose diminishes (Annison, 1960). Glucose injected into the blood persists much longer than in young or pseudoruminants (McCandless and Dye, 1950). A glucokinase in rat liver, effective in increasing glycogen deposition at physiological concentrations of glucose, is absent from adult sheep (Ballard and Oliver, 1964). Dairy cows fed 6 to 8 lb of glucose by stomach tube (Hodgson *et al.,* 1932) show a rise in blood glucose concentration; a similar rise occurs in sheep (Hungate *et al.,* 1952). Lesser quantities, which do not exceed the capacity of the rumen to ferment them, cause no increase in blood sugar (F. R. Bell and Jones, 1945).

The arteriovenous differences in blood glucose concentrations are small (R. L. Reid, 1950a, 1951; Annison *et al.,* 1957), approximately 2 mg%. The turnover constant for plasma glucose has been estimated in cattle at 0.026 and 0.045 per minute (Kleiber *et al.,* 1955) and in sheep at 0.11 per minute when fed and 0.055 per minute when fasted (Annison and White, 1961). These latter values were calculated on the basis of 2.7 and 1.5 mg of glucose used per minute per kilogram in fed and fasted sheep, respectively, with a 2-liter blood volume containing 50 mg% glucose. The 2.7-mg value would correspond with a glucose utilization of about 155 gm per day in a 40-kg sheep.

In young steers the average rate of glucose utilization was 4.71 gm per hour per 100 lb live weight (1.72 mg per minute per kilogram), and a value of 8.30 for young calves was determined (C. L. Davis and Brown, 1962). H. H. Head *et al.* (1964) found values of 1.12 and 1.46 mg per minute per kilogram by infusion and single injection, respectively. A turnover constant of 0.0188 per minute was calculated for a 457-kg cow yielding 10 kg of milk per day (Baxter *et al.,* 1955). This would amount to about 1700 gm per day, a maximal value.

Some of the glucose requirement of the ruminant is available in the form of the polysaccharide reserves of the microorganisms. This was originally postulated (Baker, 1942) as the important pathway of energy supply to the host. Quantitative analyses (Heald, 1951a,b) showed that in the sheep only small amounts of sugar could become available to the host in this way, about 1% of the requirement. Very little glucose diffuses from the rumen into the blood, the glucose concentration of portal blood being the same as in arterial blood (Schambye, 1951b). This could mean that glucose absorption from the rumen equals glucose used in the rumen epithelium. Actually, if the glucose concentration in blood is higher than in the rumen, a loss of blood glucose to the rumen might occur. At least, the finding that the glucose content of blood does not decrease during passage through the rumen tissues shows that there is no significant net loss of blood glucose into the rumen.

Exogenous glucose thus does not appear to be an important metabolite in ruminants, nor is endogenous glucose quantitatively as important as in nonruminants. Its place is taken by the VFA's in many instances. Radioactive C^{14}-carbon dioxide appears no more quickly in respired air when uniformly labeled glucose is injected than if the VFA's are introduced (Kleiber *et al.,* 1955). The VFA content in the blood of ruminants is higher than in nonruminants and remains so even after extended starvation (Annison, 1960).

1. METABOLISM OF THE VOLATILE FATTY ACIDS

A ration composed of 30% VFA salts, 30% casein, and 27% glucose supported ruminant growth (Matrone *et al.,* 1959) as well as if the acids were replaced by starch. It would be expected in this case that much of the weight increase was fat since part of the microbes produced during starch fermentation would be lacking in the acids ration.

Conversion by the host of part of the fermentation acids occurs as the acids pass through the rumen wall. The tissue contains the enzymes of the Krebs cycle (Seto and Umezu, 1959). The percentage of acetate in blood was higher (McClymont, 1951a) than in the rumen (Table IX-1), and rumen epithelial tissue was shown to utilize butyrate by converting it into β-hydroxybutyrate (Pennington, 1952). Rumen epithelium produced some acetone by decarboxylation of acetoacetic acid (Seto *et al.,* 1964). Propionate was also used by the epithelial tissue, the metabolism being dependent on carbon dioxide (Pennington, 1954). Acetate is used least. The stimulation of oxygen consumption by butyrate was greater than the stimulation by glucose. The higher VFA's are also metabolized by rumen epithelial tissue (Annison and Pennington, 1954; Annison *et al.,* 1957) (Table IX-1). Isobutyric acid is a source of glucose in the ruminant, whereas isovaleric acid is not (Menahan and Schultz, 1964b). When VFA production is low, all the butyrate and much of the propionate is metabolized in the rumen wall. At high levels of nutrition a significant fraction of the propionic and butyric acids enters the blood.

Butyrate was converted by rumen epithelium of a 75-day-old lamb at a rate of 13.7 μmoles per hour per 100 mg of dry tissue, and by epithelium from an adult sheep at a rate of 7.7 μmoles. In both types of tissue, acetate was converted at a rate of 2.1 μmoles and propionate at a rate of 2.9 μmoles per hour per 100 mg of dry tissue (D. M. Walker and Simmonds, 1962). A 7-day-old calf showed less ability to convert butyric acid than was shown by epithelial tissue from a cow (Onodera *et al.,* 1964). Analyses of rumen arterial and venous blood (Shoji *et al.,* 1964) suggest that, in addition to acetate, propionate, and ketone bodies, the rumen vein shows an increase in pyruvate, lactate, succinate, and fumarate. Formation of these acids in the epitheleium is the most likely explanation of their source.

The mammary gland tissue of ruminants differs from that of nonruminants in utilizing acetate readily to synthesize fat in tissues slices incubated *in vitro* (Kleiber *et al.,* 1952b; McClymont, 1951a). The arterio-

Table IX-1

VOLATILE FATTY ACID CONTENT OF SAMPLES OF RUMEN CONTENTS, PORTAL BLOOD, AND ARTERIAL BLOOD FROM A SHEEP ON A DIET OF HAY[a]

Time after Feeding (hr)	Sample	Total VFA concentration (mmoles/liter)	Molecular percentage of VFA's						
			Formic	Acetic	Propionic	Iso-butyric	n-Butyric	Iso-valeric + 2-methyl butyric	n-Valeric
0	Rumen contents	47	0	66	28	1	5	0	0
0.75	Portal blood	67	0	64	29	1	5	0	1
1.5		80	0	64	29	1	4	1	1
3		94	0	58	29	2	7	3	1
4.5		89	0	68	26	1	4	1	0
7		84	0	61	30	1	6	1	1
10		72	0	60	32	1	5	1	1
0	Portal blood	0.96	6	83	11	0	0	0	0
0.75		1.30	4	83	13	0	0	0	0
1.5		1.67	5	79	17	0	0	0	0
3		1.78	4	75	21	0	Trace	0	0
4.5		1.53	4	81	15	0	0	0	0
7		1.10	7	82	11	0	Trace	0	0
10		1.05	5	84	11	0	0	0	0
0	Carotid blood	0.73	5	95	0	0	0	0	0
0.75		0.82	5	95	0	0	0	0	0
1.5		0.92	4	96	0	0	0	0	0
3		0.71	5	92	3	0	0	0	0
4.5		0.58	6	92	2	0	0	0	0
7		0.52	4	96	0	0	0	0	0
10		0.51	5	95	0	0	0	0	0

[a] From Annison et al. (1957).

venous difference for blood acetate in the mammary gland is 2 to 6 mg%, 40 to 80% of the arterial level. Experiments with radioactive acetate confirmed the synthesis of acetate into fat (Popjak *et al.,* 1951a; Cowie *et al.,* 1951; Kleiber, 1954a) and showed that it was also used to form cholesterol (Rogers and Kleiber, 1955).

The concentration of acetate in peripheral blood varies with the state of nutrition more than is the case with propionate and butyrate. These latter are removed more completely in the liver. The rate of oxidation of acetate increases with an increase in acetate concentration in the peripheral blood (C. L. Davis *et al.,* 1960b,c). Infusion of propionate along with acetate decreased the total carbon dioxide production, but the fraction of carbon dioxide coming from acetate was the same as when no propionate was supplied. Increased conversion of C^{14}-acetate to fat may have occurred in the presence of propionate, due to a greater supply of glycerol. Acetate alone is not converted into fat in muscle tissue (McClymont, 1951a).

The rates at which the VFA's are used in the ruminant have been estimated. Values for acetate utilization in starved sheep range from 0.43 to 3.56 mmoles per minute per sheep (Annison and Lindsay, 1961). In fed sheep values of 4.0 moles per hour per kilogram (Annison and White, 1962a) and 5.2 mmoles per hour per kilogram (Sabine and Johnson, 1964) have been found. In later experiments (Lindsay and Ford, 1964), values from 1.00 ± 0.07 for underfed sheep to 2.44 ± 0.17 for fed sheep were found. One-fourth of the acetate was supplied from endogenous sources, the remainder from the rumen (173 gm of acetate from the rumen per day per sheep at this maximum rate). In sheep starved for 24 hours, 43 to 46% of the total acetate used was derived from endogenous sources, the remainder from the rumen.

Body fats are the chief source of endogenous acetate, and free fatty acids from fats occur in the blood. In young milk-fed goats, blood VFA's amounted to 2.7–3.8 mg/100 ml, but rose to 4.2–6.1 in the adult (Craine and Hansen, 1952). Acetate has been estimated to supply half the carbon in the carbon dioxide respired by a steer (C. L. Davis *et al.,* 1960a).

The single-injection technique for measuring acetate-C^{14} turnover gives a value for cattle of 2.1 ± 0.5 mmoles per hour per kilogram (Lee and Williams, 1962b). The rate at which acetate is utilized depends in part on the concentration of acetate in the blood (R. L. Reid, 1950c; Annison and Lindsay, 1961; C. L. Davis *et al.,* 1960b). The quantity of acetate appearing in carbon dioxide was relatively low in the experiments on starved sheep (Annison and Lindsay, 1961), only 6% of the respired carbon dioxide being from acetate, but 35% was derived from acetate during the period after feeding when this acid was most concentrated in the rumen.

Propionate utilization by the host has been estimated at 1.45 to 3.0 mmoles per hour per kilogram in sheep fed on alfalfa hay plus flaked maize; this value falls to 0.57–0.71 mmoles 24 hours after feeding (Annison and Lindsay, 1962). In view of the possible specific influence of flaked maize in increasing propionate production, the values may be derived from a microbial fermentation yielding a higher percentage of propionate than would be encountered with many rations (Jayasuriya and Hungate, 1959).

Since the rumen is presumably the only source of propionate for the ruminant, the amounts formed in the rumen can serve as a measure of the amount utilized in the host. These quantities were described in Table VI-5.

Determination of butyrate utilization by the host can also be derived from the rate of production in the rumen, since this acid is not known to be derived in any quantities through host metabolism. Since some of the butyrate is metabolized during passage through the rumen wall, tracer techniques within the ruminant circulatory system are not applicable to total production, but only to that available to tissues other than the rumen.

The utility of the individual VFA's to the host has been studied by slow infusion of large quantities of an acid into the rumen (Rook et al., 1963). All three acids increased the nitrogen retention slightly, with little difference between them; this suggests that they serve in this case as an additional source of energy rather than as precursors of cell material. This was reflected also in the relatively greater gain in fat than in fat-free material. The principal results of this study are shown in Table IX-2. The slight effect of the acids on nitrogen retention is understandable on the basis that the acids administered would not increase rumen microbial growth.

2. RESPIRATORY QUOTIENT

The respiratory quotient of the ruminants as determined by usual gasometric methods, without including methane, is approximately 1.0 (Reiset, 1863a; Zuntz, 1913), which indicates that the overall metabolism consists primarily in the oxidation of carbohydrates. In the rumen, the production of carbon dioxide and methane and the negligible consumption of oxygen make a respiratory quotient inapplicable. The theoretical respiratory quotient of the ruminant itself can be calculated from the relative proportions of the VFA's which it utilizes. On the bases of the average proportions of acids shown in Table VI-6, the respiratory quotient of the

Table IX-2

UTILIZATION OF FREE FATTY ACIDS INFUSED SLOWLY INTO THE RUMEN[a]

Acid infused	Amount infused (kcal/day)	Empty body weight gain (gm/day)	Nitrogen retention (gm/day)	Gain of fat-free material (gm/day)	Fat deposited (gm/day)	Calculated kilo calories gained (kcal/day)
Experiment 1						
Acetic	3500	377	7.4	211	166	1802
Propionic	3500	281	3.6	103	178	1787
Butyric	3500	450	5.2	148	302	3003
Experiment 2						
Acetic	5500	477	3.1	88	389	3743
Propionic	5500	554	3.9	111	443	4280
Butyric	5500	663	4.1	117	546	5246

[a] From Rook et al. (1963).

ruminant should be 0.905 (under maintenance conditions). Values approximating this have been found (Hagemann, 1899; Ustjanzew, 1911). The carbon dioxide production in the rumen fermentation would be

$$58 \text{ Hexose} \rightarrow 62 \text{ HAc} + 22 \text{ HPr} + 16 \text{ HBut} + 60.5 \text{ CO}_2 + 33.5 \text{ CH}_4 + 27 \text{ H}_2\text{O}$$

CALCULATION OF THEORETICAL RESPIRATORY QUOTIENT (R.Q.) OF RUMINANT			
62 HAc + 124 O_2 →	124 CO_2 + 124 H_2O		R.Q. = 1.0
22 HPr + 77 O_2 →	66 CO_2 + 66 H_2O		R.Q. = 0.86
16 HBut + 80 O_2 →	64 CO_2 + 64 H_2O		R.Q. = 0.80
VFA + 281 O_2 →	254 CO_2 + 254 H_2O		R.Q. = 0.905

If the 60.5 CO_2 in fermentation are added, the carbon dioxide to oxygen ratio of the ruminant plus microorganisms becomes 1.12. It exceeds 1.0 because the methane is not oxidized. Measured respiratory quotients are usually less than this. If the methane were oxidized, its respiratory quotient of 0.5 would bring the total respiratory quotient to 1.0. The fermentation carbon dioxide would be 19.2% of the total carbon dioxide produced in the animal. The carbon dioxide released from bicarbonate by acids produced in the rumen does not represent a net escape from the animal, because an equal amount returns in the bicarbonate of saliva (Zuntz, 1913).

The theoretical methane produced amounts to

$$\frac{33.5}{60.5 + 254} \times 100 = 10.6\%$$

of the carbon dioxide produced in the entire animal, including the alimentary contents. This is greater than the 7.5% found by Ritzman et al. (1936). The theoretical ratio of fermentation carbon dioxide to methane, i.e., (60.5/33.5) = 1.80, is less than the 2.6 reported by A. Krogh and Schmidt-Jenson (1920), the 2.2 for European and Zebu steers on a maintenance ration (Hungate et al., 1960), the 2.4 for producing Holstein milk cows (Hungate et al., 1961), and the 2.6 in Table VI-3. If this ratio of 60.5 CO_2/33.5 CH_4 is adjusted to give a 2.5 value, the ratio becomes 60.5 CO_2/24 CH_4, which makes the methane amount to 7.6% of the total carbon dioxide given off from the animal. This agrees with the 7.5% found by Ritzman et al. (1936) and Kleiber et al. (1945), but differs from the 8.9% of Kellner (1911) and the ca. 5% of Boycott and Damant (1907).

3. HEAT INCREMENT OF FEEDING

The heat given off when a food is metabolized is a rough measure of the inefficiency of its utilization. The less the proportion of heat given off, as compared to the energy assimilated into cell material, the more efficient the utilization of the food.

Food can be used for either of two purposes, production of adenosine triphosphate (ATP) or as a building material of cells. Since acetate does not appear to support a net synthesis in the ruminant of anything except fat, it can serve only as a precursor of either ATP or fat. Three molecules of ATP are generated in the oxidation of NADH $+$ H$^+$ (reduced nicotinamide-adenine dinucleotide) in the mitochondria, and two are generated in the oxidation of the hydrogen derived in the conversion of succinate to fumarate. Hydrogen for the reduction of NAD to NADH $+$ H$^+$ can be derived through the Krebs cycle from acetate, with the ultimate appearance of acetate carbons as carbon dioxide.

If precursors of cell material, derived from other sources than acetate, are available for cell synthesis, the ATP from acetate may be used in the growth reactions involved in cell synthesis. The heat increment will be the heat generated in the reactions of ATP genesis from acetate and the subsequent reactions of the ATP. A heat increment under these circumstances would correlate with faster growth.

If precursors of cell material are not available, acetate is converted into fat. Sterols and other lipids may be synthesized to some extent, but the bulk of acetate appears in fat. Synthesis of higher fatty acids from acetate involves the conversion of acetate to acetyl CoA, carboxylation of acetyl CoA to malonyl CoA, and the condensation of malonyl CoA with the terminal methyl group of another fatty acid to form carbon dioxide and a β-keto acyl CoA compound which is then reduced to the saturated acyl CoA compound. Acetate can provide the acetyl CoA, ATP, and NADPH for these synthetic reactions, and, under conditions in which the available acetate exceeds the quantity used in these reactions, higher fatty acids may be formed.

Glycerol is required for the conversion of fatty acids into fat. So far as is known, there is no mechanism for synthesis of glycerol from acetate. Tracer experiments (Rogers and Kleiber, 1955) show that it comes from glucose. Measurements of heat increment from acetate fed to fasted sheep have shown (Armstrong and Blaxter, 1957a) 41 cal of heat per 100 cal of acetate administered, but when it is given in a mixture of 90% acetic acid-6% propionic acid-4% butyric acid only 15 cal appeared as heat per 100 cal in the acid mixture. Disappearance of acetate injected into the blood was increased by simultaneous administration of glucose (Jarrett and Filsell, 1961).

In starved sheep, the heat increments for propionate and butyrate fed singly were 13 and 16 cal, respectively, as compared with 9 cal when fed in a mixture. Since glycerol can be derived from propionate in ruminant tissues, the high heat increment of acetate, fed singly, may be due in part to energy waste because of less availability of glycerol from the starved mammalian tissues.

The biochemical processes underlying different efficiencies in the utilization of the VFA's by ruminants are not well understood. On the basis of known pathways it might be expected that long-chain fatty acids could be synthesized from acetate with a heat increment of 11.3%, according to the following equations:

$$HAc + CoA + 8000 \text{ cal} \rightarrow AcCoA + H_2O$$
$$AcCoA + 2 O_2 + 11 ADP + 11 P_i \rightarrow 2 CO_2 + 2 H_2O + 11 ATP + CoA$$
$$2\tfrac{1}{2} HAc + 5H_2O + 20 NAD \rightarrow 5 CO_2 + 20 NADH$$
$$6 HAc + 6 CoA + 6 ATP \rightarrow 6 AcCoA + 6 AMP + 6 PP_i + 6 H_2O$$
$$5 AcCoA + 5 CO_2 + 5 ATP \rightarrow 5 \text{ malonyl CoA} + 5 ADP + 5 P_i$$
$$AcCoA + 5 \text{ malonyl CoA} + 20 NADH \rightarrow 5 CO_2 + C_{12}H_{24}O_2 + 6 CoA +$$
$$4 H_2O + 20 NAD$$

$$9\tfrac{1}{2} HAc + 2 O_2 \rightarrow 7 CO_2 + 7 H_2O + C_{12}H_{24}O_2$$
1990 kcal   1773 kcal

If glycerol is formed and condensed with fatty acids with about the same efficiency, the heat increment would be expected to be about 11%, compared to the measured 15%.

Similarly, an efficiency for propionate conversion to glucose can be calculated by assuming a pathway via succinate, fumarate, malate, oxalacetate, phosphoenolpyruvate, 1,3-diphosphoglycerate, triose phosphate, and glucose. In these reactions the hydrogen available for oxidation yields more ATP than is required for the synthesis.

$$2 \text{ propionate} + 4 ATP + 2 H_2O \rightarrow \text{glucose} + 4 ADP + 4 P_i + 4 H$$
$$4 H + O_2 + 5 ADP + 5 P_i \rightarrow 5 ATP + 2 H_2O$$
$$ATP \rightarrow ADP + P_i$$

$$2 \text{ propionate} \rightarrow \text{glucose} + 4 H$$
734 kcal   673 kcal

The measured heat increment for propionate of 13 kcal per 100 kcal used (Armstrong and Blaxter, 1957a) is more than the heat increment of 8.3% indicated by the above equations. Part of the propionate may be used for purposes other than glucose synthesis. A mixture of 3 propionate: 2 butyrate gave a heat increment of 9.3% (Armstrong *et al.,* 1957).

Calculations of efficiency for butyrate conversion into long-chain fatty acids can be made from the following pathways:

22 butyrate + 22 CoA + 22 ATP → 22 butyryl CoA + 22 AMP + 22 PP

4 butyryl CoA + 4 CoA + 24 ADP + 24 P_i + 4 O_2 → 8 acetyl CoA + 8 H_2O +
$$24 \text{ ATP}$$

2 butyryl CoA + 2 CoA + 8 NAD → 4 acetyl CoA + 8 NADH

8 acetyl CoA + 48 NAD → 16 CO_2 + 48 NADH + 8 CoA

16 butyryl CoA + 48 NADH → 4 palmityl CoA + 48 NAD + 12 CoA

4 palmityl CoA + 4 acetyl CoA + 8 NADH → 4 stearyl CoA + 4 CoA + 8 NAD

4 stearyl CoA → 4 stearate + 4 CoA

22 AMP + 22 PP → 22 ADP + 22 P_i

---

22 butyrate + 2 ADP + 2 P_i + 4 O_2 → 4 stearate + 16 CO_2 + 8 H_2O + 2 ATP
11,535 kcal   10,847 kcal   16 kcal

These theoretical equations indicate a heat increment of about 6% when butyrate is converted to fat, with the glycerol moiety disregarded.

The actual heat increments for the various acids are somewhat greater than these theoretical values based on biochemical pathways, as might be expected, since at various stages in metabolism the compounds may enter other reactions, and since the required glycerol has not been taken into account. This latter would be particularly important in conversions of acetate and butyrate to fat. The ready conversion of propionate into glycerol may explain the lower heat increments when propionate is supplied along with acetate or butyrate.

The heat increment associated with utilization of the VFA's may be an important factor in ruminant growth in hot and humid climates where heat production leads to loss in appetite (Rogerson, 1960).

4. Synthesis of Carbohydrates

Some carbohydrate is required in ruminant metabolism. Milk contains about 5% lactose. In a high-producing cow, more than a kilogram of lactose may be synthesized and secreted each day.

The small quantity of glucose absorbed from the alimentary tract as a result of digestion of the reserve polysaccharides stored in the bodies of the rumen microbes is not nearly enough to meet the carbohydrate requirement during lactation. Most of the requirement must be met by synthesis from the VFA's formed in the rumen fermentation. The ruminant differs in this metabolic feature from most other mammals. Difficulty in obtaining sufficient carbohydrate underlies the disease ketosis, which is fairly common in ruminants. Ketosis will be discussed in Chapter XII.

The mechanism for synthesis of glucose and lactose from the fatty acids has engaged the interest and activity of numerous investigators (Lindsay, 1959). Propionate has been generally accepted as a precursor of sugar (R. Clark and Malan, 1956; Kleiber *et al.,* 1953), the postulated pathway being conversion to propionyl CoA, carboxylation to methylmalonyl CoA, rearrangement into succinyl CoA (from which the CoA group is transferred back to propionic acid), oxidation of succinate to oxalacetate, decarboxylation to phosphoenolpyruvate, and reduction of phosphoenolpyruvate to triose. Two molecules of propionate can be converted into one molecule of glucose.

Conversion of propionate to lactate has been demonstrated in rumen epithelial tissue (Pennington and Sutherland, 1956b), and, in view of the rather considerable amount converted in the rumen wall (Table IX-1), conversion of propionate to lactate, and lactate to hexose, may be a quantitatively important avenue for sugar formation.

Propionic acid not absorbed during passage through the rumen epithelium is absorbed in the liver (McClymont, 1951a). Its metabolism in the liver, which does not involve oxygen uptake, does not lead to ketone bodies, whereas ketone bodies are formed from acetate and butyrate supplied to liver slices (Seto *et al.,* 1959; Leng and Annison, 1963). Succinate inhibited ketone body formation, and, in its presence, addition of acetic, propionic, or butyric acids increased the oxygen consumption of the tissue, presumably because of the additional C_4 compounds fed into the Krebs cycle.

Rumen epithelium from both normal and ketotic cows produced acetoacetate endogenously and from added butyrate (E. E. Smith *et al.,* 1961), and this production was significantly decreased if propionate was present. Greater weight gains have been found in animals given a diet favoring propionate formation (Ensor *et al.,* 1959).

Label from C^{14}-acetate appears in glucose, but this is interpreted as the result of condensation of acetyl CoA with oxalacetate to form citrate in the Krebs cycle and conversion of citrate into oxalacetate, the oxalacetate then being converted to phosphoenolpyruvate and to sugar. Although there is an uptake of acetate label into glucose in the process, there is no net glucose synthesis, since the carbons lost in oxidation from the oxalacetate moiety of citrate equal the acetate carbons gained. Insofar as is known, acetate is not converted into sugar in the ruminant host, and no enzymatic mechanisms which could catalyze such conversion, e.g., malic synthetase, have been found. Acetate appears to serve the host almost exclusively as a source of energy (Jarrett and Potter,

1950a,b). Concomitant supply of carbohydrate derivatives (pyruvic acid and C^4 dicarboxylic acids) is required for its oxidation (Jarrett and Filsell 1961; Jarrett *et al.,* 1952). An arteriovenous difference of 2.5 mg% acetate has been recorded for the head tissues of sheep (R. L. Reid, 1950b).

No biochemical pathway by which butyrate is transformed into sugar has been elucidated. Ash *et al.* (1958) found no incorporation of the label from butyrate-C^{14} into blood glucose or liver glycogen when butyrate 1-C^{14} was injected, but interpreted the accompanying increase in glucose as due to factors other than gluconeogenesis from butyrate (Ash *et al.,* 1964). This result is explained by the demonstration (R. W. Phillips, 1964) that injected labeled butyrate stimulates liver phosphorolysis of glycogen, which leads to increased blood sugar. Both the limited labeling of the glucose in these experiments and the demonstrated decrease in liver glycogen indicate that the effect of the butyrate was indirect. It was not exhibited in excised perfused livers, which suggests that mediation through some other body tissue was necessary. On the other hand insulin hypoglycemia in sheep has been relieved by injections of sodium butyrate (Potter, 1952).

When individual acids were supplied to phloridzinized ewes (Goetsch and Pritchard, 1958), acetate exerted very little effect on the level of glucose or acetone bodies. Propionate increased the concentration of glucose and decreased the acetone bodies; butyrate decreased the level of glucose and increased the level of acetone bodies. This suggests a dependence on glucose for butyrate utilization.

Labeled carbon dioxide was recovered in lactose, protein, and fat of milk (Kleiber *et al.,* 1952a). Rapid fixation of carbon dioxide into organic form in the rumen would be expected, in view of the numerous rumen bacteria using carbon dioxide in the production of propionate. Labeled carbon in butyrate appears in amino acids, via oxidation to acetyl CoA and condensation with oxalacetate in the Krebs cycle (Black *et al.,* 1961a).

Propionate inhibits acetate utilization by liver slices (Pennington and Appleton, 1958; Leng and Annison, 1963), which is interpreted as indicating that propionate inhibited production of acetyl CoA from acetate. It is possible that oxidation of propionate to pyruvate in these experiments yielded acetyl CoA. Butyrate also inhibited acetate utilization by liver, presumably by serving as a readier source of acetyl CoA. There are still some indications that acetate is assimilated into materials other than fat (Phillipson, 1958b). Infused acetate increased the nitrogen retention almost as well as propionate and butyrate (Rook *et al.,* 1963).

The various interrelationships of the VFA's have been related to the

Krebs cycle (Black *et al.,* 1961a). Liver tissue primarily metabolizes propionate and butyrate (Leng and Annison, 1963), the propionate to oxalacetate via succinate and the butyrate to β-ketobutyrate and to acetyl CoA. Acetate is not readily metabolized in the liver. The acetyl CoA and the oxalacetate are combined in the Krebs cycle, and both appear in glutamate if this material is drawn off for protein synthesis. Acetate can thus favor nitrogen assimilation. If glutamate is not formed, oxidation of α-ketoglutarate leads to oxalacetate and phosphoenolpyruvate, which can form glucose. The carbon atoms of the acetyl CoA are the ones retained in the oxalacetate, and they thus contribute significant quantities of label. However, there is no net synthesis since an acetate is oxidized during progress of the cycle to the oxalacetate stage, and also because propionate can alone give rise to oxalacetate via succinate.

Butyrate is metabolized readily to acetyl CoA in the liver and thus contributes markedly to the labeling. Acetate is not as readily used, but the label can also appear in sugar. If it is essential for the Krebs cycle to operate in production of oxalacetate, acetyl CoA derived from butyrate spares propionate which would otherwise be required as a source of acetyl CoA to feed into the Krebs cycle. Adequate supplies of butyrate can thus contribute to gluconeogenesis by diminishing the oxidation of propionate.

In experiments in which acetic, propionic, and butyric acids were simultaneously perfused through an excised goat liver very little of the labeled carbon from butyrate appeared in glucose or lactate (Holter *et al.,* 1963).

B. Nitrogen Metabolism

The utilization of proteins within the ruminant tissues is similar to the metabolism of this food in nonruminants. The amino acids essential for the ruminant, i.e., those which it cannot synthesize, are the same as those essential for the rat (Black *et al.,* 1952). Too little detailed knowledge of the quantities of essential amino acids available and required has accumulated to permit any judgment as to whether the nonessential amino acids in the rumen microorganisms are sufficient or insufficient to meet the host requirements. The fact that C^{14} from labeled acetate, propionate, and butyrate appears in nonessential amino acids (Black *et al.,* 1952) suggests that certain of the nonessential amino acids are in short supply in the rumen microbes. Glucose accounted for 26% of the alanine in the casein of milk, 23% of the serine, 10% of the glutamic acid, 9% of the asparatic acid, and 7% of the glycine (Black *et al.,* 1955).

Excess protein in the rumen causes high rumen ammonia concentrations and an increased ammonia concentration in portal blood, as is evident in Fig. IX-1. This is in turn reflected in increased blood urea levels (Fig. VII-5). The blood ammonia would presumably be available for amination of the precursors of nonessential amino acids.

Nitrogen metabolism of ruminants differs from nonruminants in the utilization of host urea by the rumen microorganisms. This is quantitatively important in ruminants on a low nitrogen ration (Schmidt-Nielsen *et al.,* 1957; Schmidt-Nielsen and Osaki, 1958). The urea enters the rumen with the saliva (Houpt, 1959) and diffuses from the blood through the rumen wall (Moir and Harris, 1962). The salivary route has been esti-

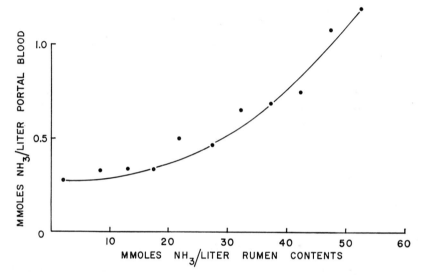

Fig. IX-1. The relationship between the concentration of ammonia in the rumen and in portal blood. From D. Lewis (1957).

mated to provide nitrogen equal to about one-tenth that in the diet (Kay, 1963), with diffusion through the rumen wall estimated to provide a somewhat greater fraction. In cases of water deprivation and a low plane of nutrition, very little urea may be excreted (Livingston *et al.,* 1962).

Extensive studies on the specific activity of amino acids in milk from cows receiving C^{14}-labeled VFA's, carbon dioxide, and glucose have been completed (Black and Kleiber, 1958; Black *et al.,* 1952, 1955, 1961a). The results are generally in accord with known routes of amino acid synthesis.

Spermine oxidase has been found in the sera of several ruminants, but not in other mammals (Blaschko and Hawes, 1959). There is no known connection between this enzyme and the rumen microbes.

It was established by Lancaster (1949a) that the percentage of nitrogen in the feces of cattle and sheep was directly related to the amount of feed consumed. The feces contained 0.83 gm of nitrogen per 100 gm of feed consumed. The explanation for this parallelism is not clear, nor is it known whether it relates to the rumen microorganisms. The nitrogen could represent in part the indigestible nitrogen of the microorganisms, but no information on this point is available.

C. Fats

Lipids do not ordinarily compose a large fraction of the feeds consumed by ruminants, but because of the large amount of feed the lipid may total as much as 450 gm (Garton, 1960). The metabolism of lipids has attracted a good deal of attention in connection with milk fats and because of the possible role of lipids in bloat (Chapter XII).

Kraus (1927) found a higher percentage of fats in rumen contents than in the hay fed to cattle and concluded that there was net synthesis of fat in the rumen. Quitteck (1936) found essentially similar results, but concluded that most of the percentage increase in fats was due to the digestion of other components of the feed.

1. Body Fat

B. H. Thomas *et al.* (1934) showed that feeding of highly unsaturated fats did not affect the degree of saturation of depot fat of young steers. Subsequently it was shown (Reiser, 1951; Reiser and Reddy, 1956; Shorland, 1953) that unsaturated acids were hydrogenated in the rumen. Bypassing the rumen by administration of unsaturated acids into the duodenum gave more double bonds (Ogilvie *et al.,* 1961) in depot fats. Unsaturated fatty acids in the fat of ruminants and herbivorous marsupials occurred in the *trans* form to the extent of 3.5 to 11.2% (Hartman *et al.,* 1954, 1955, 1958; L. M. Smith, 1961; Phatak and Patwardhan, 1953), whereas only 0.9% occur in the *trans* form in other animals, the rest being in the *cis* form. Such a difference infers hydrogenation of the *cis* isomer and enzymatic conversion to the *trans* position. Since ruminants and grazing marsupials both exhibit a fermentation in the stomach, it was presumed that microorganisms are concerned. This has been confirmed by demonstrating the same changes during *in vitro* fermentation of rumen contents. The composition of fatty acids in some forage fats is shown in Table IX-3.

The distribution of saturated and unsaturated fats in the tissues of herbivorous nonruminants such as the horse and rabbit resembles that in the forage in that considerable quantities of triethenoid C_{18} fatty acids occur. In contrast, these materials are almost absent from ruminants (Table IX-4). The increase in saturated C_{18} acids in the depot fat of the ruminants is due to saturation in the rumen of the unsaturated C_{18} acids (Shorland, 1950; Shorland *et al.*, 1955, 1957; Garton *et al.*, 1958; Hoflund *et al.*, 1955, 1956a,b; Reiser and Reddy, 1956). Fed fats did not become saturated if the rumen was bypassed (Ogilvie *et al.*, 1961) or if the fat was fed to young calves during the period when milk was fed (Hoflund *et al.*, 1956a). The lipids of the chloroplast are also hydrogenated by the rumen microbiota (D. E. Wright, 1959, 1960a).

Table IX-3

COMPOSITION OF FATTY ACIDS IN THE TRIGLYCERIDES OF FORAGES[a]

Fatty acid	Dactylis glomerata (%)	Mixed pasture grasses (%)
Saturated		
Lauric	—	2.9
Myristic	1.4	3.3
Palmitic	11.2	9.4
Stearic	2.6	1.5
Higher than C_{18}	1.5	0.7
Unsaturated		
C_{12}	—	0.3
Myristoleic	0.4	0.4
Palmitoleic	6.4	3.0
Oleic	12–19	13–19
Linoleic	19–26	20–26
Linolenic	38	39

[a] From Garton (1959).

Not only do the acids become saturated, the position of the double bonds is changed in some of the monoene acids (L. M. Smith *et al.*, 1954). The dienoic and trienoic acids of the forage are not the same as those in the rumen bacteria (Shorland *et al.*, 1955; Garton and Oxford, 1955; Lough and Garton, 1958) (Table IX-5).

In the depot fat of ruminants, and to a great extent in the milk fat, (Shorland and Hansen, 1957) are found numerous fatty acids with branched chains and an uneven number of carbon atoms, some of them unsaturated. These have been separated on the gas chromatograph after converting them to methyl esters (L. M. Smith, 1961). Comparison with the fatty acids in the rumen bacterial cells (Table IX-6) shows many similarities, and it has been suggested (Akashi and Saito, 1960) that in herbivorous animals some of the branched-chain higher acids may be derived from the intestinal microflora. They may also be synthesized from the branched-chain fatty acids resulting from the fermentation of amino acids, together with the straight-chain acids found in the rumen (Allison et al., 1962b; el-Shazly, 1952b; Keeney et al., 1962).

Inspection of Table IX-6 shows that most of the fatty acids of the milk and serum are found in the bacteria and protozoa, and that to some

Table IX-4

FATTY ACID COMPOSITION OF THE DEPOT FATS OF SEVERAL HERBIVOROUS ANIMALS[a,b]

	Nonruminants		Ruminants	
Fatty acid	Horses[c] (mesenteric fat)	Rabbit[d] (abdominal fat)	Ox[e] (perinephric fat)	Indian sheep[f] (body fat)
Saturated				
Lauric (C_{12}) and myristic (C_{14})	4.9	1.9	3.9	3.4
Palmitic (C_{16})	25.9	23.6	26.5	29.5
Stearic (C_{18})	4.9	6.1	23.1	26.6
Arachidic (C_{20})	0.2	0.9	0.7	1.3
Unsaturated				
C_{14} and C_{16} monoethenoid	6.8	5.2	3.1	3.5
Oleic (C_{18} mono-ethenoid)	33.7	12.7	40.4	3.18
Linoleic (C_{18} diethenoid)	5.2	8.9	1.8	3.3
Linolenic (C_{18} triethenoid)	16.3	39.7	—	—
C_{20-22}	2.3	1.2	0.5	0.6

[a] From Lough and Garton (1958, p. 97).
[b] All values in percent = mole percentage composition of the fatty acid.
[c] From S. S. Gupta and Hilditch (1951).
[d] From Futter and Shorland (1957).
[e] From Hilditch and Longenecker (1937).
[f] From Hilditch and Shrivastava (1949).

extent the milk acids reflect the fatty acid composition of the serum. The values for milk in the table do not include the approximately 10% of C_4 to C_{10} acids. Calculations (Keeney et al., 1962) indicate that the quantity of microbial lipid supplied to the host constitutes an appreciable fraction of the fat output in the milk and that most of the C_{15} branched acid can be accounted for by the microbial C_{15} branched-chain acids, if a turnover of once per day for the rumen contents is assumed.

Table IX-5

COMPONENT FATTY ACIDS (PERCENTAGE w/w) OF THE LIPIDS OF CLOVER-RICH PASTURE AND OF THE LIPIDS OF RUMEN CONTENTS OF SHEEP GRAZING THEREON[a]

	Saturated acids				Unsaturated acids				
	C_{14}	C_{16}	C_{18}	C_{26}	C_{14}	C_{16}	Mono-ethenoid C_{18}	Diethenoid C_{18}	Tri-ethenoid C_{18}
Clover-rich pasture	—	8.9	2.8	3.9	—	7.9	9.5	8.1	58.9
Rumen contents of sheep	1.2	16.9	48.5	5.8	0.2	1.8	19.4	2.9	3.3

[a] From Lough and Garton (1958, p. 98; after Shorland, Weenink, and Johns, 1955).

Table IX-6

PERCENTAGE COMPOSITION BY WEIGHT OF THE VARIOUS FATTY ACIDS IN BOVINE LIPIDS [a,b]

	Kind of Fatty Acid[c]																
	12:0	13:0 br	13:0	14:0 br	14:0	15:0 br	15:0	16:0 br	16:0	16:1	17:0 br	17:0	18:0	18:1	18:2	18:3	18:4 or 20:1
Milk																	
Total lipid	4.7	—	tr	—	14.1	1.1	1.4	0.2	35.3	1.3	0.8	0.8	8.9	19.5	1.5	tr	—
Bovine blood serum																	
Glyceride plus free fatty acids	tr	tr	tr	tr	1.4	1.3	1.7	0.6	23.5	5.3	0.3	1.7	17.4	27.3	18.1	—	1.7
Cholesterol ester	—	—	—	—	0.6	0.9	0.7	0.2	4.8	2.3	0.3	0.2	0.5	3.3	80.5	2.0	3.7
Phospholipid	0.3	0.2	0.9	—	1.8	tr	0.6	0.4	20.9	2.2	3.7	3.0	21.1	15.6	29.1	—	—
Rumen bacteria																	
Neutral lipid	0.6	tr	tr	0.7	2.0	6.7	4.5	2.9	26.1	1.5	1.5	0.4	10.7	16.2	18.1	—	7.9
Free fatty acids	tr	tr	tr	tr	0.9	1.9	1.2	—	16.9	0.8	0.5	1.9	58.9	12.5	4.3	—	—
Polar lipid	1.5	1.2	0.7	1.3	3.8	20.3	8.2	1.6	30.6	1.3	1.2	0.8	6.5	10.2	10.7	—	—
Rumen protozoa																	
Neutral lipid	—	—	—	—	1.0	0.5	1.1	2.2	26.5	0.7	1.2	tr	12.3	17.1	27.5	—	9.7
Free fatty acid	—	—	—	—	tr	—	0.6	tr	14.2	—	0.6	1.3	68.3	10.0	4.9	—	—
Polar lipid	tr	tr	0.7	tr	1.6	3.7	2.0	1.1	37.5	1.2	2.8	0.8	10.3	20.3	14.6	—	3.2

[a] From Keeney et al. (1962).

[b] All values in percent-percentage composition by weight.

[c] Number to the left of the colon indicates the carbon chain length; number to the right indicates the number of double bonds; br means branched; tr means trace.

The seemingly random hydrogenation and isomerization of ingested lipids by the rumen microbiota is probably not closely connected with the synthetic metabolism of the cells, but is rather an expression of the catalytic properties possessed by the total population and of the availability of molecular hydrogen. It is of interest to speculate on whether the chemical changes in the fats occur within the cells or at the cell surface.

In addition to the changes exerted on the fatty acids, the rumen microbiota hydrolyze the fats to their component fatty acids and glycerol (D. E. Wright, 1961b). The glycerol is fermented by rumen bacteria (Johns, 1953; Hobson and Mann, 1961) belonging to the genera *Selenomonas, Anaerovibrio,* and *Peptostreptococcus,* and probably additional types.

The feeding of fat for increased body storage would seem to be unnecessary if acetate ordinarily occurs in excess amounts in ruminants, the excess being converted to fat. In view of the possible diminished human nutritional usefulness of saturated fats as compared to the unsaturated ones, present standards of meat quality based on fat marbling might well be reevaluated.

2. MILK FAT

Milk fats, acids, and aldehydes reflect the relatively high concentration of oleic acid found in depot fat of cattle, 33% (L. M. Smith and Jack, 1954a,b), and in addition many odd- and even-numbered acids from C_{10} to C_{20} have been identified (L. M. Smith, 1961; Smith *et al.,* 1954; Shorland and Hansen, 1957; R. G. Jensen and Sampugna, 1962). Similar acids are found in bacteria (Scheuerbrandt and Block, 1962; Garton and Oxford, 1955; Allison *et al.,* 1962b; Keeney *et al.,* 1962). Additions of acetate to isobutyric, isovaleric, and 2-methylbutyric account for the long-chain iso- and anteiso-fatty acids characteristic of milk (Synge, 1957; Keeney *et al.,* 1962). These have been shown to be formed by *Ruminococcus* (Allison *et al.,* 1962b) and *Bacteriodes succinogenes* (G. H. Wegner and Foster, 1963).

Milk fat is synthesized chiefly from acetate (Folley and French, 1949; Popjak *et al.,* 1951a,b). The percentage of fat in milk is influenced by the relative proportions of acetate in the mixture of fermentation acids. With rations high in concentrate and low in hay, the percentage of milk fat decreases significantly (Stoddard *et al.,* 1949; Tyznik and Allen, 1951; Loosli *et al.,* 1945; McClymont and Paxton, 1947; Balch *et al.,* 1952a; Powell, 1938; Rook, 1959; J. A. B. Smith and Dastur, 1938). Flaked maize is particularly effective in this respect (Balch and Rowland, 1959; Balch *et al.,* 1954c, 1955a,b). Flaked maize caused a greater effect than maize meal (Balch *et al.,* 1954b). The rumen contents contain a greater proportion of propionate and decreased acetate, the lowered synthesis of fat correlating with reduced availability of acetate. Feeding of propionate (G. H. Schmidt and Schultz, 1958) decreased the percentage of milk fat, and acetate increased it (van Soest and Allen, 1959; Stoddard et al., 1949). Presumably the blood acetate concentration is low in low milk-fat cows, but this has not been experimentally demonstrated. In some instances the lowered fat content is correlated with an increased solids-not-fat content (Balch and Rowland, 1959; Rook, 1959); protein and lactose are presumably increased by the greater relative availability of propionate.

Addition of potassium or sodium bicarbonate to a feed causing lowered milk fat increased the percentage of fat (Emery and Brown, 1961) and caused a higher rumen pH, but no change in the proportions of VFA's in the rumen contents. This would seem to imply a relatively increased absorption of acetate at the higher pH, which seems unlikely.

D. Appetite

As indicated in Chapter V, turnover time in the rumen is influenced by the rumen volume, the quantity of saliva secreted, amount and type of food eaten, and the degree of comminution. Maximum rumen fill is obviously a factor limiting intake; the maximum fill is determined by the size of the rumen and by other factors, such as the conformation. Heredity may influence these and be a factor determining the quantity of digesta the rumen-reticulum can handle. Within limits, the faster the turnover time in the rumen the greater the benefit to the host.

Since the type of feed and the degree of comminution can be controlled and the amount of saliva is influenced by the type and amount of feed, the chief uncontrolled factor influencing turnover rate is the amount of feed. This is determined by the appetite of the animal. Appetite thus becomes extremely important in ruminant economy, as with other domestic animals. Control of appetite by a feedback mechanism has been postulated (Adler and Dye, 1957). The rate at which the digesta can leave the rumen determines the intake by releasing additional space. Much of the effect of comminution of the feed on the amount consumed is due to the fact that the smaller particles can leave the rumen without the necessity of comminution through rumination. The influence of amount of feed, percent concentrate, and time after feeding on the volume of water and digesta in the reticulo-rumen has been studied (Emery et al., 1958a); the following regression equations have been formulated:

$$Y = 12.21 - 0.05X_1 - 0.13X_2 + 2.65X_3$$

and

$$W = 11.85 - 0.05X_1 - 0.10X_2 + 1.78X_3$$

in which Y = total reticulo-rumen contents as percent of body weight, W = reticulo rumen water as percent of body weight, X_1 = the percent concentrate in the ration, X_2 = time after feeding in hours, and X_3 = the pounds of air-dry feed consumed per 100 lb of body weight.

Comparative studies of the fermentation rate in the very small (9-lb) African antelope, the suni (Hungate *et al.,* 1959), have shown that it meets its increased energy requirements by a greater fermentation rate rather than by a bigger rumen (Table VI-4). The same effect, although to a less extent, was noted in the comparison of European and Zebu-type cattle. The faster fermentation rate in the zebus correlated with a decreased retention time. Increased body weight gains have been reportd to correlate to some extent with faster turnover (Butler and Johns, 1961).

Events during pregnancy emphasize the multiplicity of factors involved in appetite. In late pregnancy the fetus takes up part of the space available in the abdominal cavity, and the rumen volume is smaller (Makela, 1956). The increased nutritional needs correlate with a greater appetite which leads to a faster rumen turnover (Graham and Williams, 1962). The factors controlling appetite are extremely complex, but their elucidation could be extremely important in achieving maximum productivity.

Faster turnover appears to correlate to a significant extent with an increased percentage of propionate in the VFA's. Since the proportion of propionate is so important in high-producing milk cows, the factors increasing the propionate deserve close study. This and other problems will be discussed in Chapter X.

CHAPTER X

Variations in the Rumen

The principles of rumen function have been presented in the preceding chapters, with particular attention to quantitative aspects. The concept of continuous fermentation in an average animal has been used to simplify the problem and to permit easier comparison of various parameters. Neither average individuals nor average conditions are common in nature; variations are universal. The factors causing variation in rumen function are the subject of this chapter. Some variations mentioned in previous chapters will be discussed more completely.

A. Influence of Kinds of Feed

One of the chief factors influencing the rumen fermentation is the variability in the components of the feed. Effects of various feeds have been the subject of a tremendous number of studies in many countries of the world. These empirical studies, coupled with nutritional discoveries of the last 50 years, are the basis for modern ruminant feeding practices. Much of the accumulated experience has been collected in Schneider's "Feeds of the World" (1947). Detailed consideration of these results is outside the scope of this volume, although some of them will be mentioned where pertinent. In the present discussion emphasis will be primarily on the effect of feed differences on microbial activity and the nature of the resulting products, both cells and fermentation acids and gases.

1. THE CHEMICAL COMPONENTS OF THE FEED

The dense rumen microbial population depends on the continuous supply of the digestible feeds included in the ration.

a. Carbohydrates

Of these digestible feeds, carbohydrates are by far the most important quantitatively because of their superiority as a source of energy under anaerobic conditions (Hungate, 1955, p. 195). The types of carbohydrates most common in forages are (1) soluble carbohydrates, chiefly sugars, (2) starch, and (3) insoluble carbohydrates composing the cell walls of forage plants.

(1) *Soluble Carbohydrates.* Sugars and other soluble carbohydrates, which may constitute 30% of the dry matter in forage (Waite and Boyd, 1953a,b), are rapidly metabolized. Fermentation is maximal almost immediately after adding sugar to *in vitro* incubated rumen contents (J. I. Quin, 1943a), which indicates that enzymes attacking these substrates are already present in the microorganisms and that the enzymes are not saturated with substrate.

The fraction assimilated is greater than is the case with many other forage components. Part of this increased assimilation represents conversion of the sugar into reserve polysaccharide in the interior of the bacteria (Baker, 1942; Gibbons *et al.*, 1955; Hoover *et al.*, 1963) and protozoa (Heald and Oxford, 1953). This is particularly true if assimilation measurements are made soon after the sugar is supplied. There is subsequent conversion of the polysaccharide into cell material and wastes.

Sugars added to a ration (Kleiber *et al.*, 1945) showed 1.9 gm of methane per 100 gm of glucose fed, or 0.21 mmoles of methane per mmole of glucose, whereas, for the total carbohydrate in forage, values of 0.48 to 0.50 (Kellner, 1911), 0.54 to 0.56 (Armsby and Fries, 1903), and 0.49 (Kleiber *et al.*, 1945) were obtained, with the glucose in these materials assumed to have a molecular weight of 171 (see also Table VI-1). The lower proportion of methane from glucose utilization suggests a greater intracellular assimilation into cell substance. A greater assimilation with glucose than with cellulose is evident from the values in Table X-1.

The greater proportion of methane in the rumen fermentation products 9–15 hours after feeding (Hungate *et al.*, 1960) as compared to the

Table X-1

NITROGEN ASSIMILATION IN RUMEN LIQUID
SUPPLIED WITH GLUCOSE OR CELLULOSE

			Nitrogen in Cells after Incubation[a]		
Experiment	Period of incubation (hr)	Initial nitrogen in cells[a]	No added substrate	400 mg of glucose added	500 mg of cellulose added
1	4	121	126	191	165
2	4	138	140	187	151
	8	138	129	233	154

[a] All values in milligrams of cell nitrogen per 100 ml of rumen liquid.

period immediately after feeding may be due to the lesser proportion of soluble sugars fermented during the later period. Similarly, the increased production of propionic acid with ground feed may be in part due to the more continuous availability of sugar at the increased turnover, although a more continuous supply of amino acids could also be a factor. Uniformly labeled glucose added to rumen contents is not preferentially converted to propionate (Otagaki *et al.*, 1963), the specific activities were in the ratio 7.4 for acetate, 1.3 for propionate, and 1.3 for butyrate, a somewhat higher proportion in acetate than would be expected from the usual proportions of these acids. Similar results were obtained (Portugal, 1963) when trace quantities of uniformly labeled glucose were added to rumen contents incubated *in vitro*.

Too much sugar in the ration (20–30%) can diminish the digestibility of the fiber (Hamilton, 1942; Hoflund *et al.*, 1948). This can be caused by depletion of noncarbohydrate nutrients by the saccharoclastic bacteria, to an excessive production of acid, or to an inhibition of cellulase by sugar (W. Smith and Hungate, unpublished experiments).

The rumen fermentation rates supported by various added sugars are not all alike. Maltose, glucose, sucrose, raffinose, fructose (J. I. Quin, 1943a), cellobiose, mannose, D-(+)-xylose, and L-(+)-arabinose were rapidly fermented by mixed rumen bacteria freed of protozoa (McNaught and Smith, 1947), but D-(−)-arabinose, sorbose, glucuronic, gluconic, and glucosaccharic acids, mannitol, sorbitol, alginic acid, glucosamine (McNaught, 1951), and L-fucose (McNaught *et al.*, 1954b) were not readily used. Analysis of the cultures with arabinose and xylose showed end products typical of the rumen fermentation. Similar products were formed from maltose, with the additional recovery of lactic acid amounting to about 10% of the substrate. Polysaccharide was also found in the bacterial suspension, an observation confirmed in other studies (R. Q. Robinson *et al.*, 1955). The action of whole rumen contents on various sugars *in vitro* has been examined by the author with manometric techniques; the results are shown in Table X-2.

Glucose and cellobiose are most rapidly utilized, fructose at almost the same rate, and maltose and sucrose less actively. Turnover rate constants of 2.5 per minute have been found for labeled glucose and fructose added to rumen contents (Portugal, 1963). D-Xylose is attacked (Pazur *et al.*, 1958), but D-arabinose turns over slowly, a rate constant of 0.12 per minute having been observed (Portugal, 1963). Lactose is not readily fermented by the mixed rumen organisms. The soluble fraction of alfalfa hay as well as whole powdered hay will usually support a fermentation as rapid as any single sugar or combination of sugars.

Maltose gives a greater percentage of gas produced than do the other

Table X-2

RATE OF GAS PRODUCTION WITH ADDED SUBSTRATE[a]

Experiment	Ration	Ground alfalfa hay	Substrate added		
			Glucose	Maltose	Other
1	Alfalfa hay	20.7	17.5	—	18.6 glucose + cellobiose
		—	—	—	19.4 glucose, cellobiose, maltose, sucrose
2	Alfalfa hay	22.9	19.4	14.8	13.7 cellobiose
3	Alfalfa hay	22.8	14.4	16.7	14.7 sucrose
4	Alfalfa hay	—	23.2	15.5	4.0 substrate
5	Alfalfa hay	—	19.4	10.5	18.8 glucose + cellobiose
6	Alfalfa hay	21.2	18.5	—	19.0 glucose + cellobiose
7	Alfalfa hay	—	21.9	15.3	—
8	Alfalfa hay	23.5	25.0	16.7	—
9	Alfalfa hay and grass pasture	18.4	12.4	9.0	—
10	Oats, barley, and alfalfa hay	15.4	10.2	8.0	10.0 sucrose
11	Wheat, barley, oats, and alfalfa hay	31.3	34.0	31.8	33.6 sucrose
12	Grains, beet pulp, and 25% alfalfa hay	41.9	56.2	56.7	—
13	Grains, beet pulp, and 25% alfalfa hay	45.0	42.7	49.8	—

[a] All values in μmoles per minute per 10 gm of rumen contents.

sugars. Analysis of the gas in the maltose fermentation vessel showed that most of the increased pressure was due to methane, as reported by others (McNeill and Jacobson, 1955). Some hydrogen was detected, but liberation of hydrogen instead of its conversion to methane could not account for the increased gas produced from maltose. The explanation for this difference with maltose is obscure. Inability of *Dasytricha* to store starch at the expense of maltose (Mould and Thomas, 1958) might allow more maltose to be metabolized by the bacteria, but it is not readily apparent why a factor such as this would increase the amount of methane formed. The percentage of hydrogen in rumen gas was increased by feeding large quantities of oatmeal mixed with water (Klein, 1926). This would increase the methane production.

A number of investigators have reported (Arias *et al.,* 1951; Hoflund *et al.,* 1948) a stimulatory effect of small amounts of sugar on fiber digestion when the forage is of poor quality. Attempts by the author to stimulate fiber digestion by adding glucose to rumen contents incubated *in vitro* slightly increased the fermentation rate after the burst of fermentation due to glucose had subsided. This could indicate increased fiber digestion, but polysaccharide reserves stored at the expense of the sugar (J. I. Quin, 1943a; Baker, 1942; Oxford, 1951; G. J. Thomas, 1960) could give a similar effect after the sugar itself had been exhausted.

The more rapid assimilation of nitrogen with glucose than when pebble-milled cellulose is the substrate (Table X-1) is due in large part to the greater rate at which glucose is fermented. It disappears from rumen contents at a rate of about 2% (w/v) per hour, as compared to a much slower average rate of digestion of the insoluble components of forage. Glucose added to a poor forage ration could facilitate the assimilation of the available protein, which might be fermented if no soluble carbohydrate were available.

The effects of various carbohydrates on the synthesis of rumen microbial protein are shown in Table X-3 (Henderickx and Martin, 1963). The most nitrogen synthesis occurred with pentoses, starch, and glycogen, the other carbohydrates giving values about as expected, with the exception of fructose. The poor assimilation on the insoluble substrates can be explained by the short incubation period of 6 hours.

(2) *Starch.* Starch is digested rapidly in the rumen, but more slowly than the sugars. Its value as a feed constituent resembles that of the soluble sugars. Since the rate of digestion of starch is faster than that of cellulose, there is a better chance that the intermediate maltose and glucose accumulate in sufficient concentration to support an accompanying nonamylolytic population. At a pH of 6.0, glucose to the extent of

Table X-3

INFLUENCE OF CARBOHYDRATES ON THE SYNTHESIS OF MICROBIAL PROTEINS [a], [b]

Carbohydrate	Result	Carbohydrate	Result	Carbohydrate	Result
Monosaccharide		Di- and trisaccharide		Polysaccharide	
Arabinose	+20.3	Cellobiose	+12.4	Dextrin	+ 9.6
Ribose	− 5.9	Lactose	+10.4	Glycogen	+16.2
Xylose	+19.6	Maltose	+10.6	Inulin	+10.0
Rhamnose	− 4.9	Melibiose	+ 5.5	Pectin	+12.0
Fructose	− 0.7	Sucrose	+ 7.6	Xylan	+ 8.9
Galactose	+ 0.9	Trehalose	+ 5.9	Starch	+18.2
Glucose	+ 7.5	Raffinose	+ 5.9	Cellulose	−15.2
Mannose	+ 7.2	Melizitose	−11.5	Hemicellulose A	− 8.8
Sorbose	−11.5			Hemicellulose B	− 1.5
				Lignin	−14.0
Acids		Alcohols			
Galacturonic acid	− 4.6	Dulcitol	− 4.6	Control	
Gluconic acid	− 6.1	Erythritol	− 3.8	Blank	−14.9
Glucuronic acid	− 7.2	Inositol	− 8.4		
Gulonic acid	− 5.1	Mannitol	− 4.6		
Mucin	− 3.2	Sorbitol	− 6.7		

[a] From Henderickx and Martin (1963).

[b] Experimental conditions: 100 ml of strained rumen fluid plus 10 ml of mineral solution (Burroughs *et al.*, 1950c) in a total volume of 200 ml. Incubated at 40°C for 6 hours under a stream of carbon dioxide flowing at 100–120 ml per minute. Carbohydrate (1 gm) and nitrogen [25 mg from $(NH_4)_2SO_4$] added.

5 μmoles per milliliter accumulates in the rumen, as compared to the normal 0.5 μmoles or less (Ryan, 1964b). Starch, in common with the soluble carbohydrates, exhibits a high percentage of conversion into fermentation products and cells. The indigestible residues are minor.

Increased acidity and a tendency for higher proportions of propionate often accompany increased starch feeding (R. L. Reid et al., 1957; Ryan, 1964b), though in some cases butyrate is increased (Donefer et al., 1963). During the course of experiments on the rates of production of the volatile fatty acids (VFA's) in lactating cows, a striking change in the ratio of propionic to acetic and butyric acids correlated with the feeding of a quantity of grain (Hungate et al., 1961). Prior to feeding, methane composed 14% of the total fermentation products, and propionate was not unusually abundant. After the grain was fed, methane constituted only 9% of the fermentation products, and the propionic acid production was high.

Increased activity of Streptococcus bovis is often postulated to explain the phenomenon, but no conclusive experimental evidence has been provided. It is more likely due to Bacteroides amylophilus, which produces much succinate that is readily decarboxylated to propionate. A gradual supply of rations high in starch would seem to be advantageous over large doses, to avoid excess acidity. No difference between adaptation with one and four daily feedings was observed (Ryan, 1964b), i.e., frequent feedings did not select a microbial population more rapidly than did a single daily feeding of the same quantity.

The rumen protozoa possess an advantage in the competition for starch in the rumen. They ingest large quantities in a very short time and thus remove it from bacterial attack. The ability of the protozoa to compete with the bacteria appears to require that the starch be in the form of raw grains. Cattle adapted to a high grain ration usually contain a very high concentration of Entodinium. Under some circumstances, other genera such as Epidinium become very numerous. In sheep fed flaked maize, the protozoa disappear (Phillipson, 1952a), possibly because of the acidity.

Various feed treatments, such as grinding and pelleting, affect the rates at which the protozoa and bacteria utilize starch. Boiled starch is much more actively fermented by rumen contents than is raw starch (McNaught, 1951), presumably due to the greater surface exposed to enzymatic attack. Flaked maize favors increased bacterial activity and a shift toward increased percentages of propionate (Balch et al., 1955b; Balch and Rowland, 1959; Eusebio et al., 1959) as judged by the effect on milk fat. Dredge corn did not produce a similar effect. Various processing treatments of corn, including heat treatments, caused a narrower acetate: propionate ratio than did the untreated corn (Newland *et al.,* 1962), although in other experiments heating the ration caused no differences in its utilization (Woods and Luther, 1962).

In sheep on a diet containing 50% wheat, starch grains observed in the abomasum (by means of an abomasal fistula) indicate that some starch escapes the rumen fermentation (Marston, 1948b), as would be expected in view of the small size of the grains, but the amount of starch escaping the rumen fermentation in this way is small. Dried potato pulp, supplemented with urea, has given better gains than an equal weight of oats (Nicholson *et al.,* 1964).

(3) *Crude Fiber.* Crude fiber is the material left after treatment with hot 1.25% sulfuric acid followed by hot 1.25% sodium hydroxide. It includes cellulose, lignin, and some hemicellulose. Hemicellulose is a heterogeneous group of materials including hexosans, pectin, pentosans and polyuronides. Some of them, e.g., pectin, are fermented rapidly, whereas others are digested at about the speed and to about the same extent as the cellulose in the forage. Many of the bacteria able to digest cellulose, such as *Butyrivibrio* and *Ruminococcus,* also digest certain of the hemicelluloses when the latter are supplied as pure materials, and, as mentioned in Chapter II, all tested pure strains of rumen bacteria able to digest a large proportion of cellulose in forage digest the hemicellulose to about the same extent.

Since the insoluble fibers of the feed digest more slowly than the starch and soluble carbohydrates, they support a more sustained fermentation and are the chief substrate just before feeding, when the other components have been largely utilized.

In the past it has been believed that some forage is essential in ruminant feeds, because of a need for fiber. The chance of nutritional disturbance is less if a considerable quantity of digestible fiber is included in the ration, but, with frequent feeding and increased understanding of the importance of obtaining a microbiota adapted to the feed, forage can be eliminated completely (R. E. Davis, 1962; Davis *et al.,* 1963).

(a) *Cellulose.* Digestion of cellulose *in vitro* leads to a high proportion of propionate in the VFA's as might be expected from the importance of succinate as a fermentation product of ruminococci and *Bacteroides succinogenes.* However, increasing the roughage in dairy cattle rations increases the proportion of acetic acid (J. M. Elliot and Loosli, 1959). The increased percentage of methane with time after feeding also suggests that acetate is more important when fiber is the substrate for the rumen fermentation. This would suggest a relatively increased activity of *Ruminococcus albus,* which produces chiefly acetate. The factors favoring a succinogenic fermentation of cellulose need study.

When pure cellulose, cellophane (Asher, 1942), or cellulose reprecipitated after solution in cuprammonium solution (Asher, 1942) is added to rumen contents incubated *in vitro,* there is a lag of 4–6 hours before increased fermentation indicates attack (McBee, 1953). If the fermentation in a control vessel without cellulose is subtracted from the experimental value, a typical growth curve results, as was shown in Fig. V-6, with an initial lag. The delay is even longer, 13–14 hours, when cotton threads are suspended in the rumen (Grosskopf, 1964).

The delay with pure cellulose could be due to its attack by a minor component of the flora which gradually increases in numbers because of the availability of its particular food, or it could be due to gradual adsorption of cellulolytic cells or enzymes onto the fresh substrate, to give an increasing rate of hydrolysis until the substrate becomes limiting. Tests with cellulose inside a sterile porous porcelain test tube suspended in the rumen (Fina *et al.,* 1958) failed to detect any cellulose digestion in the absence of microbial cells, a result suggesting that delay in attachment of cells to forage explains the lag in cellulose utilization. On the other hand, *Ruminococcus* and *Butyrivibrio* strains produce a soluble cellulase which would be expected to diffuse through the porcelain. The Whatman filter paper used in the tests is not readily digested by the soluble enzymes of the rumen, but it is attacked by *Bacteroides succinogenes.* However, the latter does not show evidence of a soluble cellulase, and the cells were unable to pass through the porcelain.

The different abilities of pure cultures (Hungate, 1957; Ghose and King, 1963) and of mixed rumen bacteria (T. I. Baker *et al.,* 1959) to attack cellulose in natural materials indicate differences in the susceptibilities of various natural "celluloses" to enzymatic attack. The digestibility of these natural materials is effectively increased by alkali treatments and also by treatment with other reagents such as acid, chlorite, and sulfite. The cellulose-splitting enzymes are produced even when bacteria (nonrumen in this case) are grown on sugar media (Hammerstrom *et al.,* 1955).

Rumen fluid does not ordinarily contain much free cellulase (King, 1956; Festenstein, 1958a, 1959; Halliwell, 1957b), although soluble cellulose derivatives are attacked. Much of the cellulase is presumably attached to the cellulose. Fiber-digesting enzymes attached to the fibers in rumen digesta have been studied by King (1956). Phosphate buffer at pH 8.5 was used to elute the cellulase from the solids, and enzyme activity was assayed with carboxymethylcellulose as substrate. One unit of enzyme was defined as the quantity causing a viscosity decrease of 0.01 in 10 minutes at 30°C when 0.1 ml of enzyme solution was added to 3.9 ml of carboxymethylcellulose-70-H in 0.05 M phosphate buffer at pH 6.75.

Between 700 and 19,000 units of enzyme activity were obtained per gram of rumen solids.

A mixture of enzymes was disclosed by fractional elution with phosphate buffers of various pH's, and electrophoresis of the fractions. Activity was high at 40°C, but absent at 50°C. Thermal inactivation at 100°C was of first order, with an inactivation constant of approximately 1 per minute. In a later study (King, 1959) evidence was obtained that rumen β-glucosidase is partly attached to the cells.

Some cellulolytic activity could be demonstrated for up to 48 hours (Meites *et al.,* 1951) when rumen fluid was incubated *in vitro* with finely divided cellulose in the presence of toluene. Of numerous additives tested, the only ones increasing cellulase activity were the ashes of rumen liquor and alfalfa. An enzyme preparation from rumen fluid failed to attack insoluble cellulose, but did digest the soluble carboxymethylcellulose (Underkofler *et al.,* 1953; Salsbury *et al.,* 1958b). Similar results with the solution obtained by blending of rumen fluid in a Waring blender, followed by centrifugation, were reported (Stanley and Kesler, 1959), with some indications that cellulose was slightly attacked. These methods have been modified with the viewpoint of obtaining rapid estimates of cellulolytic activity in the rumen (Stolk, 1959). An identity with "cellulase" of the enzymes attacking carboxymethylcellulose has not been established. All these studies indicate the need for further purification and separation of the various elements concerned with "cellulase" activity in the rumen.

Differences in digestibility of various "cellulose" preparations are partly due to differences in the associated noncellulosic material. Even defatted cotton fibers contain some noncellulosic material. If the ends of the long β-glucosidic chains in "native" cellulose are the point at which "encrusting" materials attach to cellulose, and if cellulase attack occurs only at a "free" end, a few non-β-glucosidic linkages could be very effective inhibitors. Differences in abilities of cellulolytic pure cultures to attack "native" cellulose may depend on their capacity to split linkages other than the 1,4-β-glucoside.

Solubility of forage cellulose in cupriethylenediamine correlates well with digestibility of forage grasses, but not with the digestibility of the cellulose in legumes and mixed forages (Dehority and Johnson, 1964; Johnson *et al.,* 1964). Solubility in normal sulfuric acid solution of the cellulose in both legumes and mixed forages correlated with digestibility of cellulose in these feeds, but the correlation was not perfect. These experiments indicate that the digestible and indigestible cellulose in forages may be linked in different fashions in different feeds.

Several lines of study indicate that bonds of cellulose to other materials shield it from digestion. Bombardment of wood sawdust by an intense beam of high-voltage X-rays (Lawton *et al.,* 1951) increased the digestibility of the cellulose by rumen organisms to the point that the percentage of cellulose fermented was about the same as that digested in ruminants. Untreated or ensiled beech and pine sawdust have no nutrient value for ruminants (Kirch and Jantzon, 1941). Fine grinding in a pebble mill rendered all the forage cellulose fermentable (Dehority and Johnson, 1961), whereas the unground material showed a considerable residue of unattacked cellulose even after an extended period of fermentation by rumen microorganisms. Delignification of alfalfa hay increased significantly the amount of fermentable material in the water-insoluble components (Fig. V-5). Removal of hemicellulose A and B from holocellulose caused an increased digestibility of the cellulose (Packett *et al.,* 1965).

In addition to the influence of other fiber components on the digestibility of cellulose in forages, the physical state of cellulose preparations may influence the digestibility by rumen cellulolytic bacteria (Halliwell and Bryant, 1963). Filter paper (Whatman no. 1) is not attacked to an appreciable extent by many bacterial strains which will attack the same filter paper after it is pebble-milled in water. In similar fashion, cotton fibers freed of oil by ether extraction are not as readily digested by many bacteria as are the same fibers wet ground in a pebble mill.

Some doubts as to the ability of the bacteria isolated from the rumen to digest cellulose in pure culture have been based on the premise that attack on defatted cotton fibers is the sole criterion of "cellulase." For the rumen the critical test is not whether the organisms can digest cotton, but whether they can digest the "cellulose" in the forages consumed by the ruminant. As pointed out earlier (Table II-7), a number of pure cultures, tested for their ability to decompose the water-insoluble components of alfalfa and grass hay, digest the 5% acid-soluble fraction ("hemicellulose") and the 70% sulfuric acid-soluble fraction ("cellulose") to about the same extent as they are digested during passage through the ruminant (Hungate, 1957), and enzyme experiments with pure cultures disclose cellulase activity (Halliwell and Bryant, 1963; Hungate, 1950; W. Smith and Hungate, 1965).

One of the rumen bacteria, *Bacteroides succinogenes,* is able to digest defatted cotton fibers in pure culture. Untreated Whatman filter paper added to rumen contents is rapidly digested, and abundant bacteria resembling *B. succinogenes* can be seen microscopically, massed around the disintegrating portions of the fibers. *Ruminococcus albus* predominates over *Butyrivibrio* in its incidence as colonies in cellulose-agar tubes inoculated from rumen contents of sheep fed alfalfa hay (Gouws and Kistner, 1965). Very few cellulolytic butyrivibrios were observed. When teff hay was fed, the number of cocci decreased, and the butyrivibrios increased.

The teff hay was much poorer in nitrogen, but contained about the same percentage of fiber.

The finding that some cellulolytic bacterial enzymes (Hungate, 1946b; W. Smith and Hungate, unpublished experiments) produce almost exclusively cellobiose from cellulose is explicable on the basis that only alternating β-glucosidic linkages on the surface of a polymer would be sterically identical. An enzyme would be able to attack only every second bond, and as a result cellobiose would be formed. Part of the great variability in the number of glucose units formed by enzymes acting on soluble types of cellulose results from the ability of enzyme to approach the dissolved glucose chain from any direction and therefore to achieve a proper steric orientation of any β-glucosidic bond instead of only alternate ones.

(b) *Hemicellulose.* Hemicellulose contains polymers of arabinose, xylose, glucose, galactose, rhamnose, and uronic acids (Sullivan, 1961). Legume hemicellulose contains less xylan than does grass hemicellulose.

Pectin and other polyuronides constitute 5–8% of the fresh weight of alfalfa (Conrad *et al.,* 1958b) and are about half as abundant as the cellulose in some forages (Michaux, 1951). Pectic substances disappear rapidly after forage enters the rumen (Conrad *et al.,* 1961). They are digested to the extent of 76–85% in the case of hay and 88–93% in the case of apple pulp and beet pulp (Michaux, 1950, 1951). A γ-pectin glycosidase has been found in the rumen of steers fed citrus pulp (Smart *et al.,* 1964).

Pectin-splitting enzymes have been demonstrated in rumen bacteria (Howard, 1961; D. E. Wright, 1961a) and in the protozoa (D. E. Wright, 1960b; Mah and Hungate, 1965). Methylesterase is one of the enzymes concerned, and methanol is split from the pectin (Howard, 1961; D. E. Wright, 1960b). Polygalacturonase activity has been demonstrated, and in *Ophryoscolex* is of the transeliminase type (Mah and Hungate, 1965).

A report that pectin supported a more rapid evolution of fermentation gas than did the soluble substances of legumes (Conrad *et al.,* 1961) may have resulted from the use of phosphate buffer. If the soluble carbohydrates of the alfalfa supported primarily an acidigenic fermentation, little gas would be liberated with a phosphate buffer, whereas if more carbon dioxide and methane were formed from the pectin in alfalfa fiber, these products would appear as gas even with phosphate buffer. Other investigators (Boda and Johns, 1962; Fig. V-5) find that the fermentation due to soluble materials is more rapid than that due to the insoluble constituents.

Inulin is rapidly fermented by rumen bacteria (McNaught, 1951) and many pure cultures attack this material (Table II-9). Laminarin, which composes as much as 36% of the dry weight of the littoral marine alga *Laminaria cloustoni,* was fermented by rumen microorganisms *in vitro,* but other marine algae were not (McNaught *et al.,* 1954b).

Grass forages contain 9–25% pentosans (Heald, 1952), which, when separated, are readily digested and fermented by bacteria of the sheep rumen (Heald, 1952; Howard, 1955). Analyses of feed and feces (Bondi and Meyer, 1943) show that, in sheep, pentosans are digested to the extent of 65%, and hexosans are digested 75%. The pentosan of straw was digested to the extent of 45% in the rumen (Marshall, 1949), with considerable additional digestion in the colon and cecum.

Xylans are important constituents of many grasses and are readily digested and fermented in the rumen. Xylanase is easy to demonstrate in rumen fluid, in contrast to the cellulase. It is contained also in the bacterial cells (Sorensen, 1955) of the rumen and in pure cultures of butyrivibrios (Howard *et al.,* 1960). The xylanase in 1 ml of rumen fluid is sufficient to release up to 3 mg of xylose per hour when incubated at 37°C. This would represent a daily digestion of about 7% (w/v) per day, a rate adequate for the digestion of the amounts of xylan encountered in most rations.

More than one enzyme is concerned (Howard, 1957a). Xylobiose is not hydrolyzed at the same rate as some of the higher polymers (Pazur *et al.,* 1957), and several oligoxylosides as well as xylose can be demonstrated as hydrolysis products.

Hemicelluloses separated from the natural fiber constitute an energy source intermediate in digestibility between soluble sugars and cellulose. Much of the hemicellulose in mature hays and in other plant components ferments at approximately the same rate and to the same extent as cellulose (Marshall, 1949; Michaux, 1951). When chemically separated, cellulose and hemicellulose are more completely digestible than when combined in natural forages. In some cases (Waite *et al.,* 1964) the hemicellulose in forage is significantly less digestible than the cellulose. Comminution to a very fine particle size with a pebble mill increases the digestibility also of forage hemicellulose (Dehority *et al.,* 1962). Some pure cultures of *Bacteroides* and *Ruminococcus* partially digest hemicelluloses prepared from several forages, but cannot ferment them. The partial digestion observed (Dehority, 1965) may result from the splitting of a few hexose linkages binding pentose chains together.

The products of hemicellulose fermentation are acetic, propionic, and butyric acids, with some indications that the proportion of butyrate is less than that characteristic of the rumen (Howard, 1961). Approximately equal proportions of acetic and propionic acids were formed during fermentation of a hemicellulose preparation by a washed cell suspension of rumen bacteria (Gray and Weller, 1958). In other experiments (Minato *et al.,* 1963) xylan gave rise to relatively more propionate than did pectin.

The ratios in which carbon dioxide, methane, and fermentation acids are produced in manometric vessels containing rumen contents and added natural forages differ markedly from the ratios expected when carbohydrate is fermented. The quantity of carbon dioxide formed is relatively high, and the quantity of acid is low. This type of ratio could be due to fermentation of protein or acidic substrates. In the case of protein, the ammonia formed in deamination of the amino acids is more alkaline than the amino group and consequently binds a certain amount of carbon dioxide. It is not released until acid is added at the end of the manometric run. The carbon dioxide released by fermentation acids is balanced in part by this carbon dioxide bound by the ammonia, and low values for acid production during the run are obtained. Similarly, the acid left after deamination of an amino acid, even if fermented to form another acid, will give no increase in acidity since the new acid group merely replaces the carboxyl group of the amino acid. Both these effects cause the amount of carbon dioxide released from bicarbonate during the fermentation of protein to be much smaller than the amount of acid formed. Fermentation carbon dioxide is measured accurately only if it is taken as the difference between experimental- and control-vessel carbon dioxide content after an excess of acid has been added to release all carbon dioxide from bicarbonate. The low acid recoveries from fermented organic acids in forage (Rosen *et al.,* 1956, 1961) are similarly explained.

Studies in the author's laboratory on the extent to which the various fractions of forage give aberrant ratios of fermentation products have disclosed that formic and malic acids in the soluble portion of alfalfa account for a small portion of the aberration. The amount of ammonia released during fermentation of the forage indicates that some of the effect is due to fermentation of protein. However, the chief substrate concerned with the excessive production of carbon dioxide appears to be the hemicellulose fraction.

A hemicellulose kindly prepared from alfalfa by C. A. Marsh gave a very low manometric acid value, as compared to that characteristic of carbohydrate (Table X-4). Direct analysis for VFA's shows that the relatively low values for alfalfa are due to failure of fermentation acids to release carbon dioxide (Table X-5), an expected result if acids compose an important fraction of the substrate. Only in the case of the holocellulose did the acid, measured manometrically, equal the amount produced, as measured chromatographically. In the alfalfa meal 520 μmoles of acid produced was either neutralized by ammonia or originated from an acid substrate. The uronic acids are postulated as the most important such substrate.

Fresh alfalfa has been shown to give an even lower recovery of acid in manometric experiments than is found with alfalfa hay. Grass hays show a slightly higher ratio of carbon dioxide to acid than is expected on

Table X-4

RATIOS OF FERMENTATION PRODUCTS FROM VARIOUS FRACTIONS OF ALFALFA HAY[a]

Fraction	Manometric pressure increase during experiment (mm Hg)	Percentages of Fermentation Products		
		Acid	Carbon dioxide	Methane
1. 250 mg of water-extracted meal	214	36.8	43.5	19.7
2. 212.5 mg of water-extracted meal, extracted further with concentrated NH$_4$OH	179	36.9	37.4	25.7
3. 212.5 mg of water-extracted and NH$_4$OH-extracted meal plus the NH$_4$OH extract	179	36.8	42.0	21.2
4. NH$_4$OH extract from 750 mg of water-extracted meal[b]	105	23.1	48.7	28.2

[a] Conditions of experiment: 20 ml of rumen fluid plus 250 mg of NaHCO$_3$ in each vessel (total volume, 160 ml each), gassed with 100% CO$_2$. Substrate in one sidearm and 2.5 ml of 4N H$_2$SO$_4$ in the second sidearm. Duration, 20 hours. Methane determined from the volume of residual gas after absorption of CO$_2$ with NaOH. Temperature, 39°C.
[b] The hemicellulose fraction.

Table X-5

COMPARISON OF ACID PRODUCTION FROM ALFALFA HAY MEAL BY MANOMETRIC AND CHROMATOGRAPHIC METHODS[a]

Substrate	Acid by manometry	Acid by chroma- tography	Carbon dioxide	Methane
500 mg of alfalfa meal	1340 (corrected)	1860	1270	458
500 mg of water-extracted alfalfa meal	1452	1650	928	430
500 mg of holocellulose from alfalfa	1720	1720	1108	332

[a] All values in μmoles. Conditions identical to those listed in footnote to Table X-4, except that 15 ml of rumen fluid and 25 ml of balanced salt solution were used. Duration, 42 hours. Chromatography according to Wiseman and Irwin (1957).

the basis of a carbohydrate fermentation, but the deviation is much less than with the legume.

(c) *Lignin.* Lignin is defined in a variety of ways, none of them entirely satisfactory. For most nutritional studies it is the term applied to that fraction of plant material insoluble in alcohol, ether, water, dilute alkali, and 70% (w/v) sulfuric acid. It is usually regarded as indigestible in ruminants, yet there are numerous papers (E. B. Hale *et al.,* 1940; Pigden and Stone, 1952; Sullivan, 1955; A. D. Smith *et al.,* 1956; Badawy *et al.,* 1958a; Ter-Karapetjan and Ogandzanjan, 1960b) showing that a certain amount of "lignin" disappears during passage of plant material through the ruminant, as much as 64% in some materials containing little lignin (Bondi and Meyer, 1943).

These variations in the digestibility of "lignin" can be explained by its lack of homogeneity and variations in the quantity of protein in the feed. The proportions of the various components which make up "lignin" vary from one kind of plant to another, and in the same plant at different stages of maturity (Kamstra *et al.,* 1958). Some nitrogenous materials occur in the "lignin" fraction. A "lignin" residue can even be found in *Ruminococcus albus* cells treated with alkali and 72% sulfuric acid. As plants age, lignin constitutes an increasingly large fraction of the total dry weight, and it also becomes less digestible. Since the lignin in relatively poor forages is practically indigestible, it can be used as an index or marker for the digestion of other materials. However, in young green plants lignin constitutes only a small proportion of the dry weight and is more susceptible to digestion. Furthermore, in young plants some of the other constituents which are also resistant to the same chemical agents as lignin, such as the proteins, occur in quantities large enough to interfere with the determination of lignin. A greater digestibility of the "lignin" in monocotyledonous as compared to dicotyledonous crops has been reported (Pigden and Stone, 1952).

The lignin in hays varies between 8 and 12% (Burke, 1950). In general, the more lignin contained in the forage the less the digestibility (Lancaster and Adams, 1943; Duckworth, 1946; Richards and Reid, 1953; Sullivan, 1955; Stallcup *et al.,* 1956; Sullivan and Hershberger, 1959; Glover and Dougall, 1960); conversely, the more the nitrogen the greater the digestibility (Glover and Dougall, 1960).

Lignin contains methoxyl groups ($—OCH_3$), which are split off in the rumen. In early studies (Csonka *et al.,* 1929) the methoxyl content in the lignin of bovine feces was less than that in the food, though possible contributions of protein to the fecal "lignin" could be concerned. In one study, methoxyl content of forage correlated better (-0.725) with digestibility than did lignin or the nitrogen in feed or feces (Richards *et al.,*

1958). Syringaldehyde and vanillin, derivatives of lignin, have been demonstrated in ruminant feces (Pigden and Stone, 1952). On the basis of digestibility and proximate composition of various forages, it has been concluded that lignin and methoxyl content are not the only constituents responsible for decreased digestibility in more highly lignified forages (Quicke and Bentley, 1959). In one study (Goswami and Kehar, 1956), the tannin content in leaves decreased the digestibility more than did the lignin content.

As previously mentioned in connection with cellulose digestion, the comminution of plant material in a ball mill greatly increases the digestibility (Dehority, 1961) of the cellulose in forages. Ball milling can split C—O and C—C bonds in wood and cellulose (Steurer and Hess, 1944), and break the linkages between "lignin" and "cellulose," exposing the latter to enzymatic action.

Continued X-ray irradiation or ball milling can disintegrate and solubilize insoluble plant materials. Copper-reducing substances appear after several days of pebble milling of filter paper cellulose. In the case of wood, continued X-ray treatment ultimately makes it soluble, but the material left is not fermentable by rumen microorganisms. Presumably the structure of the carbohydrate monomers has been modified. The lignin becomes soluble, but is not fermentable.

The inverse relationship between lignin content and digestibility of forages has led to the development of quantitative formulas, with regression coefficients, which represent the effect of the lignin on digestibility. These take the form of $y = a + bx$, in which y is the digestibility of a hypothetical material containing no lignin and b is the percent of lignin. The value of a is obtained by plotting lignin content against digestibility, drawing a straight line which best fits the experimental points, and extrapolating the line to the x axis. The slope of this line is b, which is always negative. These curves can be applied to a particular forage, but they may not apply to other forage species (M. H. Butterworth, 1964).

Impedence of fiber digestion by lignification applies also to the "hemicellulose" portion of the fiber, as would be expected from the similarities in the extent of digestion of the cellulose and hemicellulose components of fiber by pure cultures (Chapter II) (Dehority et al., 1962).

Lignin is not merely a relatively useless, i.e., neutral, factor in the nutritional economy of the ruminant. It separates otherwise digestible materials from digestive enzymes. It must be masticated, carried, moved along the alimentary tract, and eliminated—activities requiring expenditure of energy. The energy itself is not a serious problem, since it is probable that the ruminant has an excess of energy, but accompanying this energy expenditure is a sacrifice of alimentary space which could hold more digestible nutrients. There is also a loss of nitrogen otherwise utilizable in assimilatory processes.

b. *Organic Acids*

Acids found in feeds include malic, malonic (Fauconneau, 1954), glyceric, citric, quinic, gluconic, tartaric, oxalic, succinic, fumaric, glycolic, lactic, aconitic, oxalacetic, benzoic (Scott *et al.,* 1964), *p*-aminobenzoic, phenylacetic (Lacoste-Bastié, 1963), and phenylpropionic acids (von Tappeiner, 1886). In clover the nonvolatile acids may amount to 1 mEq per gram. The nonvolatile acids in fresh lucerne hay amounted to about 1% of the dry weight of the fresh herbage (Richardson and Hulme, 1957) and included, in relative proportions, 191 succinic, 100 L-malic, 67 malonic, 17 shikimic, 15 glyceric, 6 fumaric, 4 citramalic, 1 α-ketoglutaric, and traces of glycolic and pyrrolidonecarboxylic acids. The total nonvolatile acids in the rumen amounted to 22 ± 2.1 mg per 100 gm of rumen fluid in abbatoir ruminants starved for 24 hours, i.e., about 0.2 mEq/100 ml, but was 0.75–3.6 mEq/100 ml in fed animals, as compared to 9.4–15.4 mEq VFA per 100 ml. Succinic acid is reported (Higuchi *et al.,* 1962b) to stimulate cellulolytic activity in some pure cultures of rumen bacteria. The fermentation products of some of these nonvolatile acids are collected in Table X-6.

Caffeic (3,4-dihydroxycinnamic) acid is reduced by rumen contents to 3,4-dihydroxyphenylpropionic, *m*-hydroxyphenylpropionic, and *m*-hydroxycinnamic (*m*-coumaric) acids (Booth and Williams, 1963). To some extent, succinate, fumarate, and citrate can be used for microbial synthesis (Henderickx and Martin 1963), but sodium citrate added to the ration does not increase the rate of gain of sheep (Nicholson *et al.,* 1964). Oxalic acid, added to bovine rumen contents, is decomposed at a rate of 1.8–3.42 gm per day per lb of rumen contents, with an average of 2.59 gm (Morris and Garcia-Rivera, 1955). This is a greater rate than is required to decompose oxalate ingested as part of the forage. The fact that excess calcium diminished the rate of oxalate decomposition (Watts, 1957) suggests that the insoluble calcium oxalate is not attacked, which is in agreement with fecal elimination of an amount of oxalate about equal to the calcium oxalate in the feed (Morris and Garcia-Rivera, 1955). Calcium does not protect completely against lethal doses of oxalate (Watts, 1959a), but in animals with a sufficiently active fermentation to counteract the effects of the alkali residue from oxalate decomposition, calcium diminishes the toxicity of oxalate (Watts, 1959b).

Nucleic acids may be fermented in the rumen, but little information is available. Some activity of washed cell suspensions of rumen bacteria on purines has been reported (Doetsch and Jurtshuk, 1957). This is of interest in connection with a possible role of nucleic acids in froth stabilization in bloating ruminants (Chapter XII).

c. *Protein*

As mentioned in the discussion of minerals, protein elements in addition to nitrogen, carbon, hydrogen, and oxygen are required for fermentation and growth. Thus, if protein in the ration is limiting, addition of minerals to a ration rich in carbohydrate will stimulate microbial fermen-

Table X-6

METABOLIC PRODUCTS FROM NONVOLATILE ACIDS INCUBATED WITH WASHED BACTERIA FROM THE SHEEP RUMEN[a]

Substrate (10 mmoles per liter)	Volatile acid (mEq)	Carbon dioxide (mmoles)	Methane (mmoles)	Volatile acids (in order of importance)
Citrate	22.5	17.5	2.5	Acetic, propionic, butyric valeric
Tartrate	10.8	12	—	Acetic, propionic, butyric, valeric
Lactate	9.5	7.6	—	Propionic, butyric, valeric, acetic
Malate	13	10.5	0.6	Propionic
Fumarate	11.5	11	—	Acetic, propionic
Succinate	11.8	9	—	Propionic
Malonate	3.5	3.4	—	Acetic
Oxalate	—	3	0.5	Formic, acetic
Pyrrolidone carboxylate	2.5	1	—	Acetic, propionic, butyric

[a] The gases were measured in the Warburg respirometer, the volatile acids by gas chromatography. From Lacoste-Bastié (1963).

tation and growth. If protein is in excess, supplementation with minerals is not usually necessary. Animals fed alfalfa hay do not require supplementation with minerals (Garrett *et al.*, 1960). The importance of proteins as a source of the short branched- and straight-chain VFA's required for the growth of rumen bacteria has been discussed in Chapter VII.

High protein in the ration has been reported to delay the onset of low-fat milk production (Balch *et al.*, 1954a), but the effect could be due to use of flaked maize to replace protein in the low protein ration of these experiments. Between 5 and 10% protein supported maximum cellulolysis as measured by digestion of cotton threads (Gilchrist and Clark, 1957). Cellulose digestion was inhibited when the ration contained 18% protein, chiefly casein.

The efficiency of utilization of the protein digested by rumen microorganisms has been reported in some cases to be about the same, regardless of the kind (Dyer and Fletcher, 1958; Hamilton *et al.,* 1948; J. I. Miller and Morrison, 1942), but other studies have noted differences (Head, 1959b; I. W. McDonald, 1954). Some of these differences can be explained by differences in solubility.

As mentioned previously, the more protein in forage the greater the digestibility of the carbohydrate and protein (Glover *et al.,* 1957); the more protein in the forage the greater the concentration of nitrogen in the feces. Numerous studies (B. N. Gupta *et al.,* 1962) have shown that fecal nitrogen correlates well with digestibility of the feed. This provides a relatively easy means for assessing the nutritional quality of feeds consumed by grazing animals.

The high efficiency with which the nitrogen in fresh forage is occasionally used (Conrad *et al.,* 1960a) is unusual. Perhaps much carbohydrate is available simultaneously with the protein, which would permit assimilation of the latter, or, as indicated earlier, perhaps the protein spills over into the abomasum before it is fermented in the rumen. Such a possibility is suggested by the finding (V. J. Williams and Moir, 1951) that bacterial counts on a ration containing subterranean clover or linseed oil meal were low (18–25 $\times$ 10^9 per milliliter), whereas the biological values (83 and 80, respectively) were very similar to those for egg white (87), casein (82), and urea plus methionine (75).

Increased growth with high concentrations of protein could occur as a result of increased propionate, butyrate, and higher fatty acid production. A very slight shift (3.2%) to a higher concentration of these acids in the rumen VFA's has been noted (Woodhouse *et al.,* 1955) with increased protein in the ration. The effect of various carbohydrates on the microbial assimilation of a number of proteins has been examined (Henderickx and Martin, 1963), with results shown in Fig. X-1.

In these experiments, assimilation of S^{35} was used to measure microbial assimilation into protein, and total protein determinations were used to measure net breakdown or synthesis. The decreased efficiency of cellulose and hemicellulose in promoting synthesis is partly due to the fact that they are attacked only after some delay; the incubation period was only 6 hours. The sugars are most conducive to synthesis. The poor utilization of fibrin for synthesis is evident with nearly all the sugars. Availability of carbohydrate decreased the net digestion of most of the proteins, i.e., increased the assimilation.

d. *Fats*

In many instances, fats added to the ration diminish the digestibility of the forage (Brethour *et al.,* 1958; Grainger *et al.,* 1961); in other cases it has no effect (Esplin *et al.,* 1963). The method for preparing the fat affects the degree of inhibition. When it is added in solution in a volatile solvent, the feed is entirely covered, which decreases the hydrophily of the surface. Since digestive enzymes occur in an aqueous phase, enzymatic attack is diminished (Brooks *et al.,* 1954b). Digestion and emulsification of the fat permit gradual entrance of the enzymes, but a longer time is required for complete digestion, and the forage particles leave the rumen before digestion proceeds to the extent found in forages free of added fat. Calcium and iron counteract the surface activity of the oils

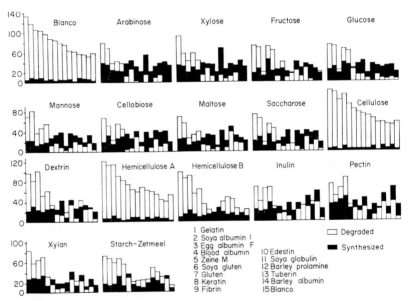

Fig. X-1. The combined effect of carbohydrate and protein sources on protein synthesis. Strained rumen fluid (100 ml) incubated for 6 hours at 40°C with 100 ml of balanced salt solution and 500 mg of the test carbohydrate and test protein containing 25 mg of nitrogen, with carbon dioxide gas. From Henderickx and Martin (1963).

In each graph of test carbohydrate there are fifteen columns, each representing the test protein with the corresponding number; the test proteins are listed by number below the graphs.

by forming insoluble soaps with the fatty acids (Grainger *et al.,* 1961; Ward *et al.,* 1957; T. W. White *et al.,* 1958).

The longer-chain fatty acids (C_{16}–C_{18}), saturated and unsaturated, only slightly diminished dry matter digestibility *in vivo* when added to a ration, but they depress cellulose digestion *in vitro* (Brooks *et al.,* 1954b). The fact that the acids were dissolved in ether before adding to the *in vitro* experiments could explain the difference.

Fats not intimately mixed with the forage are less effective in inhibiting fiber utilization and may be more or less inert insofar as the carbohydrate fermentation in the rumen is concerned. If they are highly unsaturated, they may function as acceptors of hydrogen, but in most rations the percentage of unsaturated fat is insufficient to be of quantitative importance in this respect.

Because of the abundance of acetate absorbed from the rumen, and its availability for fat systhesis, a ruminant does not normally require fat in the diet. Approximately 7% of the dry weight of the bacteria and protozoa of the rumen is lipid (P. P. Williams *et al.,* 1963). For purposes of fattening the animal, some additions of dietary fat may be economical. The fatty acids are not metabolized, except for reduction of unsaturated bonds, and would be available for absorption from the intestine.

Addition of mineral oil to the ration of milking cows decreases the carotene and vitamin A content of the milk (McGillivray, 1957; McDowell *et al.,* 1957a; McGillivray *et al.,* 1959). Infusion of oils into the rumen can markedly reduce the intake by grazing cattle (J. A. Robertson and Hawke, 1964).

Lipids are actively metabolized in the rumen (Hawke and Robertson, 1964), as was indicated in Chapter IX; they are hydrolyzed, and the unsaturated fatty acids are reduced and transformed to varying extents.

e. *Minerals*

Most natural forages contain the amounts and balance of minerals required by ruminants, though supplements often improve growth, and the intense appetite for salt displayed by wild ruminants suggests that it is a limiting nutrient. Occasionally, as in the case of cobalt and copper, an essential element is deficient in the soil and the ingested plants.

Many of the currently available processed feeds are derived from parts of plants poor in minerals and rich in digestible carbohydrate, e.g., corn cobs. With such feeds, a balance of minerals required by the rumen microorganisms must be supplied. The mineral content of a whole natural forage and a processed one may differ markedly (Tillman *et al.,* 1954a,b). Supplementation of prairie hay with the ash of alfalfa gave no response in sheep even though the alfalfa was much richer in calcium, phosphorus, potassium, chloride, and magnesium. In contrast, addition of alfalfa ash to a ration containing a large quantity of cottonseed hulls significantly improved digestibility. In the prairie hay, digestible carbohydrate rather than minerals was evidently limiting.

The importance of mineral supplementation of the ration depends on the balance between digestible nutrients and the minerals already in the feed. To a considerable extent the essential mineral content is a function of the proteinaceous materials in the feed. The minerals occur as components of many enzymes.

f. *Water*

Water is essential in the ruminant, as in all other living forms. The amount required and the means for obtaining it may differ from one species to another. A center in the hypothalamus, affected by osmotic pressure, regulates the consumption of water (Andersson and McCann, 1955). The physiological adaptation of the camel to live under adverse moisture conditions (Schmidt-Nielsen *et al.,* 1957; Schmidt-Nielsen, 1959) consists in retention of water in the circulatory system even though the tissues become partially desiccated. A considerable part of this water comes from the rumen during the first several days of water depletion in sheep (Hecker *et al.,* 1964).

Paranuclear vacuoles (Henriksson and Habel, 1961) appear in rumen epithelial cells exposed to water, but passage to the rumen capillaries has been postulated to take place between the squamous epithelial cells rather than through these vacuoles (Tamate *et al.,* 1964).

Zebu cattle (*Bos indicus*) can survive under conditions of water deprivation which kill European cattle (*Bos taurus*). The Masai tribe of East Africa may water their cattle only every third day. During the dry season water holes are few, and to obtain feed the Masai drive the herd as much as 30 miles to feed, graze them for a day, and drive them back to water on the third day. Studies (G. D. Phillips, 1960; Horrocks and Phillips, 1961) show that the zebu can use less water per body weight and feed consumed. The digestibility of fiber increases when water intake is reduced (French, 1956a; Balch *et al.,* 1953). The mechanism is in part a decreased consumption of feed (Larsen *et al.,* 1917), with consequent longer retention. Nitrogen losses are increased.

Tracer studies with tritiated water (A. L. Black *et al.,* 1964) indicate that the water pool in cattle turns over in about 3.5 days. This rapid turnover, as compared to other mammals, may be due in small part to the exchange of water with hydrogen in the rumen fermentation. Production of 330 liters of methane, with the hydrogen equilibrated with the tritiated water, would be equal to a loss of 540 gm of water per day. This is insufficient to account for the increased water turnover as compared to other mammals. The water in milk would be a much more significant factor.

Some of the native African ruminants, e.g., the dik-dik and suni, live in forests far from water, and it is doubtful that they drink water. They obtain sufficient water from the fresh forage and from metabolism. The rumen fermentation rate of the suni is more than three times the rate in the other ruminants measured (Hungate *et al.,* 1959) (Table VI-4). Water is generated not only in the aerobic respiration of the host, when the hydrogen of the substrate is combined with oxygen, but also in the fermentation, as is evident from the equations in Chapter VI for the formation of the rumen fermentation products.

The possibility that these small ruminants do not need to obtain water by drinking is purely speculative, but would be of great interest to study if they could be adapted to laboratory conditions. For many experimental purposes it would be desirable to have available smaller ruminants than those presently domesticated. A mature suni weighs only 9 lb. However, the diet of these animals is rather specialized, and the plants easily available for experimental purposes might not provide a ration capable of supporting such a rapid fermentation which was at the same time favorable to the health of the animal. It is of interest that in the agar dilution series made to detect cellulolytic bacteria, the one suni studied did not disclose any colonies with clearing of cellulose around them. This would be expected if the number of noncellulolytic bacteria greatly exceeded the number of cellulolytics. Enzymatic digestion of cellulose may be too slow to satisfy the animal's needs and to keep up with the turnover.

A faster rate of attack on cellulose in the ventral than in the dorsal rumen has been ascribed to the greater water content (Balch and Johnson, 1950), but other factors may also be important, e.g. pH and concentration of fermentation acids and ammonia. More cellulose is actually digested in the dorsal material where fiber is more abundant.

With lush green forages the amount of water may be greater than is optimal for maximum fermentation. Some animals are capable of fermenting 16 gm dry weight of substrate per 100 ml of rumen contents per day. With a turnover of once per day, a forage containing less than 16% digestible dry matter would not supply the nutrients needed by the microorganisms to support the rate of fermentation possible with a feed having a higher digestible dry matter content. Lowered intake with increased water content of silage has been noted (Dodsworth, 1954).

On the other hand, dehydration can reduce the digestibility of alfalfa (Conrad and Hibbs, 1959). Severe dehydration of barley grain (Templeton and Dyer, 1964), 48 hours at 65°C, diminished its digestibility.

In the case of soluble substrates in a continuous-feed system, the percentage of the substrate utilized increases slightly as its concentration increases. This follows from the fact that a fixed percentage of substrate is lost in the overflow. Whether similar factors cause increased utilization of insoluble substrates has not been ascertained, but, if these factors were operative, they might explain the relatively slight differences in the digestion of finely divided feeds as compared with coarse ones. The comminuted feeds are consumed at a faster rate, cause a greater percentage dry matter in rumen contents, and show a faster turnover time than the same feed in a coarse state. The shorter retention time associated with greater intake diminishes the completeness of digestion, but is counteracted in part by a more rapid digestion due to greater concentration of substrate. Since the rumen population is capable of using more food than is available, a greater concentration of substrate would give growth of more microorganisms.

Sheep consuming 750 gm of dry hay per day and weighing 75 lb drank about 2 liters of water daily (Harris and Phillipson, 1962). This is considerably less than the quantity of saliva secreted.

2. FORAGES

Since forages vary in their chemical composition according to variety, soil fertility, age, moisture, and light, they also vary in their effects on the rumen microbiota and the ruminant. Changes in the feed cause changes in the microbial population. If the feed is changed abruptly, the balance of microbial activities may be disturbed enough to show effects on the host (Annison *et al.,* 1959a,b). Slower changes in rations may not cause noticeable effects. Feeds containing larger amounts of balanced substrates readily utilizable by the rumen microbiota increase the microbial activity.

a. *Effect of Drying*

There is little evidence that drying of forage modifies its nutritional value to any great extent, as judged by digestibility in the animal (Axelsson, 1942a; Graham, 1964b) or by examinations of the rumen microorganisms (Reichl, 1960), although some effects of drying on the *in vitro* fermentation have been noted (Shinozaki *et al.,* 1957; Christian and Williams, 1957), the gas and fatty acid production from legumes being decreased by drying, whereas that from grasses was increased. A retardation of cellulose digestion in dried as compared to fresh alfalfa-brome grass (R. R. Johnson *et al.,* 1962a) was evident after 12 hours of *in vitro* incubation with rumen fluid, but the difference had disappeared at 24 hours. Slowness in hydration of the dried forage and hydrophobic surfaces of large particles may explain in part the delay.

Green grass contains a high percentage of fermentable material. Digestion coefficients of 75% have been obtained (A. T. G. McArthur, 1957a; Graham, 1964b). No effects of freezing of forage on its digestibility were noted (Raymond *et al.,* 1953a,b).

b. *Effect of Age*

Very young fresh grasses contain highly digestible dry matter. During growth, an increasing proportion of carbohydrate is formed into the structural components; this decreases the nitrogen : carbon ratio and also decreases the digestibility of the carbohydrate fraction as the structural components are laid down and "lignification" proceeds (Corbett, 1957; Noller *et al.,* 1959).

One of the chief problems in the utilization of the cellulose in natural forages is its increasing indigestibility (Axelsson, 1942b) as the forage matures. Forages containing the most crude fiber are the least digestible (Hallsworth, 1949). Chemical analyses disclose cellulose in the feces of animals fed high-fiber forages. This result would be expected to some extent due to escape from the rumen of incompletely digested small particles containing cellulose, but the amount of fecal cellulose is too great to be explained in this way (Crampton and Maynard, 1938). It has been explained by assuming that the deposition of lignin and other encrusting materials shields the cellulose from the action of the cellulase. Mature forages contain a larger percentage of lignin-like materials than do the young plants, and these lignin-like materials are less digestible (McMeekan, 1943) than is the "lignin" in young lush forage. The fiber in some materials, such as beet pulp (Eriksson, 1949) and corn cobs, is quite digestible. Wood pulps freed of lignin show digestibilities as high as 88% (Hvidsten, 1946). The decreased digestibility as forage matures is accompanied by a tendency for a rumen fermentation in which relatively more acetic acid is formed (Armstrong, 1964).

c. *Effects of Comminuting and Pelleting Forage*

Various treatments to facilitate handling of bulky forages have been made economically possible by improved engineering. One of these treatments calls for chopping the hay and compressing it into pellets with a high specific gravity, of various sizes and shapes. As pointed out in Chapter IV, rumen contents of animals fed comminuted forages are much more homogeneous and easier to sample than are digesta from animals consuming long forage (Balch, 1952).

An important effect of pelleting is to increase the concentration of dry matter in the rumen. This leads to a greater fermentation rate per unit of volume and often to greater gains in the animal. The fact that the food is in the form of pellets has relatively little effect on digestibility (Lindahl and Reynolds, 1959) although the pelleting process may raise the temperature to 49–77°C (Dow, 1959).

Feeds can be ground to varying degrees of fineness before pelleting. Grinding increases the consumption partly by increasing the turnover due to the more rapid escape from the rumen of the finely comminuted material. This correlates with decreased digestibility (Blaxter and Graham, 1956; Rodrique and Allen, 1960) (Table X-7) and counteracts the increased digestibility due to the grinding itself.

Under *ad libitum* conditions of feeding, pellets of ground material give greater gain due to greater feed consumption. The rumen volume appears to be less in animals on an equivalent quantity of pelleted feed (Meyer *et al.*, 1959b). The slower turnover noted for ground hay in one set of experiments (Balch, 1950) may be explained by inclusion of mangolds in the ground ration, which were not fed with the long hay. The almost total digestibility of the mangolds would tend to increase the retention time of dyed hay particles, as discussed in Chapter V.

Rumination is very much decreased or absent if the feed is finely ground. In one experiment in which polyethylene was added to a purified ration (Oltjen *et al.*, 1962b), the time spent ruminating varied linearly with the amount of polyethylene fed. In addition, the fine grinding and fast turnover increase the proportion of propionic acid in the fermentation products to an extent that can become important in the metabolism of the ruminant (Woods and Luther, 1962). The lowered acetate leads to a diminished percentage of fat in milk (Powell, 1938; Rodrique and Allen, 1960), and the higher propionate results in raised solids-non-fat. A ration causing low butter fat can support more rapid growth (Ensor *et al.*, 1959). Pelleting, with consequent increased rumen turnover, gave a greater proportional increase in protein than in fat, as compared to unpelleted meal (Weir *et al.*, 1959). This is the type of change expected when propionate is relatively more abundant. More efficient use of ground, purified hay as compared to chopped hay has been reported (Paladines *et al.*, 1964). In other experiments (P. I. Wright *et al.*, 1963), the pelleted gave greater gains than the ground ration.

Many of the conflicting reports on the effects of pelleting, grinding, and chopping are due to differences in the particular dimensions of the material and to other components of the ration. The effects of pelleting long hay and finely ground hay may be quite different. The plane of feeding and whether feeding is *ad libitum* or restricted may also exert an effect. On restricted feeding regimes the ground feed may be less efficiently used (Meyer *et al.*, 1959a), due to a lowered retention time.

Extremely fine comminution introduces another factor, the increase in

Table X-7

THE MEAN DIGESTIBILITY OF THE CONSTITUENTS OTHER THAN ENERGY OF THE GRASSES PREPARED IN DIFFERENT WAYS[a]

Constituent	Low level of feeding[b]			High level of feeding[c]			Standard error of means
	Chopped	Medium ground and cubed	Finely ground and cubed	Chopped	Medium ground and cubed	Finely ground and cubed	
Dry matter	80.6	73.1	73.1	75.6	67.7	64.3	±0.9
Organic matter	82.7	76.4	75.2	77.6	69.6	65.8	±0.9
Ether extracts	66.1	54.8	67.8	57.3	71.0	61.4	±6.9
Crude fiber	82.9	74.7	67.7	76.5	58.1	49.9	±1.6
Nitrogen-free extract	85.8	81.8	82.0	84.0	76.7	75.2	±1.2
Crude protein	66.8	61.9	62.6	61.0	58.7	54.8	±1.4
Cellulose	87.5	80.6	76.3	81.4	62.3	56.4	±1.7
Cellulosic pentosan	87.7	79.3	73.7	77.7	68.3	49.8	±1.3
Noncellulosic pentosan	82.5	75.9	74.9	76.9	66.4	60.1	±3.3
Dry matter[d]	80.3	76.9	75.9	79.4	69.9	65.4	—

[a] From Blaxter and Graham (1956, p. 213).
[b] 600 gm/24 hours.
[c] 1500 gm/24 hours.
[d] Results obtained with a further six sheep not confined to the calorimeter (Blaxter *et al.*, 1956).

digestibility (Koistinen, 1948; Dehority and Johnson, 1961) that results from physical separation of the cellulose from other components of the fiber. The comminution also reduces the particle size to such a degree that turnover is identical with rumen liquid and suspended bacteria. The effect of this type of comminution of fiber has been tested *in vitro,* but difficulties in preparing large quantities prevent extensive feeding trials.

Just as artificial comminution of hay increases the rate of passage through the rumen, prevention of normal comminution decreases it. Hay administered through a fistula (C. B. Bailey and Balch, 1961a) caused cows to spend 44 instead of 31% of the day ruminating and raised the dry matter content of the rumen. Salivary secretion was increased an hour after administration (C. B. Bailey and Balch, 1961b). Pearce and Moir (1964) have similarly shown that avoidance of rumination in sheep, by tying the jaws shut after feeding, markedly increased the retention and turnover time. Bicarbonate was administered with the hay to simulate salivary secretion. Lack of impaction in these experiments shows that enzymatic activity and the mixing movements of the rumen musculature are alone ultimately capable of bringing digesta to a state of division permitting its further passage, but the rate is so slow that the food consumption of the animal is reduced below maintenance.

d. *Effects of Other Treatments of the Feed*

Ensiling of forages exerts relatively little effect insofar as the nutritional value to the host is concerned, although some indications of decreased milk production have been reported (Conrad *et al.,* 1960b). Microbial ensiling processes are anaerobic, and the balances of reactions are essentially similar to the activities of the microorganisms in the rumen—except that in the absence of buffers the fermentation is inhibited by acid accumulation before much of the substrate has been fermented, and inoculation depends on the microbiota of the ensiled material. There is relatively little methane production during the ensiling process. Initially, oxygen prevents development of methane bacteria, and, by the time the redox potential is lowered to a point permitting methanogenesis, acid has accumulated. Also, few methanogenic bacteria would occur in the material to be ensiled, which would slow initiation of the methanogenesis.

The fermentation leading to silage differs from the rumen fermentation also because of the lack of the obligately anaerobic rumen bacteria. Sporeforming bacteria are relatively more important in ensiling because they occur in soil and air in large numbers, and many of them can grow anaerobically.

Various chemical treatments have been tested on forages, but none has come into widespread use. Advantages of increasing digestible carbohydrates in forages by delignification are discussed in Chapter XI. The effects of processing maize to produce flaked maize have been discussed in the section on starch in this chapter.

e. *Legume v. Grass Forages*

Legumes usually contain a larger proportion of water-soluble material than do grasses, and a greater percentage of nitrogen (Ekelund, 1949; Y. P. Gupta and Das, 1956). The mineral content is also larger. A higher proportion of polyuronides is indicated by the low recoveries of acid from *in vitro* fermentations of alfalfa hay. Fermentation gas production in the immediate period following feeding is greater for legumes than for grasses (Colvin *et al.,* 1958b; Kleiber, 1956; Cole and Kleiber, 1948). Grasses contain more crude carbohydrate and fiber than do legumes (Ekelund, 1949).

f. *Hay v. Concentrate Feeds*

A greater proportion of the bacteria observed by direct count can be cultivated when the ration is high in concentrates than when hay is the chief feed (Maki and Foster, 1957). The explanation for this observation is not clear, but it would be of interest to determine whether this relationship holds at all periods after feeding. The author has similarly obtained extremely high culture counts (5×10^{10}) from animals on a high grain ration. The greater productivity of concentrate feeds is due to their greater and more rapid digestibility as compared to forages. Young green grass can approach the utility of concentrates on a dry weight basis, but a greater dry weight of concentrates can be fed. The proportion of methane to other fermentation gases is less for concentrate than for fresh forages and hay (Hirose *et al.,* 1957).

3. FEED ADDITIVES

a. *Minerals*

Insofar as rumen microorganisms are concerned, the additives most likely to improve feed utilization are minerals and growth factors of the bacteria. Minerals are so inexpensive, particularly the trace elements, and may exert such a profound effect on both microorganisms and host, that careful consideration should always be given to the mineral balance of the ration. The quantities and availability of the elements in the ration will materially affect its adequacy; the total utilization will depend on which nutrient is limiting. The ration should be designed so that no inexpensive nutrient is limiting. This generally means that the minerals should be somewhat in excess of requirement, with due regard to the fact that, with some of the trace elements, toxic levels are reached at only low concentrations. When the ration includes forages with a good deal of protein, the trace elements are likely to be present in sufficient quantities, and supplementation is therefore unnecessary.

b. *Growth Factors for Bacteria*

Knowledge that ruminants depend to a large extent on the fermentation activities of microorganisms has led to addition of various materials to feeds in order to stimulate bacterial growth. Urea is one of the most common. Evidence has been cited in Chapter VII that, when nitrogen is limiting, urea supplementation increases microbial and host yields. Urea administration is unlikely to be economically profitable with feeds containing large quantities of protein, e.g., legumes and young grass.

The feasibility of polymerizing nitrogen sources into materials which would only slowly release ammonia into the rumen has been examined (Schoenemann, 1953), but the idea was dropped when experiments with a continuous drip of ammonia into the rumen showed no advantage over urea administration in the ration.

Ethanol has been recommended as a feed additive to provide in the rumen the reducing conditions needed by the bacteria. It is uncertain what is meant by "reducing" conditions. Ethanol is not as reduced as is methane, it is not rapidly metabolized in rumen contents (Moomaw and Hungate, 1963; Emery *et al.*, 1959), and it does not improve nutritional performance (Garrett and Meyer, 1963) when fed in trace amounts, although larger quantities may be a source of energy (Leroy, 1961).

Isovaleric and valeric acids, tested because of their known essentiality for *Bacteroides succinogenes,* have not shown an increased digestion of the ration (Hungate and Dyer, 1956; Lassiter *et al.,* 1958a), but have stimulated appetite (Hungate and Dyer, 1956) and improved performance by increasing consumption (Hemsley and Moir, 1963). In view of requirements also for isobutyric and 2-methylbutyric acids, rations high in digestible carbohydrate, but poor in protein, supplemented with all the C_4 and C_5 acids should be tested. Some response to valeric acid has been reported (Lassiter *et al.,* 1958b; K. A. Winter *et al.,* 1964).

Biotin is a likely prospect for supplementation of rations poor in organic protein. In this connection the nutritional requirements of rumen bacteria synthesizing valine, leucine, and isoleucine from carbohydrate plus ammonia would be of much interest.

Unidentified factors in rumen fluid (Garner *et al.,* 1954) stimulate growth of some rumen bacteria (Hungate, 1963a). Success in obtaining high culture counts with a defined medium (Bryant, 1963) suggests that inclusion of those nutrients in feeds containing adequate carbohydrate should support growth of the rumen bacteria and of the host on simple defined feeds.

Success in feeding milking cows on rations containing chiefly cellulose, inorganic salts, and simple nitrogen compounds (Virtanen, 1963), with no additions of branched- and straight-chain VFA's or other nutrilities, could indicate that bacteria not requiring these factors had been selected. However, it is possible that excretion of vitamins and decomposition in the rumen of microbial bodies not requiring the nutrients could yield enough of them to support adequate microbial growth and cellulose decomposition.

c. *Enzymes*

Since cellulose digestion lags behind the utilization of other forage carbohydrates, addition of cellulase with the feed to hasten the attack might seem advantageous. At present, cellulases are not available in sufficient quantity to permit feeding tests, and *in vitro* tests have not been performed. Recent application of the koji method to mass production of cellulases in Japan may reduce the cost of cellulase enough to make such tests practicable. The procedure is not likely to prove economical in commercial practice, however, as the expense of producing the enzymes could be prohibitive. Manipulation of the ration to promote cellulase production by the rumen bacteria is a more practical route to increased cellulolysis.

Addition of a proprietary preparation of amylolytic and proteolytic enzymes has been reported to improve nutrient utilization (Burroughs *et al.,* 1960; R. J. Johnson *et al.,* 1963) and to affect the ratios of VFA's (R. J. Johnson and Dyer, 1964), but the economic aspect was not analyzed. A possibly beneficial effect of fungal protease has been reported (Ralston *et al.,* 1962). Proteinases added to rumen fluid incubated *in vitro* (Theurer *et al.,* 1963) were degraded to the extent of 50% in 2–3 hours. Such materials probably are digested and used as food by certain rumen organisms.

A most interesting question would be whether continued addition of enzymes would lead to development of a microflora utilizing them as substrate. Presumably this would occur.

d. *Antibiotics*

Because of the successful use of antibiotics to increase nutrient utilization in poultry and swine, these additives have been extensively tested on ruminants. Most of the evidence obtained with calves and lambs prior to the time when forage is predominantly utilized has indicated an advantage in the use of antibiotics (Hogue *et al.,* 1956; Kesler, 1954; Mann *et al.,*

1954b; Porter, 1957). In a study with twin calves (Pritchard *et al.,* 1955), Aureomycin improved the weight gain and feed efficiency, but did not affect the digestibility of dry matter, protein, crude fiber, or nitrogen-free extract.

After the host nutrition depends on the rumen microbiota, added antibiotics have less effect (Neumann *et al.,* 1951; W. H. Hale *et al.,* 1960) or are slightly inhibitory (Horn *et al.,* 1955; Huhtanen *et al.,* 1954; Oyaért *et al.,* 1951; Shor *et al.,* 1959; Tillman and McVicar, 1956; Turner and Hodgetts, 1953), or slightly beneficial (Hatfield *et al.,* 1954; Haenlein *et al.,* 1961). Evidence for improvement in nutrition by Aureomycin has recently been obtained (Klopfenstein *et al.,* 1964), with indications of greater concentrations of protozoa.

Inhibition by antibiotics usually lasts for only a short period (Emerick and Embry, 1961; J. L. Evans *et al.,* 1957; Kesler, 1954). During this time resistant strains develop to a concentration equaling that of the inhibited strains prior to antibiotic administration. Pure cultures of rumen bacteria were found to be more susceptible to inhibition by penicillin and Terramycin than by Aureomycin, streptomycin, and Chloromycetin (Jurtshuk *et al.,* 1954). Greater inhibition by penicillin than by streptocide, sulfidine, norsulfazole, or phthalazol has been noted (Kaplan and Ljubeckaja, 1959).

Concentrations of antibiotic too low to cause a noticeable inhibition of microbial activity may nevertheless exert a selective action on the rumen microbiota. This was shown (Hungate *et al.,* 1955b; Klopfenstein *et al.,* 1964) by comparing the inhibitory effects of antibiotics *in vitro* on rumen contents of steers with and without chlortetracycline in the ration. The antibiotic added *in vitro* inhibited the fermentation of rumen contents from unsupplemented animals more than that from animals receiving the ration supplemented with a low level of the antibiotic. Destruction of certain antibiotics such as penicillin in the rumen (Wiseman *et al.,* 1960) accounts for recovery from their administration.

Though many experiments with antibiotics have been performed, relatively little is known of the mechanisms by which the effect is exerted. Splitting of urea by urease in the gut has been indicated as a possible mechanism of growth inhibition in nonruminants (Francois, 1956; Visek, 1964).

e. *Preparations of Rumen Bacteria*

Preparations of dried rumen bacteria have not been demonstrated to have any value in ruminants other than the protein and other ingredients supplied. Few of the preparations contain a high proportion of viable organisms.

B. Influence of Quantity of Feed

The chief effect of increasing the quantity of a particular feed is to increase the turnover rate in the rumen. This in turn increases the microbial activity. The rate at which a feed is consumed rarely exceeds the capacity of the rumen bacteria to attack most of it, except when extreme inanition is followed by sudden ingestion of normal rations. Under usual circumstances the rumen population develops at less than its maximum possible rate, as is evidenced by the fact that addition of feed to rumen contents removed from animals at all periods invariably causes an increase in the rate of fermentation (barring instances in which acid accumulation or other unfavorable feature stops microbial activity). A limit of growth rate is imposed by the nutrients or by the animal's appetite and ability to remove fermentation products rather than limitation by the growth rate of the rumen bacteria.

1. STARVATION

Even after extended semistarvation, the composition and numbers of bacteria per gram of rumen contents for animals on a particular ration remain relatively unchanged (Gilchrist and Kistner, 1962). The maximum rumen volume diminishes, although not in proportion to the diminution of feed. The turnover rate also diminishes (Makela, 1956). The turnover rate is reduced, i.e., turnover time is lengthened, which corresponds to the decreased intake in relation to the amount of rumen contents. This applies to the solids fraction. The liquid small particle turnover rate is also decreased; less saliva is secreted. The bacteria do not spill from the rumen as fast as with adequate rations; this keeps the microbial population at a higher concentration. The concentration of the rumen microbes reflects *the nutritive value of the feed rather than the rate of its consumption.* The viable population gradually drops as inanition continues.

In cattle dead from starvation the rumen contained 40–50 kg of digesta (Tiedemann and Gmelin, 1831), and cows fasted for 48 hours showed a significantly high capacity for cellulose digestion (J. I. Quin *et al.,* 1951). After 7 days of fast (Magee, 1932) there were still considerable quantities of digesta in the rumen. After 3–5 days of fasting the methane production falls to about 5% of the level in fed sheep (Blaxter, 1962). Rumination has been noted in sheep 41 hours after the last food was consumed (Magee and Orr, 1924). In sheep grazing on very poor forage, soil has been found to compose as much as 14% of the material consumed (Field and Purves, 1964).

The rumen microorganisms start to lose their ability to ferment substrate even after relatively short periods without food (Hungate, 1956a;

el-Shazly and Hungate, 1965), and during very long periods of inanition the concentration of bacteria falls to a very low level. When food is again provided to starved animals, several days may elapse before microorganisms regain the normal balance. The hydrogen observed in fed animals after a period of inanition (Pilgrim, 1948) indicates that the methane bacteria were limited in number. Disappearance of hydrogen as methane appears is in agreement with the hypothesis that the methanogenic organisms use hydrogen as food.

The fact that the quantity of digesta in starving animals is fairly large makes difficult the diagnosis of starvation by cursory examination of dead animals. Fermentation rate and VFA concentration are better indices. Because of the slow digestibility of the more resistant portions of plant fiber, some fermentation can continue for a long time after food was ingested. The residual material still contains viable bacteria which serve as inoculum when food is again available.

The decreased turnover rate of the coarser particles in the feed causes these elements to occur in greater proportions in the rumen than in the feed. This must be taken into account when identification of plant particles in rumen contents is used to estimate the quantities of different plants consumed by grazing or browsing ruminants (Bergerud and Russell, 1964).

2. MAINTENANCE

The level at which an animal maintains constant composition and weight is called the maintenance level, and numerous feeding experiments to determine this level have been performed. One formula for feeding at maintenance level provides feed to the extent of $0.0353 \times$ weight in kilograms (Brody *et al.,* 1934) [for a 455-kg cow this amounts to 2.3 kg of total digestible nutrients (TDN)]. Fifteen gm of 50% wheaten-50% alfalfa hay per kilogram of body weight has proven of practical value in Australian sheep (R. L. Reid, 1963, personal communication). Luitingh (1961b) in South Africa found 3.3–3.6 kg of TDN per 455 kg for maintenance of steers.

The Haecker standard for maintenance of a 455-kg cow is 3.2 kg of TDN, 0.32 kg of digestible protein, and 0.045 lb of digestible fat. Each kg of 4% milk requires, in addition, 0.24 kg of digestible carbohydrates, 0.021 kg of digestible fat, and 0.054 kg of digestible crude protein (Bull and Carroll, 1949). The National Research Council standard specifies 3.6 kg of TDN for maintenance of a 455-kg steer, with 4.1 additional kg needed for a live weight gain of 1.1 kg per day. Maintenance has been estimated by Morrison (1957) at 3.4 kg of TDN per 455 kg of body weight. The National Research Council standard for sheep is 1.7 to 2.3 kg of TDN per day per 45.5 kg, depending on the rate of gain desired. The maintenance requirement of pen-fed sheep has been estimated at 0.42 kg of digestible organic matter or 0.44 kg of TDN per 45-kg sheep, and 2.25 kg per kilogram of gain (Coop, 1962). For grazing sheep the values are higher, with 0.46 (Langlands *et al.,* 1963) to 0.67 kg (Coop and Hill, 1962) of digestible organic matter per day for the maintenance of a 45-kg sheep.

Garrett *et al.* (1959) found the following relationships for both sheep and cattle. For maintenance,

$$TDN = 0.036 \ W^{(3/4)}$$

with W as the animal weight in pounds and TDN as pounds per day. Then, for digestible energy (DE) and metabolic energy (ME) in kilocalories:

$$DE = 76 \ W^{(3/4)}$$

and

$$ME = 62 \ W^{(3/4)}$$

Marston (1948b) obtained an equation of $DE = 68 \ W^{0.73}$.

3. PRODUCTION

For gain in pounds the maintenance requirement was the same, and the required additional amount for gain is shown as a second term (Garrett *et al.,* 1959).

	Sheep	Cattle
TDN	$0.036 \ W^{(3/4)}(1 + 2.3$ gain)	$0.038 \ W^{(3/4)}(1 + 0.57$ gain)
DE	$76 \ W^{(3/4)}(1 + 2.4$ gain)	$76 \ W^{(3/4)}(1 + 0.58$ gain)
ME	$62 \ W^{(3/4)}(1 + 2.5$ gain)	$62 \ W^{(3/4)}(1 + 0.60$ gain)

These values disclose little difference between sheep and cattle. Another value (U.S. Dept. Agr., 1956) is 11.8 lb of TDN for maintenance of a 1500-lb dairy cow, and 28 lb of TDN to produce 10 lb of 3% milk, or 32 lb to produce 10 lb of 4% milk.

4. APPETITE

Since substrate concentration limits the microbial fermentation in the rumen, any factors increasing consumption will raise the plane of nutrition.

With coarse roughages the physical stretching of the rumen by the bulk of the digesta is undoubtedly a factor limiting appetite (Kay, 1963). With concentrates, the acidity developed may limit the consumption. In sheep on a high fiber diet (Badawy *et al.,* 1958b), the rumen digesta amounted to 6.5 kg, 4.6 kg on a medium fiber ration, and 4.1 kg on low fiber. Factors limiting the rate of ingestion are the rumination time and the amount of energy required for reducing the coarse digesta to a size passing the reticulo-omasal orifice (Freer *et al.,* 1962).

When boli were caught by hand at the cardia and removed through the fistula, cattle continued to eat and consumed 177% of their usual voluntary intake of hay (Campling and Balch, 1961). Water-filled balloons placed in the rumen reduced intake, the mean decrease during 10–14 days being 0.56 lb per 10 lb of water in the balloons. Each change of one lb in the rumen dry matter was accompanied by an inverse change of 0.6 lb in the voluntary intake of hay.

A diminished intake of straw as compared to hay when each was fed separately, *ad libitum* (Campling *et al.,* 1961), is interpreted as due to the lesser digestibility of the straw and a slower rate of passage. On a ration of oat straw in which nitrogen was limiting, addition of urea increased the digestibility of the oat straw and the voluntary intake (Campling *et al.,* 1962). The decreased consumption of fresh legumes by bloating ruminants (Chapter XII) (Colvin *et al.,* 1959) also indicates a relationship between rumen fill and appetite.

Crampton *et al.* (1957) have found voluntary intake better than TDN or crude fiber as an index of forage feeding value. The consumption correlated with the extent of cellulose digestion after 12 hours of *in vitro* incubation (Donefer *et al.,* 1960). A nutritive value index, derived from *in vitro* fermentations, is proposed as a measure of feeding value.

Limitation of space must ultimately depress appetite, but other factors may prevent the expression of this space limitation, and modify the feeding level at which space becomes limiting. The faster the digesta leave the rumen, by fermentation and absorption of the products or by passage to the omasum, the more material that can be consumed without reaching the point at which space is limiting. Conversely, if the food is not appetizing to the animal, space does not become limiting, and factors of appetite control the amount of food consumed.

The appetite of the animal is thus of great importance in food utilization. Factors which diminish appetite are: excess accumulation of acids in the rumen, injury, lack of suitable nutrients in the ration. This last factor is probably mediated by ingredients which impart a taste or smell. The effects on appetite of the addition of the valeric and isovaleric acids (Hungate and Dyer, 1956; Hemsley and Moir, 1963) demonstrate the importance which trace nutrients may exert on appetite.

Although conclusive evidence is lacking, several observations suggest that appetite may be linked with microbial fermentation activity. Massive transplantation of normal rumen digesta into animals off feed and displaying no appetite induces an almost immediate recovery of appetite (Dougherty, 1950, personal communication; F. M. Gilchrist and Clark, 1957). Force feeding of fermentable concentrates to feed-lot animals which have gone off feed can result in appetite restoration. The observation that animals provided with autoclaved rumen fluid exhibit less appetite than those receiving unautoclaved fluid (Bartley *et al.,* 1961) indicates that activity of the microorganisms may be necessary; stimulation is due to a continuous production rather than to the content of already formed acids. A difference between pasteurized and normal rumen fluid in their effects on rumination has also been noted (Pearce, 1965b). The possibility that metabolism of acids in the rumen wall may be associated with appetite merits consideration. An interesting increase in appetite following duodenal infusion of nitrogenous material into sheep (Egan, 1965a,b,c) indicates that an important animal factor is also involved in appetite.

C. Other Factors

1. Fistulation

It has not yet been demonstrated that a suitably prepared fistula diminishes the consumption or digestibility of feed (Drori and Loosli, 1959) or the activity of the rumen microorganisms, provided the opening is kept closed with a reasonably tight plug which causes no injury.

2. Oxygen

There is no evidence that the oxygen normally present in the gas above rumen contents seriously affects the fermentation. Oxygen raises the redox potential only in the superficial layer of the rumen contents (Broberg, 1957d). Four to fifteen milliliters of oxygen can be absorbed per liter of contents per hour (Broberg, 1957a).

3. The pH and Concentration of Fermentation Products

Retardation of acid production by accumulation of individual fermentation acids has been mentioned in connection with factors slowing fermentation by pure cultures (H. C. Lee and Moore, 1959). Modification of a mineral mix to give either an alkaline or acid balance has been successful (Oltjen *et al.,* 1962a) in modifying the rumen acidity and vitamin synthesis. Addition of alkaline buffers to the ration of steers fed timothy hay exerted no effect or was detrimental at higher levels (Nicholson and Cunningham, 1961), but with high concentrates the buffers exerted a favorable effect. Improved weight gains by addition of $NaHCO_3$ to the ration of steers fed a fiber-free ration has been reported (Preston *et al.,* 1961).

4. INDIVIDUAL ANIMAL DIFFERENCES

This is probably an important factor influencing the variations in food intake (A. Dobson and Phillipson, 1961) and rates of gain. It is related to appetite and rumen volume. Individual differences in water consumption, not correlated with body weights, have been observed (Rollinson *et al.,* 1955); observations of differences in grazing habits have also been made (Taylor *et al.,* 1955).

In addition to differences between animals, there are differences in single animals at different times (Chou and Walker, 1964a,b). Part of this is due to changes in feed, but even on a constant ration an animal will exhibit fluctuations in nutrition and rumen characteristics. Part of this may be due to fluctuations in bacterial populations attendant on the continuous adaptive, competitive, and evolutionary changes occurring within the teeming microbial population (Hungate, 1962). Mutations presumably underlie some of these variations, but they apparently occur within a restricted range, boundaries between large groups being recognizable in spite of the fluctuations within a group.

Bacteriophages may be involved in the rumen microbial fluctuations. Phages attacking rumen strains of *Streptococcus bovis* have recently been found in the rumen (Brailsford, 1965).

5. AGE AND SIZE

The demands of young ruminants for nutrients are greater than those of adults, due to the necessity for synthesis of the increasing cell material. The rumen is not proportionately larger; in fact, it is relatively smaller (Kesler *et al.,* 1951) than in the adult. The increased demand correlates with increased appetite and a consequent faster rumen turnover as compared to the adult (Columbus, 1936). Similarly, the increased concentration of protozoa in pregnant ruminants reflects an increasing turnover (Ferber, 1928).

Since the energy requirements of individuals of a particular animal species are roughly proportional to the 3/4 power of their weight (Kleiber, 1947), the energy need per unit weight of a small animal is greater than for a large one. Since the rumen is not proportionately larger, the increased needs must be met by a greater turnover.

6. SPECIES DIFFERENCES

Differences between various ruminant species in the rates of fermentation per unit of rumen contents were mentioned in Chapter IX (Hungate *et al.,* 1959, 1960). In some cases, e.g., the very high rate in the African suni, the difference was greater than could be accounted for on the basis of feed differences, although these latter were undoubtedly large, and in some cases the difference was demonstrated between animal species on the same feed (Hungate *et al.,* 1960). Zebu cattle showed slightly greater digestibilities of African feedstuffs than did European breeds, but the differences were not statistically significant (French, 1940). The protein and energy required for maintenance were found in one experiment to be only 73% and 83%, respectively, for the zebu breeds Mashona and Africander (R. C. Elliott *et al.,* 1964). Differences in palatability of different feeds for different species are probably important when areas are simultaneously occupied.

A statistically significant ($P = 0.01$) greater digestibility of grass hay by Zebu cattle as compared to Herefords (Phillips, 1961a) correlated with a greater fermentation rate ($P = 0.05$). No statistically significant difference in rumen dry matter (Phillips, 1961a,b) or rate of passage was found (Phillips *et al.,* 1960), but a greater bicarbonate and phosphorus concentration in the rumen contents of zebus suggested a greater saliva production which could account for the greater fermentation and digestibility. Greater saliva production would increase the turnover of the rumen liquid and small particle fraction. Restriction of water intake caused an increase in the forage digestibility in the Herefords, but not the zebus (Phillips, 1961a,b). In other experiments dry matter digestibility and apparent nitrogen digestibility were significantly greater in zebus than in crossbred cattle and Herefords (Ashton, 1962).

Rumen contraction rates are about the same in several zebu and European breeds (Morillo and de Alba, 1960). Zebu cattle excrete relatively more potassium in their urine and less in their feces than do European breeds (Horrocks and Phillips, 1964a,b).

Numerous reports of differences between sheep and cattle have appeared in the literature (Cipolloni *et al.,* 1951), and the fact that various microorganisms such as *Veillonella alcalescens,* Quin's oval, and large selemonads are much more abundant in sheep than in cattle indicates differences between the rumens of these two domestic ruminant species. There are many microbial species common to both, however.

Blaxter and Wainman (1961) found little difference in food utilization in sheep and cattle, yet greater efficiency in food utilization by sheep may follow from their lower maintenance energy (Blaxter and Wilson, 1962; Garrett *et al.,* 1959). Goats digest dry matter to about the same extent as sheep and cattle (Baumgardt *et al.,* 1964). A greater digestibility of feeds when daily intake was reduced has been reported for cattle (French, 1956b) but was not observed in sheep, and a difference in the effect of pelleting hay on its nutritive value index for mature sheep and cattle was noted (Buchman and Hemken, 1964). Evidence of difference in rumen ammonia and VFA concentration in the sheep, buffalo, and goat has been found (Pant *et al.*, 1963b).

The digestion of fiber in reindeer is comparable to that observed with cattle and sheep (Nordfelt *et al.*, 1961). Lichens are somewhat better digested insofar as dry matter is concerned, but the lichens did not supply sufficient protein to maintain a nitrogen balance.

Differences in vitamin content of blood according to breed and temperature has been reported for Brahman, Sauta Gertrudis, and Shorthorn heifers (Singh and Merilan, 1957), but the relationship to rumen synthesis of the vitamins was not examined.

7. Continuous vs. Infrequent Feeding

The quantity of dry matter in the rumen at various times after feeding varies according to the time between feedings and the level of feeding, i.e., maintenance or production levels. In ewes receiving 1.3 kg of feed in two feedings per day, the dry matter in the rumen before feeding was only 720 gm as compared to 1615 gm 2 hours after feeding (Boyne *et al.*, 1956).

Dry matter disappears from the rumen by absorption of fermentation products, liberation of fermentation gases, and passage to the omasum. With highly digestible feeds such as grain and concentrates the quantity leaving the rumen due to digestion and fermentation can be very large, and in these animals a considerable fluctuation in dry matter content will occur. In the above experiments the difference in dry matter is much larger than is encountered with forage rations. The ration included two-thirds ground maize.

Cattle fed twice daily at 12-hour intervals showed slightly lower average concentrations of VFA's in the rumen contents than did animals fed eight times daily at 3-hour intervals (Knox and Ward, 1961). The relative proportions of propionate and butyrate were higher with the more frequent feedings. Frequent feedings improved the weight gains in sheep (Gordon and Tribe, 1952; Moir and Somers, 1957; Rakes *et al.*, 1961). Feeding ten times daily instead of twice resulted in a 25% increased gain (Putnam *et al.*, 1961) in heifer calves. In this experiment there was no significant difference in the VFA's in the rumen of the two groups, but the number of protozoa was significantly greater in the frequently fed animals. The steady supply of ammonia nitrogen is seen as an explanation for the greater value of feed consumed continuously (Rakes *et al.*, 1961). Satter and Baumgardt (1962) have noted a better nitrogen retention and less fluctuations in rumen VFA and ammonia concentrations, but no effect on digestibility, in cows fed eight times daily on alfalfa hay, as compared to two and four times. In Portugal's (1963) experiments the average ammonia concentration in the rumen was significantly less in continuously fed sheep than in sheep fed twice daily. Rumination occurs with diurnal periodicity in sheep fed once a day, but is more or less continuous in sheep fed *ad libitum* (Pearce, 1965a,b).

These effects of frequent feeding may depend on modifications of the microbial fermentation. Change in the relative proportions of rumen fermentation products with time after feeding has been mentioned in connection with methanogenesis. The proportion of methane formed increases with time after feeding (Pilgrim, 1948; Kingwell *et al.,* 1959; Hungate *et al.,* 1960), which indicates a relatively greater production of acetate. As the total gas production in the rumen diminishes after feeding, the carbon dioxide production diminishes faster than the methane production. The early excess of carbon dioxide is due in part to the rapid fermentation of the uronic acids, which give a greater proportion of carbon dioxide to methane in their fermentation. The studies of Gray and Pilgrim (1951) indicate that relatively more propionate is formed during the period soon after feeding, the propionate diminishing with time until food is again consumed. All these observations suggest that frequent feeding improves the utilization of the nutrients.

The explanation for this improved utilization may be as follows: if proteins are digested in the rumen more rapidly than are carbohydrates, especially the cellulose and hemicellulose, amino acids and peptides will be available for assimilation during the period when soluble sugars and easily digested carbohydrates are being fermented to form cell material, but will be relatively unavailable during digestion and fermentation of the fiber in animals fed infrequently. The initial amount in excess of that which can be assimilated will be lost through fermentation, and, although part of the ammonia formed may still be available later, some will have been absorbed and lost, and the remainder may not be nutritionally equivalent to the amino acids, peptides, or other nutrients in the feed. In continuously or frequently fed animals these complex nutrients are available over a longer period. If their presence favors growth of bacteria producing propionate and succinate, the greater proportion of propionate formed in such animals would be explained. Also faster rumen turnover rate, a consequence of grinding and pelleting of the feed, would tend to increase the time during which these nutrients are available.

8. Specific Gravity

The specific gravity and size of particles influence their rate of passage through the rumen (Campling and Freer, 1962). Rate of passage increases with specific gravity up to a specific gravity of 1.2, above which it decreases. Particles with a specific gravity of less than 1.0 were retained for several weeks. These latter float and thus would not be in position to enter the omasum. The faster passage of the heavier particles is explained by their coming to rest in the reticulum and being tipped into the omasum (Grau, 1955) instead of being projected into the rumen.

9. EFFECT OF SEASON

A correlation between rumen bacterial direct counts and the season of the year has been observed (Nottle, 1956). Since the ration was kept constant, the variations were ascribed to possible effects of light and temperature. Effects of season on forages have already been discussed. Seasonal changes in the proportions of the VFA's in the rumen have been observed (Pant *et al.,* 1963a), with low proportions of acetic acid in spring and autumn when the protein content of the forage was high (Jamieson, 1959b). The effects of light and temperature can be mediated in part through the composition of the feed.

10. STEROIDS AND SIMILAR COMPOUNDS

Digestion of cellulose in sheep fed alfalfa hay and maize was improved by daily adding 10 or 20 mg of stilbestrol, cortisone, and testosterone, and in flask experiments the digestibility of cellulose was increased by these same substances and also by estrone and cholesterol (Pfander *et al.,* 1957a). Diethylstilbestrol increases the efficiency of gain in steers (M. C. Bell *et al.,* 1957; Burroughs *et al.,* 1954), and some evidence that it affected the rumen microorganisms has been reported (Brooks *et al.,* 1954a; Browning *et al.,* 1959). In experiments with various rations, diethylstilbestrol prevented the disappearance of the protozoa with some rations and increased the protozoal concentrations for all rations (Christiansen *et al.,* 1964). These results warrant study of the mechanisms involved.

D. Summary

It is evident that innumerable factors influence the kind and extent of rumen microbial activity. In addition to those mentioned, many more are known or will be found to be important. Improvement of ruminant performance for the benefit of human economy rests on an understanding of these factors and their manipulation to advantage. Some of the possibilities will be discussed in Chapter XI.

CHAPTER XI

Possible Modifications in Ruminant Feeding Practices

One of the intriguing features of ruminants is their potential to synthesize protein from carbohydrate and simple forms of nitrogen. Exploitation of this capacity requires a knowledge of the suitability of various kinds of nitrogen.

A. Nitrogen Sources for Microbial Synthesis

1. NITROGEN GAS

The possibility of rumen microbial assimilation of nonprotein nitrogen into usable protein has been considered for many years. Boussingault (1839) analyzed the food and feces of a milk cow to test for fixation of atmospheric nitrogen; none was found. Subsequent demonstrations (Toth, 1948) of nitrogen-fixing bacteria in rumen contents are unconvincing, since the organism demonstrated was *Azotobacter,* an obligate aerobe with high oxygen demands. *Azotobacter* can utilize the volatile fatty acids as a source of energy for growth on nitrogen gas, and some slight fixation of atmospheric nitrogen might occur in the upper surface layer of rumen contents exposed to the oxygen in the gas pocket at the top of the rumen. It is doubtful that it is of any quantitative significance. Growth would be much delayed by the mixing of the *Azotobacter* cells from the surface layer into the main anaerobic mass in the rumen. In view of the suppression of nitrogen fixation by ammonia, and the usual presence of ammonia in rumen fluid, conditions are unfavorable for nitrogen fixation from this standpoint as well (Toth, 1949).

Under usual conditions of ruminant feeding, nitrogen fixation will not occur, because the quantity of available fixed nitrogen is adequate. This is particularly true during the period soon after feeding, when feed nitrogen is normally in excess. During later periods, some nitrogen enters from the blood in the form of urea, and some is released from the bodies of bacteria, both by endogenous metabolism and by utilization of microbial bodies as food by other bacteria and by protozoa.

419

This does not necessarily mean that nitrogen fixation cannot be of importance in the rumen. Many anaerobic bacteria can fix atmospheric nitrogen, and although rumen bacteria have not been tested in this respect it is highly probable that imposition of suitable conditions could elicit nitrogen fixation by them or other bacteria which might be selected by the conditions. The blood circulating in the rumen wall is saturated with nitrogen at 79% of atmospheric pressure, and if nitrogen gas were used in the rumen, it could diffuse into that organ from the blood. It has been estimated that 55 liters (69 gm) of N_2 could diffuse daily into the bovine rumen (Kleiber, 1956).

2. Simple Forms of Fixed Nitrogen

Early studies (Wilckens, 1872) on the amount of protein in the rumen and in the feed indicated a net rumen synthesis of protein, but digestion of fiber and exit of microbial protein were not taken into account, which makes the results of the study inconclusive. The work shows the early interest in a "Veredelung" (upgrading) of feed nitrogen through rumen microbial activity. Wildt (1879), using silica as a marker, showed an increase in the nitrogen in rumen contents of an animal fed chopped barley straw and water.

Weiske (1879) showed that the nitrogen of asparagine was utilized in sheep. His work stimulated much interest in the possibility that amide nitrogen, abundant in plants, could be utilized in the rumen for the synthesis of protein. Müller's work (1906) further supported the idea of amide utilization. Urea became recognized as a cheap artificial source of amide nitrogen for the rumen and was shown by Voltz (1919, 1920) to be utilized by sheep. Subsequently, a great number of studies on urea utilization have been completed, and it is generally recognized that urea and ammonia can be assimilated in the rumen, provided carbohydrates and minerals are adequate (Arias *et al.,* 1951; Bloomfield *et al.,* 1964). Instances in which urea or ammonia are not utilized are usually ascribable to inclusion in the ration of a component containing sufficient organic nitrogen to meet the synthetic capacity possible with the energy source. The striking experiments of Virtanen (1963) indicate that ammonia and urea can supply the entire nitrogen needs of milking cows. Long-term experiments with sheep (Klein and Müller, 1942) similarly show that the protein requirement can be met in large part by simple forms of nitrogen.

Some methods of adding nitrogen to ruminant rations have resulted in toxicity. Ammoniated molasses (0.45 kg daily) was toxic to yearling heifers (Bartlett and Broster, 1958) and caused violent reactions (Tillman *et al.,* 1957a,b). The toxic agent was not identified.

3. Protein: Bypass of the Rumen

One of the most important possible improvements in ruminant feeds would be a device to prevent utilization of high-quality proteins (i.e., those with a high biological value for the host) in the rumen, but make them available in the abomasum and small intestine (Synge, 1952a; Blaxter and Martin, 1962). The protein of corn, zein, has already been mentioned as being resistant to microbial attack; it leaves the rumen before digestion is complete. An artificial device or treatment, e.g., a drug to stimulate closure of the esophageal groove (Mönnig and Quin, 1933, 1935), which would shield proteins from digestion in the rumen but permit digestion in the abomasum, might enable ruminants to assimilate larger quantities of protein with high biological value than is possible when the synthetic capacities of the rumen microbes are the limiting factor. The advantage of bypassing the rumen has been demonstrated (Cuthbertson and Chalmers, 1950; Chalmers *et al.,* 1954; Blaxter and Martin, 1962; Egan and Moir, 1965) by infusing casein into the abomasum of sheep, and also in the case of lysine (Devlin and Woods, 1964). This gave a net metabolizable energy of 65 as compared to 50 for casein going through the rumen. A better nitrogen retention with herring meal than with equivalent casein (Chalmers and Synge, 1954) or groundnut meal (Chalmers and Marshall, 1964) was ascribed to greater resistance to rumen digestion, less ammonia being liberated in the rumen from the herring meal. A fixed fraction (33%) of fish meal was retained by cattle at all levels up to 22% (Bowers *et al.,* 1964). Similarly, heating of soya-bean meal decreased its digestibility in the rumen of sheep, and increased its overall digestibility and the host retention of nitrogen (Tagari *et al.,* 1962). A similar effect of heating groundnut meal has been observed (Chalmers *et al.,* 1964). Denaturation of proteins with formaldehyde to shield them from attack in the rumen has been proposed (Chalmers and Synge, 1954).

A lowered ammonia level in the rumen when groundnut oil was added to the protein supplement suggests that coating of proteins with oil might shield them from the rumen fermentation (Jayasinghe, 1961). Digestion could occur in the intestine through the combined action of emulsifying agents and proteinases. This bypass of the rumen would improve the nutrition of the host only if the protein had a reasonably high biological value. Poor proteins would be more valuable if converted to microbial form.

4. Proteins: Feeding Practices

A modification in feeding practice that might improve forage utilization would be the administration of protein concentrates in small doses during the day, even though the forage might be consumed in a few large feeds. This would avoid the initial excess of protein or other form of nitrogen, which results in loss of nitrogen as ammonia, and would supply needed nitrogen at later stages when fiber is largely undergoing digestion and nitrogen is needed. This could spare some of the previously synthesized microbial cells, the decomposition of which releases ammonia at periods some time after food is ingested.

Stimulation of the rumen fermentation by a yeast preparation has been noted (Butz et al., 1958a). The extent would undoubtedly depend on the nutrients in the yeast rather than on yeast growth. Reports that enzyme supplements improve ruminant performance have not been substantiated (Strong et al., 1960), and it would not be expected that they would be economical since the rumen microorganisms in a properly fed animal elaborate them at little cost. Ligninases and cellulases would appear to be the enzymes most likely to improve forage utilization if added to the feed.

5. PROTEINS: RUMEN EFFECTS ON BIOLOGICAL VALUE

The fact that nitrogenous constituents of the feed undergo microbial conversions before they are assimilated by the host makes it possible to feed ruminants on many nitrogen sources unsuited to monogastric animals (Synge, 1952a).

The literature shows ruminant utilization of such bizarre nitrogen sources as sarcosine and creatine (C. D. Campbell et al., 1959; G. C. Anderson et al., 1959), orotic acid (Casati and Garuti, 1961), biuret (G. C. Anderson et al., 1959, Ewan et al., 1958; Hatfield et al., 1959; T. C. Campbell et al., 1963; Clark et al., 1963; Gutowski et al., 1958b; MacKenzie and Altona, 1964a,b; R. S. Gray and Clark, 1964; questioned by R. R. Johnson and McClure, 1964), cyanuric acid (Altona and MacKenzie, 1964), Penicillium (Woodman and Evans, 1947), feather meal (Jordan and Croom, 1957), sewage sludge (Hackler et al., 1957; G. C. Anderson et al., 1959; McClymont, 1946), chicken litter (Noland et al., 1955), entrails (Kehar and Chanda, 1947), yeast (Hrenov, 1961; Ryś et al., 1962), and even rumen contents (Moon and Varley, 1942). This merely skims the surface of the possibilities, but it indicates the wide variety of nitrogenous materials which can be utilized by the ruminant.

The equivalence of proteins in ruminant feeds (Dyer and Fletcher, 1958; Ellis et al., 1956; Glover et al., 1957) is explained by the microbial digestion and assimilation which precedes utilization by the host. Approximately the same kinds of microbes are formed regardless of the type of protein fed. This follows because of the capacity of the microorganisms to synthesize amino acids.

The amino acids in some forages are chiefly those that are not essential in mammals, which explains the low value of plant juices for monogastric animals (Synge, 1951). The leaf proteins, chiefly within the chloroplasts (Synge, 1952a), more nearly resemble in composition the proteins in animal tissues.

Even if proteins are essentially similar in amino acid composition, losses of protein nitrogen in the rumen from a soluble protein such as casein may be greater than from less soluble plant proteins (C. Little *et al.,* 1963; Oyaért and Bouckaert, 1960), due to more rapid digestion (Kay and Hobson, 1963). If the soluble protein has a greater biological value, as with casein, part of this advantage is lost during rumen conversions. Proteins of low biological value may be upgraded by microbial conversions in the rumen, but those of high biological value can only suffer loss. The biological value of the protein can be greatly affected by other nitrogenous and nonnitrogenous constituents of the food.

B. Use of Carbohydrates for Microbial Synthesis

1. DIVERSITY OF SOURCES

Many substances have been tested as ruminant feeds (Schneider, 1947), e.g., wood cellulose, ground prunes, winery pomace, avocado meal, asparagus butts (Folger, 1940), heather (Armstrong and Thomas, 1953), live oak (*Quercus agrifolia*) leaves, chamise (*Adenostoma fasciculatum*) (Bissell and Weir, 1957), elephant grass (Bredon, 1957), pineapple residues (Rogerson, 1956), cassava roots (French, 1937a), coffee hulls (Rogerson, 1955), thorn (*Acacia*) trees (Steyn, 1943), chervil seed (Dijkstra, 1956), mesquite wood (Marion, 1956), and date stones (Richter and Becker, 1956). Some of them, such as extracted tobacco seed (Maymone and Tiberio, 1948), appear to be relatively indigestible; others, such as ivy leaves (Maymone and Tiberio, 1949), are fairly digestible.

Many plant species differ in palatability for ruminants, particularly browse plants. Examinations of the factors underlying palatability and digestibility of a wide variety of feeds could yield interesting and possibly profitable results. Use of seaweeds by Orkney sheep is a case in point (Eadie, 1957).

2. PROBLEMS IN CARBOHYDRATE UTILIZATION

Poor forages usually include those mature portions of the vegetative parts of the plant from which nitrogen and minerals have been mobilized into the fruit and in which lignification has diminished the digestibility of the carbohydrate. In some cases the nitrogen content of the forage has been about as accurate as the lignin content as a basis for predicting digestibility (Armstrong *et al.,* 1964a; Bateman and Blaxter, 1964). The straws of cereal grains, corn cobs (Burroughs *et al.,* 1945), and wood cellulose are examples. Much of the carbohydrate material in cereal straws is not digestible, which makes them unsatisfactory as feed not so much because of their lack of nitrogen and other essential minerals, which can be supplemented, but because of their lack of digestible carbohydrate (R. C. Elliott and Topps, 1964). Supplementation of roughage with a simple nitrogen source (MacKenzie and Altona, 1964b) can assist in maintaining sheep on a poor roughage diet, but growth is absent or slight. The bulk which must be handled is large in comparison with the yield in food. Addition of nitrogenous concentrates to a wheat straw ration does not increase cellulose digestion (Abou Akkada and el-Shazly, 1958). Along the same lines, the small quantities of nitrogen and minerals contained in the straws are inadequate if the straw is supplemented with digestible carbohydrate.

3. TREATMENTS TO INCREASE DIGESTIBILITY

Attempts to improve the utilization of poor forages could well start with chemical treatments to increase the amount of digestible carbohydrate. Studies such as those of Ferguson (1942, 1943) have demonstrated the efficacy of sodium hydroxide treatments to increase the digestibility of straw. Unfortunately, the method employed removes the soluble materials and in addition is economically impractical. Treatments which remove lignin but lose the soluble materials in forage are less useful than those salvaging the latter. Encouraging results with a method to remove lignin without taking out digestible components have been derived from *in vitro* experiments (Sullivan and Hershberger, 1959) with chlorite- and sulfite-treated forage. A sodium hydroxide treatment of hammer-milled wheat straw and poplar wood markedly increased digestibility (R. K. Wilson and Pigden, 1964); the mixture was palatable even though it contained much residual sodium hydroxide.

Ensiling of beech sawdust did not increase the nutritional value (Kirsch and Jantzon, 1941). Essentially similar results were obtained with olive twigs (Maymome and Cece-Ginestrelli, 1947), i.e., the material remained indigestible. This is not surprising since the rumen bacteria would be expected to be fully as capable as those in silage to digest wood cellulose.

In contrast, the physical comminution of wood to small particles renders it digestible (Koistenen, 1948) by breaking the bonds connecting "encrusting" materials to the cellulose. The fine grinding of wood by the mandibles of termites may explain the ability of these insects to digest the cellulose in wood.

The increases in digestibility of wheat straw (Pritchard *et al.,* 1962) and of basswood sawdust exposed to high-voltage X-rays (Lawton *et al.,* 1951) indicate that high-energy irradiation can break the bonds preventing digestion of the cellulose. This process is not economically feasible.

In view of the ability of the "white rot" fungi to decompose lignin under aerobic conditions, it would be of interest to examine the effect of their attack on wood on its rumen digestibility.

A theoretically sound chemical treatment can be formulated to increase the digestibility of forage, without losing soluble materials, and at the same time provide minerals. A mixture of cations characteristic of rumen contents could be applied in the form of the hydroxide to the forage; a spray might be used to obtain even application. The mixture would include sodium, potassium, calcium, and magnesium hydroxides. If the treatment were in a closed chamber, ammonium hydroxide could also be used. The material would then be held at elevated temperature, e.g., by application of steam, for a period found optimal in increasing digestibility, and then neutralized with a mixture of acids representing the anions in the rumen, i.e., phosphoric, hydrochloric, and sulfuric acids to neutralize the bases and at the same time provide essential minerals. Urea or other nitrogen source could be added later if volatilization of ammonia entailed losses during the chemical treatment. Carbon dioxide from the air could serve to complete the neutralization of the treated forage, although, in view of the large quantities of carbon dioxide generated in the rumen, such neutralization would perhaps be superfluous, provided the animals found the somewhat alkaline material palatable. Traces of straight- and branched-chain volatile fatty acids could be added to improve palatability.

Chlorite-sulfite treatment to remove lignin without losing organic matter increased the digestibility of several poor forages (Sullivan and Hershberger, 1959); even the fiber of alfalfa hay is rendered more digestible by treatments removing lignin (Fig. V-5).

Delignified wood cellulose has been extensively employed for feeding ruminants (Hvidsten, 1946; Moskovits, 1942), particularly in northern European countries during war periods when usual fodders are unavailable. The material is highly digestible and has proven practicable, though not equal to good forage. As previously mentioned, milk production can be supported by wood cellulose (Rook and Campling, 1959; Virtanen, 1963). One of the problems in feeding wood cellulose is the rapid passage of the finely divided materials through the rumen (Rook and Campling, 1959). Methods for aggregating into larger pieces (Virtanen, 1963) which would still expose a large surface to cellulolytic enzymes may increase retention and give more complete utilization.

4. PRODUCTS FROM CELLULOSE

The rumen bacterium which most readily digests wood cellulose is *Bacteroides succinogenes*. It requires a mixture of straight- and branched-chain C_4–C_6 fatty acids, factors usually supplied during the fermentation of protein. In the absence of protein it is necessary to supply these factors if active growth of *Bacteroides succinogenes* is to be obtained. Further investigation of the value of wood cellulose (and other celluloses from delignified plant materials) seems well worthwhile, particularly in view of the evidence that cellulose fermentation can occasionally be associated with the production of a relatively high proportion of propionic acid (Elsden, 1945b; Dehority *et al.*, 1960; Salsbury *et al.*, 1961; Gray and Pilgrim, 1952a). This high propionate concentration is understandable if most of the cellulose digestion is due to *Bacteroides succinogenes,* since succinic acid is an important fermentation product and is rapidly decarboxylated to propionate. On the other hand, the finding that milk fat percentage increased when wood cellulose was fed (Hvidsten, 1946; Virtanen, 1963) suggests that acetate production was relatively greater. Possibly the other nutrients in these experiments supported a different population, with an acetigenic rather than a propiogenic fermentation.

5. READILY DIGESTIBLE CARBOHYDRATES

Starch and glucose are well suited to supply digestible carbohydrate since they contain negligible quantities of other nutrients. In too high concentrations they lead to acid accumulation and indigestion and are therefore not well suited to compose a major fraction of the diet under usual feeding regimes. In five out of seven calves fed a purified ration containing 87% starch, glucose, and cellophane, the pH fell below 5.0 (Agrawala *et al.*, 1953a). Cane molasses as one-third of the ration caused no marked ill effects (Henke *et al.*, 1940). A synthetic ration containing 61% starch gave some growth in lambs (Thomas *et al.*, 1951). As much as 95% concentrates has been fed successfully to fattening lambs (Meyer and Nelson, 1961).

If acid accumulation could be avoided, there seems to be no a priori reason why starch and sugar should not be well suited for extensive utilization by ruminants. Other readily fermentable materials such as xylan and pectin also fall in this category (Belasco, 1956; Conrad *et al.*, 1958b). The acid accumulation can be avoided by feeding slowly over an extended period and by adding buffers to the feed. If the readily fermentable material is supplied only as fast as the fermentation products can be absorbed, acid should not accumulate. In addition to avoidance of acid accumulation, frequent feeding increases the efficiency of microbial growth and in practice has been shown to give greater weight gains (J. G. Gordon and Tribe, 1952; Preston, 1963).

Purified carbohydrates such as glucose, if employed as exclusive substrate, could select a rumen biota quite different from that characterizing animals receiving forage. It might be possible to obtain marked differences in ratios of fermentation products available to the host. An extensive study of ruminant nutrition with glucose as the chief carbon source could be a means for testing the effect of other nutrients.

C. Other Nutrients Influencing Microbial Synthesis in the Rumen

Most of the trace elements are concerned with reactions in which proteins play a part, and they therefore occur in conjunction with protein. This factor must be considered and carefully provided for in experiments substituting simple forms of nitrogen for the forms in forage. The improvement in cellulose digestion noted when green legumes forage was substituted for a supplement of starch, casein, dried yeast, and minerals (Louw *et al.,* 1948) suggests that forages may contain microbial nutrients in addition to the known mineral and organic growth factors, although the possibility that the favorable effect of the legume was in its known mineral, nitrogen, and carbohydrate components cannot be excluded. Similarly, 1–2% of an acid extract of clover leaves has stimulated microbial development and gas production (Takahashi *et al.,* 1963, 1964).

The importance of protein fermentation in the rumen for providing the C_4–C_5 branched- and straight-chain fatty acids shows how protein may influence nutrition in a fashion not directly connected with suitability of nitrogen sources for assimilation. The concentrates employed in ruminant feeds undoubtedly function in part in a similar fashion. Infusion of isobutyrate, isovalerate, and 2-methylbutyrate into the rumen caused 1 or 2 gm more nitrogen to enter the sheep rumen each day (Kay and Phillipson, 1964). Under some circumstances, a microbiota not requiring these factors or generating them from simple nutrients can apparently be selected (Virtanen, 1963).

Successful utilization of simple forms of nitrogen requires more than a knowledge of the kinds and proportions of minerals essential for the microbes and the host; it also requires a knowledge of the organic nutrilites stimulatory to growth of the rumen microorganisms. Nutrients such as those listed in Table II-10 should be tested in a ration containing purified carbohydrate (Glucose, starch, or cellulose). Unidentified factors (Luizzo *et al.,* 1961) need further study in order to test them in similar fashion.

Although not immediately practicable, experiments on feeding ruminants simple defined nutrients is of much theoretical interest, and preliminary tests (Hungate and Dyer, 1956; Hemsley and Moir, 1963) show promise of future practical applications. The variations in the proportions of volatile fatty acids in the rumen over a long period (Jamieson, 1959a) suggest that attention to forage constituents could disclose nutritional methods for increasing the relative quantity of propionic acid. In some cases urea causes acetic acid to predominate more than if casein was the source of nitrogen (G. V. Davis and Stallcup, 1964).

D. Importance of Balanced Nutrients

Balance of nutrients in ruminant feeds is extremely important (Summers *et al.,* 1957). If proteinaceous feeds are the source of essential minerals (Lassiter *et al.,* 1958b) and organic growth factors in some rations, their content of the organic growth factors, rather than the actual protein, may be the factor determining the quantity of "protein" required. With such rations, supplementation with minerals, organic nutrilities, and carbohydrates may permit use of less of the feed protein.

This is the difficult problem of practical nutrition, to balance the feed ingredients so that they are all adequate. As any component is increased relative to other required foods, it is used with decreasing efficiency, which means that its cost in relation to the gain is greater. However, if too little is used, other components are in excess and are less efficiently utilized. Between the extremes of surplus and inadequacy is a point of greatest efficiency, which the feeder seeks under the particular conditions of operation. It is obvious that very cheap feeds can be supplied in greater excess than can the expensive ones. Interaction between various feed ingredients has been demonstrated *in vitro* (Francois *et al.,* 1954) and *in vivo* (Swift *et al.,* 1947).

An illustration of the importance of balance is afforded by experiments (Oltjen *et al.,* 1959) testing the effects of trace mineral supplements in a ration containing sorghum grain, prairie hay, protein concentrate, ground limestone, and salt. Cobalt, copper, iron, iodine, manganese, and zinc did not improve the utilization of this ration, but when maize was substituted for the sorghum, the trace minerals improved the weight gain. An increase in minerals to 6.5% of the ration gave a linear increase in the rate of gain (Oltjen *et al.,* 1962c). Corn cobs contain a high fraction of digestible carbohydrates, and supplementation of a corn cob ration with balanced minerals and nitrogen (Burroughs *et al.,* 1949b, 1950a) has proven highly advantageous in practice.

Another example of the importance of balanced nutrients for microbial growth is provided by comparisons between the percentage of nitrogen in the ration and the nitrogen passing as protein to the abomasum (Oyaért and Bouckaert, 1960) when various proportions of protein and carbohydrate are fed (see Table XI-1). With clover hay alone or with added casein, increasing the available carbohydrate increased the nitrogen retained. The effect was not as marked with corn gluten, linseed meal, and peanut meal. These were less available for microbial growth, perhaps because they are less soluble. They were perhaps digested in the intestine, their lowered retention being due to deficiencies in essential amino acids.

Linseed oil meal can have a high biological value (V. J. Williams *et al.,* 1953) under some feeding conditions; in addition, however, the lower bacterial counts in these experiments, as compared to those with casein, indicate that some of the protein was not digested and fermented in the rumen.

The value of a feed can vary according to the level at which it is fed (Blaxter and Wainman, 1964) and according to the other feeds in the ration. From the discussion in Chapter VII it is obvious that much of the protein fed in excess of that required for rumen synthesis of microbial nitrogenous constituents will be wasted through fermentation to ammonia, carbon dioxide, and acids. Many feeds, such as lush spring grass, contain more nitrogen as protein than can be assimilated into the host (Sjollema, 1950). Supplementation with carbohydrate can improve the protein utilization (Watanabe and Umezu, 1963). The balance of nutrients can influence the ratios in which the volatile fatty acids are produced. A ration of known components, with starch and glucose as the carbohydrate source, gave an increased proportion of propionate when the casein was reduced from 20 to 2% of the ration and 4.6% urea was added (Matrone *et al.,* 1964). With urea but no casein, butyrate was much increased.

Increased knowledge of nutrilites required by the rumen microbes could indicate a means to save a significant fraction of the feed protein, if trace substances in it are limiting, and permit a different optimal balance of feed ingredients.

E. Rumen Inoculation

Recognition of the importance of rumen microorganisms for the functioning of the host has led to the concept that malfunction of the host may be due to a defect in the microbial population. The most obvious defect would be the absence of a particular type of bacterium or protozoan performing an essential function. The equally obvious remedy would be

Table XI-1

EFFECTS OF NITROGENOUS AND STARCH SUPPLEMENTS ON NITROGEN METABOLISM IN SHEEP[a]

Ration		Total Nitrogen			Protein Nitrogen			
Description	Dry matter (gm)	Ration (gm)	Omasum (gm)	Difference (gm)	Ration (gm)	Omasum (gm)	Difference (gm)	Nitrogen retention (gm)
800 gm of clover hay	704	18.4	17.13	−1.27	14.72	15.56	+0.84	0.20
+ 300 gm of potato starch	959	18.4	18.48	+0.08	14.70	17.13	+2.41	2.10
+ 100 gm of casein	794	30.4	19.76	−10.64	26.60	15.60	−11.00	3.70
+ casein and starch	1050	30.4	24.19	−6.21	26.60	19.78	−6.82	7.20
+ 200 gm of corn gluten	880	32.8	25.98	−6.82	28.60	23.15	−5.45	6.40
+ gluten and starch	1139	32.8	28.95	−3.85	28.60	26.95	−1.65	6.30
+ 300 gm of linseed meal	977	31.42	25.87	−5.55	26.90	22.74	−4.16	5.00
+ linseed and starch	1232	31.42	28.67	−2.75	26.90	26.49	−0.91	6.50
+ 180 gm of peanut meal	867	30.33	21.91	−8.42	26.19	19.50	−6.69	2.50
+ peanut and starch	1132	30.33	26.18	−4.15	26.19	24.14	−2.05	3.85

[a] From Oyaért and Bouckaert (1960).

the addition of the missing organisms. This has led to numerous experiments with inoculation of calves with rumen contents from adult cattle (Conrad *et al.,* 1950; Hibbs *et al.,* 1953a,b, 1954, 1956, 1957; Hibbs and Conrad, 1958; Hibbs and Pounden, 1948, 1949, 1950; Pounden and Hibbs, 1948a, 1949a,b; A. T. G. McArthur, 1957b) and to widespread advertising of the value of various pills and additives reported to contain the bacteria typical of normal rumen function.

It is doubtful that rumen inoculations are effective in changing the nature of the microbiota if the ration remains the same, although some exceptions can be noted. Two factors determine whether a particular microbe will occur in the rumen: first, whether the conditions are favorable to its growth, and, second, whether it gains access to these favorable conditions. Inoculation is essential only if the organism is absent from the rumen, although if it is present in very small numbers inoculation can hasten its development, particularly if the feed is changed.

Where ruminants are herded together, rumen microbes spread throughout the herd. It is improbable that a particular organism will be completely missing from the entire group. Rumen inoculation can be beneficial to calves separated from the herd at an early age. Such animals remain free of protozoa unless inoculated (Conrad *et al.,* 1958b).

A paucity in numbers of a particular microbe indicates conditions unfavorable for its development. Augmentation of numbers occurs only if conditions supporting additional growth are provided. The feeding regimen and the host physiology determine the composition of the rumen microbial population. Exceptions to this generalization may occur, but it is difficult to document them, to prove rigorously that organisms capable of improved performance are absent and that the environment is not the chief factor determining the composition of that particular population.

The exact composition of the microbial population in an animal on a particular ration is not necessarily the same at different times. A particular serotype of *Butyrivibro* in a Jersey heifer was rare 2 years after it had been common in the same animal, even though the feed throughout the period had been exclusively alfalfa hay (Margherita *et al.,* 1964). However, isolations from many parts of the earth and from many individual animals have disclosed a fair similarity in the genera and species encountered, and complete absence of a particular kind of microorganism is probably rare. Changes in the protozoan population have been shown to follow introduction of new species (Eadie, 1962b). The changes in morphologically identifiable rumen microbes after exchange of digesta between paired sheep with populations differing in relative composition (Warner, 1962b) resulted in a seeming modification of the numbers of entodinia but not of oscillospira. The initial microbial differences in these sheep were presumably due to differences in host physiology, but the modifications after transfer of microbiota are not explained.

In New Zealand cattle, Clarke (1964a) has recently observed specimens of *Bütschlia,* the ciliate described from the rumen by Schuberg (1888) and others, but not reported in the recent literature from any part of the world. Clarke has also observed *Charon equi,* previously described only from the cecum of the horse. New Zealand is isolated from other parts of the world, and the mammalian fauna consists almost exclusively of introduced species. Chief among introduced eutherian mammals, insofar as numbers are concerned, are sheep, horses, and cattle. Whether isolation of cattle and horses on the islands is concerned with the establishment of *Charon* in cattle is a matter for conjecture, but it suggests that, in some cases, failure of a microbe to gain access to the rumen explains its absence there.

A fermentation somewhat similar to that in the rumen would probably develop if rumen conditions were artificially created and the system inoculated with soil. According to the concept of maximum biochemical work, selection would give a population of bacteria producing approximately the proportions of fermentation products characteristic of the rumen. A test of this concept could be one of the most valuable uses for a true artificial rumen, i.e., a gnotobiotic (known organisms) system simulating the rumen, except for inoculum, which could be controlled.

Many of the advertised rumen bacteria preparations contain bacteria obtained directly from the rumen and dried under conditions unlikely to preserve viability. The author tested such a preparation for rumen cellulolytic bacteria and found only *Clostridium lochheadii,* presumably because its spores resisted the desiccation method employed. Similar results have been observed by others (Harbers *et al.,* 1961; Kamstra *et al.,* 1959; Pounden *et al.,* 1954; Richter *et al.,* 1961). Calves isolated from birth differ from normal calves in appearance and in the incidence of scours, and these differences persist after dried rumen contents are administered to the isolated group (Wieringa, 1956). Even when rumen bacteria are lyophilized under conditions most conducive to survival, a large number of the cells are killed by the process.

If rumen inoculation is to be practiced, fresh rumen contents from a healthy animal consuming the same type of feed as the animal to be inoculated should be used (Allison *et al.,* 1964b). A method for catching the cud of a ruminating animal in order to obtain an inoculum has been widely used (Pounden and Hibbs, 1949b). This practice is very old and is mentioned by Brugnone (1809), ". . . as one encounters it [the bolus] in ruminants, a bolus which is removed very easily from the mouth, and which some veterinarians give as a sure remedy to induce rumination in animals, in which this function is suspended due to illness" (translation).

After some digestive disturbances the animal no longer feeds. Massive transfers of rumen contents of a healthy animal can restore appetite and normal rumen function. The massive transplant exerts an effect in fashions other than simple inoculation. Possibly the large quantity of fermentation acids in the transferred material stimulates rumen activity and restores appetite. If this is the explanation, similar results should be obtained by artificially introducing a considerable quantity of fresh readily fermentable feed into the rumen by stomach tube, which would provide a substrate for the bacteria still present. The loss of appetite is more likely due to lack of an active fermentation than to absence of suitable bacteria. If some drastic treatment has killed the bacteria, a combination of inoculation and force feeding may be required to restore rumen function.

An instance in which rumen inoculation has been shown to improve performance is during the transition from a forage ration to the high grain and concentrate regimen of the feed lot. The microflora differ for these two feeds (Pounden and Hibbs, 1948b; Maki and Foster, 1957). Fairly massive transplants of rumen contents, from already adapted animals to those starting the ration, give sustained weight increases during an adaptive period in which there is a slower growth in uninoculated controls, (Allison *et al.,* 1964b). Feeding starch four times instead of once a day does not hasten selection of an adapted population (Ryan, 1964b), although it may lessen the chance of acute indigestion during the adaptation period.

Similarly, in the adaptation of calves to a roughage ration, copious inoculation can hasten the acquisition of the normal adult biota and improve performance when roughage is an important component of the ration (Conrad and Hibbs, 1953). In long-term feeding experiments inoculation has little effect (Ewan *et al.,* 1958; Harrison *et al.,* 1957) other than the nutrients included. There may be instances in which adaptation of the microbiota extends over a long period, e.g., with biuret as a nitrogen source (T. C. Campbell *et al.,* 1963), in which case inoculation might hasten the process. An animal already well adapted would be a necessity as a source of inoculum.

The most conclusive evidence for an effect of rumen inoculation is to be found in the experiments with sheep fasted for a period (4 days) long enough to reduce the numbers of methane bacteria to a very low level (Pilgrim, 1948). Hydrogen composed 50% of the gas in the rumen of an uninoculated animal on the first day feeding was resumed, and 5% on the second day. A similar sheep, drenched with normal rumen fluid prior to the first resumed feeding, showed only 10% hydrogen 6 hours later and none after 24 hours. Methane production was delayed almost 2 days in the uninoculated animal. Even after 15 days of starvation, copious methane production had developed in uninoculated animals by the sixth day of resumed feeding, following 4 days in which hydrogen was the chief combustible gas. A few viable methanogenic bacteria are evidently retained in the rumen even after severe starvation. They use the hydrogen and carbon dioxide generated by other bacteria fermenting the slight quantities of digestible substances (microbial bodies, rumen epithelial cells, organic constituents of saliva, and digesta residues) in the rather large volume of digesta retained in the rumen. After 3 days of starvation the rumen bacteria in a sheep recovered their former concentration within 24 hours after food was again available (Coop, 1949).

F. Importance of Turnover

The interrelationships between rumen volume, fermentation rate, rate of turnover, and completeness of digestion (H. H. Mitchell, 1942) indicate an importance of turnover rate in the economy of feed utilization by the host. Since appetite is an important determiner of turnover, it also becomes extremely important in ruminant nutrition (Campling, 1964).

The most complete digestion of forages will be obtained with the longest retention time in the rumen, i.e., with the slowest turnover rate, and this will be correlated with the largest rumen volume, slowest fermentation rate, and greatest digestibility. There obviously comes a point at which the work of carrying the increased volume of rumen contents exceeds the advantage to be derived from more complete digestion. In a number of studies the advantages due to increased turnover balance the loss in digestibility, with no net difference in the value of the ration (Blaxter and Graham, 1956; Forbes et al., 1925). Under other conditions, e.g., much activity required, as during grazing, the faster turnover may be advantageous (Phillips et al., 1960).

The exact point at which increased rumen volume becomes detrimental depends on factors such as the relative costs of labor and feed, and also on the physiology of the animal. It would be expected that the greater the activity of the animal the less the volume of rumen contents that can be efficiently transported. If labor costs are high and feed is cheap, it may be more economical to sacrifice completeness of utilization to gain increased growth with the same labor investment. Under practical feeding conditions, increased consumption of feed is not always closely correlated with total digestibility (P. E. Anderson et al., 1959), particularly with small differences in feeding level. With differences of two to three times the quantity of food consumed, a negative correlation between digestibility and quantity was observed. Since the rate of digestion of a given piece of forage diminishes with time in the rumen, increases in turnover rate will not cause a proportional decrease in digestibility. Also, different intake levels do not necessarily lead to proportionally faster turnover, since the volume of rumen contents may increase with higher levels. However, large differences in intake speed up turnover and decrease total digestibility, other things being equal. This is illustrated in Fig. XI-1.

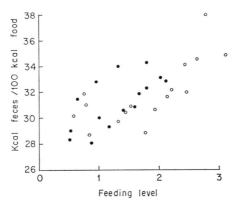

Fig. XI-1. Relationship between feeding level and digestibility in sheep (○) and cattle (●). From Blaxter and Wainman (1961, p. 422).

More rapid turnover within the same rumen volume leads to an increased productivity of microorganisms, but the increase is not proportional to the increased turnover, due to lessened digestibility. Fermentation in the rumen steadily decreases after the first 1 to 6 hours following ingestion. A greater proportion of less digestible material accumulates. Loss of this material by increased feed rate is more than offset by the increased entrance rate of the readily utilizable components.

The fermentation rate increases with increased turnover rate until the enzyme systems become saturated with substrate. Before this point is reached, the capacity of the host to consume food, secrete saliva, and eliminate residues becomes limiting. This occurs also before the turnover rate exceeds the microbial growth rate. The microorganisms would wash out of the rumen if the turnover rate exceeded their growth rate.

Methane wastes are less at higher levels of feeding, i.e., at faster turnover, as is evident in Fig. XI-2. Zuntz (1913) and Bratzler and Forbes (1940) also found a diminished proportion of methane as the feed consumption increased. In Marston's (1948b) experiments, of 508 Kcal of digestible energy, 14.6% appeared as methane; 1130 kcal yielded 9.4%, 1430 kcal gave 8.2%, 1760 gave 9.5%, and 2475 kcal showed 6.5% of the energy as methane. During the time after feeding the percentage of methane formed in the rumen increases (Washburn and Brody, 1937; Hungate *et al.*, 1960).

In comparisons of four different pastures, weight gains at slaughter were greatest in sheep showing the smallest rumen volumes (Butler and Johns, 1961). With feeds of greater digestibility, consumption is greater and turnover and fermentation are faster which gives increased growth (Johns *et al.*, 1963). Rations giving higher proportions of propionate

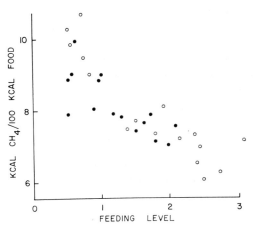

Fig. XI-2. Relationship between feeding level and production of methane in sheep (○) and cattle (●). From Blaxter and Wainman (1961, p. 422).

and butyrate are more efficient in milk production (Coppock *et al.*, 1964). The amount of protein and total solids exclusive of fat increases with higher levels of feeding (Rook and Line, 1961; Ueyama and Hirose, 1964). This may result from the higher proportions of propionate formed when the turnover rate is faster.

These considerations again emphasize the value of the readily fermentable carbohydrates in the ruminant. Starch and sugar are so completely digestible that fast feed rates do not cause a large accumulation of undigested material in the rumen. Much of the advantage of concentrates and young forages appears to be explicable in terms of these factors. Careful experiments with rations high in sugar and starch, with all accessory nutrients and with continuous feeding of the animal, might very well provide evidence for much increased rates of gain when maximum useful turnover rate is achieved through the use of readily digestible materials with little residual bulk.

Rations consisting entirely of concentrates have been fed to dairy cows (Moe *et al.,* 1963) and to steers (Nicholson *et al.,* 1962a,b; R. E. Davis, 1962), though the latter showed some kidney and liver damage. Additions of $NaHCO_3$ to the ration did not improve the nutritional response in some instances (Nicholson *et al.,* 1962a), but in others a favorable effect was observed (Nicholson and Cunningham, 1961; Oltjen *et al.,* 1962a; Preston *et al.,* 1961). Regulation of rumen acidity by ration constituents needs further study.

A smaller average volume of rumen contents is associated with continuous feeding of a given quantity of ration, as compared to consumption at infrequent intervals (Murray *et al.,* 1962). Continuous feeding assists in achieving maximum turnover by eliminating large fluctuations in rumen volume. It may also increase the proportion of propionate.

Some of the problems associated with high concentrate feeds may be overcome by continuous feeding with the aid of mechanical devices, many of which are employed in large feeding operations (Dobie and Curley, 1963; Finner and Baumgardt, 1963). Alternate feeding of hay and concentrate can affect the ratios of volatile fatty acids as compared to simultaneous feeding of these materials (Palmquist *et al.,* 1964).

G. *In Vitro* Evaluation of Feedstuffs

Waentig and Gierisch (1919) appear to have been the first to conceive of an evaluation of digestibility by *in vitro* incubation of feed with rumen contents. Rumen fluid was incubated for several days with the test feed, and the loss in weight was compared to that in a boiled control. Digestibility of up to 30% of the insoluble material was obtained. This was less than the 60% observed in feeding trials, due to the use of too high a concentration of feed in the cultures with a resultant high acidity.

Much of the information on relative digestibility now gained through feeding trials could probably be obtained with *in vitro* experiments set up to simulate rumen conditions. It must be ascertained experimentally that the quantity of substrate (feed) provided in the *in vitro* culture is small enough so that its fermentation does not develop conditions unfavorable to microbial growth (Church and Peterson, 1960). This quantity will depend on the buffering capacity of the medium and the digestibility of the substrate. With bicarbonate (0.5%)-carbon dioxide (at atmospheric pressure) as buffer, the digestible material should probably not exceed 0.5% (w/v). It is also important to provide in the culture the required accessory nutrients (Marquardt and Asplund, 1964a). When

30% rumen fluid is used as inoculum, the quantity of required nutrients in the added fluid is probably sufficient, but if a washed cell suspension is used, additional nutrients are essential unless those required are present in the feed to be tested. Provision of all nutrients except the one to be tested for digestibility gives the maximum digestibility value. Quantity of inoculum is also important (Marquardt and Asplund, 1964b).

The *in vitro* digestibility with rumen fluid can serve as a rough index to the total digestibility of the feed (Naga and el-Shazly, 1963; el-Shazly *et al.,* 1963; Hershberger *et al.,* 1959b; Bowden and Church, 1962b; Asplund *et al.,* 1958; Donefer *et al.,* 1962; Armstrong *et al.,* 1964a,b), but since the relative proportion of readily digestible versus slowly digestible material is an important factor in the value of the feed, a single determination of digestibility after a certain time does not provide all the needed information. Determinations of digestibility after various periods of incubation will give information on the quantity of speedily versus slowly digestible components. In practice it is advantageous to measure the rate of utilization (fermentation) at various times to pick up significant differences in feed value that would be missed in a single terminal measurement. Manometric methods are valuable for this purpose.

Experiments using rumen washed cell suspensions have shown that the fermentation of a given feed is most rapid when the cell suspension is obtained from an animal on that feed (Hinders and Ward, 1961). It must be emphasized that *in vitro* experiments cannot be substituted for actual feeding trials. They are useful for obtaining clues to possible differences in feeds, but the final measure of a feed is its effect on the ruminant. Use of *in vitro* methods is particularly apt in studying fermentability of possible forage and browse plants of wild ruminants (A. D. Smith, 1957). *In vitro* techniques (Nagy *et al.,* 1964) have disclosed a toxicity of the essential oils in sagebrush (*Artemisia tridentata*) on the microbes from the rumen of mule deer (*Odocoileus hermionus*).

The ratios of propionic and acetic acid from various feeds have been tested (Stewart and Schultz, 1958) and significant differences found, grass or beet pulp giving significantly more acetate, whereas molasses or corn meal produced more propionate. Asplund *et al.* (1958) compared the digestibility of grass and legume hays and oat straw.

One of the most useful applications of *in vitro* techniques could well be to test in a preliminary fashion the numerous feed additives. In many cases these are quite worthless, or priced above the nutritional value. Many of these feed additives are advocated as particularly stimulatory to the rumen bacteria. It should be possible to test claims of this sort validly and inexpensively by appropriate use of *in vitro* cultures of rumen contents.

Another possible use of *in vitro* fermentations is to test the nutritive qualities of forages unpalatable to ruminants. This would show whether the forage contained digestible and fermentable materials and, added to fermentable substrates, whether toxic materials inhibited microbial growth. Activity *in vitro* can serve as an inexpensive means for preliminary testing prior to actual feeding trials.

It is commonly assumed that rumen bacteria must be used as inoculum for *in vitro* incubations of forages to test their digestibility. Similar digestibilities can be obtained, however, by using soil as an inoculum and incubating under aerobic conditions (Birch, 1958). Digestibility is a feature of the chemical and physical composition of the feed, and, given a sufficiently large mixed microbial population from almost any source, organisms capable of utilizing the digestible components would be present. Long incubation gives digestibility values with a low variability coefficient for a particular feed (Baumgardt and Oh, 1964; R. F. Barnes *et al.*, 1964) because chemical resistance to digestion becomes limiting, and the variation in rate of digestion at earlier incubation periods becomes less important. With long aerobic incubation, utilization would be more complete than anerobic incubation, due to a capacity of certain aerobes to decompose lignin, which is not fermentable anaerobically.

Rumen bacteria are easier to obtain than most other microbial preparations likely to be active in forage decomposition, which makes their use in *in vitro* experiments relatively easy. Pending proofs of other equally satisfactory sources of inoculum, it is most sensible to study potential rumen digestibility with microbes from the rumen.

H. Rumen Neutralization of Toxins and Production of Antibiotics

In vitro experiments (Cook, 1957) suggest that some of the insecticides highly toxic to animals may be destroyed in the rumen. In view of the increasing use of pesticides, this factor could well permit utilization of some treated products which would be toxic to nonruminants. A detoxification of the gossypol in cotton seed occurs in the rumen through combination of protein with gossypol in a bond which resists subsequent proteolysis (Reiser and Fu, 1962). An inactivation of digitalis during *in vitro* incubation with rumen contents has been observed (Westermarck, 1959). An important agent not metabolized in the rumen is 2,4-dichlorophenoxyacetic acid (2,4-D) (D. E. Clark *et al.*, 1964).

The possible formation of antibiotics in the rumen has been investigated, and some indications of antibiotics have been noted, but neither the materials concerned nor the causative organisms have been identified (Hoflund *et al.*, 1957).

I. Problem of Excess Energy

In general, nutrients are used most efficiently by living organisms when they are limiting, as in maintenance. The increased growth yield of bacteria in continuous cultures with sugar as limiting nutrient is an example (Hungate, 1963a; Hobson and Smith, 1963), as is the greater efficiency of feed in preventing loss of weight as compared to gain in sheep and cattle (Blaxter and Wainman, 1961).

The maintenance energy needs include a large fraction for performance of mechanical work, as in locomotion, circulation, and movement of digesta through the alimentary tract (Graham, 1964a). For this, acetate, propionate, or butyrate can be used. The proportion of the acids is less important, and the feeds are used efficiently (Blaxter and Wainman, 1961). In production, the proportion of the acids becomes more important, since propionate is the chief precursor of carbohydrates.

Some of the problems with ruminants in hot climates are concerned with the excess heat evolved as the level of production increases (de Alba and Sampaio, 1957; McClymont, 1952; Allen, 1962). In addition, if considerable quantities of acetate are formed, the heat increment is high. A lowered acetate : propionate ratio has been observed in heat-stressed cattle (Weldy et al., 1964). Excess heat decreases intake, with resultant decreased weight gains. Cold drinking water assists in accomodating the excess heat (Noffsinger et al., 1961). If nutritional practice can be used to increase the proportion of propionate formed, ruminant productivity in hot climates can be increased (Ittner et al., 1958). Also, in cooler climates, increased propionate ratios could conceivably reduce the amount of acetate to a value low enough to make the energy for growing ruminants limiting, in which case reduction of maintenance energy expenditure by eliminating grazing, by grinding and pelleting feed to reduce rumination and increase consumption, and by restricting animal location should increase productivity (Graham, 1964a).

J. Selection of Heritable Nutritional Traits

It is highly probable that hereditary factors are concerned with differences in turnover rate in individual animals. Genetic selection for high turnover rate might appreciably increase average productivity.

Some reports in the literature (Blaxter and Wainman, 1961; M. A. MacDonald, 1957; A. Robertson, 1963) indicate differences in utilization of energy by different individuals of the same ruminant species. In some instances these differences suggest a greater net gain from a given quantity of food. Differences in productivity between inbred cattle lines have been noted (Flower et al., 1963). Characteristics such as blood supply to the rumen, quantity of saliva secreted, chemical composition of the saliva, efficiency of mixing and of rumination, and undoubtedly other factors vary among individuals and could easily influence the kind and magnitude of microbial fermentation and synthesis in the rumen. Selection for combinations of characters promoting feed conversion could significantly improve ruminant productivity (Mather, 1959). Selection for high milk and butterfat production has provided marked improvements in these characteristics. Selection for rate of growth in beef breeds might be similarly fruitful.

Crosses between bovine species yield intermediates carrying carcass and other characteristics intermediate between both parent stocks, which indicates heritability (W. H. Black *et al.,* 1934; F. D. Carroll *et al.,* 1955) of characteristics of economic importance. Detailed analysis of nutritional characteristics could disclose valuable selectable traits.

Because of heritable nutritional differences between domestic and wild ruminant species, maximum productivity of grazing land may be achieved by the use of mixture of ruminant species rather than a single type. This is particularly important in marginal grazing lands. In some areas of Africa, it is estimated that harvest of the game would yield more food than would domestic ruminants. There is evidence that the food habits are not identical (McMahan, 1964).

K. Rumen Contents as a Tissue

For many purposes it is unnecessary to know the kinds and proportions of microorganisms in the rumen if interest centers on a particular activity (Doetsch, 1957). It may be feasible to study differences in total enzyme concentration, for example, without being concerned with the kinds and proportions of microbes synthesizing the enzymes. The isolation of β-glucosidase (Conchie, 1954), enzymes hydrolyzing lysolecithin (Dawson, 1959), demonstration of a system activating acetate to acetyl CoA (van Campen and Matrone, 1962), and studies of the succinic enzymes (Baldwin 1965) are examples of this type of approach.

L. Conclusion

Knowledge of the relationship between the ruminant and its symbiotic microorganisms has been an important factor in developing these ideas for increasing the productivity of ruminants. There are undoubtedly many additional possibilities for improvement of ruminant feeding practices. The suggestions in this chapter can only serve as an introduction to the problem and as a survey which may suggest additional profitable avenues for investigation.

CHAPTER XII

Abnormalities in the Rumen

Because of the dependence of ruminant nutrition on microbial fermentation, it is subject to malfunctions not present in other animals. Chief among these are bloat and acute indigestion. In addition, enterotoxemia, nitrate poisoning, ketosis, ammonia toxicity, and possibly hypomagnesemia are related to the activities of the microorganisms of the rumen.

A. Bloat

Bloat is characterized by a build-up of pressure in the rumen. There are two sorts of ruminant bloat, frothy bloat and free-gas bloat. In both, the animal is unable to eructate the fermentation gases, but for different reasons.

Various devices have been developed for measuring the degree of bloat, including instruments (tympanometers, Cole and Kleiber, 1945) to record the pressure against the body wall, changes in girth (C. S. W. Reid, 1957b), and intraruminal pressure by means of balloons (Andersson et al., 1958) or by telemetering of pressures from small transmitters placed in the rumen (Dracy and Essler, 1964; Payne, 1960; Wallace et al., 1959).

By direct measurement in the just-succumbed animal, Reiset (1868b) found a rumen gas pressure of 63 mm Hg. Subsequent studies have corroborated this approximate value, but have demonstrated variability among animals in the effect of increased pressures (Dougherty, 1955) and have shown that the type of gas used artifically to raise rumen pressures influences the magnitude of the pressure withstood, higher pressures with oxygen being less effective than with carbon dioxide. Experimental carbon dioxide pressures of 60 mm Hg can be fatal to sheep (Dougherty et al., 1955). These pressures, although insufficient to cut off arterial blood (which has a systolic pressure of well above 100 mm Hg), are sufficient to close the large veins returning blood to the heart and thus cause acute circulatory impairment.

1. FREE-GAS BLOAT

An hereditary bloat in certain types of dwarf Hereford cattle (Hafez et al., 1959) is due to a physiological defect which prevents eructation. This statement is based on a single experience of the author with a Hereford dwarf bloater at Pullman, Washington. It consumed very little food, yet the ribs were distended and malformed by gas pressure in the rumen. The animal was examined to see if it appeared to contain a normal rumen flora and fauna. A stomach tube was inserted, but difficulty was experienced in forcing the tube past the cardia. When this was accomplished by applying considerable force, there was a rush of gas out of the tube, and the distended sides of the animal fell to a nearly normal position. No rumen contents could be obtained by applying suction to the stomach tube. A few liters of water were pumped in, and after about 5 minutes a second attempt to obtain rumen contents was successful. A small quantity of liquid and solid digesta was pumped out and examined microscopically. Protozoa characteristic of normal animals were found, which indicated that the microbiota was normal.

Subsequent studies on similar animals (Fletcher and Hafez, 1960) indicated that the microbial population was more concentrated than in controls, as evidenced by a more rapid fermentation with added substrate. The digesta appeared to have undergone more complete decomposition than in the controls, in keeping with the more concentrated microbial population. The microbes apparently were not diluted by salivary flow, which was therefore presumably less than normal. Because of the inability to eructate, the animals bloated as food was eaten, due to the fermentation gases formed. The resulting distention inhibited appetite (fortunately).

This type of bloat may be caused by any factor interfering with eructation. The author has observed another case of free-gas bloat. At Pullman, the Animal Husbandry Department was feeding a group of steers a ration high in grain when one animal started to bloat and appeared in danger of collapse. The attendant inserted a stomach tube. After the tube was moved about in the rumen, a large volume of gas suddenly escaped, and the distension receded to normal.

A sample of rumen contents was taken and tested for acidity. The pH was 4.2. In order to be sure that acidity was the factor responsible for the bloat, stomach tubes were inserted in two other steers on the same ration. The pH's of the rumen contents were 4.1 and 4.3, respectively. There was thus no conclusive evidence that the bloat was due only to the high acidity. In spite of this, the author is inclined to ascribe this case of bloat to high acidity coupled with greater susceptibility of the animal, or with some unknown difference in microbial activity. The high acidities (pH 4.0–4.5) in sheep, which characterize acute indigestion due to overfeeding on grain (Hungate *et al.*, 1952), were not commonly associated with bloat, although a tendency to bloat was observed in one instance.

Many cases of free-gas bloat may be due to injury, abnormal growths (J. A. Benson, 1957), or other causes, but, to the best of the author's knowledge, high acidity is the only known mechanism by which rumen microbial activity can cause an abnormal accumulation of free gas in the rumen.

2. Frothy Bloat

Frothy bloat results when the fermentation gases, instead of rising in the digesta to coalesce with the pocket of free gas in the top of the rumen, form small bubbles which increase in size within the mass of digesta (J. I. Quin, 1943b). The situation is comparable to that in rising bread. The gas bubbles are held in the solid matrix, and the entire mass increases in size. A specific gravity of 0.46 has been reported (W. E. Thomas, 1956) for digesta in bloated cattle. Various methods of measuring the froth-producing capacity, froth strength, and stability of rumen digesta have been developed (Mangan, 1958).

The eructation mechanism of the host is not disturbed during the initial stages of frothy bloat. The animal eructates the gas initially collected in the top of the rumen (Hancock, 1954) and can void gas much faster than it forms (Cole *et al.*, 1942), if it collects as free-gas, but as the expanding mass of digesta increases in volume, the animal is finally unable to clear the cardia of solids. The receptor system sensitive to contact of digesta, described in Chapter IV, prevents the cardia and the pharyngo-esophageal sphincter from opening, and the digesta cannot escape (Dougherty and Habel, 1955). A small gas pocket may still remain in the top of the rumen, but it cannot be maneuvered into a position near the cardia and eructated. Impairment of eructation in the later stages of frothy bloat includes failure of the second or eructation rumen contraction to occur (E. I. Williams, 1955a).

In severe cases the rumen distension continues until the pressure cuts off the circulation and the animal collapses and dies within a minute or two unless steps are taken to relieve the pressure (D. M. Walker, 1960). In acute cases the body wall can be perforated, which permits a gush of digesta to escape. Upon this release of pressure the animal almost immediately recovers from the acute symptoms and in many cases recovers from the peritonitis attendant on the invasion of the peritoneal cavity by rumen and other microorganisms when the rumen and body wall are punctured.

Many types of fresh lush legumes may cause frothy bloat, and it is also observed in some feed-lot animals fed rations containing a high proportion of grain.

a. *Legume Bloat*

Alhough the formation of a froth which does not rise in the rumen was postulated by J. I. Quin (1943a,b) as the cause of legume bloat and was substantiated by Clark (1948), the hypothesis was not immediately accepted. Later, through the efforts of veterinary supply houses and practicing veterinarians, and through substantiation by experiments in several parts of the world (Clark, 1950; Weiss, 1953a; Hungate *et al.,* 1955a; Johns, 1954), the importance of a stable foam in legume bloat became well established. The causes underlying the froth formation are still not completely understood.

(1) *Multiplicity of Factors in Legume Bloat.* A great many factors have been considered as causes of legume bloat (Johns, 1954; Cole *et al.,* 1945, 1956). As the name implies, an essential factor is the consumption of fresh legumes or a few other fresh lush feeds. However, although these must be consumed if the bloat is to occur, it does not necessarily occur. This makes experimental investigations difficult. It cannot be stated that legumes cause bloat; in a pasture consisting exclusively of legumes, some animals bloat and others do not. Of the animals which bloat on one day, some may not bloat on the day following, and a few new bloaters may be found. Over a long period of observation of a large group of cattle, a few animals can be picked which bloat regularly, and others which bloat rarely (Barrentine *et al.,* 1954; Mendel and Boda, 1961; Johns, 1954).

These observations focus attention on animal factors and on changed conditions within an animal at different times; these conditions make the animal liable to bloat during some periods and resistant at others. This feature increases the difficulty in pinning down the underlying causes. Whatever factor is postulated (and a tremendous number have been postulated) it must be shown to operate in the bloating animal and not in the nonbloater in the same pasture at the same time. Many good explanations have failed on this particular aspect of the proof.

Differences in microorganisms have been invoked as the cause of bloat. The following activities have been suggested as critical: increased rate of gas production, formation of toxins, production of slime (Hungate *et al.,* 1955a; Gutierrez *et al.,* 1958), failure to destroy saponin, release of nucleic acids (Gutierrez *et al.,* 1961), fermentation of pectin (Conrad *et al.,* 1958c), microbial phosphatase (Kolb, 1957), destruction of chloroplast lipids (Mangan, 1959), destruction of mucin (Fina *et al.,* 1961a), and bursting of holotrich protozoa (Clarke, 1965b).

If differences in microorganisms are a factor in bloat, the differences are not sufficient to change significantly the ratios of fermentation products (Hungate *et al.,* 1955a; Reiset, 1868b), and isolation studies (Bryant *et al.,* 1960) have not disclosed clear cut bacterial differences. Bloat can be caused by administering alfalfa juice (Ferguson and Terry, 1955) or egg white (Boda *et al.,* 1957).

There is little evidence that excess production of gas is the main factor in bloat (J. I. Quin, 1943b; Hungate *et al.,* 1955a). Much more gas is produced from ingested alfalfa, 17.8 liters per kilogram in 4 hours after feeding as compared to 6.25 liters from Sudan grass (Cole and Kleiber, 1948), but this larger quantity could easily be eructated if the gas could coalesce. Similarly, there is little evidence that excess acid production on legumes is a factor in legume bloat, although it has been considered (Shinozaki, 1959). The legumes contain so much protein and polyuronides, with their alkalinizing effect during fermentation, that microbial conversions would not tend to form excess acids from this feed.

(2) *Lipid Foam Breaking Theory.* This hypothesis has been developed by the group in the Plant Chemistry Division at Palmerston North, New Zealand, under the direction of A. T. Johns. It was noted that the stability of the froth was less in the presence of the alfalfa chloroplasts (Mangan, 1959). The chloroplasts contain lipids, galactosyl glycerol esters of linolenic acid (Weenink, 1959, 1962; A. A. Benson *et al.,* 1959). When the lipids were removed from the chloroplasts, the latter were no longer effective in breaking the froth. Hydrogenation of the lipid by bacteria and protozoa of the rumen was demonstrated. The tendency of the lipid to break the froth diminished as the lipid was hydrogenated. Both the bacteria (D. E. Wright, 1960a) and protozoa (D. E. Wright, 1959) of the rumen were shown to be capable of hydrogenating the lipids. Examination of the microorganisms in the rumen of nonbloating animals disclosed that lipids were hydrogenated about as effectively as in the bloating animals, which makes necessary the postulation of some additional factor to explain the absence of bloat. The rate of release of galactolipids from forage varied, but did not correlate with bloat (R. W. Bailey, 1964).

(3) *Microbial Slime.* The fact that the rumen microbes are concerned in some fashion with the production of the stable froth of legume bloat is shown by the effectiveness of oral penicillin in preventing the disease.

The hypothesis of microbial slime production was formulated because of the observation that certain butyrivibrios, cultured on sugar, produced so much slime that the medium gelled (Hungate *et al.,* 1955a). Butyrivibrios are abundant in the rumen of cattle fed on fresh alfalfa (Gutierrez *et al.,* 1958, 1959b), and an increase in slime has been demonstrated (Gutierrez *et al.,* 1963), but no differences in slime or butyrivibrios (Bryant *et al.,* 1960) in bloating and nonbloating individuals have been found. The slime formed from alfalfa saponins by *Butyrivibrio* strains was different from the slime in animals bloating on a high grain ration (Gutierrez and Davis, 1962b).

Another microbial hypothesis (Clarke, 1965a,b) has been developed on the basis of observations on numbers of holotrich protozoa during bloat. These show unexpected decreases in concentration, explained by bursting of the holotrichs due to excess starch storage from the abundant soluble carbohydrates in legumes. The bursting releases nucleic acids which greatly increase the viscosity. The hypothesis is attractive, particularly since nucleic acids might persist in the rumen, as does the froth-promoting factor, during the extended period between feedings. Earlier observations of Koffman (1937) indicated fewer protozoa in the rumen of bloated as compared to normal sheep.

The rate of gas production by digesta from bloating animals was found to be slightly greater than in the nonbloaters, all consuming ladino clover pasture (Hungate *et al.,* 1955a). The difference in gas production between bloaters and nonbloaters is not enough to be important, if the gas bubbles coalesce with the large gas phase in the dorsal rumen. The indication of a faster rate in the bloaters is perhaps significant in suggesting that there is a greater turnover rate in the bloating than in the nonbloating individuals. A production of gas is essential for the expression of bloat, but the gas *per se,* unless the bubbles are stabilized as a froth, rises through the rumen contents and can be eructated.

There is no evidence that the rumen fermentation is abnormal in bloating animals. Reiset (1868b) collected rumen gas at the instant of death from a cow which had started to bloat 2 hours earlier and found 74.33% carbon dioxide, 23.46% methane, 2.21% nitrogen, and no oxygen or hydrogen sulfide. In a bloated sheep he found 76% carbon dioxide. These values are well within the normal range (Kleiber *et al.,* 1943). The fermentation products in bloating and nonbloating steers on a ladino clover pasture (Hungate *et al.,* 1955a) were identical in the percentage of acid, carbon dioxide, and methane, even though the total quantities differed.

There have been so many proposals to explain frothy bloat, each with some experimental support, that uncertainty will exist until a number of different laboratories in different parts of the world confirm a particular hypothesis. Unfortunately, experiments on bloat are expensive because of the large number of animals needed to test the many possible factors. It is difficult to obtain a bloat-provoking pasture and to maintain it for a considerable period in which experiments can be performed.

(4) *Host Factors.* Explanation of bloat on the basis of microbial differences does not rule out host influences. In fact, it is difficult to conceive of maintenance of divergent microbial populations in ruminants in close contact with each other, unless the two hosts differ in some respect favoring the microbial difference. Importance of microbes is consistent with an importance of host features, including some that may be heritable.

Identical twin cows show marked similarities in bloating behavior (Johns, 1954), and susceptibility to bloat differs according to sire (Knapp *et al.*, 1943; Hancock, 1954).

Host features of possible import in bloat include (Mendel and Boda, 1961) rumen motility and the eructation reflex, rate of eating, and salivary composition and flow. Of these host features, saliva production—both kind and amount—and turnover rate have been implicated in the genesis of frothy bloat (Weiss, 1953a). Bloaters secrete less saliva, with a greater bicarbonate concentration, than do nonbloaters (Mendel and Boda, 1961). Correlated with this is the fact that the digesta of bloaters contained a greater concentration of solids. No differences in rumen contractions, water consumption, or pH of rumen contents were observed. The increased salivary flow in nonsusceptible animals reduced the concentration of the froth stabilizing material. Other workers have also reported that saliva reduced the frothiness of digesta (Van Horn and Bartley, 1961), and mucin has been cited as the effective component (Bartley, 1957). In other studies (R. M. Meyer *et al.*, 1964), variability in saliva production masked any correlation with bloat susceptibility.

The frothing tendency in the contents of animals susceptible to bloat is just as great before the animals are turned out to pasture in the morning as just after they have fed (Hungate *et al.*, 1955a). Whatever the factor concerned with froth stabilization, it is not destroyed by the rumen microorganisms during the 16 hours between feedings.

The appearance of the rumen digesta in a bloating steer is quite different from that in a nonbloating animal on the same pasture. Casual inspection shows the normal digesta to be darker in color and apparently less digested than in the bloater, and the liquid separates from the solids more readily. If normal digesta is vigorously shaken it develops a distinct top layer of foam which gradually subsides.

Digesta from a bloating animal shows a lighter green color, tinged with more yellow. The recently ingested forage has become smeared with digesta and appears to be more changed from the fresh material than in the nonbloating animal. The digesta from the bloater may actually *appear* to be less foamy. No large gas bubbles are evident, and no surface layer of froth appears on shaking. However close inspection shows myriads of small gas bubbles trapped in the digesta, too small to rise to the top through the viscous mass. They gradually enlarge, and new ones are formed as fermentation proceeds.

The difference in appearance of the digesta in the rumen of bloating and nonbloating animals is due in part to the physical effect of the gas bubbles within the solids; the bubbles give it the lighter color. Chemical analyses show (Bartley and Bassette, 1961) a high percentage of lipid, carotenoid pigments, and nucleic acid protein in the material surrounding the bubbles. The thicker layer of viscous digesta smeared on the surface of recently ingested forage gives the impression of a more rapid rate of attack in bloaters, which is consistent with the greater fermentation rate.

The rumen contents of cattle bloating on clover pasture showed more ethanol-precipitable slime as the clover was ingested (Gutierrez *et al.,* 1963). The slime contained 61–64% protein, 8–14% carbohydrate, and 7–10% ribonucleic acid.

Toxins do not appear to be an important factor in legume bloat. A bloating steer in the experimental herd at Mississippi State University was able repeatedly to eructate large quantities of gas (Hungate *et al.,* 1955a), and several investigators have observed frequent eructation during the early stages. The important factor is not the failure of the normal eructation mechanism, but the superposition of the inhibitory mechanism preventing escape of solid digesta.

In view of the common capacity of mammals to vomit, it is rather peculiar that this mechanism is lacking (Brugnone, 1809) as an effective means of voiding excess solids digesta during legume bloat. Vomiting does occasionally occur in ruminants (Clark, 1956) and vomition and rumination centers in the medulla have been found (Andersson *et al.,* 1959). Vomition can be induced with drugs (Dougherty *et al.,* 1965), but is not invoked during legume bloat. The underlying reasons constitute an intriguing problem in physiology.

Although the word "foamy" is often used to describe the gas bubbles in rumen contents of bloating animals, "frothy" is the more accurate term. The surface layer enclosing a gas bubble in a foam is elastic, and the bubble can increase or shrink in size, the bounding layer expanding or shrinking according to the resultant force from gas pressure and the surface tension. The layer bounding a bubble of froth, on the contrary, is not elastic. It breaks if the gas pressure exceeds the strength of the film, unless new film forms, and wrinkles and collapses if the gas pressure falls. This is observed in the froth from rumen contents of steers bloating on legumes.

Froth formation can be visualized as the appearance of a minute center of gas which rapidly increases in size because of diffusion of gas from the liquid. For the bubble to have formed, the pressure of the gas in solution must have exceeded that in the newly formed bubble, and there will therefore be an initially high diffusion gradient. The materials in the gas-liquid interface reorient, and new bonds create a nonelastic structure. Growth of the bubble may occur during the period before new bonds have completely formed, and perhaps by addition of new surface after the nonelastic surface structure has formed. However, as mentioned in the section on manometry, nuclei for gas bubbles readily form in rumen contents. Instead of a few large gas bubbles forming from a few nuclei, a great many form, which makes the bubbles very small and less likely to coalesce and rise against the restraining viscosity and rigidity of the digesta and the gas-liquid interfacial layers.

The interfacial layer of material surrounding a gas bubble formed in the digesta of a bloating animal is thick compared to that in a nonbloater. This layer was caught as a film on the fine wire grid used in electron microscopy, and when dried withstood the force of the electron beam. No details of ultrastructure could be seen.

So many constituents of digesta are suited to form stable films that detection of those actually concerned is difficult. Proteins, lipids, bacterial slimes, saponins, mucin, and nucleic acids form viscous solutions under certain conditions. Bacteria are caught in the film and increase the difficulty in detecting the important ingredients in the film itself. The proteins would seem to be the most suited for stable froth production. Denaturation and precipitation could give an inelastic film. A protein with a sedimentation constant of 18 S, isolated from alfalfa leaves, is reported to be responsible for the froth stabilization characteristic of legume bloat (J. M. McArthur *et al.*, 1964), and in one investigation the nitrogen content of alfalfa correlated with the degree of bloat (Miltimore *et al.*, 1964b). Enhancement of bloat severity by Ca^{++} and Mg^{++} (K. J. Smith and Woods, 1962) is consistent with formation of an inelastic froth. Actual evidence is much needed to disclose the role played by various constituents in froth formation during bloat.

(5) *Toxins.* The possibility that toxins might be concerned in acute cases of frothy bloat has been considered (Dougherty, 1940, 1941). The large quantities of easily hydrolyzed cyanogens in some legumes make cyanide a potential danger (Clark and Quin, 1945; Clark, 1951; Dougherty and Christensen, 1953); in the case of lotaustralin (Blakely and Coop, 1949; Coop and Blakely, 1949, 1950) and birds-foot trefoil (Parsons *et al.*, 1952) it may be lethal. The hydrocyanic acid in clover juice has been shown to be toxic to smooth muscle (W. C. Evans and E. T. R. Evans, 1949), but comparisons of various legumes for their cyanide content has shown little correlation with the tendency to cause bloat (Dougherty, 1956a). Carbon monoxide (Dougherty, 1940, 1941) to the extent of 0.2% has been reported to occur in the rumen gas of cattle fed ladino clover.

Histamine (Dougherty, 1942b) has been considered a possible factor in bloat, and occurrence of histamine and histamine-like materials (Shinozaki, 1957) in rumen contents indicates that decarboxylation of histidine occurs in the rumen. However, the histamine in rumen contents of ruminants bloating on legumes has not been shown to be greater than in those not bloating. Histamine is not rapidly absorbed (Shinozaki, 1957).

The rapidity with which recovery from bloat occurs after the pressure is released (D. M. Walker, 1960) is not compatible with a toxin as the cause of the acute symptoms. Indirect actions of toxins cannot be ex-cluded as contributory causes of bloat (as with saponins), but there is no positive evidence that they are directly concerned.

Hydrogen sulfide is a constituent of rumen gas (Dougherty, 1941, 1942a; J. M. McArthur and Miltimore, 1961), and the quantity increases when legumes are fed, due to the greater quantities of sulfur (Dougherty, 1941). Since it can cause paralysis of the rumen, it has been considered a factor of possible importance in legume bloat, but apparently the quan-tities absorbed do not exceed the capacity of the ruminant to oxidize them, and hydrogen sulfide does not reach a toxic concentration in the blood, except when eructation introduces large concentrations into the lungs.

A difference in the effects of legume juice and grass juice on isolated smooth muscle of the rabbit gut has been noted (Ferguson, 1948); the legume juice diminished the frequency and amplitude of the movements, whereas the grass juice usually stimulated them. A flavone was isolated from the legumes (Ferguson *et al.*, 1949, 1950), but no role in bloat was demonstrated.

A factor in alfalfa is toxic to rat diaphragm muscle (Jackson *et al.*, 1959). Digesta from a cow dying of bloat was more inhibitory of motility in isolated rabbit intestine than were several clovers, with the exception of red clover (Parsons *et al.*, 1955). The relation of these effects to legume bloat is unknown.

(6) *Forage Factors.* The salient feature of frothy bloat is its depend-ence on ingestion of fresh legumes or a few other plants such as crucifers and young rye grass (Johns, 1958). The well-known capacity for saponins to stabilize foams, and the abundance of saponins in many legumes, has stimulated much interest in a possible role of saponins in bloat (Lindahl *et al.*, 1954). It is evident that if saponins alone were responsi-ble there should be no nonbloaters in a herd pastured on legumes. Since nonbloaters commonly occur, it must be concluded that, although sapo-nin may play a part in bloat and may even be essential for its appearance, there is an additional factor or factors required for expression of the phenomenon. Froth due to saponins is most stable at pH 5.0, a more acid reaction than that of the rumen digesta in legume-bloated animals (Mangan, 1959). Mucogenic strains of *Butyrivibrio* which attack sap-onins have been found in the rumen of steers fed fresh alfalfa (Gutier-rez *et al.*, 1958, 1959b).

Pectins (Conrad *et al.*, 1961) and readily digested hemicelluloses (Con-rad *et al.*, 1958c; Head, 1959a) have also been suggested as important in legume bloat. The increased viscosity of digesta in legume bloat has been ascribed to activity of pectin methylesterase (J. Gupta and Nichols, 1962).

The high sugar content in fresh legumes (Healy and Nutter, 1915; Henrici, 1949, 1952) and the nonvolatile acids (Nichols and Penn, 1959) have been proposed as substrates responsible for foam stability. Both are undoubtedly concerned with gas production, but an effect on foam stability is dubious except as microbial activity might produce slime.

Dried legume hay almost never causes frothy bloat when fed alone. The intraruminal pressure of animals consuming alfalfa hay was 5 mm Hg, whereas with the same alfalfa in a fresh state the pressure was 27 mm Hg (Boda, 1958; Boda and Johns, 1962). Dry matter consumption was similar, but the rate of gas production 90–200 minutes after ingestion was greater for the fresh alfalfa.

Denaturation of the proteins appears to be the most likely effect of drying, which suggests that denaturation of protein at the bubble surface in bloating animals contributes stability. Addition of egg white to the rumen increases the degree of bloat (Boda et al., 1957), and protein has been identified as an important component of the alcohol-insoluble material in the rumen of bloating cattle (Bartley and Bassette, 1961). Froth formed from protein is most stable at pH 6.0, near the usual rumen pH.

The unusually high efficiencies with which the nitrogen in fresh legumes can be utilized by dairy cows (Conrad et al., 1960a) may be due to a decreased rumen digestion of the protein. If so, the protein may remain in the rumen over a long period, a necessary feature in explaining the persistence of frothiness in the digesta (Hungate et al., 1955a).

(7) Control. Knowledge of control methods is more satisfactory than knowledge of the fundamental causes of legume bloat. Substances with froth-breaking characteristics (Clark, 1948, 1950; A. H. Quin et al., 1949) promote coalescence of the small gas bubbles into the large bubbles which separate from the solids to produce a continuous gas phase which can be eliminated by eructation.

If symptoms of bloat are noted soon enough, the acute attack can often be alleviated by treatment. The specific treatments employed are beyond the scope of this volume, but in general they are designed (1) to break the foam within the rumen and assist coalescence of the small bubbles or (2) to diminish the production of gas. The speed with which bloat sometimes develops, 30–90 minutes after animals start to graze a legume pasture, may make it difficult to catch the disease in time for these treatments to be effective.

Once severe symptoms of bloat have developed, foam breakers are less effective than if applied prior to the onset of bloat, due to the difficulty in rapidly mixing the administered material with the rumen contents (Clark,

1950). During the onset of bloat, there is an increased activity of the rumen musculature, but when distention has increased to the point of distress there is little mixing of the digesta. Injection of oil or other foam-breaking material through the rumen wall (L. R. Brown *et al.,* 1958) or administration by stomach tube may be effective (R. H. Johnson *et al.,* 1958, 1960a).

Methyl silicone prevented bloat when administered in large doses to steers before grazing them on ladino clover pasture (Barrentine *et al.,* 1954). Various oils have been used, alkyl-aryl sodium sulfonate (Blake *et al.,* 1957b), and polyoxypropylene polyoxyethylene (Bartley, 1965). One of the first demonstrations of foam-breaking was with turpentine and coal-tar reagents (Clark, 1948, 1950), long used by veterinarians to treat legume-bloated animals. Mucin in linseed meal has been reported to diminish frothing (Van Horn and Bartley, 1961). There is variability in the effectiveness of some foam-breaking materials, presumably because of variability in the animals and rations (Miltimore *et al.,* 1964a).

Prevention of legume bloat by spraying pastures with oils has been demonstrated by New Zealand investigators to be a practicable method of control (Johns, 1954; C. S. W. Reid, 1958, 1959; Reid and Johns, 1957; Hutchings, 1961). The quantities of oils consumed do not affect milk production (McDowall *et al.,* 1957c). Vegetable oils increased the unsaturated fatty acids in the butter fat, and some of them changed the odor of the butter, but total butter fat and total nonfat milk solids were unaffected. Light oils reduced the intake of the treated member of monozygotic twins (C. S. W. Reid, 1957a); only heavy paraffin oil showed no effect on appetite. Paraffin oil decreased the absorption of carotene and vitamin E (C. S. W. Reid, 1959).

Administration of penicillin can prevent legume bloat (Barrentine *et al.,* 1956; Johns *et al.,* 1959; Mangan *et al.,* 1959). The effect ultimately wears off (Shawver and Williams, 1960; Schellenberger *et al.,* 1964). The decreased effectiveness after several weeks of administration suggests a selection of resistant strains or development of bacteria which attack penicillin (Wiseman *et al.,* 1960). Rotation of the antibiotics used can prolong the effectiveness (R. H. Johnson *et al.,* 1960b; Van Horn *et al.,* 1961), or several can be simultaneously administered (Van Horn *et al.,* 1963).

b. *Feed-Lot Bloat*

Factors concerned in feed-lot bloat are less known than those in legume bloat, chiefly because until recently it has not been an important economic problem. With increased intensive feeding of rations high in grain, the problem is more common. Study has been facilitated by the development in Michigan of an experimental ration conducive to feed-lot bloat (C. K. Smith *et al.,* 1953); this ration consists of 61% barley, 16% soybean oil meal, 22% alfalfa meal, and 1% NaCl. Other laboratories (Lindahl *et al.,* 1957) have obtained comparable bloat with the ration.

A statement in one publication (D. R. Jacobson and Lindahl, 1955) suggests that feed-lot bloat may sometimes be of a free-gas type: "In taking rumen samples of animals with frothy bloat, gas repeatedly escaped when the curvature of the stomach tube was up during placement of the tube into the rumen." It is not stated whether the distention of the animal subsided with the escape of this gas, but, if so, the phenomenon in those particular animals could be considered as an example of the free-gas type. Few data are available on the acidity in the rumen contents of bloating feed-lot animals, but with the high starch ration high acidities occasionally develop (R. Jensen *et al.,* 1954a). Some of the animals studied by Bryant *et al.* (1961) showed a pH as low as 5.0 and contained few or no protozoa. In other experiments (Gutierrez *et al.,* 1959a; Elam *et al.,* 1960), the pH of the rumen was not unusually acidic, since numerous entodinia occurred. These observations suggest that in feed-lot bloat the free-gas or the frothy type may be concerned.

Encapsulated bacteria become more abundant in the rumen of bloating animals (D. R. Jacobson *et al.,* 1957). An ethanol-precipitable slime can be found in the supernate from centrifuged digesta. The amount correlates to some extent with the degree of bloat. The slime contained 34% crude protein, acid-hydrolyzable polysaccharides, and nucleic acids (Gutierrez *et al.,* 1961). Decomposition of mucin by rumen bacteria has been postulated as a factor preventing foam stability (Fina *et al.,* 1961b), but no conclusive evidence has been reported.

Streptococcus bovis and *Peptostreptococcus elsdenii* have been implicated as agents responsible for feed-lot bloat (Gutierrez *et al.,* 1959a). *Peptostreptococcus elsdenii* disappears following penicillin treatment (Bryant *et al.,* 1961). After long adaptation to the ration the high propionate : acetate ratio returns to normal (D. R. Jacobson and Lindahl, 1955) and *Streptococcus bovis* is not particularly abundant (Bryant *et al.,* 1961).

In feed-lot bloated animals the intensity of bloating was increased by addition of 4–8% soya bean oil (Elam *et al.,* 1960). The saliva of bloating and nonbloating cattle is not sufficiently different to implicate salivary differences as important factors in the etiology of feed-lot bloat (Emery *et al.,* 1960).

B. Acute Indigestion

Ruminants not accustomed to a grain diet often suffer acute digestive disturbances, and in many cases death, within 24 hours after consumption of a large quantity of grain (Reiset, 1863b; M. J. Masson, 1951;

Hungate *et al.,* 1952; F. M. Gilchrist and Clark, 1957). Rumen atony and loss of appetite are symptoms of the disease. The rumen atony as well as death is also associated with overeating of mangolds (Scarisbrick, 1954; Penny, 1954), fruits (Irwin, 1956; Portway, 1957; Merrill, 1952) brewers dried grains (Owens, 1959), and kiawe beans (H. Adler, 1949), and even when semistarved animals gain access to lush feeds (Coop, 1949). In most cases there is a sudden bloom of *Streptococcus bovis* in the rumen and accompanying production of lactic acid (Hungate *et al.,* 1952; Krogh, 1959, 1960, 1961a,b; P. K. Briggs *et al.,* 1957; Ryan, 1964a). Acidity increases to a pH of 4.0–4.5 and in some cases is high for a week to 10 days. The acidity inhibits muscular activity of the rumen wall (Ash, 1956), and, if sufficient carbohydrate is available in the rumen, death usually results. The volatile fatty acids (VFA's) disappear from the rumen within 4 to 24 hours (Ryan, 1964a). If the sheep survive more than 4 days, succinic acid begins to appear in fairly high concentrations (Ryan, 1964a). The glucose concentration in the rumen may be high during the early periods of exposure to excess grain (Ryan, 1964a). During gradual adaptation to grain, glucose and succinic acid accumulate when the pH falls to ca. 6.0 (Ryan, 1964b).

The lactic acid in the rumen is almost all L-(+) prior to feeding grain (Ryan, 1964b; Turner and Hodgetts, 1949–1959), and changes to an approximately equal concentration of the D- and L-forms during adaptation to the high grain ration. The concentration of rumen lactate does not change appreciably during gradual adaptation (Ryan, 1964b) to a grain diet.

In some experiments (Ash, 1956) the VFA's inhibited rumen motility more than did citric, phosphoric, and lactic acids, presumably because of greater permeability of the rumen wall. The undissociated VFA's may be the actual acids inhibiting motility in acute indigestion when lactic acid or other strong acid is administered. Induction of low pH through addition of the VFA's not only lowers the pH, but also increases the total amounts of undissociated VFA's and is therefore more effective than added strong acids alone. Formic acid added to the rumen is very effective in decreasing rumen motility (Shinozaki, 1959). Butyric is less effective, followed by propionic, with acetic acid relatively ineffective. These were all compared at identical molar concentrations; thus, the percentage increase over the usual rumen concentration was least for acetic acid and greater for propionic, butyric, and formic acid, since their normal rumen concentrations decrease in that order. The effect of acids is directly on the rumen musculature and not on the central nervous system (Ash, 1956).

The numbers of *Streptococcus bovis* diminish almost immediately after the maximum concentration of several billion per milliliter is reached, and lactobacilli predominate (Hungate *et al.,* 1952; Krogh, 1959,

1963a,b). Numbers of lactobacilli in the rumen of animals on high concentrate diet are greater than in those fed grass (C. A. E. Briggs, 1955). This agrees with the ability of lactobacilli to grow in media too acid for streptococci. Abundance of lactobacilli in the rumen has in some instances been interpreted as implicating lactobacilli in the development of the acidity, but the examination may be made too late to catch the burst of *S. bovis* (Perry *et al.*, 1955). *Streptococcus bovis* occurs in larger numbers in hay-fed ruminants than do the lactobacilli and has an extremely rapid potential growth rate. These factors permit it in most cases to outgrow the lactobacilli initially, but *S. bovis* is inhibited at the high acidities it causes, whereas the lactobacilli are not. Lactobacilli are quite numerous in animals with chronic high rumen acidity. In some cases (Hungate *et al.*, 1952; Krogh, 1961a), *S. bovis* does not appear to be the chief agent causing the initial high acidity. Other factors which may explain the failure of *S. bovis* to maintain high numbers in animals fed rations high in starch have been discussed in Chapter II.

Ulcerative rumen lesions and liver abscesses occur in cattle fed high grain rations (H. G. Smith, 1944; T. J. Robinson *et al.*, 1951); the occurrence correlates with the proportion of grain fed (R. Jensen *et al.*, 1954a). The speed of the shift from forage to grain influences the severity of the symptoms (R. Jensen *et al.*, 1954b), a change in 12 days causing significantly more damage than adaptation over 30 days. Dosing with $CaCO_3$ was ineffective in reducing the damage. Increased rumen parakeratosis was noted in animals receiving finely ground feed (Garrett *et al.*, 1961).

Dosing with small quantities of alkali does not affect the development of acute indigestion due to excess of readily fermentable substrate. The alkali assists in neutralizing the acid and permits extensive growth of *S. bovis*. When the neutralizing capacity of the alkali is exceeded, the greater *S. bovis* population drops the acidity even faster than with no added alkali, provided feed is still in excess. On the other hand, if the buffer is sufficient to take care of the fermentation products from the total quantity of substrate in the rumen, the symptoms might be prevented. The resulting high concentration of cation could cause a damaging alkalosis. The effect of dosing with large quantities of a buffer such as $CaCO_3$ has not been tested.

Feeding 3% bicarbonate in the ration can increase the acetate: propionate ratio and maintain the milk-fat content when high grain rations are fed (C. L. Davis *et al.*, 1964; Emery *et al.*, 1964b). If the high acidities brought on by high grain rations are the cause, damage and deaths of steers in feed lots can probably be reduced by extending the feeding of the ration over a longer period.

Histamine and tyramine concentrations increase in the sheep rumen during development of acidity following high grain feeding (Dain *et al.*, 1955), and lactobacilli decarboxylating histidine have been isolated (Rodwell, 1953). This accords with the known more rapid decarboxylation of amino acids at pH's between 4 and 5 than at neutrality. Some killing of cells as acidity increases, with release of histidine, may contribute to the quantities of histamine released. Direct effects of introducing histamine into the rumen, at high acidities, have not been examined. Histamine and methylamine have been found in the rumen of sheep and goats fed on ladino clover (Shinozaki, 1957) and in sheep at 6 hours after administration of 300 gm of glucose (Sanford, 1963). Added histamine was not rapidly absorbed. Histamine has been implicated also in horses foundered on grain (Akerblom, 1934).

The feasibility of removing toxins in the rumen by means of lavage with a stomach tube has been demonstrated (Pounden, 1954). If followed by heavy dosing with rumen liquid from a normal animal, plus some ground feed, this procedure should lead to recovery of appetite unless the rumen wall has been too badly damaged. Thiamine has been reported to be effective in alleviating the symptoms of acute indigestion (Broberg, 1960).

The high acidity following overeating on grain is accompanied by accumulation of ethanol (Allison *et al.,* 1964a). It is doubtful that the alcohol itself is responsible for the symptoms, since none are observed in cattle fed 170 gm of ethanol per 100 kg per day in fermented apple pomace (Leroy and Zelter, 1955).

C. Nitrate Poisoning

Nitrate or nitrite can serve as a source of nitrogen for assimilatory reactions of rumen bacteria after being reduced to ammonia. With some rations, the rumen nitrogen in the form of nitrate may exceed the assimilatory capacity of the microbial population. The excess is reduced to nitrite and ammonia. Less than 5% of administered high doses of nitrate is excreted as such in the urine (D. Lewis, 1951a).

The rate of nitrate reduction to nitrite exceeds the rate at which nitrite is further reduced, and nitrite accumulates when the feed or water (J. B. Campbell *et al.,* 1954) contains more than a few tenths of a percent of nitrate. Grazing sheep contained 0 to 15.7 mg of nitrate nitrogen per 100 ml of rumen contents (Jamieson, 1959a).

Nitrite enters the blood and reacts with hemoglobin to form methemoglobin, which cannot transport oxygen. In cattle killed by nitrate, 50–60% of the blood hemoglobin is in the form of methemoglobin (Dodd and

Coup, 1957). Two grams of KNO_3 in a sheep caused 60% conversion of the hemoglobin into methemoglobin (D. Lewis, 1951a), although in other experiments (Jamieson, 1959a) 10–25 gm per day was not toxic.

As much as 8% nitrate (as KNO_3) has been reported in corn stalks and turnip tops (Brakenridge, 1956; Dodd and Coup, 1957), which makes it easy under some conditions for ruminants to consume lethal amounts. The hemoglobinuria in Japanese cattle fed on beets (Ota, 1958) presumably resulted from the high nitrate content. In experimental calves on good rations, the threshold toxicity level has been estimated at 30 gm/ 100 lb live weight (Prewitt and Merilan, 1958). Rather similar thresholds were found in sheep (Setchell and Williams, 1962). Over a rumen pH range between 5.7 and 7.8, the least nitrite accumulation occurred at pH 6.6 (Holtenius, 1957).

Plant nitrate concentrations are higher in seasons of low rainfall or drought, and nitrate poisoning of livestock is more common during such periods. Part of this increased toxicity in dry seasons is due to the poor feed conditions. Increased nitrate concentrations in the soil give higher concentrations in the crop (Ericksson and Vestervall, 1960). Greater quantities of nitrate can be consumed without ill effects by animals on a good ration (Holtenius and Nielsen, 1957). The favorable effects of a high feeding level may be due to a more rapid reduction of nitrite when fermentation is more active and decreases the amount absorbed. Individual animals differ in their susceptibility to nitrate (A. J. Winter, 1962).

The nitrogenous substances formed when nitrite is reduced in the rumen have not been studied extensively. By analogy with soil denitrification, hydroxylamine, nitrous oxide, and nitrogen would be expected. An early finding of possible significance in this respect is the report (Reiset, 1863a) of an excess of nitrogen in his metabolism chamber after the sheep was fed on beet chips. As more chips were consumed, greater amounts of nitrogen were found. Evidence for adaptation to a high nitrate ration has been observed (Sinclair and Jones, 1964).

Inspection of Table II-9B shows that a number of rumen bacteria can reduce nitrate to nitrite. Nitrate-reducing fungi isolated from the rumen (Holtenius and Nielsen, 1957) were probably ingested with the feed; filamentous fungi are rarely seen (Sapiro *et al.,* 1949) in rumen contents. Hydrogen, formate, succinate, lactate, citrate, and glucose can serve as sources of hydrogen for the rumen reduction of nitrate to nitrite (D. Lewis, 1951b).

D. Ammonia Toxicity

A toxicity of ruminal ammonia for sheep was demonstrated (Clark *et al.,* 1951; Clark and Lombard, 1951) by administering graded levels of urea and examining the pH and ammonia concentration in the rumen. The rumen is not as well buffered on the alkaline side of neutrality as on the acid side (Turner and Hodgetts, 1955b). At higher pH's, ammonia toxicity is greater than at low, and additives which affect the acidity influence the toxicity. The high pH is not in itself toxic (Ash, 1959b), since pH's of 9–10 in a rumen filled with an artificial salt solution did not inhibit rumen mobility.

Carbohydrate starvation contributes to ammonia toxicity (Juhasz, 1962) because it diminishes the production of fermentation acids. In the absence of carbohydrate, fermentation of proteins, including the microbial cells, yields ammonia. Succulent green forages and those supplemented with urea yield higher ammonia levels in the rumen than do hay and grain (Ryś *et al.,* 1957).

A somnolence in sheep has been observed when the blood ammonia exceeded 0.4 mmoles per liter (D. Lewis *et al.,* 1957). At levels above 30 mmoles of ammonia per liter in the rumen, the ammonia level in blood increased, and symptoms of toxicity appeared when the blood ammonia level reached 0.4–0.5 mg/100 ml of blood (Lewis, 1960; Repp *et al.,* 1955b). Six grams of ammonium carbamate given intravenously to sheep (Hale and King, 1955) were fatal. No ammonia could be detected in the blood, and the authors suggest that the symptoms of urea toxicity are due to ammonium carbamate.

The toxicity of urea is presumably due to release of ammonia through urease activity in the rumen. As little as 0.5 gm of urea per kilogram of body weight caused inhibition of rumen motility and was a lethal dose on some occasions (Annicolas *et al.,* 1956c); as much as 2 gm per kilogram did not kill at other times. In one experiment, feeding of 1 gm per kilogram daily for 3–4 weeks produced no untoward effects. Some of the variability in response has been ascribed to differences in the acidity of the rumen contents (Clark *et al.,* 1951; Clark and Lombard, 1951; Annicolas *et al.,* 1956a). On poor rations less urea can be tolerated (Annicolas *et al.,* 1956b; Clark *et al.,* 1951). These observations are in accord with those on ammonia toxicity. A diminution in ammonia absorption has been obtained by supplying nitrogen to the rumen as diammonium phosphate rather than urea (Russell *et al.,* 1962).

E. Hypomagnesemia

Observed interactions between ammonia levels and magnesium (Head and Rook, 1955) suggest that rumen precipitation of Mg^{++} as $MgNH_4PO_4$ may be a factor in hypomagnesemia. The magnesium in lush spring grass is less available than that in usual rations (Rook *et al.,* 1964). Additional

$MgSO_4$ (5 gm per day) did not relieve the deficiency in a goat, but massive 30–50-gm daily doses did relieve it (Uesaka *et al.,* 1962a). Decreased retention of feed phosphorus accompanied the decreased Mg^{++} retention. The forage in these cases contained as much as 25% protein. It would be expected that compounds such as $MgNH_4PO_4$ would dissolve in the acid abomasum, unless acid secreted in the abomasum is buffered by feed protein and ammonia sufficiently to prevent solution of the precipitate. The lush spring grass commonly associated with hypomagnesemia is rich in protein. This would cause high ammonia concentrations in the rumen. Decreased magnesium retention with increased nitrogen content of the forage has been noted (Stillings *et al.,* 1964). The metabolism of magnesium under various feeding conditions has been examined (Rook *et al.,* 1958; Rook and Balch, 1958, 1962; Rook and Wood, 1960; Evered, 1961; Head and Rook, 1955), but the causes of hypomagnesemia have not yet been established. A study of abomasal and intestinal pH during hypomagnesemia might be rewarding.

F. Enterotoxemia

Sheep consuming much grain sometimes exhibit symptoms described as enterotoxemia. Immunization of the animals against the epsilon toxin of *Clostridium welchii* (*perfringens*) Type D (Whitlock and Fabricant, 1947) prevents the symptoms. The disease can occur with a number of readily fermentable diets. It has been shown (Bullen and Batty, 1957) that the permeability of the small intestine to protein is increased in animals suffering from enterotoxemia and that detectable amounts of toxin are absorbed (Batty and Bullen, 1961). In some instances natural absorption of the toxin in sublethal quantities appears to immunize the animal (Bullen and Batty, 1957).

The conditions under which the disease occurs suggest that food is consumed so rapidly that some of it passes through the stomach compartments to the small intestine where it supports growth of *Clostridium welchii*. Information on the relative numbers of *Clostridium welchii* in the small intestine of sheep suffering from enterotoxemia, as compared to healthy sheep, has not been obtained. A direct relationship of this disease to the rumen has not been demonstrated.

G. Ketosis

Milking ruminants are subject to periods of lack of appetite and lowered milk production accompanied by appearance of ketone bodies (β-hydroxybutyric acid, acetoacetic acid, and acetone) in the blood and urine (Bach and Hibbitt, 1959). The concentration of acetate in the blood is unusually high (Aafjes, 1964), but in the rumen is lower than normal. Ketosis is most common in high-producing dairy cattle and during late pregnancy (Hoflund and Hedstrom, 1948).

Ketone bodies appear also in the blood and urine of starved ruminants, and since ketosis is almost always accompanied by varying degrees of inanition it is difficult to distinguish symptoms of ketosis from those of starvation. A. Robertson and Thin (1953) studied starvation ketosis and noted that food deprivation increased the acetone bodies in animals on high planes of nutrition associated with milk production or pregnancy, but had relatively little effect on dry cows. In both fasted and ketotic milking cows less acetate-1-C^{14} is assimilated into water-insoluble fatty acids than into water-soluble acids (Thin *et al.,* 1962).

Analyses of rumen contents in ketotic animals after appetite loss disclose very low concentrations of total VFA's (R. E. Brown and Shaw, 1957; Hungate, 1959, unpublished observations). The proportion of acetic acid is higher than normal, and the propionate is lower (Shaw *et al.,* 1955; Baaij, 1959; R. E. Brown and Shaw, 1957; Schultz, 1954). Starved normal cows showed the identical proportions of VFA, with acetate high. Ketotic cows with maintained appetite exhibited normal VFA patterns. In cows prone to ketosis, during periods when clinical symptoms are absent, the concentrations of propionate and butyrate are low (Aafjes, 1964).

Isopropanol can be detected in rumen contents from ketotic cows (Thin and Robertson, 1953), but is absent from the normal rumen. Acetone diffusing into the rumen is presumed to be reduced to form the isopropanol found in animals suffering from ketosis. In normal animals β-hydroxybutyric acid could be detected in the blood, but neither acetoacetate nor acetone was found. Acetoacetate is absent from the rumen of normal cattle (van der Horst, 1961).

A rumen microbial genesis of acetone and isopropanol could perhaps occur under acid rumen conditions. The finding of ethanol (Allison *et al.,* 1964a) suggests that these other neutral products could also form. An inverse correlation between blood acetone bodies and propionate concentration in the rumen has been found (van Soest and Allen, 1959).

Silage contains as much as 410 mg of ketone bodies per kilogram (J. H. Adler, 1956), a quantity which was believed to contribute to the ketone bodies found in the blood in amounts characteristic of ketotic cows. None of the animals tested showed clinical symptoms of ketosis, and all were producing milk in large quantities. The results show that the clinical symptoms of ketosis are not due to accumulation of ketone bodies in the blood. In these experiments the ketone bodies presumably passed through the rumen wall, but the possibility that they were formed within the animal is not excluded. The concentration of ketone bodies in the rumen was not investigated.

It is generally believed that ketosis results from insufficiency of carbohydrate needed for the oxidation of the fatty acids (Sampson and Hayden, 1936; Dye and McCandless, 1948; Hoflund and Hedstrom, 1948). Administration of glucose relieved symptoms of ketosis (Hayden et al., 1946).

The fact that ketosis often occurs in the cows producing the highest milk yields and that it commonly comes after calving, when the food consumption and milk production are the highest, suggests that occasionally a period of high acidity in the rumen may cause loss of appetite, with a resulting starvation ketosis. The period of high acidity would precede the symptoms and would be hard to detect, since prediction of the disease would be necessary in order to examine for acidity at the right time.

In cases of ketosis in which appetite remains good and production continues, the high blood ketone bodies may result from an inadequacy of carbohydrate. Normally the proteins, propionate, and other acids are adequate for synthesis of amino acids and sugars in the host, but under the increased requirements during pregnancy and high milk production the amount may be insufficient. Insofar as the rumen is concerned, an increased production of acetate and/or butyrate relative to propionate appears to be the most likely explanation for the lack of carbohydrate, but information on the relative rates of production of these acids in ketotic cows is not available. The lowered synthesis of milk fat from acetate in ketotic cows (Kronfeld et al., 1959) might have been due to a greater availability of butyrate.

In some cases, feeding of sugar increased the incidence of ketosis as compared to animals receiving cellulose (Breirem et al., 1949). Since small amounts of sugar would not reach the blood (Schambye, 1951c) and could be fermented to a combination of VFA's in which propionate was low, this evidence is not inconsistent with the view that ketosis results from a lack of carbohydrate. It does show that *feeding* carbohydrate will not necessarily cure ketosis. Other gluconeogenic materials not affected by the rumen would be expected to be more effective. Dosing with sources of carbohydrate, either intravenously or, with some of them, into the rumen, partly alleviates the symptoms of ketosis, but does not insure that they will not recur. Propionate, normally the source of ruminant carbohydrate, would be expected to relieve the symptoms of ketosis, if administered in sufficient amount, and is found in practice to be effective (Schultz, 1958).

The success of a glycerol drench in relieving ketosis (R. B. Johnson,

1951a, 1954) may be due either to direct absorption or to a propiogenic fermentation. Glycerol fermentation by bacteria usually leads to propionic acid as an important product. Propylene glycol is also effective in relieving symptoms of ketosis (R. B. Johnson, 1954). The effectiveness of oral lactate (Seekles, 1951) has been interpreted as due to propionic acid formation (Hueter *et al.,* 1956), but as indicated in Chapter VI, direct evidence for this conversion is lacking, and in some cases lactate leads to an increase in blood ketone bodies (Radloff and Schultz, 1963) and does not reduce the incidence of ketosis.

The low blood glucose and high ketone body concentration have been interpreted as due to a deficiency in coenzyme A and adenosine triphosphate (ATP). Bovine ketosis was treated successfully with vitamin B_{12} and with oral cobalt. In view of the role of B_{12} as coenzyme for the isomerization between methylmalonyl CoA and succinyl CoA, B_{12} abundance in the rumen could be a factor in the relative proportion of propionate and acetate in the rumen. Ketosis of cattle in cobalt-deficient areas was alleviated in nine out of twelve animals given cobalt orally (J. G. Henderson, 1948). No controls were reported.

Lack of protein in the ration has also been implicated as a predisposing factor in acetonemia (Hoflund and Hedstrom, 1948). Treatment with methionine alleviated neural disturbances accompanying ketosis (Lecomte, 1954). Host factors are important in ketosis; certain individuals develop the disease whereas others under the same conditions do not. It can recur many times in the same individual and be almost chronic.

A deficiency of nicotine-adenine dinucleotides (coenzymes I and II) and an impaired lipogenesis have been suggested as causes of ketosis (Kronfeld and Kleiber, 1959; Kronfeld, 1961). The metabolic reactions of acetate in normal and ketotic cows are alike (Simesen *et al.,* 1961). Pyruvate and oxoglutarate occurred in higher concentrations in the blood of ketotic cows than in normal controls (Bach and Hibbitt, 1959), and lower levels of citrate and succinate were found.

Acetoacetic acid has been shown to be converted to β-hydroxybutyric acid in the blood of sheep. Injection of butyrate increased blood ketones. Acetate increased them slightly, but propionate exerted no effect. An opposite result has been found (Schultz and Smith, 1951), with no rise in blood acetone bodies when acetate was fed, but an increase with propionate, butyrate, or caproate.

Propionate decreased the ketone body production by rumen epithelium or even caused an uptake by these tissues (Pennington and Pfander, 1957). A decrease in the oxygen consumption of liver homogenate cells obtained by biopsy has been observed (Sauer *et al.,* 1958) in ketotic cows as compared to normal. Comparisons with normals at the same reduced intake as the ketotic cows would be of interest in this connection.

Intravenously injected acetate is cleared faster by ewes in early stages of pregnancy than by those in the last 2 weeks (Pugh and Scarisbrick, 1952). The latter showed 9.3–25.5 mg of ketone bodies (calculated as acetone) per 100 ml of blood prior to injection of the acetate. The results indicate either a lessened ability of ketotic animals to oxidize acetate or that the metabolism was already flooded with acetate, the additional injected acetate therefore causing less effect. The results of others (Simesen *et al.*, 1961) can be similarly interpreted.

Some investigators (W. J. Miller *et al.*, 1954a) have reported alleviation of ketosis through feeding of acetate, but without controls. The reported relief of ketosis by ammonium lactate (Seekles, 1951) could be due to absorption from the alimentary tract if the lactate was not rapidly fermented.

Resumption of appetite in ketotic cows has been obtained by injecting cortisone and ACTH (Dye *et al.*, 1953; Shaw *et al.*, 1951). Whether this indicates a specific connection between ketosis and hormonal balance or a generalized influence in restoring appetite has not been determined. A concept of adrenal insufficiency as a cause of ketosis has been advanced (E. A. White, 1955) with the suggestion that additional vitamin B_{12} in the rumen might stimulate the adrenal gland.

Ketotic cows have been treated with anterior pituitary lobe hormone preparations (Fincher and Hayden, 1940), but simultaneous administration of glucose, molasses, and chloral obscured any action of the hormone. In other experiments (A. G. Hall, 1940), glucose improved the effectiveness of hormone treatment.

H. Poisonous Constituents of Plants

Many plants contain materials toxic to the ruminant. These include tannic acid (Begovic *et al.*, 1957), hydrocyanic acid (Dougherty and Christensen, 1953), and terpenes (Crane *et al.*, 1957).

In most cases a toxicity mediated through the rumen has not been distinguished from direct action on the host. Further study from this point of view might disclose means for utilization of additional forages through chemical or other treatments to eliminate toxic factors. Hydrogen cyanide added to rumen contents, or ingested in combination with glucosides in plants, is rapidly destroyed in the rumen by microbial activity (Coop and Blakely, 1949). A poisonous South African plant, *Geigeria,* is reported (Clark, 1956) to cause vomiting, further evidence that this reflex has not been entirely lost in ruminants. Bracken (*Pteris aquilima*) is toxic to cattle and apparently acts on the bone marrow (I. A. Evans *et al.,* 1958). It is reported to contain an enzyme which attacks thiamine (J. A. Henderson *et al.,* 1952).

Chlorophyll is converted to phylloerythrin in the rumen and absorbed by the host. Normally it is excreted in bile by the liver, but some plants— such as *Tribulus* (Zygophyllaceae), *Lippia* (Verbenaceae), and *Panicum* (Graminae)—inhibit bile secretion (Quin, 1933), and the circulating phylloerythrin photosensitizes the ruminant (Simmonet and LeBars, 1953a).

The literature on plants poisonous to ruminants is much too abundant to consider completely here, and much of it is not peculiarly associated with the rumen. However investigation of the factors concerned in de-toxification might in some instances suggest methods for diminishing losses or increasing gains when such feeds are a problem.

References

Aafjes, J. H. (1964). *Life Sci.* **3**, 1327–1334.

Abelson, P. H., and H. H. Darby (1949). *Science* **110**, 566.

Abou Akkada, A. R., and T. H. Blackburn (1963). *J. Gen. Microbiol.* **31**, 461–469.

Abou Akkada, A. R., and B. H. Howard (1960). *Biochem. J.* **76**, 445–451.

Abou Akkada, A. R., and B. H. Howard (1961). *Biochem. J.* **78**, 512–517.

Abou Akkada, A. R., and B. H. Howard (1962). *Biochem. J.* **82**, 313–320.

Abou Akkada, A. R., and K. el-Shazly (1958). *J. Agr. Sci.* **51**, 157–163.

Abou Akkada, A. R., and K. el-Shazly (1964). *Appl. Microbiol.* **12**, 384–390.

Abou Akkada, A. R., and K. el-Shazly (1965). *J. Agr. Sci.* **64**, 251–255.

Abou Akkada, A. R., P. N. Hobson, and B. H. Howard (1959). *Biochem. J.* **73**, 44–45P.

Abou Akkada, A. R., J. M. Eadie, and B. H. Howard (1963). *Biochem. J.* **89**, 268–272.

Adams, F. W., and J. R. Haag (1957). *J. Nutr.* **63**, 585–590.

Adams, S. L., and R. E. Hungate (1950). *Ind. Eng. Chem.* **42**, 1815–1818.

Adler, H. (1949). *J. Am. Vet. Med. Assoc.* **115**, 263.

Adler, J. H., and J. A. Dye (1957). *Cornell Vet.* **47**, 506–514.

Adler, J. H., and N. Cohen (1960). *Bull. Res. Council Israel* **8E**, 162–166.

Adler, J. H., J. A. Dye, D. E. Boggs, and H. H. Williams (1958) *Cornell Vet.* **48**, 53–56.

Adler, J. H., N. Cohen, and H. Rodrig (1960a). *Refuah Vet.* **17**, 55–51.

Adler, J. H., E. Mayer, and M. Egyed (1960b). *Refuah Vet.* **17**, 145–144.

Aggazzotti, A. (1910). *Clin. Vet.* **33**, 54–75.

Agrawala, I. P., C. W. Duncan, and C. F. Huffman (1953a). *J. Nutr.* **49**, 29–40.

Agrawala, I. P., C. F. Huffman, R. W. Luecke, and C. W. Duncan (1953b). *J. Nutr.* **49**, 631–638.

Akashi, S., and K. Saito (1960). *J. Biochem. (Tokyo)* **47**, 222–229.

Akerblom, E. (1934). *Skand. Arch. Physiol.* **68**, Suppl.

Akssenowa, M. J. (1932). *Arch. Tierernaehr. Tierzucht* **7**, 295–304.

Albaugh, R., J. H. Meyer, and S. E. Smith (1963). *Calif. Agr.* **17**, 4–5.

Albertson, P. A., and G. D. Baird (1962). *Exptl. Cell Res.* **28**, 296–322.

Alexander, R., M. J. Macpherson, and A. E. Oxford (1952). *J. Comp. Pathol. Therap.* **62**, 252–259.

Aliev, A. A. (1958). *Sechenov Physiol. J. USSR (English Transl.)* **44**, 49–51.

Allen, T. E. (1962). *Australian J. Agr. Res.* **13**, 165–179.

Allison, M. J. (1965). *Biochem. Biophys. Res. Commun.* **18**, 30–35.

Allison, M. J., and M. P. Bryant (1963). *Arch. Biochem. Biophys.* **101**, 269–277.

Allison, M. J., M. P. Bryant, and R. N. Doetsch (1958). *Science* **128**, 474–475.

Allison, M. J., M. P. Bryant, and R. N. Doetsch (1959). *Arch. Biochem. Biophys.* **84**, 246–247.

Allison, M. J., M. P. Bryant, and R. N. Doetsch (1962a). *J. Bacteriol.* **83**, 523–532.

Allison, M. J., M. P. Bryant, I. Katz, and M. Keeney (1962b). *J. Bacteriol.* **83**, 1084–1093.

Allison, M. J., R. W. Dougherty, J. A. Bucklin, and E. E. Snyder (1964a). *Science* **144**, 54–55.

Allison, M. J., J. A. Bucklin, and R. W. Dougherty (1964b). *J. Animal Sci.* **23**, 1164–1171.

Altona, R. E., and H. I. MacKenzie (1964). *J. S. African Vet. Med. Assoc.* **35**, 203–205.

Ambo, K., S. Matsuda, and M. Umezu (1963). *Tohoku J. Agr. Res.* **14**, 25–28.

Ammerman, C. B., and W. E. Thomas (1953). *J. Animal Sci.* **11**, 754–755.

Ammerman, C. B., and W. E. Thomas (1955). *Cornell Vet.* **45**, 443–450.

Ammerman, C. B., R. M. Forbes, U. S. Garrigus, A. L. Newmann, H. W. Norton, and E. E. Hatfield (1957). *J. Animal Sci.* **16**, 796–810.

Anderson, A. C. (1934). *Skand. Arch. Physiol.* **69**, 33–58.

Anderson, C. M. (1956). *New Zealand J. Sci. Technol.* **A37**, 379–394.

Anderson, G. C., G. A. McLaren, J. A. Welch, C. D. Campbell, and G. S. Smith (1959). *J. Animal Sci.* **18**, 134–140.

Anderson, J. P., and E. D. Andrews (1952). *Nature* **170**, 807.

Anderson, P. E., J. T. Reid, M. J. Anderson, and J. W. Stroud (1959). *J. Animal Sci.* **18**, 1299–1307.

Anderson, R., E. Cheng, and W. Burroughs (1956). *J. Animal Sci.* **15**, 489–495.

Anderson, R. L., and E. J. Ordal (1961). *J. Bacteriol.* **81**, 130–138.

Andersson, B. (1951). *Acta Physiol. Scand.* **23**, 8–23.

Andersson, B., and S. M. McCann (1955). *Acta Physiol. Scand.* **35**, 191–201.

Andersson, B., R. Kitchell, and N. Persson (1958). *Acta Physiol. Scand.* **44**, 92–102.

Andersson, B., R. Kitchell, and N. Persson (1959). *J. Physiol. (London)* **147**, 11P.

Andrews, E. D., and J. P. Anderson (1954). *New Zealand J. Sci. Technol.* **A35**, 483–488.

Ankersmit, P. (1905). *Zentr. Bakteriol. Parasitenk., Abt. I Orig.* **39**, 359–369, 574–584, and 687–695; (1906) **40**, 100–118.

Annicolas, D., H. LeBars, J. Nugues, and H. Simmonet (1956a). *Bull. Acad. Vet. France* **29**, 257–261.

Annicolas, D., H. Lebars, J. Nugues, and H. Simmonet (1956b). *Bull. Acad. Vet. France* **29**, 263–265.

Annicolas, D., H. Lebars, J. Nugues, and H. Simmonet (1956c). *Bull. Acad. Vet. France* **29**, 225–232.

Annison, E. F. (1954a). *Biochem. J.* **57**, 400–405.

Annison, E. F. (1954b). *Biochem. J.* **58**, 670–680.

Annison, E. F. (1956). *Biochem. J.* **64**, 705–714.

Annison, E. F. (1960). *Australian J. Agr. Res.* **11**, 58–64.

Annison, E. F. (1964). *Proc. 2nd Intern. Symp. Physiol. Digestion Ruminant, 1964, Ames, Iowa.* Oral communication.

Annison, E. F., and D. Lewis (1959). "Metabolism in the Rumen," 184 pp. Methuen, London.

Annison, E. F., and D. B. Lindsay (1961). *Biochem. J.* **78**, 777–785.

Annison, E. F., and D. B. Lindsay (1962). *Biochem. J.* **85**, 474–479.

Annison, E. F., and R. J. Pennington (1954). *Biochem. J.* **57**, 685–692.

Annison, E. F., and R. R. White (1961). *Biochem. J.* **80**, 162–169.

Annison, E. F., and R. R. White (1962a). *Biochem. J.* **84**, 546–552.

Annison, E. F., and R. R. White (1962b). *Biochem. J.* **84**, 552–557.

Annison, E. F., M. I. Chalmers, S. B. M. Marshall, and R. L. M. Synge (1954). *J. Agr. Sci.* **44**, 270–273.

Annison, E. F., K. J. Hill, and D. Lewis (1957). *Biochem. J.* **66**, 592–599.

Annison, E. F., D. Lewis, and D. B. Lindsay (1959a). *J. Agr. Sci.* **53**, 34–41.

Annison, E. F., D. Lewis, and D. B. Lindsay (1959b). *J. Agr. Sci.* **53**, 42–45.

Ardo, K. M., K. W. King, and R. W. Engel (1964). *J. Animal Sci.* **23**, 734–736.

Arduini, A., and V. Dagnino (1953). *Ateneo Parmense* **24**, 22–45.

Arias, C., W. Burroughs, P. Gerlaugh, and R. M. Bethke (1951). *J. Animal Sci.* **10**, 683–692.

Armsby, H. P. (1917). "The Nutrition of Farm Animals." Macmillan, New York.

Armsby, H. P., and J. A. Fries (1903). *U.S. Bur. Animal Ind. Bull.* **51**, 77 p.

Armsby, H. P., and C. R. Moulton (1925). "The Animal as Converter of Matter and Energy," 236 pp. Chemical Catalogue Co., New York.

Armstrong, D. G. (1964). *J. Agr. Sci.* **62**, 399–416.

Armstrong, D. G., and K. L. Blaxter (1957a). *Brit. J. Nutr.* **11**, 247–272.

Armstrong, D. G., and K. L. Blaxter (1957b). *Brit. J. Nutr.* **11**, 413–425.

Armstrong, D. G., and B. Thomas (1953). *J. Agr. Sci.* **43**, 223–228.

Armstrong, D. G., K. L. Blaxter, and N. M. Graham (1957). *Brit. J. Nutr.* **11**, 392–408.

Armstrong, D. G., K. L. Blaxter, N. M. Graham, and F. W. Wainman (1958). *Brit. J. Nutr.* **12**, 177–188.

Armstrong, D. G., K. L. Blaxter, and R. Waite (1964a). *J. Agr. Sci.* **62**, 417–424.

Armstrong, D. G., R. H. Alexander, and M. McGowan (1964b). *Proc. Nutr. Soc.* (*Engl. Scot.*) **23**, xxvi–xxvii.

Arnaudi, C. (1931). *Boll. Sez. Ital. Soc. Intern. Microbiol.* **3**, 35–40.

Ash, R. M. (1956). *J. Physiol.* (*London*) **133**, 75–76P.

Ash, R. W. (1959a). *J. Physiol.* (*London*) **149**, 72–73P.

Ash, R. W. (1959b). *J. Physiol.* (*London*) **147**, 58–73.

Ash, R. W. (1961a). *J. Physiol.* (*London*) **156**, 93–111.

Ash, R. W. (1961b). *J. Physiol.* (*London*) **157**, 185–207.

Ash, R. W., and R. B. N. Kay (1957). *J. Physiol.* (*London*) **139**, 23–24P.

Ash, R. W., and R. N. B. Kay (1959). *J. Physiol.* (*London*) **149**, 43–57.

Ash, R. W., R. J. Pennington, and R. S. Reid (1958). *Biochem. J.* **71**, 9–10P.

Ash, R. W., R. J. Pennington, and R. S. Reid (1964). *Biochem. J.* **90**, 353–360.

Asher, T. (1942). Mikroscopische Untersuchungen über die Veränderungen der Zellulose bei der bakteriellen Verdauung im Magendarmkanal der Wiederkäuer. Thesis, Hanover.

Ashton, G. C. (1962). *J. Agr. Sci.* **58**, 333–342.

Asplund, J. M., R. T. Berg, L. W. McElroy, and W. J. Pigden (1958). *Can. J. Animal Sci.* **38**, 171–180.

Atkinson, R. L., F. H. Kratzer, and G. F. Stewart (1957). *J. Dairy Sci.* **40**, 1114–1132.

Awerinzew, S., and R. Mutafowa (1914). *Arch. Protistenk.* **33**, 109–118.

Axelsson, J. (1942a). *Biedermann's Zentr., B. Tierernaehr.* **14**, 198–211.

Axelsson, J. (1942b). *Biedermann's Zentr., B. Tierernaehr.* **14**, 212–232.

Ayers, W. A. (1958). *J. Bacteriol.* **76**, 504–509.

Ayers, W. A. (1959). *J. Biol. Chem.* **234**, 2819–2822.

Baaij, P. K. (1959). "Enkele aspecten van de pensdigestie bij runderen in verband met acetonemie," 96 pp. Uitgeversmaatschappij Neerlandia, Utrecht.

Bach, J. S., and K. G. Hibbitt (1959). *Biochem. J.* **72**, 87–92.

Bachmann, B. J. (1955). *J. Gen. Microbiol.* **13**, 541–551.

Badawy, A. M., R. M. Campbell, D. P. Cuthbertson, and B. F. Fell (1957). *Nature* **180,** 756–757.

Badawy, A. M., R. M. Campbell, D. P. Cuthbertson, B. F. Fell, and W. S. Mackie (1958a). *Brit. J. Nutr.* **12,** 367–383.

Badawy, A. M., R. M. Campbell, D. P. Cuthbertson, and W. S. Mackie (1958b). *Brit. J. Nutr.* **12,** 384–390.

Badawy, A. M., R. M. Campbell, D. P. Cuthbertson, and W. S. Mackie (1958c). *Brit. J. Nutr.* **12,** 391–403.

Bailey, C. B. (1961a). *Brit. J. Nutr.* **15,** 443–451.

Bailey, C. B. (1961b). *Brit. J. Nutr.* **15,** 489–498.

Bailey, C. B., and C. C. Balch (1961a). *Brit. J. Nutr.* **15,** 183–188.

Bailey, C. B., and C. C. Balch (1961b). *Brit. J. Nutr.* **15,** 371–402.

Bailey, R. W. (1958a). *J. Sci. Food Agr.* **9,** 743–747.

Bailey, R. W. (1958b). *J. Sci. Food Agr.* **9,** 748–753.

Bailey, R. W. (1958c). *New Zealand J. Agr. Res.* **1,** 825–833.

Bailey, R. W. (1959a). *Biochem. J.* **71,** 23–26.

Bailey, R. W. (1959b). *Biochem. J.* **72,** 42–49.

Bailey, R. W. (1959c). *New Zealand J. Agr. Res.* **2,** 355–364.

Bailey, R. W. (1962). *Nature* **195,** 79–80.

Bailey, R. W. (1963). *Biochem. J.* **86,** 509–514.

Bailey, R. W. (1964). *New Zealand J. Agr. Res.* **7,** 417–426.

Bailey, R. W., and E. J. Bourne (1961). *Nature* **191,** 277–278.

Bailey, R. W., and R. T. J. Clarke (1959). *Biochem. J.* **72,** 49–54.

Bailey, R. W., and R. T. J. Clarke (1963a). *Nature* **198,** 787.

Bailey, R. W., and R. T. J. Clarke (1963b). *Nature* **199,** 1291–1292.

Bailey, R. W., and B. H. Howard (1962). *Arch. Bioch. Biophys.* **99,** 299–303.

Bailey, R. W., and B. H. Howard (1963a). *Biochem. J.* **86,** 446–452.

Bailey, R. W., and B. H. Howard (1963b). *Biochem. J.* **87,** 146–151.

Bailey, R. W., and A. E. Oxford (1958a). *Nature* **182,** 185–186.

Bailey, R. W., and A. E. Oxford (1958b). *J. Gen. Microbiol.* **19,** 130–145.

Bailey, R. W., and A. E. Oxford (1959). *J. Gen. Microbiol.* **20,** 258–266.

Bailey, R. W., and A. M. Roberton (1962). *Biochem. J.* **82,** 272–277.

Bailey, R. W., R. T. J. Clarke, and D. E. Wright (1962). *Biochem. J.* **83,** 517–523.

Baker, F. (1942). *Nature* **150,** 479–481.

Baker, F. (1943). *Ann. Appl. Biol.* **30,** 230–239.

Baker, F. (1945). *Proc. Nutr. Soc. (Engl. Scot.)* **3,** 199–201.

Baker, F., and S. T. Harriss (1947). *Nutr. Abstr. Rev.* **17,** 3–12.

Baker, F., and R. Martin (1938). *Nature* **141,** 877–878.

Baker, F., and H. Nasr (1947). *J. Roy. Microscop. Soc.* **67,** 27–42.

Baker, T. I., G. V. Quicke, O. G. Bentley, R. R. Johnson, and A. L. Moxon (1959). *J. Animal Sci.* **18,** 655–662.

Balch, C. C. (1950). *Brit. J. Nutr.* **4,** 361–388.

Balch, C. C. (1952). *Brit. J. Nutr.* **6,** 366–375.

Balch, C. C. (1958). *Brit. J. Nutr.* **12,** 330–345.

Balch, C. C., and V. W. Johnson (1950). *Brit. J. Nutr.* **4,** 389–394.

Balch, C. C., and A. Kelly (1950). *Brit. J. Nutr.* **4,** 395–398.

Balch, C. C., and C. Line (1957). *J. Dairy Res.* **24,** 11–19.

Balch, C. C., and S. J. Rowland (1959). *J. Dairy Res.* **26,** 162–172.

Balch, C. C., A. Kelly, and G. Heim (1951). *Brit. J. Nutr.* **5,** 207–216.

Balch, C. C., D, A Balch, S. Bartlett, C. P. Cox, and S. J. Rowland (1952a). *J. Dairy Res.* **19,** 39–50.

Balch, C. C., D. A. Balch, S. Bartlett, V. W. Johnson, and S. J. Rowland (1952b). *Brit. J. Nutr.* **6**, 356–365.

Balch, C. C., D. A. Balch, V. W. Johnson, and J. Turner (1953). *Brit. J. Nutr.* **7**, 212–224.

Balch, C. C., D. A. Balch, S. Bartlett, C. P. Cox, S. J. Rowland, and J. Turner (1954a). *J. Dairy Res.* **21**, 165–171.

Balch, C. C., D. A. Balch, S. Bartlett, Z. D. Hosking, V. W. Johnson, S. J. Rowland, and J. Turner (1954b). *J. Dairy Res.* **21**, 172–177.

Balch, C. C., D. A. Balch, S. Bartlett, V. W. Johnson, S. J. Rowland, and J. Turner (1954c). *J. Dairy Res.* **21**, 305–317.

Balch, C. C., D. A. Balch, S. Bartlett, Z. D. Hosking, V. W. Johnson, S. J. Rowland, and J. Turner (1955a). *J. Diary Res.* **22**, 10–15.

Balch, C. C., D. A. Balch, S. Bartlett, M. P. Bartrum, V. W. Johnson, S. J. Rowland, and J. Turner (1955b) *J Dairy Res.* **22**, 270–289.

Balch, C. C., J. T. Reid, and J. W. Stroud (1957). *Brit. J. Nutr.* **11**, 184–197.

Balch, D. A. (1958). *Brit. J. Nutr.* **12**, 18–24

Balch, D. A., and S. J. Rowland (1957). *Brit. J. Nutr.* **11**, 288–298.

Baldissera, N. C. (1951). *Boll. Soc. Ital. Biol. Sper.* **27**, 954–956.

Baldwin, R. L. (1965). *In* "Physiology of Digestion in the Ruminant" (R. W. Dougherty, ed.), pp. 379–389. Butterworth, London and Washington, D.C.

Baldwin, R. L., and L. P. Milligan (1965). *Biochim. Biophys. Acta* **92**, 421–432.

Baldwin, R. L., and D. L. Palmquist (1965). *Appl. Microbiol.* **13**, 194–200.

Baldwin, R. L., W. A. Wood, and R. S. Emery (1962). *J. Bacteriol.* **83**, 907–913.

Baldwin, R. L., W. A. Wood, and R. S. Emery (1963). *J. Bacteriol.* **85**, 1346–1349.

Baldwin, R. L., W. A. Wood and R. S. Emery (1965). *Biochim. Biophys. Acta* **97**, 202–213.

Ballard, F. J., and I. T. Oliver (1964). *Biochem. J.* **92**, 131–136.

Barcroft, J. (1945). *Proc. Nutr. Soc. (Engl. Scot.)* **3**, 247–251.

Barcroft, J., R. A. McAnally, and A. T. Phillipson (1944). *J. Exptl. Biol.* **20**, 120–129.

Barnes, I. J., H. W. Seeley, and P. J. van Demark (1961). *J. Bacteriol.* **82**, 85–93.

Barnes, R. F., G. O. Mott, L. B. Packett, and M. P. Plumlee (1964) *J. Animal Sci.* **23**, 1061–1064.

Barnett, A. J. G., and R. E. B. Duncan (1953). *J. Agr. Sci.* **43**, 260–264.

Barnett, A. J. G., and R. L. Reid (1957a). *J Agr. Sci.* **48**, 315–321.

Barnett, A. J. G., and R. L. Reid (1957b). *J. Agr. Sci.* **49**, 171–179.

Barnett, A. J. G., and R. L. Reid (1957c). *J. Agr. Sci.* **49**, 180–183.

Barnett, A. J. G., and R. L. Reid (1957d). *Biochem. J.* **65**, 13P.

Barrentine, B. F., C. B. Shawver, and L. W. Williams (1954). *J. Animal Sci.* **13**, 1006.

Barrentine, B. F., C. B. Shawver, and L. W. Williams (1956). *J. Animal Sci.* **15**, 440–446.

Barth, J., and S. L. Hansard (1962). *Proc. Soc. Exptl. Biol. Med.* **109**, 448–451.

Bartlett, S., and K. L. Blaxter (1947). *J. Agr. Sci.* **37**, 32–44.

Bartlett, S., and W. H. Broster (1958). *J. Agr. Sci.* **50**, 60–63.

Bartley, E. E. (1957). *J. Animal Sci.* **16**, 1084.

Bartley, E. E. (1965). *J. Dairy Sci.* **48**, 102–104.

Bartley, E. E., and R. Bassette (1961). *J. Dairy Sci.* **44**, 1365–1366.

Bartley, E. E., T. J. Claydon, L. R. Fina, C. Hay and I. S. Yadava (1961). *J. Dairy Sci.* **44**, 553–555.

de Bary, A. (1879). *Rev. Intern. Sci.* **3**, 301–309.

Bastié, A. M. (1957). *Arch. Sci. Physiol.* **11,** 87–100.

Bateman, J. V., and K. L. Blaxter (1964). *J. Agr. Sci.* **63,** 129–131.

Bath, D. L., W. C. Weir, and D. T. Torell (1956). *J. Animal Sci.* **15,** 1166–1171.

Bath, I. H., and M. J. Head (1961). *J. Agr. Sci.* **56,** 131–136.

Bath, I. H., C. C. Balch, and J. A. F. Rook (1962). *Proc. Nutr. Soc. (Engl. Scot.)* **21,** ix.

Batty, I., and J. J. Bullen (1961). *J. Pathol. Bacteriol.* **81,** 447–458.

Bauchop, T., and S. R. Elsden (1960). *J. Gen. Microbiol.* **23,** 457–469.

Bauman, H. E., and E. M. Foster (1956). *J. Bacteriol.* **71,** 333–338.

Baumgardt, B. R., and H. K. Oh (1964). *J. Dairy Sci.* **47,** 263–266.

Baumgardt, B. R., W. J. Byer, H. F. Jumah, and C. R. Krueger (1964). *J. Dairy Sci.* **47,** 160–164.

Baxter, C. F., M. Kleiber, and A. L. Black (1955). *Biochem. Biophys. Acta* **17,** 354–361.

Bechdel, S. I., H. E. Honeywell, R. A. Dutcher, and M. H. Knutsen (1928). *J. Biol. Chem.* **80,** 231–238.

Becker, D. E., and S. E. Smith (1951a). *J. Animal Sci.* **10,** 266–271.

Becker, D. E., and S. E. Smith (1951b). *J. Nutr.* **43,** 87–100.

Becker, D. E., S. E. Smith and J. K. Loosli (1949). *Science* **110,** 71–72.

Becker, E. R. (1933). *Trans. Am. Microscop. Soc.* **52,** 217–219.

Becker, E. R., and R. C. Everett (1930). *Am. J. Hyg.* **11,** 362–370.

Becker, E. R., and T. S. Hsuing (1929). *Proc. Natl. Acad. Sci. U.S.* **15,** 684–690.

Becker, E. R., and M. Talbott (1927). *Iowa State Coll. J. Sci.* **1,** 345–372.

Becker, E. R., J. A. Schulz, and M. A. Emmerson (1929). *Iowa State Coll. J. Sci.* **4,** 215–241.

Becker, R. B., S. P. Marshall, and P. T. D. Arnold (1963). *J. Dairy Sci.* **46,** 835–839.

Begovic, S., M. Janjatovic, L. J. Kozic, and P. Nesic (1957). *Vet. Glasnik* **11,** 761–767.

Beijer, W. H. (1952). *Nature* **170,** 576–577.

Belasco, I. J. (1954a). *J. Animal Sci.* **13,** 739–747.

Belasco, I. J. (1954b). *J. Animal Sci.* **13,** 748–757.

Belasco, I. J. (1954c). *J. Animal Sci.* **13,** 601–610.

Belasco, I. J. (1956). *J. Animal Sci.* **15,** 496–508.

Bell, F. R. (1958a). *J. Physiol. (London)* **142,** 503–515.

Bell, F. R. (1958b). *J. Physiol. (London)* **143,** 46–47P.

Bell, F. R. (1958c). *Nature* **181,** 494.

Bell, F. R., and E. R. Jones (1945). *J. Comp. Pathol. Therap.* **55,** 117–124.

Bell, F. R., and A. M. Lawn (1955). *J. Physiol. (London)* **128,** 577–592.

Bell, M. C., J. R. Taylor, and R. L. Murphree (1957). *J. Animal Sci.* **16,** 821–827.

Bell, M. C., B. G. Diggs, R. S. Lowrey, and P. L. Wright (1964) *J. Nutr.* **84,** 367–372.

Bensadoun, A., and J. T. Reid (1962). *J. Dairy Sci.* **45,** 540–543.

Bensadoun, A., O. L. Paladines, and J. T. Reid (1962). *J. Dairy Sci.* **45,** 1203–1210.

Benson, A. A., J. F. Wintermans, and R. Weiser (1959). *Plant Physiol.* **34,** 315–317.

Benson, J. A. (1957). *Vet. Record* **69,** 412.

Bentley, O. G., R. R. Johnson, S. Vanecko, and C. H. Hunt (1954a). *J. Animal Sci.* **13,** 581–593.

Bentley, O. G., M. Moinuddin, T. V. Hershberger, E. W. Klosterman, and A. L. Moxon (1954b). *J. Animal Sci.* **13,** 789–801.

Bentley, O. G., A. Lehmkuhl, R. R. Johnson, T. V. Hershberger, and A. L. Moxon (1954c). *J. Am. Chem. Soc.* **76,** 5000–5001.

Bentley, O. G., R. R. Johnson, T. V. Hershberger, J. H. Cline, and A. L. Moxon (1955). *J. Nutr.* **57,** 389–400.

Benzie, D., and A. T. Phillipson (1957). "The Alimentary Tract of the Ruminant." Oliver & Boyd, Edinburgh and London.

Béranger (1961). *Ann. Nutr. Aliment.* **15,** 287–290.

Bergerud, A. T., and L. Russell (1964). *J. Wildlife Management* **28,** 809–814.

Bergmann, H. D., and H. H. Dukes (1925). *J. Am. Vet. Med. Assoc.* **67,** 364–366.

Bergmann, H. D., and H. H. Dukes (1926). *J. Am. Vet. Med. Assoc.* **69,** 600–612.

Birch, H. F. (1958). *J. Agr. Sci.* **51,** 377–379.

Bissell, H. D., and W. C. Weir (1957). *J. Animal Sci.* **16,** 476–480.

Black, A. L., and M. Kleiber (1958). *J. Biol. Chem.* **232,** 203–209.

Black, A. L., M. Kleiber, and A. H. Smith (1952). *J. Biol. Chem.* **197,** 365–370.

Black, A. L., M. Kleiber and C. F. Baxter (1955). *Biochim. Biophys. Acta* **17,** 346–353.

Black, A. L., M. Kleiber, and A. M. Brown (1961a). *J. Biol. Chem.* **236,** 2399–2403

Black, A. L., J. R. Luick, M. Kleiber, and C. F. Baxter (1961b). *Am. J. Physiol.* **201,** 74–76.

Black, A. L., N. F. Baker, J. C. Bartley, T. E. Chapman, and R. W. Phillips (1964). *Science* **144,** 876–878.

Black, W. H., A. T. Semple, and J. L. Lush (1934). *U.S. Dept. Agr. Tech. Bull.* **417,** 54 pp.

Blackburn, T. H. (1965). *In* "Physiology of Digestion in the Ruminant" (R. W. Dougherty, ed.), pp. 322–334. Butterworth, London and Washington, D. C.

Blackburn, T. H., and P. N. Hobson (1960a). *J. Gen. Microbiol.* **22,** 272–281.

Blackburn, T. H., and P. N. Hobson (1960b). *J. Gen. Microbiol.* **22,** 282–289.

Blackburn, T. H., and P. N. Hobson (1960c). *J. Gen. Microbiol.* **22,** 290–294.

Blackburn, T. H., and P. N. Hobson (1960d). *Brit. J. Nutr.* **14,** 445–456.

Blackburn, T. H., and P. N. Hobson (1962). *J. Gen. Microbiol.* **29,** 69–81.

Blackburn, T. H., and R. E. Hungate (1963). *Appl. Microbiol.* **11,** 132–135.

Bladen, H. A., and R. N. Doetsch (1959). *J. Agr. Food Chem.* **7,** 791–794.

Bladen, H. A., M. P. Bryant, and R. N. Doetsch (1961a). *Appl. Microbiol.* **9,** 175–180.

Bladen, H. A., M. P. Bryant, and R. N. Doetsch (1961b). *J. Dairy Sci.* **44,** 173–174.

Blaizot, J., and P. Raynaud (1957). *Compt. Rend.* **244,** 802–803.

Blaizot, J., and P. Raynaud (1958). *Arch. Sci. Physiol.* **12,** 285–292.

Blake, J. T., R. S. Allen, and N. L. Jacobson (1957a). *J. Animal Sci.* **16,** 190–200.

Blake, J. T., N. L. Jacobson, and R. S. Allen (1957b). *Am. J. Vet. Res.* **18,** 756–760.

Blakely, R. L., and J. E. Coop (1949). *New Zealand J. Sci. Technol.* **A31,** No. 3, 1–16.

Blaschko, H., and R. Hawes (1959). *J. Physiol. (London)* **145,** 124–131.

Blaxter, K. L. (1952). *Brit. J. Nutr.* **6,** 110–117.

Blaxter, K. L. (1962). *Brit. J. Nutr.* **16,** 615–626.

Blaxter, K. L. (1963). *Brit. J. Nutr.* **17,** 105–117.

Blaxter, K. L., and N. M. Graham (1955). *J. Agr. Sci.* **46,** 292–306.

Blaxter, K. L., and N. M. Graham (1956). *J. Agr. Sci.* **47,** 207–217.

Blaxter, K. L., and A. K. Martin (1962). *Brit. J. Nutr.* **16,** 397–407.

Blaxter, K. L., and J. A. F. Rook (1955). *Brit. J. Nutr.* **9,** 121–132.

Blaxter, K. L., and F. W. Wainman (1961). *J. Agr. Sci.* **57,** 419–425.

Blaxter, K. L., and F. W. Wainman (1964). *J. Agr. Sci.* **63,** 113–128.

Blaxter, K. L., and R. S. Wilson (1962). *Proc. Nutr. Soc.* (*Engl. Scot.*) **21,** xxii.

Blaxter, K. L., M. K. Hutcheson, J. M. Robertson, and A. L. Wilson (1952). *Brit. J. Nutr.* **6,** i–ii.

Blaxter, K. L., N. M. Graham, and F. W. Wainman (1956). *Brit. J. Nutr.* **10,** 69–91.

Block, R. J., J. A. Stekol, and J. K. Loosli (1951). *Arch. Biochem. Biophys.* **33,** 353–363.

Bloomfield, R. A., R. P. Wilson, and G. B. Thompson (1964). *J. Animal Sci.* **23,** 868.

Boda, J. M. (1958). *J. Dairy Sci.* **41,** 295–301.

Boda, J. M., and A. T. Johns (1962). *Nature* **193,** 195–196.

Boda, J. M., B. S. Silver, W. H. Colvin, and H. H. Cole (1957). *J. Dairy Sci.* **40,** 759–767.

Boggs, D. E. (1959). An *in vivo* N^{15} tracer study of amino acid metabolism in the rumen of sheep on a purified diet. Ph.D. Thesis, Cornell University, Ithaca, New York.

Bond, J., D. O. Everson, J. Gutierrez, and E. J. Warwick (1962). *J. Animal Sci.* **21,** 728–733.

Bondi, A. H., and H. Meyer (1943). *J. Agr. Sci.* **33,** 123–128.

Booth, A. N., and R. T. Williams (1963). *Nature* **198,** 684–685.

Bouckaert, J. H., and W. Oyaért (1952). *Zootechnia* **1,** 21–29.

Bouckaert, J. H., and W. Oyaért (1954). *Nature* **174,** 1195.

Boussingault, J. B. (1839). *Ann. Chim. Phys.* [2] **71,** 113–127.

Bowden, D. M., and D. C. Church (1962a). *J. Dairy Sci.* **45,** 972–979.

Bowden, D. M., and D. C. Church (1962b). *J. Dairy Sci.* **45,** 980–985.

Bowen, J. M. (1962). *Am. J. Vet. Res.* **23,** 685–686.

Bowers, H. H., T. R. Preston, I. McDonald, N. A. MacLeod, and E. B. Philip (1964). *Proc. Nutr. Soc.* (*Engl. Scot.*) **23,** xxxi–xxxii.

Bowie, W. C. (1962). *Am. J. Vet. Res.* **23,** 858–867.

Boycott, A. E., and G. C. C. Damant (1907). *J. Physiol.* (*London*) **36,** 283–287.

Boyne, A. W., R. M. Campbell, R. M. Davidson, and D. P. Cuthbertson (1956). *Brit. J. Nutr.* **10,** 325–333.

Boyne, A. W., J. M. Eadie, and K. Raitt (1957). *J. Gen. Microbiol.* **17,** 414–423.

Brailsford, M. D. (1965). Isolation and characterization of *Streptococcus bovis* bacteriophages. M. S. Thesis, Iowa State University, Ames, Iowa.

Brakenridge, D. T. (1956). *New Zealand Vet. J.* **4,** 165–166.

Brandt, C. S., and E. J. Thacker (1958). *J. Animal Sci.* **17,** 218–223.

Bratzler, J. W., and E. B. Forbes (1940). *J. Nutr.* **19,** 611–613.

Braune, R. (1914). *Arch. Protistenk.* **32,** 111–170.

Bredon, R. M. (1957). "Feeding of Livestock in Uganda." *Occasional Bull., Dept. Vet. Sci. Animal Ind. Uganda,* No. **1,** 34 pp.

Breirem, K., F. Ender, K. Halse, and L. Skagsvold (1949). *Acta Agr. Suecana* **3,** 89–120.

Brethour, J. R., R. J. Sirny, and A. D. Tillman (1958). *J. Animal Sci.* **17,** 171–179.

Bretschneider, L. H. (1931). *Verhandl. Deut. Zool. Ges.* Suppl. 5, 324–330.

Bretschneider, L. H. (1934). *Arch. Protistenk.* **82,** 298–330.

Briggs, C. A. E. (1955). *Dairy Sci. Abstr.* **17,** 714–722.

Briggs, M. (1953). *J. Gen. Microbiol.* **9,** 234–248.

Briggs, P. K., J. P. Hogan, and R. L. Reid (1957). *Australian J. Agr. Res.* **8,** 674–690.

Broberg, G. (1957a). *Nord. Veterinarmed.* **9,** 57–60.

Broberg, G. (1957b). *Nord. Veterinarmed.* **9,** 918–930.

Broberg, G. (1957c). *Nord. Veterinarmed.* **9,** 931–941.

Broberg, G. (1957d). *Nord. Veterinarmed.* **9,** 942–950.

Broberg, G. (1960). Acute overeating with cereals in ruminants. Thesis, Lovisa Nya Tryckeri, Lovisa.

Brody, S., R. C. Procter, and U. S. Ashworth (1934). *Missouri, Univ., Agr. Expt. Sta. Res. Bull.* **220,** 4–40.

Brooks, C. C., G. B. Garner, M. E. Muhrer, and W. H. Pfander (1954a). *Science* **120,** 455–456.

Brooks, C. C., G. B. Garner. C. W. Gehrke, M. E. Muhrer, and W. H. Pfander (1954b). *J. Animal Sci.* **13,** 758–764.

Brouwer, E. (1961). Mineral relationships of the ruminant. *In* "Digestive Physiology and Nutrition of the Ruminant" (D. Lewis, ed.), pp. 154–166. Butterworth, London and Washington, D.C.

Brown, D. W., and W. E. C. Moore (1960). *J. Dairy Sci.* **43,** 1570–1574.

Brown, F. B., and E. L. Smith (1954). *Biochem. J.* **56,** xxxiv–xxxv.

Brown, L. R., R. H. Johnson, N. L. Jacobson, and P. G. Homeyer (1958). *J. Animal Sci.* **17,** 374–385.

Brown, R. E., and C. L. Davis (1962). *Federation Proc.* **21,** 396.

Brown, R. E., and J. C. Shaw (1957). *J. Dairy Sci.* **40,** 667–671.

Brown, R. E., C. L. Davis, J. R. Staubus, and W. O. Nelson (1960). *J. Dairy Sci.* **43,** 1788–1795.

Browning, C. B., F. C. Fountaine, and E. E. Bartley (1959). *J. Dairy Sci.* **42,** 1857–1866.

Brüggemann, J., and D. Giesecke (1963). *Z. Tierphysiol. Tierernalhr. Futtermittelk.* **18,** 215–226.

Brüggemann, J., D. Giesecke, and K. Drepper (1962). *Z. Tierphysiol. Tierernalhr. Futtermittelk.* **17,** 162–188.

Brugnone (1809). *Mem. Roy. Acad. Turin* 1–56 and 309–346.

Brumpt, E., and Joyeux, C. (1912). *Bull. Soc. Pathol. Exotique* **5,** 499.

Brunaud, M. (1954). *Rev. Med. Vet.* **105,** 535–580.

Brünnich, J. C., and W. R. Winks (1931). *Queensland Agr. J.* **35,** 183–188.

Bruno, C. F., and W. E. C. Moore (1962). *J. Dairy Sci.* **45,** 109–115.

Bryant, M. P. (1952). *J. Bacteriol.* **64,** 325–335.

Bryant, M. P. (1956). *J. Bacteriol.* **72,** 162–167.

Bryant, M. P. (1959). *Bacteriol. Rev.* **23,** 125–153.

Bryant, M. P. (1963). *J. Animal Sci.* **22,** 801–813.

Bryant, M. P. (1965). *In* "Physiology of Digestion in the Ruminant" (R. W. Dougherty, ed.), pp. 411–418. Butterworth, London and Washington, D. C.

Bryant, M. P., and L. A. Burkey (1953a). *J. Dairy Sci.* **36,** 205–217.

Bryant, M .P., and L. A. Burkey (1953b). *J. Dairy Sci.* **36,** 218–224.

Bryant, M. P., and R. N. Doetsch (1954a). *Science* **120,** 944–945.

Bryant, M. P., and R. N. Doetsch (1954b). *J .Dairy Sci.* **37,** 1176–1183.

Bryant, M. P., and I. M. Robinson (1961a). *Appl. Microbiol.* **9,** 96–103.

Bryant, M. P., and I. M. Robinson (1961b). *J. Dairy Sci.* **44,** 1446–1456.

Bryant, M. P., and I. M. Robinson (1961c). *Appl. Microbiol.* **9,** 91–95.

Bryant, M. P., and I. M. Robinson (1962a). *Abstr. 8th Intern. Cong. Microbiol., Montreal, 1962* p. 42.

Bryant, M. P., and I. M. Robinson (1962b). *J. Bacteriol.* **84,** 605–614.

Bryant, M. P., and I. M. Robinson (1963). *J. Dairy Sci.* **46,** 150–154.

Bryant, M. P., and N. Small (1956a). *J. Bacteriol.* **72,** 16–21.

Bryant, M. P., and N. Small (1956b). *J. Bacteriol.* **72,** 22–26.

Bryant, M. P., and N. Small (1960). *J. Dairy Sci.* **43,** 654–667.

Bryant, M. P., N. Small, C. Bouma, and I. M. Robinson (1958a). *J. Dairy Sci.* **41,** 1747–1767.

Bryant, M. P., N. Small, C. Bouma, and H. Chu (1958b). *J. Bacteriol.* **76,** 15–23.

Bryant, M. P., N. Small, C. Bouma, and I. M. Robinson (1958c). *J. Bacteriol.* **76,** 529–537.

Bryant, M. P., I. M. Robinson, and H. Chu (1959). *J. Dairy Sci.* **42,** 1831–1847.

Bryant, M. P., B. F. Barrentine, J. F. Sykes, I. M. Robinson, C. V. Shawver, and L. W. Williams (1960). *J. Dairy Sci.* **43,** 1435–1444.

Bryant, M. P., I. M. Robinson, and I. L. Lindahl (1961). *Appl. Microbiol.* **9,** 155–515.

Buchman, D. T., and R. W. Hemken (1964). *J. Dairy Sci.* **47,** 861–864.

Buisson, J. (1923a). *Compt. Rend. Soc. Biol.* **89,** 1217–1219.

Buisson, J. (1923b). *Ann. Parasitol. Humaine Comparee* **1,** 209–246.

Buisson, J. (1924). *Ann. Parasitol. Humaine Comparee* **2,** 155–160.

Bull, S., and Carrol, W. E. (1949). "Principles of Feeding Farm Animals." Interstate Publ., Danville, Illinois.

Bullen, J. J., and I. Batty (1957). *J. Pathol. Bacteriol.* **73,** 511–518.

Bullen, J. J., R. Scarisbrick, and A. Maddock (1953). *J. Pathol. Bacteriol.* **65,** 209–219.

Buller, C. S., and J. M. Akagi (1964). *J. Bacteriol.* **88,** 440–443.

Burke, J. D. (1950). *Proc. 1950 Cornell Nutr. Conf. Feed Manuf.* pp. 5–10.

Burroughs, W. (1950). *Proc. 1950 Cornell Nutr. Conf. Feed Manuf.* pp. 11–23.

Burroughs, W., and P. Gerlaugh (1949). *J. Animal Sci.* **8,** 3–8.

Burroughs, W., P. Gerlaugh, A. F. Schalk, E. A. Silver, and L. E. Kunkle (1945). *J. Animal Sci.* **4,** 373–386.

Burroughs, W., P. Gerlaugh, E. A. Silver, and A. F. Schalk (1946a). *J. Animal Sci.* **5,** 272–278.

Burroughs, W., P. Gerlaugh, E. A. Silver, and A. F. Schalk (1946b). *J. Animal Sci.* **5,** 338–349.

Burroughs, W., P. Gerlaugh, B. H. Edgington, and R. M. Bethke (1949a). *J. Animal Sci.* **8,** 271–278.

Burroughs, W., P. Gerlaugh, B. H. Edgington, and R. M. Bethke (1949b). *J. Animal Sci.* **8,** 9–18.

Burroughs, W., P. Gerlaugh, anl R. M. Bethke (1950a). *J. Animal Sci.* **9,** 207–213.

Burroughs, W., J. Long, P. Gerlaugh, and R. M. Bethke (1950b). *J. Animal Sci.* **9,** 523–530.

Burroughs, W., N. Frank, P. Gerlaugh, and R. M. Bethke (1950c). *J. Nutr.* **40,** 9–24.

Burroughs, W., L. S. Gall, P. Gerlaugh, and R. M. Bethke (1950d). *J. Animal Sci.* **9,** 214–220.

Burroughs, W., H. G. Headley, R. M. Bethke, and P. Gerlaugh (1950e). *J. Animal Sci.* **9,** 513–522.

Burroughs, W., C. Arias, P. DePaul, P. Gerlaugh, and R. M. Bethke (1951a). *J. Animal Sci.* **10,** 672–682.

Burroughs, W., A. Latona, P. DePaul, P. Gerlaugh, and R. M. Bethke (1951b). *J. Animal Sci.* **10,** 693–705.

Burroughs, W., C. C. Culbertson, J. Kastelic, E. Cheng, and W. H. Hale (1954). *Science* **120,** 66–67.

Burroughs, W., W. Woods, S. A. Ewing, J. Greig, and B. Theurer (1960). *J. Animal Sci.* **19,** 458–464.

Butler, G. W., and A. T. Johns (1961). *J. Australian Inst. Agr. Sci.* **27,** 123–133.

Butler, G. W., and P. J. Peterson (1961). *New Zealand J. Agr. Res.* **4,** 484–491.

Butterworth, J. P., S. E. Bell, and M. G. Garvock (1960). *Biochem. J.* **74,** 180–182.

Butterworth, M. H. (1964). *J. Agr. Sci.* **63,** 319–321.

Butz, H., H. Meyer, and K. Korber (1958a). *Zuechtungskunde* **30,** 39–43.

Butz, H., H. Meyer, and C. Shulken (1958b). *Berlin. Muench. Tieraertzl. Wochschr.* **71,** 163–168.

Buziassy, C., and D. E. Tribe (1960a). *Australian J. Agr. Res.* **11,** 989–1001.

Buziassy, C., and D. E. Tribe (1960b). *Australian J. Agr. Res.* **11,** 1002–1008.

Calaby, J. H. (1958). *Australian J. Biol. Sci.* **11,** 571–580.

Caldwell, D. R., M. P. Bryant, and D. C. White (1962). *Proc. Am. Dairy Sci. Assoc.* Paper No. 105. Oral communication.

Campbell, A. S. (1929). *Arch. Protistenk.* **66,** 331–339.

Campbell, C. D., G. A. McLaren, G. C. Anderson, J. A. Welch, G. S. Smith, and J. Brooks (1959). *J. Animal Sci.* **18,** 780–789.

Campbell, J. B., A. N. Davis, and P. J.Myhr (1954). *Can. J. Comp. Med. Vet. Sci.* **18,** 93–101.

Campbell, L. L., and J. R. Postgate (1965). *Bacteriol. Rev.* **29,** 359–363.

Campbell, T. C., J. K. Loosli, R. G. Warner, and I. Tosaki (1963). *J. Animal Sci.* **22,** 139–145.

Camper, P. (1803). "Ouvres de P. Camper, qui ont pour objet l'histoire naturelle, la physiologie et l'anatomie comparée" (Hansen, ed.), 3 vols. Paris.

Campling, R. C. (1964). *Proc. Nutr. Soc. (Engl. Scot.)* **23,** 80–88.

Campling, R. C., and C. C. Balch (1961). *Brit. J. Nutr.* **15,** 523–530.

Campling, R. C., and M. Freer (1962). *Brit. J. Nutr.* **16,** 507–518.

Campling, R. C., M. Freer, and C. C. Balch (1961). *Brit. J. Nutr.* **15,** 531–540.

Campling, R. C., M. Freer, and C. C. Balch (1962). *Brit. J. Nutr.* **16,** 115–125.

Card, C. S., and L. H. Schultz (1953). *J. Dairy Sci.* **36,** 599.

Cardon, B. P. (1953). *J. Animal Sci.* **12,** 536–540.

Cardoso, D. M., and W. F. Almeida (1958). *Arquiv. Inst. Biol. (Sao Paulo)* **25,** 237–252.

Carnie, J. A., and Porteous, J. W. (1959). *Biochem. J.* **73,** 47–48P.

Carroll, E. J. (1960). Urea utilizing organisms from the rumen. Ph.D. Thesis, University of Missouri, Columbia, Missouri.

Carroll, E. J., and R. E. Hungate (1954). *Appl. Microbiol.* **2,** 205–214.

Carroll, E. J., and R. E. Hungate (1955). *Arch. Biochem. Biophys.* **56,** 525–536.

Carroll, F. D., W. C. Rollins, and N. R. Ittner (1955) *J. Animal Sci.* **14,** 218–223.

Casati, R. M., and A. Garuti (1961). *Atti Soc. Ital. Sci. Vet.* **15,** 440–442.

Cason, J. L., E. S. Ruby, and O. T. Stallcup (1954). *J. Nutr.* **52,** 457–465.

Castle, E. J. (1956a). *Brit. J. Nutr.* **10,** 15–23.

Castle, E. J. (1956b). *Brit. J. Nutr.* **10,** 115–125.

Cavallo, G., and F. D'Onofrio (1956). *Giorn. Microbiol.* **1,** 349–356.

Certes, A. (1889). *J. Micrographie* **13**, 277–279.

Chalmers, M. I., and S. B. M. Marshall (1964). *J. Agr. Sci.* **63**, 277–282.

Chalmers, M. I., and R. L. M. Synge (1954). *J. Agr. Sci.* **44**, 263–269.

Chalmers, M. I., D. P. Cuthbertson, and R. L. M. Synge (1954). *J. Agr. Sci.* **44**, 254–262.

Chalmers, M. I., J. B. Jayasinghe, and S. B. M. Marshall (1964). *J. Agr. Sci.* **63**, 283–288.

Chalupa, W., J. L. Evans, and M. C. Stillions (1964). *J. Animal Sci.* **23**, 802–807.

Chamberlain, C. C., and W. Burroughs (1962). *J. Animal Sci.* **21**, 428–432.

Chance, C. M., C. F. Huffman, and C. W. Duncan (1953a). *J. Dairy Sci.* **36**, 253–259.

Chance, C. M., C. W. Duncan, C. F. Huffman, and R. W. Luecke (1953b). *J. Dairy Sci.* **36**, 495–503.

Chance, C. M., C. R. Smith, C. F. Huffman, and C. W. Duncan (1953c). *J. Dairy Sci.* **36**, 743–751.

Chappel, C. F. (1952). Factors affecting the digestibility of low quality roughages by lambs. Ph.D. Thesis, Oklahoma A.&M. College, Stillwater, Oklahoma.

Cheng, E. W., G. Hall, and Burroughs, W. (1955). *J. Dairy Sci.* **38**, 1225–1230.

Chou, K. C., and D. M. Walker (1964a). *J. Agr. Sci.* **62**, 7–13.

Chou, K. C., and D. M. Walker (1964b). *J. Agr. Sci.* **62**, 15–25.

Christian, K. R., and V. J. Williams (1957). *New Zealand J. Sci. Technol.* **A38**, 1003–1010.

Christiansen, W. C. (1963). Nutrient and other environmental influences upon rumen protozoal growth and metabolism. Ph.D. Thesis, Iowa State University, Ames, Iowa.

Christiansen, W. C., L. Y. Quinn, and W. Burroughs (1962). *J. Animal Sci.* **21**, 706–710.

Christiansen, W. C., W. Woods, and W. Burroughs (1964). *J. Animal Sci.* **23**, 984–988.

Christie, A. O., and J. W. Porteous (1957). *Biochem. J.* **67**, 19P.

Christie, L. E., and G. A. Lassiter (1958). *Mich. State Univ., Agr. Expt. Sta., Quart. Bull.* **40**, 714–723.

Christl, H. (1958). *Z. Parasitenk.* **18**, 297–307.

Church, D. E., and R. G. Petersen (1960). *J. Dairy Sci.* **43**, 81–92.

Cipolloni, M. A., B. H. Schneider, H. L. Lucas, and H. M. Parlech (1951). *J. Animal Sci.* **10**, 337–343.

Claren, O. B. (1942). *Z. Physiol. Chem.* **276**, 97–107.

Clark, D. E., J. E. Young, R. L. Younger, L. M. Hunt, and J. K. McLaren (1964). *J. Agr. Food. Chem.* **12**, 43–47.

Clark, R. (1948). *Onderstepoort J. Vet. Sci. Animal Ind.* **23**, 389–393.

Clark, R. (1950). *J. S. African Vet. Med. Assoc.* **21**, 173–178.

Clark, R. (1951). *Onderstepoort J. Vet. Res.* **25**, 67–72.

Clark, R. (1953). *Onderstepoort J. Vet. Res.* **26**, 137–140.

Clark, R. (1956). *J. S. African Vet. Med. Assoc.* **27**, 79–104.

Clark, R., and W. A. Lombard (1951). *Onderstepoort J. Vet. Res.* **25**, 79–92.

Clark, R., and J. R. Malan (1956). *Onderstepoort J. Vet. Res.* **27**, 101–109.

Clark, R., and J. I. Quin (1945). *Onderstepoort J. Vet. Sci. Animal Ind.* **20**, 209–212.

Clark, R., and J. I. Quin (1951). *Onderstepoort J. Vet. Res.* **25**, 93–103.

Clark, R., and K. E. Weiss (1952a). *J. S. Afr. Vet. Med. Assoc.* **23**, 163–165.

Clark, R., and K. E. Weiss (1952b). *J. S. Afr. Vet. Med. Assoc.* **23**, 103–106.

Clark, R., W. Oyaert, and J. I. Quin (1951). *Onderstepoort J. Vet. Res.* **25,** 73–78.

Clark, R., E. L. Barrett, and J. H. Kellerman (1963). *J. S. African Vet. Med. Assoc.* **34,** 419–423.

Clarke, R. T. J. (1959). *J. Gen. Microbiol.* **20,** 549–553.

Clarke, R. T. J. (1963). *J. Gen. Microbiol.* **33,** 401–408.

Clarke, R. T. J. (1964a). *New Zealand J. Agr. Res.* **7.** 248–257.

Clarke, R. T. J. (1964b). *New Zealand J. Agr. Res.* **7,** 525–530.

Clarke, R. T. J. (1964c). Studies on the flora and fauna of the bovine rumen. Ph.D. Thesis, Massey University of Manawatu, New Zealand.

Clarke, R. T. J. (1965a). *Nature* **205,** 95–96.

Clarke, R. T. J. (1965b). *New Zealand J. Agr. Res.* **8,** 1–9.

Clarke, R. T. J., and M. E. di Menna (1961). *J. Gen. Microbiol.* **25,** 113–117.

Clarke, R. T. J., and R. E. Hungate (1966). *Appl. Microbiol.* Submitted.

Claydon, T. J., and G. W. Teresa (1959). *J. Dairy Sci.* **42,** 712–714.

Claypool, D. W., D. R. Jacobson, and R. F. Wiseman (1961). *J. Dairy Sci.* **44,** 174–175.

Cleveland, L. R. (1923). *Proc. Natl. Acad. Sci. U.S.* **9,** 424–428.

Cleveland, L. R. (1924). *Biol. Bull.* **46,** 177–225.

Cleveland, L. R. (1926). *Quart. Rev. Biol.* **1,** 51–60.

Cline, J. H. (1952). Ruminant nutrition. Studies on sulfur and mineral metabolism of rumen microorganisms *in vitro.* M.S. Thesis, Ohio State University, Columbus, Ohio.

Cline, J. H., T. V. Hershberger, and O. G. Bentley (1958). *J. Animal Sci.* **17,** 284–292.

Coats, D. A., and R. D. Wright (1957). *J. Physiol. (London)* **135,** 611–622.

Coats, D. A., D. A. Denton, J. R. Goding, and R. D. Wright (1956). *J. Physiol. (London)* **131,** 13–31.

Coats, D. A., D. A. Denton, and R. D. Wright (1958). *J. Physiol. (London)* **144,** 108–122.

Cole, H. H., and M. Kleiber (1945). *Am. J. Vet. Res.* **6,** 188–193.

Cole, H. H., and M. Kleiber (1948). *J. Dairy Sci.* **34,** 1016–1023.

Cole, H. H., S. W. Mead, and M. Kleiber (1942). *Calif., Univ., Agr. Expt. Sta. Bull.* **662.**

Cole, H. H., C. F. Huffman, M. Kleiber, T. M. Olson, and A. F. Schalk (1945). *J. Animal Sci.* **4,** 183–236.

Cole, H. H., R. W. Dougherty, C. F. Huffman, R. E. Hungate, M. Kleiber, and W. D. Maclay (1956). A review of bloat in ruminants. *Natl. Acad. Sci.—Natl. Res. Council Publ.* **388.**

Coleman, G. S. (1958). *Nature* **182,** 1104–1105.

Coleman, G. S. (1960a). *Nature* **187,** 518–520.

Coleman, G. S. (1960b). *J. Gen. Microbiol.* **22,** 555–563.

Coleman, G. S. (1960c). *J. Gen. Microbiol.* **22,** 423–436.

Coleman, G. S. (1962). *J. Gen. Microbiol.* **28,** 271–281.

Coleman, G. S. (1964). *J. Gen. Microbiol.* **37,** 209–223.

Colin, G. (1852). *Compt. Rend.* **34,** 681–683.

Colin, G. (1854). "Traité de physiologie comparée des animaux domestiques," Vol. I. ballière et Fils, Paris.

Colin, G. (1873). "Traité de physiologie comparée," 2nd ed., Vol. I. Ballière et Fils, Paris.

Colin, G. (1886). "Traité de physiologie comparée," 3rd ed., Vol. I. Ballière et Fils, Paris.

Columbus, A. (1934). Der quantitative Verlauf der Entleerung des Pansens bei Schafen und Ziegen, mit Berücksichtigung der Gesamtentleerung des Magen-Darmkanals. 49 pp. Thesis, Berlin.

Columbus, A. (1936). *Forschungsdienst* **2,** 208–213.

Colvin, H. W., J. D. Wheat, E. A. Rhode, and J. M. Boda (1957). *J. Dairy Sci.* **40,** 492–502.

Colvin, H. W., P. T. Cupps, and H. H. Cole (1958a). *J. Dairy Sci.* **41,** 1557–1564.

Colvin, H. W., P. T. Cupps, and H. H. Cole (1958b). *J. Dairy Sci.* **41,** 1565–1579.

Colvin, H. W., J. M. Boda, and T. Wegner (1959). *J. Dairy Sci.* **42,** 333–345.

Comar, C. L., G. K. Davis, R. F. Taylor, C. F. Huffman, and R. E. Ely (1946). *J. Nutr.* **32,** 61–68.

Comline, R. S., and R. N. B. Kay (1955). *J. Physiol. (London)* **129,** 55–56P.

Comline, R. S., and D. A. Titchen (1951). *J. Physiol. (London)* **115,** 210–226.

Comline, R. S., and D. A. Titchen (1957). *J. Physiol. (London)* **139,** 24–25P.

Conchie, J. (1954). *Biochem. J.* **58,** 552–560.

Conrad, H. R., and J. W. Hibbs (1953). *J. Dairy Sci.* **36,** 1326–1334.

Conrad, H. R., and J. W. Hibbs (1955). *J. Dairy Sci.* **38,** 548.

Conrad, H. R., and J. W. Hibbs (1959). *Ohio Farm Home Res.* **44,** 75–77.

Conrad, H. R., and J. W. Hibbs (1961). *J. Dairy Sci.* **44,** 1903–1909.

Conrad, H. R., J. W. Hibbs, W. D. Pounden, and T. S. Sutton (1950). *J. Dairy Sci.* **33,** 585–592.

Conrad, H. R., H. R. Smith, J. R. Vandersall, W. D. Pounden, and J. W. Hibbs (1958a). *J. Dairy Sci.* **41,** 1094–1099.

Conrad, H. R., J. W. Hibbs, and N. Frank (1958b). *J. Dairy Sci.* **41,** 1248–1261.

Conrad, H. R., W. D. Pounden, O. G. Bentley, and A. W. Fetter (1958c). *J. Dairy Sci.* **41,** 1586–1592.

Conrad, H. R., J. W. Hibbs, and A. D. Pratt (1960a). *Ohio Agr. Expt. Sta. Bull.* **861.**

Conrad, H. R., J. W. Hibbs, A. D. Pratt, and J. R. Vandersall (1960b). *Ohio Agr. Expt. Sta. Bull.* **867.**

Conrad, H. R., W. D. Pounden, and B. A. Dehority (1961). *J. Dairy Sci.* **44,** 2015–2026.

Cook, J. W. (1957). *J. Agr. Food Chem.* **5,** 859–863.

Cook, R. M., and R. H. Ross (1964). *J. Animal Sci.* **23,** 601.

Coop, I. E. (1949). *New Zealand J. Sci. Technol.* **A31,** 1–12.

Coop, I. E. (1962). *J. Agr. Res.* **58,** 179–186.

Coop, I. E., and R. L. Blakely (1949). *New Zealand J. Sci. Technol.* **A30,** 277–291.

Coop, I. E., and R. L. Blakely (1950). *New Zealand J. Sci. Technol.* **A31** No. 5, 44–58.

Coop, I. E., and M. K. Hill (1962). *J. Agr. Sci.* **58,** 187–199.

Coppock, C. E., W. P. Flatt, L. A. Moore, and W. E. Stewart (1964). *J. Dairy Sci.* **47,** 1359–1364.

Corbett, J. L. (1957). *Natl. Agr. Advisory Serv. Quart. Rev.* **37,** 9 pp.

Corbett, J. L., J. F. D. Greenhalgh, and P. MacDonald (1958a). *Nature* **182,** 1014–1016.

Corbett, J. L., J. F. D. Greenhalgh, P. E. Gwynn, and D. Walker (1958b). *Brit. J. Nutr.* **12,** 266–276.

Corbett, J. L., J. F. D. Greenhalgh, and E. Florence (1959). *Brit. J. Nutr.* **13,** 337–344.

Coup, M. R., and A. G. Campbell (1964). *New Zealand J. Agr. Res.* **7,** 624–638.

Cowie, A. T., W. G. Duncombe, S. J. Folley, T. H. French, R. F. Glascock, L. Massart, G. J. Peeters, and G. Popjak (1951). *Biochem. J.* **49**, 610–615.

Craine, E. M., and R. G. Hansen (1952). *J. Dairy Sci.* **35**, 631–636.

Crampton, E. W., and L. A. Maynard (1938). *J. Nutr.* **15**, 383–395.

Crampton, E. W., E. E. Lister, and L. E. Lloyd (1957). *J. Animal Sci.* **16**, 1056.

Crane, A., W. O. Nelson, and R. E. Brown (1957). *J. Dairy Sci.* **40**, 1317–1324.

Crawley, H. (1923). *Proc. Acad. Natl. Sci. Phila.* **75**, 393–412.

Cresswell, E. (1960). *Nature* **186**, 560–561.

Crossland, A., E. C. Owen, and R. Proudfoot (1958). *Brit. J. Nutr.* **12**, 312–329.

Csonka, F. A., M. Phillips, and D. B. Jones (1929). *J. Biol. Chem.* **85**, 65–75.

Cunningham, H. M., and G. J. Brisson (1955). *Can. J. Agr. Sci.* **35**, 511–512.

Cunningham, I. J., and K. G. Hogan (1959). *New Zealand J. Agr. Res.* **2**, 134–144.

Cunningham, M. D., F. A. Martz, and C. P. Merilan (1964). *J. Dairy Sci.* **47**, 382–385.

Cuthbertson, D. P., and M. I. Chalmers (1950). *Biochem. J.* **46**, xvii–xviii.

Czepa, A., and R. Stigler (1926). *Arch. Ges. Physiol.* **212**, 300–356.

Dain, J. A., A. L. Neal, and R. W. Dougherty (1955). *J. Animal Sci.* **14**, 930–935.

Dain, J. A., A. L. Neal, and H. W. Seeley (1956). *J. Bacteriol.* **72**, 209–213.

Dale, H. E., R. E. Stewart, and S. Brody (1954). *Cornell Vet.* **44**, 368–374.

Danielli, J. F., M. W. S. Hitchcock, R. A. Marshall, and A. T. Phillipson (1945). *J. Exptl. Biol.* **22**, 75–84.

Davey, D. G. (1936). *J. Agr. Sci.* **26**, 328–330.

Davey, L. A., G. C. Chesseman, and C. A. E. Briggs (1960). *J. Agr. Sci.* **55**, 155–163.

Davidson, J. (1954). *J. Sci. Food Agr.* **5**, 86–92.

Davis, C. L., and R. E. Brown (1962). *J. Dairy Sci.* **45**, 513–516.

Davis, C. L., R. E. Brown, J. R. Staubus, and W. O. Nelson (1960a). *J. Dairy Sci.* **43**, 231–240.

Davis, C. L., R. E. Brown, J. R. Staubus, and W. O. Nelson (1960b). *J. Dairy Sci.* **43**, 241–249.

Davis, C. L., R. E. Brown, and J. R. Staubus (1960c). *J. Dairy Sci.* **43**, 1783–1787.

Davis, C. L., R. E. Brown, and D. E. Beitz (1964). *J. Dairy Sci.* **47**, 1217–1219.

Davis, G. V., and O. T. Stallcup (1964). *J. Dairy Sci.* **47**, 1237–1242.

Davis, R. E. (1962). *Agr. Res.* **10**, 7.

Davis, R. E., R. R. Oltjen, and J. Bond (1963). *J. Animal Sci.* **22**, 640–643.

Davison, K. L., and J. Seo (1963). *J. Dairy Sci.* **46**, 862–863.

Dawbarn, M. C., D. C. Hine, and J. Smith (1957a). *Australian J. Exptl. Biol. Med. Sci.* **35**, 97–102.

Dawbarn, M. C., D. C. Hine, and J. Smith (1957b). *Australian J. Exptl. Biol. Med. Sci.* **35**, 273–276.

Dawson, R. M. C. (1959). *Nature* **183**, 1822–1823.

Dawson, R. M. C., P. F. V. Ward, and T. W. Scott (1964). *Biochem. J.* **90**, 9–12.

de Alba, J. and J. M. C. Sampaio (1957). *J. Animal Sci.* **16**, 725–731.

Dehority, B. A. (1961). *J. Dairy Sci.* **44**, 687–692.

Dehority, B. A. (1963). *J. Dairy Sci.* **46**, 217–222.

Dehority, B. A. (1965). *J. Bacteriol.* **89**, 1515–1520.

Dehority, B. A., and R. R. Johnson (1961). *J. Dairy Sci.* **44**, 2242–2249.

Dehority, B. A., and R. R. Johnson (1964). *J. Animal Sci.* **23**, 203–207.

Dehority, B. A., and H. W. Scott (1963). *J. Animal Sci.* **22**, 1118–1119.

Dehority, B. A., O. G. Bentley, R. R. Johnson, and A. L. Moxon (1957). *J. Animal Sci.* **16,** 502–514.

Dehority, B. A., R. R. Johnson, O. G. Bentley, and A. L. Moxon (1958). *Arch. Biochem. Biophys.* **78,** 15–27.

Dehority, B. A., K. el-Shazly, and R. R. Johnson (1960). *J. Animal Sci.* **19,** 1098–1109.

Dehority, B. A., R. R. Johnson, and H. R. Conrad (1962). *J. Dairy Sci.* **45,** 508–512.

Demaux, G., H. LeBars, J. Molle, A. Rérat, and H. Simonnet (1961). *Bull. Acad. Vet. France* **34,** 85–88.

Denton, D. A. (1957). *Quart. J. Exptl. Physiol.* **42,** 72–95.

Devlin, T. J., and W. K. Roberts (1963). *J. Animal Sci.* **22,** 648–653.

Devlin, T. J., and W. Woods (1964). *J. Animal Sci.* **23,** 872–873.

de Waele, A., and G. Genie (1943a). *Verhandel. Koninkl. Vlaam. Acad. Wetenschap. Belg.* **5,** No. 7.

de Waele, A., and G. Genie (1943b). *Verhandel. Koninkl. Vlaam. Acad. Wetenschap. Belg.* **5,** No. 11.

Dewey, D. W., H. J. Lee, and H. R. Marston (1958). *Nature* **181,** 1367–1371.

Dick, A. T. (1952). *Australian Vet. J.* **28,** 30–33.

Dick, A. T. (1954). *Australian J. Agr. Res.* **5,** 511–544.

Dick, A. T. (1956). *In* "Inorganic Nitrogen Metabolism" (W. D. McElroy and B. Glass, eds.), pp. 445–473. Johns Hopkins Press, Baltimore, Maryland.

Dijkstra, N. D. (1956). *Landbouwk. Tijdschr.* **68,** 609–611.

Dion, H. W., D. G. Calkins, and J. J. Pfiffner (1952). *J. Am. Chem. Soc.* **74,** 1108.

Dirkson, G., and L. Wolf (1963). *Tieralrtzl. Umschau* **18,** 282–285.

Dobie, J. B., and R. G. Curley (1963). *Univ. Calif. Agr. Expt. Sta. Extension Circ.* **517,** 38 pp.

Dobson, A. (1959). *J. Physiol. (London)* **146,** 235–251.

Dobson, A. (1961). *In* "Digestive Physiology and Nutrition of the Ruminant" (D. Lewis, ed.), pp. 68–78. Butterworth, London and Washington, D.C.

Dobson, A., and A. T. Phillipson (1956). *J. Physiol. (London)* **133,** 76P–77P.

Dobson, A., and A. T. Phillipson (1958). *J. Physiol. (London)* **140,** 94–104.

Dobson, A., and A. T. Phillipson (1961). *Proc. Symp. Colston Res. Soc.* **13,** 1–12.

Dobson, A., R. N. B. Kay, and I. McDonald (1960). *Res. Vet. Sci.* **1,** 103–110.

Dobson, M. J. (1955). *J. Physiol. (London)* **128,** 25P.

Dobson, M. J., W. C. B. Brown, A. Dobson, and A. T. Phillipson (1956). *Quart. J. Exptl. Physiol.* **41,** 247–253.

Dodd, D. C., and M. R. Coup (1957). *New Zealand Vet. J.* **5,** 51–54.

Dodsworth, T. L. (1954). *J. Agr. Sci.* **44,** 383–393.

Doetsch, R. N. (1957). *J. Dairy Sci.* **40,** 1204–1207.

Doetsch, R. N., and P. Jurtshuk (1957). *Univ. Maryland Agr. Exptl. Sta. Misc. Publ.* **291,** 8–11.

Doetsch, R. N., R. Q. Robinson, R. E. Brown, and J. C. Shaw (1953). *J. Dairy Sci.* **36,** 825–831.

Doetsch, R. N., J. C. Shaw, J. J. McNeill, and P. Jurtshuk (1955). *Univ. Maryland Agr. Exptl. Sta. Misc. Publ.* **238,** 1–2.

Doetsch, R. N., B. H. Howard, S. O. Mann, and A. E. Oxford (1957). *J. Gen. Microbiol.* **16,** 156–168.

Dogiel, V. A. (1925a). *Ann. Parasitol. Humaine Comparée* **3,** 116–142.

Dogiel, V. A. (1925b). *Arch. Protistenk.* **50,** 283–442.

Dogiel, V. A. (1925c). *Arch. Russes Protistol.* **4,** 43–65.

Dogiel, V. A. (1926a). *Ann. Parasitol. Humaine Comparée* **4,** 61–64.

Dogiel, V. A. (1926b). *Ann. Parasitol. Humaine Comparée* **4,** 241–271.

Dogiel, V. A. (1927a). *Arch. Entwicklungsmech. Organ.* **109,** 380–389.

Dogiel, V. A. (1927b). *Arch. Protistenk.* **59,** 1–288.

Dogiel, V. A. (1928). *Ann. Parasitol. Humaine Comparée* **6,** 323–338.

Dogiel, V. A. (1947). *Quart. J. Microscop. Sci.* **88,** 337–343.

Dogiel, V. A., and T. Fedorowa (1925). *Zool. Anz.* **62,** 97–107.

Dogiel, V. A., and T. Fedorowa (1926). *J. Genet.* **16,** 257–268.

Dogiel, V. A., and T. Winogradowa-Fedorowa (1930). *Wiss. Arch. Landwirtsch. Abt.* B, *Arch. Tierenaehr Tierzucht* **3,** 172–188.

Dohner, P. M., and B. P. Cardon (1954). *J. Bacteriol.* **67,** 608–611.

Donefer, E., E. W. Crampton, and L. E. Lloyd (1960). *J. Animal Sci.* **19,** 545–552.

Donefer, E., L. E. Lloyd, and E. W. Crampton (1962). *J. Animal Sci.* **21,** 815–818.

Donefer, E., L. E. Lloyd, and E. W. Crampton (1963). *J. Animal Sci.* **22,** 425–428.

Dougherty, R. W. (1940). *J. Am. Vet. Med. Assoc.* **96,** 43–46.

Dougherty, R. W. (1941). *J. Am. Vet. Med. Assoc.* **99,** 110–114.

Dougherty, R. W. (1942a). *Am. J. Vet. Res.* **3,** 401–402.

Dougherty, R. W. (1942b). *Cornell Vet.* **32,** 269–280.

Dougherty, R. W. (1955). *Cornell Vet.* **45,** 331–357.

Dougherty, R. W. (1956a). *Natl. Acad. Sci.—Natl. Res. Council Publ.* **388,** 26–32.

Dougherty, R. W. (1956b). *Proc. 7th Intern. Grasslands Congr., Palmerston, New Zealand, 1956* Paper No. 24, pp. 262–272.

Dougherty, R. W., and R. B. Christensen (1953). *Cornell Vet.* **43,** 481–486.

Dougherty, R. W., and R. E. Habel (1955). *Cornell Vet.* **45,** 459–464.

Dougherty, R. W., and C. D. Meredith (1955). *Am. J. Vet. Res.* **16,** 96–100.

Dougherty, R. W., C. D. Meredith, and R. B. Barrett (1955). *Am. J. Vet. Res.* **16,** 79–90.

Dougherty, R. W., R. E. Habel, and H. E. Bond (1958). *Am. J. Vet. Res.* **19,** 115–128.

Dougherty, R. W., W. E. Stewart, M. M. Nold, I. L. Lindahl, C. H. Mullenax, and B. F. Leek (1962a). *Am. J. Vet. Res.* **23,** 205–212.

Dougherty, R. W., K. J. Hill, F. L. Campetti, R. C. McClure, and R. E. Habel (1962b). *Am. J. Vet. Res.* **23,** 213–219.

Dougherty, R. W., C. H. Mullenax and M. J. Allison (1965). *In* "Physiology of Digestion in the Ruminant" (R. W. Dougherty, ed.), pp. 159–170. Butterworth, London and Washington, D.C.

Dow, J. K. D. (1959). *Proc. Nutr. Soc.* (*Engl. Scot.*) **18,** 134–139.

Downes, A. M., and I. W. McDonald (1964). *Brit. J. Nutr.* **18,** 153–162.

Downie, H. G. (1954). *Am. J. Vet. Res.* **15,** 217–223.

Dracy, A. E., and W. O. Essler (1964). *J. Dairy Sci.* **47,** 1428–1429.

Drawert, F., H. J. Kuhn, and A. Rapp (1962). *Z. Physiol. Chem.* **329,** 84–89.

Drori, D., and J. K. Loosli (1959). *J. Animal Sci.* **18,** 206–210.

Drori, D., and J. K. Loosli (1961). *J. Animal Sci.* **20,** 233–238.

Dryden, L. P., A. M. Hartman, M. P. Bryant, I. M. Robinson, and L. A. Moore (1962). *Nature* **195,** 201–202.

Duckworth, J. (1946). *Tropical Agriculturist* **23,** 4–8.

Duncan, C. W., I. P. Agrawala, C. F. Huffman, and R. W. Luecke (1953). *J. Nutr.* **49,** 41–49.

Duncan, D. L. (1953). *J. Physiol.* (*London*) **119,** 157–169.

Duncan, D. L., and A. T. Phillipson (1951). *J. Exptl. Biol.* **28,** 32–40.

Dussardier, M. (1954a). *Compt. Rend. Soc. Biol.* **148,** 446.

Dussardier, M. (1954b). *J. Physiol.* (*London*) **46,** 777–797.

Dussardier, M., J. L. Durel, and C. Lavenet (1961). *Ann. Biol. Animale Biochim. Biophys.* **1,** 113–116.

Dye, J. A., and E. L. McCandless (1948). *Cornell Vet.* **38,** 331–340.

Dye, J. A., S. J. Roberts, N. Blampied, and M. G. Fincher (1953). *Cornell Vet.* **43,** 128–160.

Dyer, I. A., and D. W. Fletcher (1958). *J. Animal Sci.* **17,** 391–397.

Dzuick, H. E., and A. F. Sellers (1955a). *Am. J. Vet. Res.* **16,** 499–504.

Dzuick, H. E., and A. F. Sellers (1955b). *Am. J. Vet. Res.* **16,** 411–417.

Eadie, J. M. (1957). *Proc. Roy. Soc. Edinburgh* **B66,** 276–287.

Eadie, J. M. (1962a). *J. Gen. Microbiol.* **29,** 563–578.

Eadie, J. M. (1962b). *J. Gen. Microbiol.* **29,** 579–588.

Eadie, J. M., and P. N. Hobson (1962). *Nature* **193,** 503–505.

Eadie, J. M., and A. E. Oxford (1954). *Nature* **174,** 973.

Eadie, J. M., and A. E. Oxford (1955). *J. Gen. Microbiol.* **12,** 298–310.

Eadie, J. M., and A. E. Oxford (1957). *Nature* **179,** 485.

Eadie, J. M., S. O. Mann, and A. E. Oxford (1956). *J. Gen. Microbiol.* **14,** 122–133.

Eadie, J. M., P. N. Hobson, and S. O. Mann (1959). *Nature* **183,** 624–625.

Eadie, J. M., D. J. Manners, and J. B. Stark (1963). *Biochem. J.* **89,** 91P.

Eberlein, R. (1895). *Z. Wiss. Zool.* **59,** 233–304.

Edin, H. (1918). *Medd. Centralanstalt. Forsoksvaesendet Jordbruks.* **165.** Cited by Anderson, A. C. (1934).

Edin, H. (1926). *Medd. Centralanstalt. Forsoksvaesendet Jordbruks.* **309.** Cited by Anderson, A. C. (1934).

Edlefsen, J. L., C. W. Cook, and J. T. Blake (1960). *J. Animal Sci.* **19,** 561–567.

Edwards, D. C., and R. A. Darroch (1956). *Brit. J. Nutr.* **10,** 286–292.

Egan, A. R. (1965a). *Australian J. Agr. Res.* **16,** 451–462.

Egan, A. R. (1965b). *Australian J. Agr. Res.* **16,** 463–472.

Egan, A. R. (1965c). *Australian J. Agr. Res.* **16,** 473–483.

Egan, A. R., and R. J. Moir (1965). *Australian J. Agr. Res.* **16,** 437–449.

Einszporn, T. (1961). *Acta Parasitol. Polon.* **9,** 195–210.

Ekelund, S. (1949). *Ann. Agr. Coll. Sweden* **16,** 179–326.

Ekern, A., and J. T. Reid (1963). *J. Dairy Sci.* **46,** 522–529.

Ekman, J., and I. Sperber (1953). *Ann. Agr. Coll. Sweden* **19,** 227–231.

Elam, C. J., P. A. Putnam, and R. E. Davis (1959). *J. Animal Sci.* **18,** 718–725.

Elam, C. J., J. Gutierrez, and R. E. Davis (1960). *J. Animal Sci.* **19,** 1089–1097.

Ellenberger, W. (1883). *Arch. Wiss. Prakt. Tierheilk.* **9,** 128–147.

Ellenberger, W., and V. Hofmeister (1887a). *Landwirtsch. Jahrb.* **16,** 201–280.

Ellenberger, W., and V. Hofmeister (1887b). *Arch. Anat. Physiol. Leipzig, Physiol. Abt.* Suppl., pp. 138–147.

Ellenberger, W., and A. Scheunert (1925). "Lehrbuch der vergleichenden Physiologie der Haussäugetiere", pp. 215–285. P. Parey, Berlin.

Elliot, J. M., and J. K. Loosli (1959). *J. Dairy Sci.* **42,** 843–848.

Elliott, R. C., and J. H. Topps (1964). *Animal Prod.* **6,** 345–355.

Elliott, R. C., W. D. C. Reed, and J. H. Topps (1964). *Brit. J. Nutr.* **18,** 519–528.

Ellis, W. C., and W. H. Pfander (1957). *J. Animal Sci.* **16,** 1053.

Ellis, W. C., and W. H. Pfander (1958). *J. Nutr.* **65,** 235–250.

Ellis, W. C., G. B. Garner, M. E. Muhrer, and W. H. Pfander (1956). *J. Nutr.* **60,** 413–425.

Ellis, W. C., W. H. Pfander, M. E. Muhrer, and E. E. Pickett (1958). *J. Animal Sci.* **17,** 180–188.

Ellis, W. C., M. G. Murray, R. S. Reid, and T. M. Sutherland (1962). *J. Physiol. (London)* **164,** 3P–4P.

Elsden, S. R. (1945a). *J. Exptl. Biol.* **22,** 51–62.

Elsden, S. R. (1945b). *Proc. Nutr. Soc. (Engl. Scot.)* **3,** 243–247.

Elsden, S. R. (1946). *Biochem. J.* **40,** 252–256

Elsden, S. R., and D. Lewis (1953). *Biochem. J.* **55,** 183–189.

Elsden, S. R., and A. K. Sijpesteijn (1950). *J. Gen. Microbiol.* **4,** xi.

Elsden, S. L., M. W. S. Hitchcock, R. A Marshall, and A. T. Phillipson (1946). *J. Exptl. Biol.* **22,** 191–202.

Elsden, S. R., B. E. Volcani, F. M. C. Gilchrist, and D. Lewis (1956). *J. Bacteriol.* **72,** 681–689.

el-Shazly, K. (1952a). *Biochem. J.* **51,** 640–647.

el-Shazly, K. (1952b). *Biochem. J.* **51,** 647–653.

el-Shazly, K. (1958). *J. Agr. Sci.* **51,** 149–156.

el-Shazly, K., and R. E. Hungate (1965). *Appl. Microbiol.* **13,** 62–69.

el-Shazly, K., B. A. Dehority, and R. R. Johnson (1961a). *J. Animal Sci.* **20,** 268–273.

el-Shazly, K., R. R. Johnson, B. A. Dehority, and A. L. Moxon (1961b). *J. Animal Sci.* **20,** 839–843.

el-Shazly, K., A. R. Abou Akkada, and M. M. A. Naga (1963). *J. Agr. Sci.* **61,** 109–114.

Emerick, R. J., and L. B. Embry (1961). *J. Animal Sci.* **20,** 844–848.

Emery, R. S., and L. D. Brown (1961). *J. Dairy Sci.* **44,** 1899–1892.

Emery, R. S., C. K. Smith, and C. F. Huffman (1957a). *Appl. Microbiol.* **5,** 360–362.

Emery, R. S., C. K. Smith, and L. Faito (1957b). *Appl. Microbiol.* **5,** 363–366.

Emery, R. S., C. K. Smith, and T. R. Lewis (1958a). *J. Dairy Sci.* **41,** 647–650.

Emery, R. S., C. K. Smith, and C. F. Huffman (1958b). *Mich. State Univ., Agr. Expt. Sta., Quart. Bull.* **40,** 460–467.

Emery, R. S., T. R. Lewis, J. P. Everett, and C. A. Lassiter (1959). *J. Dairy Sci.* **42,** 1182–1186.

Emery, R. S., C. K. Smith, R. M. Grimes, C. F. Huffman, and C. W. Duncan (1960). *J. Dairy Sci.* **43,** 76–80.

Emery, R. S., L. D. Brown, and J. W. Thomas (1964a). *J. Dairy Sci.* **47,** 1322–1324.

Emery, R. S., L. D. Brown, and J. W. Thomas (1964b). *J. Dairy Sci.* **47,** 1325–1329.

Eng, K. S., M. E. Riewe, J. H. Craig, and J. C. Smith (1964). *J. Animal Sci.* **23,** 1129–1132.

Ensor, W. L., J. C. Shaw, and H. F. Tellechea (1959). *J. Dairy Sci.* **42,** 189–191.

Eriksson, S. (1949). *Kgl. Lantbruks-Hogskol. Ann.* **16,** 167–168.

Eriksson, S., and F. Vestervall (1960). *Nord. Jordbrugsforskn.* **42,** 83–91.

Esplin, G., W. H. Hale, F. Hubbert, and B. Taylor (1963). *J. Animal Sci.* **22,** 695–698.

Eusebio, A. N., J. C. Shaw, E. C. Leffel, S. Lakshmanan, and R. N. Doetsch (1959). *J. Dairy Sci.* **42,** 692–697.

Evans, J. L., R. B. Grainger, and C. M. Thompson (1957). *J. Animal Sci.* **16,** 110–117.

Evans, I. A., A. J. Thomas, W. C. Evans, and C. M. Edwards (1958). *Brit. Vet. J.* **114,** 253–267.

Evans, W. C., and E. T. R. Evans (1949). *Nature* **163,** 373–375.

Evered, D. F. (1961). *Nature* **189,** 228–229.

Ewan, R. C., E. E. Hatfield, and U. S. Garrigus (1958). *J. Animal Sci.* **17,** 298–303.

Faber, J. (1651). "Nova plantarum, animalium, et mineralium Mexicanorum historia." Rome. Quoted by Brugnone (1809) as "in-folio, p. 625."

Fahraeus, G. (1944). *Ann. Agr. Coll. Sweden* **12,** 1–22.

Falaschini, A. (1934). *Boll. Lab. Zool. Agrar. Bachicoltura Rov. Inst. Super. Agrar. Milano* **4,** 151–160.

Fantham, H. B. (1922). *S. African J. Sci.* **19,** 332–339.

Farr, W. K., and S. H. Eckerson (1934). *Contrib. Boyce Thompson Inst.* **6,** 309–313.

Fauconneau, G. (1958). *Ann. Agron. (Paris)* **9,** 1–13.

Fauconneau, G. (1961). *Ann. Nutr. Aliment.* **15,** 291–299.

Fauconneau, G., and L. Chevillard (1953). *Compt. Rend. Soc. Biol.* **147,** 69–72.

Felinski, L., and S. Baranow-Baranowski (1959a). *Roczniki Nauk Rolniczych, Ser. B* **74,** 379–405.

Felinski, L., and S. Baranow-Baranowski (1959b). *Roczniki Nauk Rolniczych, Ser. B* **75,** 1–21.

Fenner, H., and J. M. Elliot (1963). *J. Animal Sci.* **22,** 624–627.

Ferber, K. E. (1928). *Z. Tierzuecht. Zuechtungsbiol.* **12,** 31–63.

Ferber, K. E. (1929a). *Z. Tierzuecht. Zuechtungsbiol.* **15,** 375–390.

Ferber, K. E. (1929b). *Wiss. Arch. Landwirtsch. Abt. B, Arch. Tierernaehr. Tierzucht* **1,** 597–600.

Ferber, K. E., and T. Winogradowa-Fedorowa (1929). *Biol. Zentr.* **49,** 321–328.

Ferguson, W. S. (1942). *Biochem. J.* **36,** 786–789.

Ferguson, W. S. (1943). *J. Agr. Sci.* **33,** 174–177.

Ferguson, W. S. (1948). *Nature* **161,** 816.

Ferguson, W. S., and R. A. Terry (1955). *J. Agr. Sci.* **46,** 257–266.

Ferguson, W. S., B. de Ashworth, and R. A. Terry (1949). *Nature* **163,** 606.

Ferguson, W. S., B. de Ashworth, and R. A. Terry (1950). *Nature* **166,** 116–117.

Festenstein, G. N. (1958a). *Biochem. J.* **69,** 562–567.

Festenstein, G. N. (1958b). *Biochem. J.* **70,** 49–51.

Festenstein, G. N. (1959). *Biochem. J.* **72,** 75–79.

Field, A. C., and D. Purves (1964). *Proc. Nutr. Soc. (Engl. Scot.)* **23,** xxiv–xxv.

Fina, L. R., G. W. Teresa, and E. E. Bartley (1958). *J. Animal Sci.* **17,** 667–674.

Fina, L. R., C. A. Hay, E. E, Bartley, and B. Mishra (1961a). *J. Animal Sci.* **20,** 654–658.

Fina, L. R., E. E. Bartley, and R. Mishra (1961b). *J. Dairy Sci.* **44,** 1202.

Fina, L. R., C. L. Keith, E. E. Bartley, P. A. Hartman, and N. L. Jacobson (1962). *J. Animal Sci.* **21,** 930–934.

Fincher, M. G., and C. E. Hayden (1940). *Cornell Vet.* **30,** 197–216.

Findlay, J. D. (1958). *Proc. Nutr. Soc. (Engl. Scot.)* **17,** 186–191.

Finner, M. F., and B. R. Baumgardt (1963). *J. Dairy Sci.* **46,** 341.

Fiorentini, A. (1889). "Intorno ai protisti dello stomaco dei bovini," 27 pp. Pavia.

Fiorentini, A. (1890). *J. Micrographie* **14,** 23–28, 79–83, and 178–183.

Flatt, W. P., R. G. Warner, and J. K. Loosli (1958). *J. Dairy Sci.* **41,** 1593–1600.

Fletcher, D. W., and E. S. E. Hafez (1960). *Appl. Microbiol.* **8,** 22–27.

Florentin, P. (1952). *Rev. Med. Vet.* **103,** 530–542.

Flower, A. E., J. S. Brinks, J. J. Urick, and F. S. Willson (1963). *J. Animal Sci.* **22,** 914–918.

Fluorens, M. (1833). *Mem. Acad. Sci. Paris* **12,** 483–506 and 531–550.

Folger, A. H. (1940). *Calif. Univ., Agr. Expt. Sta. Bull.* **635,** 11 pp.

Folley, S. J., and T. H. French (1949). *Nature* **163,** 174–175.

Forbes, E. B., J. A. Fries, and W. W. Braman (1925). *J. Agr. Res.* **31,** 987–995.

Forbes, E. B., W. W. Braman, M. Kriss, C. D. Jeffries, R. W. Swift, R. B. French, R. C. Miller, and C. V. Smythe (1928). *J. Agr. Res.* **37,** 253–300.

Ford, J. E., and S. H. Hutner (1955). *Vitamins Hormones* **13,** 101–133.

Ford, J. E., E. S. Holdsworth, and J. W. G. Porter (1953). *Proc. Nutr. Soc. (Engl. Scot.)* **12,** xi–xii.

Ford, J. E., K. D. Perry, and C. A. E. Briggs (1958). *J. Gen. Microbiol.* **18,** 273–284.

Forman, S. A., and F. Sauer (1962). *Can. J. Animal Sci.* **42,** 9–17.

Forsyth, G., and E. L. Hirst (1953). *J. Chem. Soc.* pp. 2132–2135.

Foubert, E. L., and H. C. Douglas (1948). *J. Bacteriol.* **56,** 25–34.

Francois, A. (1956). *Natl. Acad. Sci.—Natl. Res. Council Bull.* **397, 85.**

Francois, A., A. M. Leroy, and S. Z. Zelter (1954). *Ann. Zootech.* **4,** 103–110.

Freer, M., and R. C. Campling (1963). *Brit. J. Nutr.* **17,** 79–88.

Freer, M., R. C. Campling, and C. C. Balch (1962). *Brit. J. Nutr.* **16,** 279–295.

French, M. H. (1937a). *Ann. Rept. Dept. Vet. Sci. Animal Husbandry Tanganyika Territ.* **1937–38,** 81–82.

French, M. H. (1937b). *Ann. Rept. Dept. Vet. Sci. Animal Husbandry Tanganyika, Territ.* **1937–38,** 83–85.

French, M. H. (1940). *J. Agr. Sci.* **30,** 503–510.

French, M. H. (1956a). *Empire J. Exptl. Agr.* **24,** 128–136.

French, M. H. (1956b). *Empire J. Exptl. Agr.* **24,** 235–244.

Friend, D. W., H. M. Cunningham, and J. W. G. Nicholson (1962). *Can. J. Animal Sci.* **42,** 55–62.

Friend, D. W., H. M. Cunningham, and J. W. G. Nicholson (1963a). *Can. J. Animal Sci.* **43,** 156–168.

Friend, D. W., H. M. Cunningham, and J. W. G. Nicholson (1963b). *Can. J. Animal Sci.* **43,** 174–181.

Friend, D. W., J. W. G. Nicholson, and H. M. Cunningham (1964). *Can. J. Animal Sci.* **44,** 303–308.

Fulghum, R. S., and W. E. C. Moore (1963). *J. Bacteriol.* **85,** 808–815.

Furstenberg, M. H., and Rohde (1872). *Die Rindviehzucht* **1,** 207. Cited in Krzywanek and Quast (1937).

Futter, J. R., and F. B. Shorland (1957). *Biochem. J.* **65,** 689–693.

Galgan, M. W., and B. H. Schneider (1951). *Wash., State Coll., Agr. Expt. Sta. Circ.* **166.**

Gall, L. S., and C. N. Huhtanen (1951). *J. Dairy Sci.* **34,** 353–362.

Gall, L. S., C. N. Stark, and J. K. Loosli (1947). *J. Dairy Sci.* **30,** 891–899.

Gall, L. S., W. Burroughs, P. Gerlaugh, and B. H. Edgington (1949a). *J. Animal Sci.* **8,** 433–440.

Gall, L. S., S. E. Smith, D. E. Becker, C. N. Stark, and J. K. Loosli (1949b). *Science* **109,** 468–469.

Gall, L. S., W. E. Thomas, J. K. Loosli, and C. N. Huhtanen (1951). *J. Nutr.* **44,** 113–122.

Gallup, W. D., L. S. Pope, and C. K. Whitehair (1952). *J. Animal Sci.* **11,** 621–630.

Ganimedov, L. A., and L. F. Zajac (1961). *Veterinariya* **38,** 70–72.

Gant, D. E., E. L. Smith, and L. F. J. Parker (1954). *Biochem. J.* **56,** xxxiv.

Garner, G. B., M. E. Muhrer, W. C. Ellis, and W. H. Pfander (1954). *Science* **120,** 435–436.

Garrett, W. N., and J. H. Meyer (1963). *Calif. Agr.* **17,** 11–12.

Garrett, W. N., J. H. Meyer, and G. L. Lofgreen (1959). *J. Animal Sci.* **18,** 528–547.

Garrett, W. N., G. P. Lofgreen, and J. H. Meyer (1960). *Calif. Agr.* **14,** 8.

Garrett, W. N., J. H. Meyer, and G. P. Lofgreen (1961). *J. Animal Sci.* **20,** 833–838.

Garton, G. A. (1951). *J. Exptl. Biol.* **28,** 358–368.

Garton, G. A. (1959). *Proc. Nutr. Soc. (Engl. Scot.)* **18,** 112–117.

Garton, G. A. (1960). *Nutr. Abstr. Rev.* **30,** 1–16.

Garton, G. A., and A. E. Oxford (1955). *J. Sci. Food Agr.* **6,** 142–148.

Garton, G. A., P. N. Hobson, and A. K. Lough (1958). *Nature* **182,** 1511–1512.

Garton, G. A., A. K. Lough, and E. Vioque (1961). *J. Gen. Microbiol.* **25,** 215–225.

Gelei, J., and O. Sebestyen (1932). *Acta Biol.* **2,** 141–161.

Ghose, S., and K. W. King (1963). *Textile Res. J.* **33,** 392–398.

Gibbons, R. J., and R. N. Doetsch (1955). *Univ. Maryland Agr. Expt. Sta. Misc. Publ.* **238.**

Gibbons, R. J., and R. N. Doetsch (1959). *J. Bacteriol.* **77,** 417–428.

Gibbons, R. J., and L. P. Engle (1964). *Science* **146,** 1307–1309.

Gibbons, R. J., and R. D. McCarthy (1957). *Univ. Maryland Agr. Expt. Sta. Misc. Publ.* **291,** 12–16.

Gibbons, R. J., R. N. Doetsch, and J. C. Shaw (1955). *J. Dairy Sci.* **38,** 1147–1154.

Giesecke, D. (1960a). *Naturwissenschaften* **47,** 475–476.

Giesecke, D. (1960b). *Zentralb Bakteriol., Parasitenk. Abt. I, Orig.* **179,** 448–455.

Giesecke, D. (1962). *Zentralb. Bakteriol., Parasitenk. Abt. I, Orig.* **186,** 170–178.

Gilchrist, F. M. C., and R. Clark (1957). *S. African Vet. Med. Assoc.* **28,** 295–309.

Gilchrist, F. M. C., and A. Kistner (1962). *J. Agr. Sci.* **59,** 77–83.

Gill, J. W., and K. W. King (1958). *J. Bacteriol.* **75,** 666–673.

Gilroy, J. J., and F. G. Hueter (1957). *Univ. Maryland Agr. Expt. Sta. Misc. Publ.* **291,** 4–8.

Glawischnig, E. (1962). *Wien. Tieraerztl. Monatsschr.* **49,** 230–235.

Glover, J., and H. W. Dougall (1960). *J. Agr. Sci.* **55,** 391–394.

Glover, J., and D. W. Duthie (1960). *J. Agr. Sci.* **55,** 403–408.

Glover, J., D. W. Duthie, and M. H. French (1957). *J. Agr. Sci.* **48,** 373–378.

Godfrey, N. W. (1961). *J. Agr. Sci.* **57,** 173–175 and 177–183.

Goetsch, G. D., and W. R. Pritchard (1958). *Am. J. Vet. Res.* **19,** 637–641.

Gordon, G. R., and W. E. C. Moore (1961). *J. Dairy Sci.* **44,** 1772–1773.

Gordon, J. G. (1958a). *J. Agr. Sci.* **51,** 78–80.

Gordon, J. G. (1958b). *J. Agr. Sci.* **51,** 81–83.

Gordon, J. G. (1958c). *J. Agr. Sci.* **50,** 34–42.

Gordon, J. G., and D. E. Tribe (1952). *Brit. J. Nutr.* **6,** 89–93.

Goswami, M. N. D., and N. D. Kehar (1956). *Proc. Indian Sci. Congr.* **3,** 359–360. Cited by Kehar (1956).

Gottschalk, A., and E. R. B. Graham (1958). *Z. Naturforsch.* **13,** 821–822.

Gouws, L., and A. Kistner (1965). *J. Agr. Sci.* **64,** 51–57.

Gradzka-Majewska, I. (1961). *Acta Parasitol. Polon.* **9,** 169–191.

Graham, N. M. (1964a). *Australian J. Agr. Res.* **15,** 969–973.

Graham, N. M. (1964b). *Australian J. Agr. Res.* **15,** 974–981.

Graham, N. M. and A. J. Williams (1962). *Australian J. Agr. Res.* **13**, 894–900.

Grainger, R. B., M. C. Bell, J. W. Stroud, and F. H. Baker (1961). *J. Animal Sci.* **20**, 319–322.

Grau, H. (1955). *Berlin Muench. Tieraerztl. Wochschr.* **68**, 271–275.

Gray, F. V. (1947a). *J. Exptl. Biol.* **24**, 1–10.

Gray, F. V. (1947b). *J. Exptl. Biol.* **24**, 15–19.

Gray, F. V. (1948). *J. Exptl. Biol.* **25**, 135–144.

Gray, F. V., and A. F. Pilgrim (1950). *Nature* **166**, 478.

Gray, F. V., and A. F. Pilgrim (1951). *J. Exptl. Biol.* **28**, 83–90.

Gray, F. V., and A. F. Pilgrim (1952a). *Nature* **170**, 375.

Gray, F. V., and A. F. Pilgrim (1952b). *J. Exptl. Biol.* **29**, 54–56.

Gray, F. V., and R. A. Weller (1958). *Australian J. Agr. Res.* **9**, 797–801.

Gray, F. V., A. F. Pilgrim, H. J. Rodda, and R. A. Weller (1951a). *Nature* **167**, 954.

Gray, F. V., A. F. Pilgrim, and R. A. Weller (1951b). *J. Exptl. Biol.* **28**, 74–82.

Gray, F. V., A. F. Pilgrim, H. J. Rodda, and R. A. Weller (1952). *J. Exptl. Biol.* **29**, 57–65.

Gray, F. V., A. F. Pilgrim, and R. A. Weller (1953). *Nature* **172**, 347.

Gray, F. V., A. F. Pilgrim, and R. A. Weller (1954). *J. Exptl. Biol.* **31**, 49–55.

Gray, F. V., A. F. Pilgrim, and R. A. Weller (1958a). *Brit. J. Nutr.* **12**, 404–413.

Gray, F. V., A. F. Pilgrim, and R. A. Weller (1958b). *Brit. J. Nutr.* **12**, 413–421.

Gray, F. V., G. B. Jones, and A. F. Pilgrim (1960). *Australian J. Agr. Res.* **11**, 383–388.

Gray, F. V., R. A. Weller, A. F. Pilgrim, and G. B. Jones (1962). *Australian J. Agr. Res.* **13**, 343–349.

Gray, R. S., and R. Clark (1964). *Onderstepoort J. Vet. Res.* **31**, 91–95.

Greenwood, C. T. (1954). *Biochem. J.* **57**, 151–153.

Grieve, C. M., A. R. Robblee, and L. W. McElroy (1963) *Can. J. Animal Sci.* **43**, 196–201.

Grosskopf, J. F. W. (1964). *Onderstepoort J. Vet. Res.* **31**, 69–75.

Gruby, D., and O. Delafond (1843). *Compt. Rend.* **17**, 1305–1308.

Günther, A. (1899). *Z. Wiss. Zool.* **65**, 529–572.

Günther, A. (1900). *Z. Wiss. Zool.* **67**, 640–662.

Gunsalus, I. C., and C. W. Shuster (1961). *In* "The Bacteria" (I. C. Gunsalus and R. Y. Stanier, eds.), Vol. 2. Chapt. 1. Academic Press, New York.

Gupta, B. N., B. N. Majumdar, and N. D. Kehar (1962). *Ann. Biochem. Exptl. Med.* (*Calcutta*) **22**, 105–112.

Gupta, J., and R. E. Nichols (1962). *Am. J. Vet. Res.* **23**, 128–133.

Gupta, S. S., and T. P. Hilditch (1951). *Biochem. J.* **48**, 137–146.

Gupta, Y. P., and N. B. Das (1956). *Indian J. Agr. Sci.* **26**, 373–379.

Gutierrez, J. (1953). *J. Bacteriol.* **66**, 123–128.

Gutierrez, J. (1955). *Biochem. J.* **60**, 516–522.

Gutierrez, J. (1958). *J. Protozool.* **5**, 122–126.

Gutierrez, J. (1959). *J. Protozool.* **6**, Suppl., 21.

Gutierrez, J., and R. E. Davis (1959). *J. Protozool.* **6**, 222–226.

Gutierrez, J., and R. E. Davis (1962a). *Appl. Microbiol.* **10**, 305–308.

Gutierrez, J., and R. E. Davis (1962b). *J.Animal Sci.* **21**, 819–823.

Gutierrez, J., and R. E. Hungate (1957). *Science* **126**, 511.

Gutierrez, J., R. E. Davis, and I. L. Lindahl (1958). *Science* **127**, 335.

Gutierrez, J., R. E. Davis, I. L. Lindahl, and E. J. Warwick (1959a) *Appl. Microbiol.* **7**, 16–22.

Gutierrez, J., R. E. Davis, and I. L. Lindahl (1959b). *Appl. Microbiol.* **7**, 304–308.

Gutierrez, J., R. E. Davis, and I. L. Lindahl (1961). *Appl. Microbiol.* **9**, 209–212.

Gutierrez, J., P. P. Williams, R. E. Davis, and E. J. Warwick (1962). *Appl. Microbiol.* **10**, 548–551.

Gutierrez, J., H. W. Essig, P. P. Williams, and R. E. Davis (1963). *J. Animal Sci.* **22**, 506–509.

Gutowski, B. (1960). *Acta Physiol. Polon.* **11**, 105–118.

Gutowski, B., W. Barej, A. Temler, and I. Nowosielska (1958a). *Acta Physiol. Polon.* **9**, 341–350.

Gutowski, B., A. Temler, W. Barej, and I. Nowosielska (1958b). *Acta Physiol. Polon.* **9**, 461–476.

Gutowski, B., W. Barej, and A. Temler (1958c). *Acta Physiol. Polon.* **9**, 669–675.

Gutowski, B., W. Barej, A. Temler, and I. Nowosielska (1960). *Acta Physiol. Polon.* **11**, 119–128.

Gyllenberg, H., and M. Lampila (1955). *Maataloustieteellinen Aikakauskirja* **27**, 53–56.

Habel, R. E. (1956). *Cornell Vet.* **46**, 555–633.

Hackler, L. R., A. L. Neumann, and B. C. Johnson (1957). *J. Animal Sci.* **16**, 125–129.

Haenlein, G. F. W., C. R. Richards, and I. Huff, Jr. (1961). *J. Dairy Sci.* **44**, 2004–2014.

Hafez, E. S. E., M. E. Ensminger, and W. E. Ham (1959). *J. Agr. Sci.* **53**, 339–346.

Hagemann, O. (1891). *Landwirtsch. Jahrb.* **20**, 261–291.

Hagemann, O. (1899). *Arch. Anat. Physiol., Physiol. Abt.,* Suppl., 111–140.

Hale, E. B., C. W. Duncan, and C. F. Huffman (1940). *J. Dairy Sci.* **23**, 953–967.

Hale, E. B., C. W. Duncan, and C. F. Huffman (1947a). *J. Nutr.* **34**, 733–746.

Hale, E. B., C. W. Duncan, and C. F. Huffman (1947b). *J. Nutr.* **34**, 747–758.

Hale, W. H., and U. S. Garrigus (1953). *J. Animal Sci.* **12**, 492–496.

Hale, W. H., and R. P. King (1955). *Proc. Soc. Exptl. Biol. Med.* **89**, 112–114.

Hale, W. H., A. L. Pope, P. H. Phillips, and G. Bohstedt (1950). *J. Animal Sci.* **9**, 414–419.

Hale, W. H., W. B. Hardie, W. C. Sherman, W. P. Crawford, and W. M. Reynolds (1960). *J. Animal Sci.* **19**, 648.

Hall, A. G. (1940). *Cornell Vet.* **30**, 216.

Hall, E. R. (1952). *J. Gen. Microbiol.* **7**, 350–357.

Hall, O. G., H. D. Baxter, and C. S. Hobbs (1961). *J. Animal Sci.* **20**, 817–819.

Halliwell, G., and M. P. Bryant (1963). *J. Gen. Microbiol.* **32**, 441–448.

Hallsworth, E. G. (1949). *J. Agr. Sci.* **39**, 254–258.

Halse, K., and W. Velle (1956). *Acta Physiol. Scand.* **37**, 380–390.

Hamilton, T. S. (1942). *J. Nutr.* **23**, 101–110.

Hamilton, T. S., W. B. Robinson, and B. C. Johnson (1948). *J. Animal Sci.* **7**, 26–33.

Hamlin, L. J., and R. E. Hungate (1956). *J. Bacteriol.* **72**, 548–554.

Hammerstrom, R. A., K. D. Claus, J. W. Coghlan, and R. H. McBee (1955). *Arch. Biochem. Biophys.* **56**, 123–129.

Hancock, J. (1954). *J. Agr. Sci.* **45**, 80–95.

Hanold, F. J., E. E. Bartley, and F. W. Atkeson (1957). *J. Dairy Sci.* **40**, 369–376.

Harbers, L. H., and A. D. Tillman (1962). *J. Animal Sci.* **21**, 575–582.

Harbers, L. H., J. M. Prescott, and C. E. Johnson (1961). *J. Animal Sci.* **20**, 6–9.

Hardie, W. B. (1952). Vitamin B_{12} production by *Streptococcus bovis* from the bovine rumen. M.S. thesis, Washington State University, Pullman, Washington.

Hardison, W. A., W. N. Linkous, R. W. Engel, and G. C. Graf (1959). *J. Dairy Sci.* **42,** 346–352.

Harker, K. W., J. I. Taylor, and D. H. L. Rollinson (1954). *J. Agr. Sci.* **44,** 193–198.

Harris, L. E., and H. H. Mitchell (1941a). *J. Nutr.* **22,** 167–182.

Harris, L. E., and H. H. Mitchell (1941b). *J. Nutr.* **22,** 183–196.

Harris, L. E., and A. T. Phillipson (1962). *Animal Prod.* **4,** 97–116.

Harrison, W. A., G. A. Miller, and G. C. Graf (1957). *J. Dairy Sci.* **40,** 363–368.

Hart, E. B., G. Bohstedt, H. J. Deobald, and M. I. Wegner (1938). *Proc. Am. Soc. Animal Prod.* **31,** 333–336.

Hartley, W. J., and A. B. Grant (1961). *Federation Proc.* **20,** 679–688.

Hartman, L., F. B. Shorland, and I. R. C. McDonald (1954). *Nature* **174,** 185–186.

Hartman, L., F. B. Shorland, and I. R. C. McDonald (1955). *Biochem. J.* **61,** 603–607.

Hartman, L., F. B. Shorland, and B. Cleverley (1958). *Biochem. J.* **69,** 1–5.

Harvey, J. M. (1952). *Australian Vet. J.* **28,** 209–215.

Hatfield, E. E., U. S. Garrigus, and H. W. Norton (1954). *J. Animal Sci.* **13,** 715–725.

Hatfield, E. E., U. S. Garrigus, F. M. Forbes, A. L. Neumann, and W. Gaither (1959). *J. Animal Sci.* **18,** 1208–1219.

Haubner, G. C. (1837). "Über die Magenverdauung der Wiederkäuer nach Versuchen, nebst einer Prüfung der Flourenschen Versuche über das Wiederkäuen." W. Dietze, Anclam.

Haubner, G. K. (1854). *Ver. Koenigreichs Sachsen* **2,** 56–57.

Haubner, G. K. (1855). *Z. Deut. Landwirthe* [N.F.] **6,** 177–182.

Havre, G. N., O. Dynna, and F. Ender (1960). *Acta Vet. Scand.* **1,** 250–276.

Hawke, J. C., and J. A. Robertson (1964). *J. Sci. Food Agr.* **15,** 283–289.

Hayden, C. E., M. G. Fincher, S. J. Roberts, W. J. Gibbons, and A. G. Danks (1946). *Cornell Vet.* **36,** 71–84.

Hayes, B. W., C. O. Little, and G. E. Mitchell (1964a). *J. Animal Sci.* **23,** 764–766.

Hayes, B. W., G. E. Mitchell, C. O. Little, J. R. Thompson, and N. W. Bradley (1964b). *J. Animal Sci.* **23,** 877.

Head, H. H., J. D. Connolly, and W. F. Williams (1964). *J. Dairy Sci.* **47,** 1371–1377.

Head, M. J. (1953). *J. Agr. Sci.* **43,** 281–293.

Head, M. J. (1959a). *Nature* **183,** 757.

Head, M. J. (1959b). *Proc. Nutr. Soc.* (*Engl. Scot.*) **18,** 108–112.

Head, M. J., and J. A. F. Rook (1955). *Nature* **176,** 262–263.

Heald, P. J. (1951a). *Brit. J. Nutr.* **5,** 75–83.

Heald, P. J. (1951b). *Brit. J. Nutr.* **5,** 84–93.

Heald, P. J. (1952). *Biochem. J.* **50,** 503–508.

Heald, P. J. (1953). *Brit. J. Nutr.* **7,** 124–130.

Heald, P. J., and A. E. Oxford (1953). *Biochem. J.* **53,** 506–512.

Heald, P. J., A. E. Oxford, and B. Sugden (1952). *Nature* **169,** 1055.

Heald, P. J., N. Krogh, S. O. Mann, J. C. Appleby, F. M. Masson, and A. E. Oxford (1953). *J. Gen. Microbiol.* **9,** 207–215.

Healy, D. J., and J. W. Nutter (1915). *Kentucky Agr. Expt. Sta. Circ.* **5.**

Hecker, J. F., D. E. Budtz-Olsen, and M. Ostwald (1964). *Australian J. Agr. Res.* **15,** 961–968.

Helwig, R. (1960). *Z. Tierphysiol. Tierernaehr. Futtermittelk.* **15**, 127–135.

Hemsley, J. A., and R. J. Moir (1963). *Australian J. Agr. Res.* **14**, 509–517.

Henderickx, H. (1960a). *Mededel. Landbouwhogeschool Opzoekingssta. Staat Gent* **25**, 859–868.

Henderickx, H. (1960b). *Z. Tierphysiol. Tierernaehr. Futtermittelk.* **15**, 218–227.

Henderickx, H. (1961a). *Arch. Intern. Physiol. Biochim.* **69**, 443–448.

Henderickx, H. (1961b). *Arch. Intern. Physiol. Biochim.* **69**, 449–458.

Henderickx, H., and J. Martin (1963). *Compt. Rend. Rech., Inst. Rech. Sci. Ind. Agr., Bruxelles* **31**, 110 pp.

Henderson, J. A., E. V. Evans, and R. A. McIntosh (1952). *J. Am. Vet. Med. Assoc.* **120**, 375–378.

Henderson, J. G. (1948). *Cornell Vet.* **38**, 292–304.

Henderson, R., R. E. Horvat, and R. J. Block (1954). *Contrib. Boyce Thompson Inst.* **17**, 337–341.

Henke, L. A., S. H. Work, and A. W. Burt (1940). *Hawaii Agr. Expt. Sta. Bull.* **85**, 37 pp.

Henneberg, W. (1919). *Berlin. Klin. Wochschr.* **56**, 693–694.

Henneberg, W. (1922). *Zentr. Bakteriol. Parasitenk. Abt. II* **55**, 242–281.

Henneberg, W., and F. Stohmann (1860, 1864). "Beiträge zur Begründung einer rationellen Fütterung der Wiederkäuer," 2 vol. Schwetschke and Son, Braunschweig.

Henneberg, W., and F. Stohmann (1885). *Z. Biol.* **21**, 613–624.

Henrici, M. (1949). *Onderstepoort J. Vet. Sci. Animal Ind.* **22**, 373–413.

Henrici, M. (1952). *Onderstepoort J. Vet. Res.* **25**, 45–92.

Henriksson, K. B., and R. E. Habel (1961). *Anat. Record* **139**, 499–507.

Hershberger, T. V., O. G. Bentley, and A. L. Moxon (1959a). *J. Animal Sci.* **18**, 663–670.

Hershberger, T. V., T. A. Long, E. W. Hartsook, and R. W. Swift (1959b). *J. Animal Sci.* **18**, 770–779.

Hibbs, J. W., and H. R. Conrad (1958). *J. Dairy Sci.* **41**, 1230–1247.

Hibbs, J. W., and W. D. Pounden (1948). *J. Dairy Sci.* **31**, 1055–1061.

Hibbs, J. W., and W. D. Pounden (1949). *J. Dairy Sci.* **32**, 1016–1024.

Hibbs, J. W., and W. D. Pounden (1950). *Ohio Farm Home Res.* **35**, 30–31.

Hibbs, J. W., W. D. Pounden, and H. R. Conrad (1953a). *J. Dairy Sci.* **36**, 717–727.

Hibbs, J. W., H. R. Conrad, and W. D. Pounden (1953b). *J. Dairy Sci.* **36**, 1319–1325.

Hibbs, J. W., H. R. Conrad, and W. D. Pounden (1954). *J. Dairy Sci.* **37**, 724–736.

Hibbs, J. W., H. R. Conrad, W. D. Pounden, and N. Frank (1956). *J. Dairy Sci.* **39**, 171–179.

Hibbs, J. W., H. R. Conrad, and W. D. Pounden (1957). *Ohio Farm Home Res.* **41**, 4–6.

Higginbottom, C., and D. W. F. Wheater (1954). *J. Agr. Sci.* **44**, 434–442.

Higuchi, M., T. Uemura, and C. Furusaka (1962a). *J. Agr. Chem. Soc. Japan* **36**, 451–454.

Higuchi, M., T. Uemura, and C. Furusaka (1962b). *J. Agr. Chem. Soc. Japan* **36**, 455–459.

Higuchi, M., and T. Uemura (1965). *J. Agr. Chem. Soc. Japan* **39**, 95–101.

Hilditch, T. P., and H. E. Longenecker (1937). *Biochem. J.* **31**, 1805–1819.

Hilditch, T. P., and R. K. Shrivastava (1949). *J. Am. Oil Chemists' Soc.* **26**, 1–4.

Hill, F. D., J. H. Saylor, R. S. Allen, and N. L. Jacobson (1960). *J. Animal Sci.* **19**, 1266.

Hill, K. J. (1955). *Quart. J. Exptl. Physiol.* **40,** 32–39.

Hill, K. J. (1957). *J. Physiol. (London)* **139,** 4–5P.

Hill, K. J. (1958). *J. Physiol. (London)* **143,** 30–31P.

Hill, K. J., and J. L. Mangan (1964). *Biochem. J.* **93,** 39–45.

Hinders, R. G., and G. W. Ward (1961). *J. Dairy Sci.* **44,** 1129–1133.

Hine, D. C., and M. C. Dawbarn (1954). *Australian J. Exptl. Biol. Med. Sci.* **32,** 641–652.

Hinkson, R. S., A. W. Mahoney, and B. R. Poulton (1964). *J. Dairy Sci.* **47,** 1461.

Hird, F. J. R., and M. J. Wiedemann (1964). *Biochem. J.* **92,** 585–589.

Hirose, Y., S. Suzuki, and Y. Asahida (1957). *Japan J. Zootech. Sci.* **28,** 177–180.

Hobson, P. N. (1965a). *J. Gen. Microbiol.* **38,** 161–166.

Hobson, P. N. (1965b). *J. Gen. Microbiol.* **38,** 167–180.

Hobson, P. N., and M. J. MacPherson (1952). *Biochem. J.* **52,** 671–679.

Hobson, P. N., and M. J. MacPherson (1954). *Biochem. J.* **57,** 145–151.

Hobson, P. N., and S. O. Mann (1955). *J. Gen. Microbiol.* **13,** 420–435.

Hobson, P. N., and S. O. Mann (1957). *J. Gen. Microbiol.* **16,** 463–471.

Hobson, P. N., and S. O. Mann (1961). *J. Gen. Microbiol.* **25,** 227–240.

Hobson, P. N., and M. R. Purdom (1959). *Nature* **183,** 904–905.

Hobson, P. N., and M. R. Purdom (1961). *J. Appl. Bacteriol.* **24,** 188–193.

Hobson, P. N., and W. Smith (1963). *Nature* **200,** 607–608.

Hobson, P. N., E. S. M. Mackay, and S. O. Mann (1955). *Res. Correspondence* **8,** 2 pp.

Hobson, P. N., S. O. Mann, and A. E. Oxford (1958). *J. Gen. Microbiol.* **19,** 462–472.

Hobson, P. N., S. O. Mann, and W. Smith (1962). *J. Gen. Microbiol.* **29,** 265–270.

Hobson, P. N., S. O. Mann, and W. Smith (1963). *Nature* **198,** 213.

Hodgson, R. E., W. H. Riddell, and J. S. Hughes (1932). *J. Agr. Res.* **44,** 357–365.

Hoekstra, W. G., A. L. Pope, and P. H. Phillips (1952a). *J. Nutr.* **48,** 431–441.

Hoekstra, W. G., A. L. Pope, and P. H. Phillips (1952b). *J. Nutr.* **48,** 421–430.

Hoernicke, H., W. F. Williams, D. R. Waldo, and W. P. Flatt (1964). *Z. Tierphysiologie, Tierernährung Futtermittelkunde* **19,** 118–119.

Hoflund, S. (1940). *Svensk Veterinartidskr.* **45,** Suppl. 15, 322 pp.

Hoflund, S., and H. Hedstrom (1948). *Cornell Vet.* **38,** 405–417.

Hoflund, S., J. I. Quin, and R. Clark (1948). *Onderstepoort J. Vet. Sci. Animal Ind.* **23,** 395–409.

Hoflund, S., J. Holmberg, and G. Sellmann (1955). *Cornell Vet.* **45,** 254–261.

Hoflund, S., J. Holmberg, and G. Sellmann (1956a). *Cornell Vet.* **46,** 51–53.

Hoflund, S., J. Holmberg, and G. Sellmann (1956b). *Cornell Vet.* **46,** 53–57.

Hoflund, S., P. Andersson, G. Broberg, and L. Hässler (1957). *Nord. Veterinarmed.* **9,** 257–273.

Hofmeister, V. (1881). *Arch. Wiss. Prakt. Tierheilk.* **7,** 169–197.

Hogan, J. P. (1961). *Australian J. Biol. Sci* **14,** 448–460.

Hogan, J. P. (1964a). *Australian J. Agr. Res.* **15,** 384–396.

Hogan, J. P. (1964b). *Australian J. Agr. Res.* **15,** 397–407.

Hogan, J. P., and A. T. Phillipson (1960). *Brit. J. Nutr.* **14,** 147–156.

Hogue, D. E., R. G. Warner, C. H. Griffin, and J. K. Loosli (1956). *J. Animal Sci.* **15,** 788–793.

Holdsworth, E. S. (1953). *Nature* **171,** 149–150.

Hollis, L., C. F. Chappel, R. MacVicar, and C. K. Whitehair (1954). *J. Animal Sci.* **13,** 732–738.

Holmes, P., R. J. Moir, and E. J. Underwood (1953). *Australian J. Biol. Sci.* **6**, 637–644.

Holtenius, P. (1957). *Acta Agr. Scand.* **7**, 113–163.

Holtenius, P., and N. Nielsen (1957). *Nord. Vetinarmed.* **9**, 210–213.

Holter, J. B., R. D. McCarthy, and E. M. Kesler (1963). *J. Dairy Sci.* **46**, 1256–1259.

Honigberg, B. M., W. Balamuth, E. C. Bovee, J. O. Corliss, M. Gojdics, R. P. Hall, R. R. Kudo, N. D. Levine, A. R. Loeblich, J. Weiser, and D. H. Wenrich (1964). *J. Protozool.* **11**, 7–20.

Hoover, W. H., E. M. Kesler, and R. J. Flipse (1963). *J. Dairy Sci.* **46**, 733–739.

Hopffe, A. (1919). *Zentr. Bakteriol. Parasitenk. Abt. 1, Orig.* **83**, 374–386.

Hopkins, L. L., A. L. Pope, and C. A. Baumann (1964). *J. Animal Sci.* **23**, 674–681.

Hopson, J. D., R. R. Johnson, and B. A. Dehority (1963). *J. Animal Sci.* **22**, 448–453.

Horiguchi, M., and M. Kandatsu (1960). *Bull. Agr. Chem. Soc. Japan* **24**, 565–570.

Horn, L. H., R. R. Snapp, and L. S. Gall (1955). *J. Animal Sci.* **14**, 243–248.

Horrocks, D. (1964a). *J. Agr. Sci.* **63**, 369–372.

Horrocks, D. (1964b). *J. Agr. Sci.* **63**, 373–375.

Horrocks, D., and G. D. Phillips (1961). *J. Agr. Sci.* **56**, 379–382.

Horrocks, D., and G. D. Phillips (1964a). *J. Agr. Sci.* **63**, 359–363.

Horrocks, D., and G. D. Phillips (1964b). *J. Agr. Sci.* **63**, 365–367.

Hoshino, S., and Y. Hirose (1963). *J. Dairy Sci.* **46**, 323–326.

Houpt, T. R. (1959). *Am. J. Physiol.* **197**, 115–120.

Howard, B. H. (1955). *Biochem. J.* **60**, i.

Howard, B. H. (1957a). *Biochem. J.* **67**, 643–651.

Howard, B. H. (1957b). *Biochem. J.* **67**, 18P.

Howard, B. H. (1959a). *Biochem. J.* **71**, 671–675.

Howard, B. H. (1959b). *Proc. Nutr. Soc. (Engl. Scot.)* **18**, 103–108.

Howard, B. H. (1959c). *Biochem. J.* **71**, 675–680.

Howard, B. H. (1961). *Proc. Nutr. Soc. (Engl. Scot.)* **20**, xxix.

Howard, B. H. (1963a). *Biochem. J.* **89**, 89P.

Howard, B. H. (1963b). *Biochem. J.* **89**, 90–91P.

Howard, B. H., G. Jones, and M. R. Purdom (1960). *Biochem. J.* **74**, 173–180.

Hrenov, I. I. (1961). *Vestn. Sel'skokhoz. Nauki,* **11**, 56–62. Cited in *Nutr. Abst. Rev.* **32**, p. 587.

Hsiung, T. S. (1931). *Bull. Fan Mem. Inst. Biol.* **2**, 29–43.

Hsiung, T. S. (1932). *Bull. Fan Mem. Inst. Biol.* **2**, 87–107.

Hubbert, F., E. Cheng, and W. Burroughs (1958a). *J. Animal Sci.* **17**, 559–568.

Hubbert, F., E. Cheng, and W. Burroughs (1958b). *J. Animal Sci.* **17**, 576–585.

Hudman, D. B., and H. O. Kunkel (1953). *J. Agr. Food Chem.* **1**, 1060–1062.

Hueter, F. G., J. C. Shaw, and R. N. Doetsch (1956). *J. Dairy Sci.* **39**, 1430–1437.

Hueter, F. G., R. J. Gibbons, J. C. Shaw, and R. N. Doetsch (1958). *J. Dairy Sci.* **41**, 651–661.

Huhtanen, C. N., and L. S. Gall (1953a). *J. Bacteriol.* **65**, 548–553.

Huhtanen, C. N., and L. S. Gall (1953b). *J. Bacteriol.* **65**, 554–560.

Huhtanen, C. N., and L. S. Gall (1955). *J. Bacteriol.* **69**, 102–103.

Huhtanen, C. N., M. R. Rogers, and L. S. Gall (1952). *J. Bacteriol.* **64**, 17–23.

Huhtanen, C. N., R. K. Saunders, and L. S. Gall (1954a). *J. Dairy Sci.* **37**, 328–335.

Huhtanen, C. N., F. J. Carleton, and H. R. Roberts (1954b). *J. Bacteriol.* **68**, 749–751.

Hungate, R. E. (1939). *Ecology* **20**, 230–245.

Hungate, R. E. (1942). *Biol. Bull.* **83**, 303–319.

Hungate, R. E. (1943). *Biol. Bull.* **84**, 157–163.

Hungate, R. E. (1946a). *J. Elisha Mitchell Sci. Soc.* **62**, 9–24.

Hungate, R. E. (1946b). *J. Bacteriol.* **48**, 499–513.

Hungate, R. E. (1947). *J. Bacteriol.* **53**, 631–645.

Hungate, R. E. (1950). *Bacteriol. Rev.* **14**, 1–49.

Hungate, R. E. (1955). In "Biochemistry and Physiology of the Protozoa" (S. H. Hutner and A. Lwoff, eds.), Vol. 2, pp. 159–200. Academic Press, New York.

Hungate, R. E. (1956a). *J. Agr. Food Chem.* **4**, 701–703.

Hungate, R. E. (1956b). *Natl. Acad. Sci.—Natl. Res. Council Pub.* **388**, 4–9.

Hungate, R. E. (1957). *Can. J. Microbiol.* **3**, 289–311.

Hungate, R. E. (1960). *Bacteriol. Rev.* **24**, 353–364.

Hungate, R. E. (1962). In "The Bacteria" (I. C. Gunsalus and R. Y. Stanier, eds.) Vol. IV, pp. 95–119. Academic Press, New York.

Hungate, R. E. (1963a). *J. Bacteriol.* **86**, 848–854.

Hungate, R. E. (1963b). *Symp. Soc. Gen. Microbiol.* **13**, 266–297.

Hungate, R. E. (1965a). "Cellulose in Animal Nutrition." *Biol. Sci. Curric. Study,* Pamphlet 22. Heath, Boston, Massachusetts.

Hungate, R. E. (1965b). In "Physiology of Digestion in the Ruminant" (R. W. Dougherty, ed.), pp. 311–321. Butterworth, London and Washington, D.C.

Hungate, R. E., and I. A. Dyer (1956). *J. Animal Sci.* **15**, 485–488.

Hungate, R. E., R. W. Dougherty, M. P. Bryant, and R. M. Cello (1952). *Cornell Vet.* **42**, 423–449.

Hungate, R. E., D. W. Fletcher, R. W. Dougherty, and B. F. Barrentine (1955a). *Appl. Microbiol.* **3**, 161–173.

Hungate, R. E., D. W. Fletcher, and I. A. Dyer (1955b). *J. Animal Sci.* **14**, 997–1001.

Hungate, R. E., G. D. Phillips, A. MacGregor, D. P. Hungate, and H. K. Buechner (1959). *Science* **130**, 1192–1194.

Hungate, R. E., G. D. Phillips, D. P. Hungate, and A. MacGregor (1960). *J. Agr. Sci.* **54**, 196–201.

Hungate, R. E., R. A. Mah, and M. Simesen (1961). *Appl. Microbiol.* **9**, 554–561.

Hungate, R. E., M. P. Bryant, and R. A. Mah (1964). *Ann. Rev. Microbiol.* **18**, 131–166.

Hunt, C. H., C. H. Kick, E. W. Burroughs, R. M. Bethke, A. F. Schalk, and P. Gerlaugh (1941). *J. Nutr.* **21**, 85–92.

Hunt, C. H., E. W. Burroughs, R. M. Bethke, A. F. Schalk, and P. Gerlaugh (1943). *J. Nutr.* **25**, 207–216.

Hunt, C. H., O. G. Bentley, and T. Hershberger (1952). *Federation Proc.* **11**, 233–234.

Hunt, C. H., O. G. Bentley, T. V. Hershberger, and J. H. Cline (1954). *J. Animal Sci.* **13**, 570–580.

Hunt, W. G., and R. P. Moore (1958). *Appl. Microbiol.* **6**, 36–39.

Hutchings, A. (1961). *Queensland Agr. J.* **87**, 19–21.

Hvidsten, H. (1946). *Norg. Landbrukshøgskol. Foringsforsøkene.* **60**, 172. Cited from *Nutr. Abst. Rev.* **19**, abst. 2857.

Hydén, S. (1961a). *Kgl. Lantbruks-Hogskol. Ann.* **27,** 51–79.

Hydén, S. (1961b). *Kgl. Lantbruks-Hogskol. Ann.* **27,** 273–285.

Iggo, A. (1951). *J. Physiol. (London)* **115,** 74P–75P.

Iggo, A. (1955). *J. Physiol. (London)* **128,** 593–607.

Iggo, A. (1956). *J. Physiol. (London)* **131,** 248–256.

Irwin, D. H. (1956). *J. S. African Vet. Med. Assoc.* **27,** 9–13.

Ittner, N. R., T. E. Bond, and C. F. Kelley (1958). *Univ. Calif., Agr. Expt. Sta. Bull.* **761.**

Jackson, H. D., R. A. Shaw, W. R. Pritchard, and B. W. Hatcher (1959). *J. Animal Sci.* **18,** 158–162.

Jacobs, N. J., and M. J. Wolin (1963a). *Biochim. Biophys. Acta* **69,** 18–28.

Jacobs, N. J., and M. J. Wolin (1963b). *Biochim. Biophys. Acta* **69,** 29–39.

Jacobson, D. R. and I. L. Lindahl (1955). *Univ. Maryland Agr. Expt. Sta. Misc. Pub.* **238,** 9–15.

Jacobson, D. R., I. L. Lindahl, J. J. McNeill, J. C. Shaw, R. N. Doetsch, and R. E. Davis (1957). *J. Animal Sci.* **16,** 515–524.

Jacobson, N. L., D. Espe, and C. Y. Cannon (1942). *J. Dairy Sci.* **25,** 785–799.

James, A. T., and A. J. P. Martin (1952). *Biochem. J.* **50,** 679–690.

Jameson, A. P. (1925). *Parasitology* **17,** 403–405.

Jamieson, N. D. (1959a). *New Zealand J. Agr. Res.* **2,** 96–106.

Jamieson, N. D. (1959b). *New Zealand J. Agr. Res.* **2,** 314–328.

Jamieson, N. D., and T. M. Loftus (1958). *New Zealand J. Agr. Res.* **1,** 17–30.

Jarrett, I. G. (1948a). *J. Council Sci. Ind. Res.* **21,** 311–315.

Jarrett, I. G., and O. H. Filsell (1961). *Nature* **190,** 1114–1115.

Jarrett, I. G., and B. J. Potter (1950a). *Nature* **166,** 515–517.

Jarrett, I. G., and B. J. Potter (1950b). *Australian J. Exptl. Biol. Med. Sci.* **28,** 595–602.

Jarrett, I. G., B. J. Potter, and O. H. Filsell (1952). *Australian J. Exptl. Biol. Med. Sci.* **30,** 197–206.

Jasper, D. E. (1953). *Am. J. Vet. Res.* **14,** 184–191.

Jayasinghe, J. B. (1961). *Ceylon Vet. J.* **9,** 135–140.

Jayasuriya, G. C. N., and R. E. Hungate (1959). *Arch. Biochem. Biophys.* **82,** 274–287.

Jensen, E. A., and D. M. Hammond (1964). *J. Protozool.* **11,** 386–394.

Jensen, R., H. M. Deane, L. J. Cooper, V. A. Miller, and W. R. Graham (1954a). *Am. J. Vet. Res.* **15,** 202–216.

Jensen, R., W. E. Connell, and A. W. Deem (1954b) *Am J. Vet. Res.* **15,** 425–428.

Jensen, R. G., and J. Sampugna (1962). *J. Dairy Sci.* **45,** 435–437.

Jensen, R. G., K. L. Smith, J. E. Edmondson, and C. P. Merilan (1956). *J. Bacteriol.* **72,** 253–258.

Jírovec, O. (1933). *Z. Parasitenk.* **5,** 584–591.

Johanson, R., R. J. Moir, and E. J. Underwood (1949). *Nature* **163,** 101.

Johns, A. T. (1948). *Biochem. J.* **42,** ii–iii.

Johns, A. T. (1951a). *J. Gen. Microbiol.* **5,** 317–325.

Johns, A. T. (1951b). *J. Gen. Microbiol.* **5,** 326–336.

Johns, A. T. (1953). *New Zealand J. Sci. Technol.* **A35,** 262–269.

Johns, A. T. (1954). *New Zealand J. Sci. Technol.* **A36,** 289–320.

Johns, A. T. (1955a). *New Zealand J. Sci. Technol.* **A37,** 301–311.

Johns, A. T. (1955b). *New Zealand J. Sci. Technol.* **A37,** 323–331.

Johns, A. T. (1958). *Vet. Rev. Annotations* **4,** 17–31.

Johns, A. T., J. L. Mangan, and C. S. W. Reid (1958). *Proc. New Zealand Soc. Animal Prod.* **18,** 21–30.

Johns, A. T., F. H. McDowall, and W. A. McGillivray (1959). *New Zealand J. Agr. Res.* **2,** 62–71.

Johns, A. T., M. J. Ulyatt, and A. C. Glenday (1963). *J. Agr. Sci.* **61,** 201–207.

Johnson, B. C., T. S. Hamilton, H. H. Mitchell, and W. B. Robinson (1942). *J. Animal Sci.* **1,** 236–245.

Johnson, B. C., T. S. Hamilton, W. B. Robinson, and J. C. Garey (1944). *J. Animal Sci.* **3,** 287–298.

Johnson, C. E., and J. M. Prescott (1959). *J. Animal Sci.* **18,** 830–835.

Johnson, C. E., L. H. Harbers, and J. M. Prescott (1959). *J. Animal Sci.* **18,** 599–606.

Johnson, D. E., W. E. Denirsson, and D. W. Bolin (1964). *J. Animal Sci.* **23,** 499–505.

Johnson, R. B. (1951a). *North Am. Veterinarian* **32,** 327–332.

Johnson, R. B. (1951b). *Cornell Vet.* **41,** 115–121.

Johnson, R. B. (1954). *Cornell Vet.* **44,** 6–21.

Johnson, R. B., O. G. Bentley, and A. L. Moxon (1956). *J. Biol. Chem.* **218,** 379–390.

Johnson, R. H., L. R. Brown, N. L. Jacobson, and P. G. Homeyer (1958). *J. Animal Sci.* **17,** 893–902.

Johnson, R. H., R. S. Allen, N. L. Jacobson, W. R. Woods, and D. R. Warner (1960a). *J. Dairy Sci.* **43,** 1341–1342.

Johnson, R. H., P. A. Hartman, N. L. Jacobson, L. R. Brown, and H. H. Van Horn, Jr. (1960b). *J. Animal Sci.* **19,** 735–744.

Johnson, R. J., and I. A. Dyer (1964). *J. Animal Sci.* **23,** 602.

Johnson, R. J., I. A. Dyer, and J. A. Templeton (1963). *J. Animal Sci.* **22,** 838.

Johnson, R. R., and K. E. McClure (1964). *J. Animal Sci.* **23,** 208–213.

Johnson, R. R., B. A. Dehority, and O. G. Bentley (1958). *J. Animal Sci.* **17,** 841–850.

Johnson, R. R., B. A. Dehority, H. R. Conrad, and R. R. Davis (1962a). *J. Dairy Sci.* **45,** 250–252.

Johnson, R. R., B. A. Dehority, A. Burk, J. L. Parsons, and H. W. Scott (1962b). *J. Animal Sci.* **21,** 892–896.

Johnson, R. R., B. A. Dehority, A. Burk, K. E. McClure, and J. L. Parsons (1964). *J. Animal Sci.* **23,** 1124–1128.

Johnston, R. P., E. M. Kesler, and R. D. McCarthy (1961). *J. Dairy Sci.* **44,** 331–339.

Jordan, R. M., and H. G. Croom (1957). *J. Animal Sci.* **16,** 118–124.

Joyner, A. E., E. M. Kesler, and J. R. Holter (1963). *J. Dairy Sci.* **46,** 1108–1113.

Juhasz, B. (1962). *Acta Vet. Acad. Sci. Hung.* **12,** 383–395.

Jurtshuk, P., R. N. Doetsch, J. J. McNeill, and J. C. Shaw (1954) *J. Dairy Sci.* **37,** 1466–1472.

Jurtshuk, P., R. N. Doetsch, and J. C. Shaw (1958). *J. Dairy Sci.* **41,** 190–202.

Kaishio, Y., S. Higaki, S. Horii, and Y. Awai (1951). *Bull. Natl. Inst. Agr. Sci., Yahagi, Japan, Ser. G* **2,** 131–140. Cited from *Nutr. Abst. Rev.* **22,** abst. 3040.

Kameoka, K., and H. Morimoto (1959). *J. Dairy Sci.* **42,** 1187–1197.

Kameoka, K., S. Takahashi, and H. Morimoto (1956). *J. Dairy Sci.* **39,** 462–467.

Kamstra, L. D., A. L. Moxon, and O. G. Bentley (1958). *J. Animal Sci.* **17,** 199–208.

Kamstra, L. D., P. R. Zimmer, and L. B. Embry (1959). *J. Animal Sci.* **18,** 849–854.

Kandatsu, M., and N. Takahashi (1959). *Japan. J. Zootech. Sci.* **30,** 166–170.

Kandatsu, M., T. Matsumoto, Y. Kazama, K. Kikuno, and Y. Ichinose (1955). *J. Agr. Chem. Soc. Japan* **29,** 759–764.

Kane, E. A., W. C. Jacobson, and L. A. Moore (1952). *J. Nutr.* **47,** 263–273.

Kaplan, V. A., and A. V. Ljubeckaja (1959). *Sb. Tr. Har'kov. Zootec. Inst.* **10,** 131–135.

Karunairatnam, M. C., and G. A. Levvy (1951). *Biochem. J.* **49,** 210–215.

Kawashima, R., Y. Hashimoto, Y. Kamada, and S. Uesaka (1959). *Bull. Res. Inst. Food Sci. Kyoto Univ.* no. 22, 26–32.

Kay, R. N. B. (1954). *J. Physiol. (London)* **125,** 24P–25P.

Kay, R. N. B. (1955). *J. Physiol. (London)* **130,** 15P–16P.

Kay, R. N. B. (1958a). *J. Physiol. (London)* **144,** 463–475.

Kay, R. N. B. (1958b). *J. Physiol. (London)* **144,** 476–489.

Kay, R. N. B. (1958c). *J. Physiol. (London)* **143,** 63P–64P.

Kay, R. N. B. (1959). *Nature* **183,** 552–553.

Kay, R. N. B. (1960a). *J. Physiol. (London)* **150,** 515–537.

Kay, R. N. B. (1960b). *J. Physiol. (London)* **150,** 538–545.

Kay, R. N. B. (1963). *J. Dairy Res.* **30,** 261–288.

Kay, R. N. B., and P. N. Hobson (1963). *J. Dairy Res.* **30,** 261–313.

Kay, R. N. B., and A. T. Phillipson (1957). *J. Physiol. (London)* **139,** 7P.

Kay, R. N. B., and A. T. Phillipson (1959). *J. Physiol. (London)* **148,** 507–523.

Kay, R. N. B., and A. T. Phillipson (1964). *Proc. Nutr. Soc. (Engl. Scot.)* **23,** xlvi.

Keating, E. K., W. H. Hale, and F. Hubbert (1964). *J. Animal Sci.* **23,** 111–117.

Keeney, M., I. Katz, and M. J. Allison (1962). *J. Am. Oil Chemists' Soc.* **39,** 198–201.

Kehar, N. D. (1956). *Proc. 7th Intern. Congr. Animal Husbandry, Madrid* Subject 8 pp. 121–135.

Kehar, N. D., and R. Chanda (1947). *Indian J. Vet. Sci.* **17,** 189–192.

Kellner, O. (1911). *Proc. 5th Intern. Congr. Dairy Sci., Stockholm, 1911.* From Nehring (1956). pp. 133–137.

Kemble, A. R., and H. T. McPherson (1954). *Biochem. J.* **58,** 44–46.

Kercher, C. J., and S. E. Smith (1955). *J. Animal Sci.* **14,** 458–464.

Kesler, E. M. (1954). *J. Animal Sci.* **13,** 10–19.

Kesler, E. M., M. Ronning, and C. B. Knodt (1951). *J. Animal Sci.* **10,** 969–974.

Khouvine, Y. (1923). "Digestion de la cellulose par la flore intestinale de l'homme, *B. cellulosae dissolvens, n.sp.*" These. La Cour d'Appel, Paris.

Kick, C. H., P. Gerlaugh, A. F. Schalk, and E. A. Silver (1937a). *J. Agr. Res.* **55,** 587–597.

Kick, C. H., P. Gerlaugh, and A. F. Schalk (1937b). *Proc. Am. Soc. Animal Prod.* **1937,** 95–97.

Kiddle, P., R. A. Marshall, and A. T. Phillipson (1951). *J. Physiol. (London)* **113,** 207–217.

Kimura, F. T., and V. L. Miller (1957). *J. Agr. Food Chem.* **5,** 216.

King, K. W. (1956). *Virginia Agr. Expt. Sta. Tech. Bull.* **127,** 3–16.

King, K. W. (1959). *J. Dairy Sci.* **42,** 1848–1856.

King, K. W., and P. H. Smith (1955). *J. Bacteriol.* **70,** 726–729.

Kingwell, R. G., R. A. Oppermann, W. O. Nelson, and R. E. Brown (1959). *J. Dairy Sci.* **42**, 912–913.

Kirsch, W., and H. Jantzon (1941). *Z. Tierernaehrung Futtermittelk.* **5**, 244–259.

Kistner, A. (1960). *J. Gen. Microbiol.* **23**, 565–576.

Kistner, A. (1965). *In* "Physiology of Digestion in the Ruminant" (R. W. Dougherty, ed.), pp. 419–432. Butterworth, London and Washington, D.C.

Kistner, A., L. Gouws, and F. M. C. Gilchrist (1962). *J. Agr. Sci.* **59**, 85–91.

Klatte, F. J., G. E. Mitchell, and C. D. Little (1964). *J. Agr. Food Chem.* **12**, 420–421.

Kleiber, M. (1947). *Physiol. Rev.* **27**, 511–541.

Kleiber, M. (1954a). *Rev. Can. Biol.* **13**, 333–348.

Kleiber, M. (1954b). *Wiss. Abhandl. Deut. Akad. Landwirtschaftswiss. Berlin* **5**, 117–139.

Kleiber, M. (1956). *Natl. Acad. Sci.—Natl. Res. Council Publ.* **388**, 10–20.

Kleiber, M., H. H. Cole, and S. W. Mead (1943). *J. Dairy Sci.* **26**, 929–933.

Kleiber, M., W. M. Regan, and S. W. Mead (1945). *Hilgardia* **16**, 511–571.

Kleiber, M., A. H. Smith, and A. L. Black (1952a). *J. Biol. Chem.* **195**, 707–714.

Kleiber, M., A. H. Smith, A. L. Black, M. A. Brown, and B. M. Tolbert (1952b). *J. Biol. Chem.* **197**, 371–379.

Kleiber, M., A. L. Black, M. A. Brown, and B. M. Tolbert (1953). *J. Biol. Chem.* **203**, 339–346.

Kleiber, M., A. L. Black, M. A. Brown, C. F. Baxter, J. R. Luick, and F. H. Stadtman (1955). *Biochim. Biophys. Acta* **17**, 252–260.

Klein, W. (1915). *Biochem. Z.* **72**, 169–252.

Klein, W. (1926). *Z. Tierzuecht. Zuechtungsbiol.* **6**, 56.

Klein, W., and R. Müller (1942). *Z. Tierzuecht. Zuechtungsbiol.* **51**, 201–212.

Klein, W., H. Schmid, and E. Studt (1937). *Z. Tierzuecht. Zuechtungsbiol.* **39**, 135–161.

Klein, W., H. Schmid, E. Studt, and R. Müller (1939). *Z. Tierzuecht. Zuechtungsbiol.* **43**, 76–119.

Klopfenstein, T. J., D. B. Purser, and W. J. Tyznik (1964). *J. Animal Sci.* **23**, 490–495.

Kluyver, A. J., and C. B. van Niel (1956). "The Microbe's Contribution to Biology," 182 pp. Harvard Univ. Press, Cambridge, Massachusetts.

Knapp, R., A. L. Baker, and R. W. Phillips (1943). *J. Animal Sci.* **2**, 221–225.

Knight, C. A., R. A. Dutcher, N. B. Guerrant, and S. I. Bechdel (1940). *Proc. Soc. Exptl. Biol. Med.* **44**, 90–93.

Knoth, M. (1928). *Z. Parasitenk.* **1**, 262–282.

Knott, J. C., H. K. Murer, and R. E. Hodgson (1936). *J. Agr. Res.* **53**, 553–556.

Knox, K. L., and G. M. Ward (1961). *J. Dairy Sci.* **44**, 1550–1553.

Köhler, W. (1940). *Arch. Mikrobiol.* **11**, 432–469.

Koffman, M. (1937). *Skand. Veterinartidskr.* **27**, 45; Cited from *Vet. Bull.* **7**, 592.

Koffman, M. (1938). *Lantbrukshogskol. Ann.* **5**, 201–247.

Kofoid, C. A., and J. F. Christensen (1934). *Univ. Calif. (Berkeley) Publ. Zool.* **39**, 341–392.

Kofoid, C. A., and R. F. MacLennan (1930). *Univ. Calif. (Berkeley) Publ. Zool.* **33**, 471–544.

Kofoid, C. A., and R. F. MacLennan (1932). *Univ. Calif. (Berkeley) Publ. Zool.* **37**, 53–152.

Kofoid, C. A., and R. F. MacLennan (1933). *Univ. Calif. (Berkeley) Publ. Zool.* **39,** 1–34.

Koistinen, O. A. (1948). *Ann. Acad. Sci. Fenn. Ser. A, II Chem.* **31,** 9–73.

Kolb, E. (1957). *Berlin. Muench. Tieraerztl. Wochschr.* **70,** 242–246.

Komarck, R. J., and E. C. Leffel (1961). *J. Animal Sci.* **20,** 782–784.

Kon, S. K. (1945). *Proc. Nutr. Soc. (Engl. Scot.)* **3,** 217–238.

Kon, S. K., and J. W. G. Porter (1947). *Nutr. Abstr. Rev.* **17,** 31–37.

Kon, S. K., and J. W. G. Porter (1953a). *In* "Biochemistry and Physiology of Nutrition" (G. H. Bourne and G. W. Kidder, eds.), Vol. 1, pp. 291–327. Academic Press, New York.

Kon, S. K., and J. W. G. Porter (1953b). *Proc. Nutr. Soc. (Engl. Scot.)* **12,** xii.

Kon, S. K., and J. W. G. Porter (1954). *Vitamins Hormones.* **12,** 53–65.

Koval'skiö, V. V., and Y. I. Raetskaya (1955). *Dokl. Akad. Nauk SSSR* **100,** 1131–1134.

Kraschenninikow, S. (1929). *Z. Zellforsch. Mikroskop. Anat.* **8,** 470–483.

Kraus, F. K. (1927). *Biol. Generalis* **3,** 347.

Kreipe, H. (1927). Untersuchungen Über die Milchsäurebakterienflora des Kuh-pansens. Dissertation, Kiel.

Krogh, A., and H. O. Schmidt-Jensen (1920). *Biochem. J.* **14,** 686–696.

Krogh, N. (1959). *Acta Vet. Scand.* **1,** 74–97.

Krogh, N. (1960). *Acta Vet. Scand.* **1,** 383–410.

Krogh, N. (1961a). *Acta Vet. Scand.* **2,** 103–119.

Krogh, N. (1961b). *Acta Vet. Scand.* **2,** 357–374.

Krogh, N. (1963a). *Acta Vet. Scand.* **4,** 27–40.

Krogh, N. (1963b). *Acta Vet. Scand.* **4,** 41–51.

Kronfeld, D. S. (1961). *Am. J. Vet. Res.* **22,** 496–501.

Kronfeld, D. S., and M. Kleiber (1959). *J. Appl. Physiol.* **14,** 1033–1035.

Kronfeld, D. S., E. G. Tombropoulus, and M. Kleiber (1959). *J. Appl. Physiol.* **14,** 1029–1032.

Krzywanek, F. W. (1929). *Arch. Ges. Physiol.* **222,** 89–96.

Krzywanek, F. W., and P. Quast (1937). *Arch. Ges. Physiol.* **238,** 333–340.

Kubikova, M. (1935). *Zool. Anz.* **111,** 175–177.

Kuhn, H. (1964). *Folia Primatol. Abt. Senckenberg. Anat. Inst., Univ. Frank-furt-am-Main* **2,** 193–221.

Lacoste, A. M. (1961). *Compt. Rend.* **252,** 1233–1235.

Lacoste-Bastié, A. M. (1963). Recherches sur le métabolisme des acides organiques non volatils et des acides aminés cycliques chez les bactéries du rumen. Thèse Docteur des Sciences. Université de Toulouse.

Ladd, J. N. (1959). *Biochem. J.* **71,** 16–22.

Ladd, J. N., and D. J. Walker (1959). *Biochem. J.* **71,** 364–373.

Lagace, A. (1961). *J. Am. Vet. Med. Assoc.* **138,** 188–190.

Lambourne, L. J. (1957a). *J. Agr. Sci.* **48,** 273–285.

Lambourne, L. J. (1957b). *J. Agr. Sci.* **49,** 515–425.

Lampila, M. (1955). *Maataloustieteellinen Aikakauskirja* **27,** 142–153.

Lancaster, R. J. (1949a). *New Zealand J. Sci. Technol.* **A31,** 31–38.

Lancaster, R. J. (1949b). *Nature* **163,** 330–331.

Lancaster, R. J., and A. F. R. Adams (1943). *New Zealand J. Sci. Technol.* **A25,** 131–151.

Langlands, J. P., J. L. Corbett, I. McDonald, and G. W. Reid (1963) *Animal Prod.* **5,** 11–16.

Lardinois, C. C., R. C. Mills, C. A. Elvehjem, and E. B. Hart (1944). *J. Dairy Sci.* **27,** 579–583.

Larsen, C., E. H. Hungerford, and D. E. Bailey (1917). *S. Dakota Agr. Expt. Sta.,* **175,** 648–693.

Lassiter, C. A., R. S. Emery, and C. W. Duncan (1958a). *J. Animal Sci.* **17,** 358–362.

Lassiter, C. A., R. S. Emery, and C. W. Duncan (1958b). *J. Dairy Sci.* **41,** 552–553.

Lawton, E. J., W. D. Bellamy, and R. E. Hungate (1951). *Science* **113,** 380–382.

LeBars, H., and H. Simonnet (1954a). *Rec. Med. Vet.* **130,** 689–700.

LeBars, H., and H. Simonnet (1954b). *Rec. Med. Vet.* **130,** 777–788.

LeBars, H., R. Nitescu, and H. Simonnet (1953a). *Bull. Acad. Vet. France* **26,** 287–300.

LeBars, H., R. Nitescu, and H. Simonnet (1953b). *Bull. Acad. Vet. France* **26,** 351–360.

LeBars, H., R. Nitescu, and H. Simonnet (1953c). *Bull. Acad. Vet. France* **26,** 445–447.

LeBars, H., J. Lebrument, R. Nitescu, and H. Simonnet (1954a). *Bull. Acad. Vet. France* **27,** 53–67.

LeBars, H., J. Lebrument, and H. Simonnet (1954b). *Bull. Acad. Vet. France* **27,** 69–73.

Lecomte, A. (1954). *Bull. Acad. Vet. France* **27,** 75–77.

Lee, H. C., and W. E. C. Moore (1959). *J. Bacteriol.* **77,** 741–747.

Lee, S. D., and W. F. Williams (1962a). *J. Dairy Sci.* **45,** 517–521.

Lee, S. D., and W. F. Williams (1962b). *J. Dairy Sci.* **45,** 893–896.

Legg, S. P., and L. Sears (1960). *Nature* **186,** 1061–1062.

Leifson, E. (1960). "Atlas of Bacterial Flagellation." Academic Press, New York.

Leng, R. A., and E. F. Annison (1963). *Biochem. J.* **86,** 319–327.

Lenkeit, W. (1930). *Wiss. Arch. Landwirtsch. Abt. B, Arch. Tierernaehr. Tierzucht* **3,** 631–638.

Lenkeit, W., and A. Columbus (1934). *Arch. Wiss. Prakt. Tierheilk.* **68,** 126–133.

Leroy, A. M. (1958). *Ann. Nutr. Aliment.* **12,** 283–286.

Leroy, A. M. (1961). *Ann. Nutr. Aliment.* **15,** 280–287.

Leroy, A. M., and S. Z. Zelter (1955). *Ann. Zootech.* **4,** 69–91.

Lesperance, A. L., and V. R. Bohman (1963). *J. Animal Sci.* **22,** 686–694.

Lessel, E. F., and R. S. Breed (1954). *Bacteriol. Rev.* **18,** 165–168.

Lev, M. (1959). *J. Gen. Microbiol.* **20,** 697–703.

Lewis, D. (1951a). *Biochem. J.* **48,** 175–180.

Lewis, D. (1951b). *Biochem. J.* **49,** 149–153.

Lewis, D. (1954). *Biochem. J.* **56,** 391–399.

Lewis, D. (1955). *Brit. J. Nutr.* **9,** 215–230.

Lewis, D. (1957). *J. Agr. Sci.* **48,** 438–446.

Lewis, D. (1960). *J. Agr. Sci.* **55,** 111–117.

Lewis, D. (1962). *J. Agr. Sci.* **58,** 73–79.

Lewis, D., and S. R. Elsden (1955). *Biochem. J.* **60,** 683–692.

Lewis, D. and W. I. McDonald (1958). *J. Agr. Sci.* **51,** 108–118.

Lewis, D., K. J. Hill, and E. F. Annison (1957). *Biochem. J.* **66,** 587–592.

Lewis, F. T. (1915). *Anat. Record* **9,** 102–103.

Lewis, T. R., and R. S. Emery (1962a). *J. Dairy Sci.* **45,** 765–768.

Lewis, T. R., and R. S. Emery (1962b). *J. Dairy Sci.* **45,** 1487–1492.

Lewis, T. R., and R. S. Emery (1962c). *J. Dairy Sci.* **45**, 1363–1368.

Liebetanz, E. (1910). *Arch. Protistenk.* **19**, 19–80.

Lindahl, I. L., and P. J. Reynolds (1959). *J. Animal Sci.* **18**, 1074–1079.

Lindahl, I. L., A. C. Cook, R. E. Davis, and W. D. Maclay (1954). *Science* **119**, 157–158.

Lindahl, I. L., R. E. Davis, D. R. Jacobson, and J. C. Shaw (1957). *J. Animal Sci.* **16**, 165–178.

Lindsay, D. B. (1959). *Vet. Rev. Annotations* **5**, 103–128.

Lindsay, D. B., and E. J. H. Ford (1964). *Biochem. J.* **90**, 24–30.

Lines, E. J. (1935). *J. Council Sci. Ind. Res.* **8**, 117–119.

List, A. (1885). Untersuchungen über die in und auf dem Körper des gesunden Schafes vorkommenden niederen Pilze. Dissertation, Leipzig.

Little, O., E. Cheng, and W. Burroughs (1958). *J. Animal Sci.* **17**, 1190.

Little, C. O., W. Burroughs, and W. Woods (1963). *J. Animal Sci.* **22**, 358–363.

Livingston, H. G., W. J. A. Payne, and M. T. Friend (1962). *Nature* **194**, 1057–1058.

Lofgreen, G. P., and M. Kleiber (1954). *J. Animal Sci.* **13**, 258–264.

Lofgreen, G. P., W. C. Weir, and J. F. Wilson (1953). *J. Animal Sci.* **12**, 347–352.

Loosli, J. K., H. L. Lucas, and L. A. Maynard (1945). *J. Dairy Sci.* **28**, 147–153.

Loosli, J. K., H. H. Williams, W. E. Thomas, F. H. Ferris, and L. A. Maynard (1949). *Science* **110**, 144–145.

Loosli, T. K., and L. E. Harris (1945). *J. Animal Sci.* **4**, 435–437.

Lough, A. K., and G. A. Garton (1958). *In* "Essential Fatty Acids" (M. Sinclair, ed.), p. 97–100. Butterworth, London and Washington, D.C.

Louw, J. G., S. I. Bodenstein, and J. I. Quin (1948). *Onderstepoort J. Vet. Sci. Animal Ind.* **23**, 239–259.

Louw, J. G., H. H. Williams, and L. A. Maynard (1949). *Science* **110**, 478–480.

Lubinsky, G. (1955a). *Can. J. Microbiol.* **1**, 440–450.

Lubinsky, G. (1955b). *Can. J. Microbiol.* **1**, 675–684.

Lubinsky, G. (1957a). *Can. J. Zool.* **35**, 111–133.

Lubinsky, G. (1957b). *Can. J. Zool.* **35**, 135–140.

Lubinsky, G. (1957c). *Can. J. Zool.* **35**, 141–159.

Lubinsky, G. (1957d). *Can. J. Zool.* **35**, 579–580.

Luecke, R. W., R. Culik, F. Thorp, L. H. Blakeslee, and R. H. Nelson (1950). *J. Animal Sci.* **9**, 420–425.

Luedke, A. J., J. W. Bratzler, and H. W. Dunne (1959). *Am. J. Vet. Res.* **20**, 690–696.

Lugg, J. W. H. (1938). *J. Agr. Sci.* **28**, 688–694.

Lugg, J. W. H. (1949). *Advan. Protein Chem.* **5**, 230–295.

Luick, J. R., D. T. Torell, and W. Sirc (1959). *Intern. J. Appl. Radiation Isotopes* **4**, 169–172.

Luitingh, H. C. (1961a). *J. Agr. Sci.* **56**, 333–342.

Luitingh, H. C. (1961b). *J. Agr. Sci.* **56**, 389–396.

Luizzo, J. A., S. L. Hansard, J. G. Lee, and A. F. Novak (1961). *J. Nutr.* **75**, 231.

Lungwitz, M. (1891). *Arch. Wiss. Prak. Tierheilk.* **18**, 80–110.

Lupien, P. J., F. Sauer, and G. V. Hatina (1962). *J. Dairy Sci.* **45**, 210–217.

Lusk, J. W., C. B. Browning, and J. T. Miles (1962). *J. Dairy Sci.* **45**, 69–73.

Lyttleton, J. W. (1960). *New Zealand J. Agr. Res.* **3**, 63–68.

McAnally, R. A. (1942). *Biochem. J.* **36**, 392–399.

McAnally, R. A. (1943). *Onderstepoort J. Vet. Sci. Animal Ind.* **18**, 131–138.

McAnally, R. A., and A. T. Phillipson (1944). *Biol. Rev. Cambridge Phil. Soc.* **19**, 41–54.

McArthur, A. T. G. (1957a). *New Zealand J. Sci. Technol.* **A38**, 696–699.

McArthur, A. T. G. (1957b). *New Zealand J. Sci. Technol.* **A38**, 700–701.

McArthur, J. M., and J. E. Miltimore (1961). *Can. J. Animal Sci.* **41**, 187–196.

McArthur, J. M., J. E. Miltimore, and M. J. Pratt (1964). *Can. J. Animal Sci.* **44**, 200–206.

McBee, R. H. (1953). *Appl. Microbiol.* **1**, 106–110.

McCance, R. A., and E. M. Widdowson (1944). *Ann. Rev. Biochem.* **13**, 315–346.

McCandless, E. L., and J. A. Dye (1950). *Am. J. Physiol.* **162**, 434–446.

McCarthy, R. D., J. B. Holter, J. C. Shaw, F. G. Hueter, and J. McCarthy (1957). *Univ. Maryland Agr. Expt. Sta. Misc. Publ.* **291**, 33–36.

McClymont, G. L. (1946). *Australian Vet. J.* **22**, 95–98.

McClymont, G. L. (1951a). *Australian J. Agr. Res.* **2**, 92–103.

McClymont, G. L. (1951b). *Australian J. Agr. Res.* **2**, 158–180.

McClymont, G. L. (1952). *Australian J. Sci. Res., Ser B* **5**, 374–383.

McClymont, G. L., and R. Paxton (1947). *Agr. Gaz. N. S. Wales* **58**, 551–553.

McCormick, N. G., E. R. Ordal, and H. R. Whiteley (1962a). *J. Bacteriol.* **83**, 887–898.

McCormick, N. G., E. J. Ordal, and H. R. Whiteley (1962b). *J. Bacteriol.* **83**, 899–906.

McDonald, I. R., and D. A. Denton (1956). *Nature* **177**, 1035–1036.

McDonald, I. W. (1948a). *Biochem. J.* **42**, 584–587.

McDonald, I. W. (1948b). *J. Physiol. (London)* **107**, 21P.

McDonald, I. W. (1948c). *Biochem. J.* **42**, xiii.

McDonald, I. W. (1952). *Biochem. J.* **51**, 86–90.

McDonald, I. W. (1954). *Biochem. J.* **56**, 120–125.

McDonald, I. W. (1958). *Proc. Australian Soc. Animal Prod.* **2**, 46–51.

McDonald, I. W., and R. J. Hall (1957). *Biochem. J.* **67**, 400–403.

MacDonald, J. B., and E. M. Madlener (1957). *Can. J. Microbiol.* **3**, 679–686.

MacDonald, M. A. (1957). *New Zealand J. Sci. Technol.* **A38**, 987–996.

McDougall, E. I. (1948). *Biochem. J.* **43**, 99–109.

McDowall, F. H., W. A. McGillivray, and C. S. W. Reid (1957a). *New Zealand J. Sci. Technol.* **A38**, 839–852.

McDowall, F. H., M. R. Patchell, and C. S. W. Reid (1957b). *New Zealand J. Sci. Technol.* **A38**, 1036–1053.

McDowall, F. H., C. S. W. Reid, and M. R. Patchell (1957c). *New Zealand J. Sci. Technol.* **A38**, 1054–1080.

McElroy, L. W., and H. Goss (1939). *J. Biol. Chem.* **130**, 437–438.

McElroy, L. W., and H. Goss (1940a). *J. Nutr.* **20**, 527–540.

McElroy, L. W., and H. Goss (1940b). *J. Nutr.* **20**, 541–550.

McElroy, L. W., and H. Goss (1941a). *J. Nutr.* **21**, 163–173.

McElroy, L. W., and H. Goss (1941b). *J. Nutr.* **21**, 405–409.

McElroy, L. W., and T. H. Jukes (1940). *Proc. Soc. Exptl. Biol. Med.* **45**, 296–297.

McGaughey, C. A., and K. C. Sellers (1948). *Nature* **161**, 1014–1015.

McGillivray, W. A. (1957). *New Zealand J. Sci. Technol.* **A38,** 878–886.

McGillivray, W. A., F. H. McDowall, and C. S. W. Reid (1959). *New Zealand J. Agr. Res.* **2,** 35–46.

MacKenzie, H. I., and R. E. Altona (1964a). *J. S. African Vet. Med. Assoc.* **35,** 301–306.

MacKenzie, H. I., and R. E. Altona (1964b). *J. S. African Vet. Med. Assoc.* **35,** 309–312.

McLean, J. W., G. G. Thomson, and J. H. Claxton (1959). *Nature* **184,** 251–252.

MacLennan, R. F. (1933). *Univ. Calif. (Berkeley) Publ. Zool.* **39,** 205–250.

MacLennan, R. F. (1934). *Arch. Protistenk.* **81,** 412–419.

MacLeod, R. A., and J. F. Murray (1956). *J. Nutr.* **60,** 245–259.

McMahan, C. A. (1964). *J. Wildlife Management* **28,** 798–808.

McManus, W. R. (1961). *Nature* **192,** 1161–1163.

McManus, W. R. (1962). *Australian J. Agr. Res.* **13,** 907–923.

McMeekan, C. P. (1943). *New Zealand J. Sci. Technol.* **A25,** 152–153.

McNaught, M. L. (1951). *Biochem. J.* **49,** 325–332.

McNaught, M. L., and J. A. B. Smith (1947). *Nutr. Abstr. Rev.* **17,** 18–31.

McNaught, M. L., J. A. B. Smith, K. M. Henry, and S. K. Kon (1950a). *Biochem. J.* **46,** 32–36.

McNaught, M. L., E. C. Owen, and J. A. B. Smith (1950b). *Biochem. J.* **46,** 36–43.

McNaught, M. L., E. C. Owen, K. M. Henry, and S. K. Kon (1954a). *Biochem. J.* **56,** 151–156.

McNaught, M. L., J. A. B. Smith, and W. A. P. Black (1954b). *J. Sci. Food Agr.* **5,** 350–352.

McNeill, J. J., and D. R. Jacobson (1955). *Univ. Maryland Agr. Expt. Sta. Misc. Publ.* **238,** 7–8.

McNeill, J. J., and D. R. Jacobson (1955). *J. Dairy Sci.* **38,** 608.

McNeill, J. J., R. N. Doetsch, and J. C. Shaw (1954). *J. Dairy Sci.* **37,** 81–88.

Macpherson, M. J. (1953). *J. Pathol. Bacteriol.* **66,** 95–102.

Magee, H. E. (1932). *J. Exptl. Biol.* **9,** 409–426.

Magee, H. E., and J. B. Orr (1924). *J. Agr. Sci.* **14,** 619–625.

Mah, R. A. (1962). Experiments on the culture and physiology of *Ophryoscolex purkynei*. Ph.D. Thesis, University of California, Davis, California.

Mah, R. A. (1964). *J. Protozool.* **11,** 546–552.

Mah, R. A., and R. E. Hungate (1965). *J. Protozool.* **12,** 131–136.

Makela, A. (1956). *Suomen Maataloustieteellisen Seuran Julkaisuja* **85,** 1–139.

Maki, L. R., and E. M. Foster (1957). *J. Dairy Sci.* **40,** 905–913.

Mallèvre, A. (1891). *Arch. Ges. Physiol.* **49,** 460–477.

Mangan, J. L. (1958). *New Zealand J. Agr. Res.* **1,** 140–147.

Mangan, J. L. (1959). *New Zealand J. Agr. Res.* **2,** 47–61.

Mangan, J. L., A. T. Johns, and R. W. Bailey (1959). *New Zealand J. Agr. Res.* **2,** 342–354.

Mangold, E. (1929). *Handbuch Ernaehr. Stoffwechsels Landwirtschlaftl. Nutztiere* **2,** 160.

Mangold, E. (1943). *Ergeb. Biol.* **19,** 1–81.

Mangold, E., and W. Klein (1927). "Bewegungen und Innervation des Wiederkäuermagens." Thieme, Leipzig.

Mangold, E., and W. Lenkeit (1931). *Wiss. Arch. Landwirtsch. Abt. B* **5,** 201–205.

Mangold, E., and T. Radeff (1930). *Wiss. Arch. Landwirtsch. Abt. B* **4,** 173–199.

Mangold, E., and K. Schmitt-Krahmer (1927). *Biochem. Z.* **191**, 411–422.

Mangold, E., and F. Usuelli (1930). *Arch. Tierernaehr. Tierzucht* **3**, 189–201.

Mann, S. O., and A. E. Oxford (1954). *J. Gen. Microbiol.* **11**, 83–90.

Mann, S. O., and A. E. Oxford (1955). *J. Gen. Microbiol.* **12**, 140–146.

Mann, S. O., F. M. Masson, and A. E. Oxford (1954a). *J. Gen. Microbiol.* **10**, 142–149.

Mann, S. O., F. M. Masson, and A. E. Oxford (1954b). *Brit. J. Nutr.* **8**, 246–252.

Manusardi, L. (1931). *Boll. Lab. Zool. Agrar. Bachicoltura Milano* **4**, 140–148.

Margherita, S. S., and R. E. Hungate (1963). *J. Bacteriol.* **86**, 855–860.

Margherita, S. S., R. E. Hungate, and H. Storz (1964). *J. Bacteriol.* **87**, 1304–1308.

Margolin, S. (1930). *Biol. Bull.* **59**, 301–305.

Markoff, I. (1911). *Biochem. Z.* **34**, 211–232.

Markoff, I. (1913). *Biochem. Z.* **57**, 1–70.

Marquardt, R. R., and J. M. Asplund (1964a). *Can. J. Animal Sci.* **44**, 16–23.

Marquardt, R. R., and J. M. Asplund (1964b). *Can. J. Animal Sci.* **44**, 24–28.

Marsh, C. A. (1954). *Biochem. J.* **58**, 609–617.

Marsh, C. A. (1955). *Biochem. J.* **59**, 375–382.

Marshall, R. A. (1949). *Brit. J. Nutr.* **3**, 1–2.

Marshall, R. A., and A. T. Phillipson (1945). *Proc. Nutr. Soc. (Engl. Scot.)* **3**, 238–243.

Marston, H. R. (1935). *J. Council Sci. Ind. Res.* **8**, 111–116.

Marston, H. R. (1948a). *Biochem. J.* **42**, 564–574

Marston, H. R. (1948b). *Australian J. Sci. Res.* **B1**, 93–129.

Marston, H. R., and R. M. Smith (1952). *Nature* **170**, 792–793.

Martin, J. E., L. R. Arrington, J. E. Moore, C. B. Ammermann, G. K. Davis, and R. L. Shirley (1964). *J. Nutr.* **83**, 60–64.

Masson, F. M., and A. E. Oxford (1951). *J. Gen. Microbiol.* **5**, 664–672.

Masson, M. J. (1951). *Research (London)* **4**, 73–78.

Masson, M. J., and A. T. Phillipson (1951). *J. Physiol. (London)* **113**, 189–206.

Masson, M. J., and A. T. Phillipson (1952). *J. Physiol. (London)* **116**, 98–111.

Masson, M. J., A. Iggo, R. S. Reid, and G. F. Mann (1952). *J. Roy. Microscop. Soc.* **72**, 67–69.

Masters, C. J. (1964a). *Australian J. Biol. Sci.* **17**, 183–189.

Masters, C. J. (1964b). *Australian J. Biol. Sci.* **17**, 190–199.

Masters, C. J. (1964c). *Australian J. Biol. Sci.* **17**, 200–207.

Mather, R. E. (1959). *J. Dairy Sci.* **42**, 878–885.

Matrone, G., H. A. Ramsey, and G. H. Wise (1959). *Proc. Soc. Exptl. Biol. Med.* **100**, 8–11.

Matrone, G., C. R. Gunn, and J. J. McNeill (1964). *J. Nutr.* **84**, 215–219.

Matsumoto, T. (1961). *Tohoku J. Agr. Res.* **12**, 213–220.

Matsumoto, T., and Y. Shimoda (1962). *Tohoku J. Agr. Res.* **13**, 135–140.

Matteuzzi, D. (1964). *Ric. Sci.* **34** (II–B), 703–710.

Maymone, B., and D. Cece-Ginestrelli (1947). *Ann. Sper. Agrar.* **1**, 7.

Maymone, B., and M. Tiberio (1948). *Ann. Sper. Agrar.* **2**, 123–132.

Maymone, B., and M. Tiberio (1949). *Ann. Sper. Agrar.* **3**, 37–51.

Maynard, L. A., and J. K. Loosli (1956). "Animal Nutrition," McGraw-Hill, New York.

Meiske, J. C., R. L. Salsbury, J. A. Hoefer, and R. W. Luecke (1958). *J. Animal Sci.* **17**, 774–781.

Meister, A. (1934). Untersuchungen über den Eisengehalt fremdkörperfreier und fremdkörperhaltiger Vormageninhalte beim Rinde zwecks diagnostischer Verwertung bei der Fremdkörpererkrankung. Dissertation, Hanover. 44 pp. Abst. in *Jahresber. Vet.-Med.* **56,** 453.

Meites, S., R. C. Buttell, and T. S. Sutton (1951). *J. Animal Sci.* **10,** 203–210.

Menahan, L. A., and L. H. Schultz (1964a). *J. Dairy Sci.* **47,** 1080–1085.

Menahan, L. A., and L. H. Schultz (1964b). *J. Dairy Sci.* **47,** 1086–1091.

Mendel, V. E. (1961). *J. Dairy Sci.* **44,** 679–686.

Mendel, V. E., and J. M. Boda (1961). *J. Dairy Sci.* **44,** 1881–1897.

Merrill, S. D. (1952). *Vet. Med.* **47,** 405–406.

Meyer, J. H., and A. O. Nelson (1961). *Calif. Agr.* **15,** 14.

Meyer, J. H., and A. O. Nelson (1963). *J. Nutr.* **80,** 343–349.

Meyer, J. H., W. C. Weir, and J. D. Smith (1955). *J. Animal Sci.* **14,** 160–172.

Meyer, J. H., R. L. Gaskill, G. S. Stoewsand, and W. C. Weir (1959a). *J. Animal Sci.* **18,** 336–346.

Meyer, J. H., W. C. Weir, J. B. Dobie, and J. L. Hull (1959b). *J. Animal Sci.* **18,** 976–982.

Meyer, R. M., E. E. Bartley, J. L. Morrill, and W. E. Stewart (1964). *J. Dairy Sci.* **47,** 1339–1345.

Michaux, A. (1950). *Compt. Rend.* **230,** 2051–2053.

Michaux, A. (1951). *Compt. Rend.* **232,** 121–123.

Miller, J. I., and F. B. Morrison (1942). *J. Animal Sci.* **1,** 352.

Miller, J. K., and W. J. Miller (1960). *J. Dairy Sci.* **43,** 1854–1856.

Miller, W. J., W. J. Tyznick, N. N. Allen, and D. K. Sorenson (1954a). *J. Am. Vet. Med. Assoc.* **124,** 291–294.

Miller, W. J., R. K. Waugh, and G. Matrone (1954b). *J. Animal Sci.* **13,** 283–288.

Mills, R. C., A. N. Booth, G. Bohstedt, and E. B. Hart (1942). *J. Dairy Sci.* **25,** 925–929.

Mills, R. C., C. C. Lardinois, I. W. Rupel, and E. B. Hart (1944). *J. Dairy Sci.* **27,** 571–578.

Miltimore, J. E., W. J. Pigden, J. M. McArthur, and T. H. Antey (1964a). *Can. J. Animal Sci.* **44,** 96–101.

Miltimore, J. E., J. L. Mason and J. M. McArthur (1964b). *Can. J. Animal Sci.* **44,** 309–314.

Minato, H., A. Endo, T. Koriyama, and T. Uemura (1962). *J. Agr. Chem. Soc. Japan* **36,** 106–110.

Minato, H., T. Sai, A. Endo, T. Murakami, and T. Uemura (1963). *J. Agric. Chem. Soc. Japan* **37,** 379–384.

Mitchell, H. H. (1942). *J. Animal Sci.* **1,** 159–173.

Mitchell, R. L., and J. Tosic (1949). *J. Gen. Microbiol.* **3,** xvi–xvii.

Moe, P. W., H. F. Tyrrell, and J. T. Reid (1963). *Proc. 1963 Cornell Nutr. Conf. Feed Manuf.* pp. 66–70. Cornell Univ., Ithaca, New York.

Mönnig, H. O., and J. I. Quin (1933). *Onderstepoort J. Vet. Sci. Animal Ind.* **1,** 117–133.

Mönnig, H. O., and J. I. Quin (1935). *Onderstepoort J. Vet. Sci. Animal Ind.* **5,** 485–499.

Moir, R. J. (1951). *Australian J. Agr. Res.* **2,** 322–330.

Moir, R. J. (1965). *In* "Physiology of Digestion in the Ruminant" (R. W. Dougherty, ed.), pp. 1–14. Butterworth, London and Washington, D.C.

Moir, R. J., and L. E. Harris (1962). *J. Nutr.* **77,** 285–298.

Moir, R. J., and M. J. Masson (1952). *J. Pathol. Bacteriol.* **64,** 343–350.

Moir, R. J., and M. Somers (1956). *Nature* **178,** 1472.

Moir, R. J., and M. Somers (1957). *Australian J. Agr. Res.* **8,** 253–265.

Moir, R. J., and V. J. Williams (1950). *Australian J. Sci. Res.* **B3,** 381–392.

Moir, R. J., M. Somers, G. Sharman, and H. Waring (1954). *Nature* **173,** 269–270.

Moir, R. J., M. Somers, and H. Waring (1956). *Australian J. Biol. Sci.* **9,** 293–304.

Monroe, R. A., H. E. Sauberlich, C. L. Comar, and S. L. Hood (1952). *Proc. Soc. Exptl. Biol. Med.* **80,** 250–259.

Montgomery, M. J., L. H. Schultz, and B. R. Baumgardt (1963). *J. Dairy Sci.* **46,** 1380–1384.

Moomaw, C. R., and R. E. Hungate (1963). *J. Bacteriol.* **85,** 721–722.

Moon, F. E., and N. Varley (1942). *Vet. Record* **54,** 359–360.

Moore, L. A., and O. B. Winter (1934). *J. Dairy Sci.* **17,** 297–305.

Moore, W. E. C., and E. P. Cato (1965). *Intern. Bull. Bacteriol. Nomenclature Taxonomy* **15,** 69–80.

Moore, W. E. C., and K. W. King (1958). *J. Dairy Sci.* **41,** 1451–1455.

Morillo, F. J., and J. de Alba (1960). *Cornell Vet.* **50,** 66–72.

Morris, M. P., and R. Garcia-Rivera (1955). *J. Dairy Sci.* **38,** 1169.

Morrison, F. B. (1957). "Feeds and Feeding." Morrison, Ithaca, New York.

Moskovits, I. (1942). *Monthly Bull. Agr. Sci. Pract.* **33,** 408T–420T.

Mould, D. L., and G. J. Thomas (1958). *Biochem. J.* **69,** 327–337.

Moussu, G. (1888). *Compt. Rend. Soc. Biol.* **40,** 280–281.

Mowry, H. A., and E. R. Becker (1930). *Iowa State Coll. J. Sci.* **5,** 35–60.

Müller, M. (1906). *Arch. Ges. Physiol.* **112,** 245–291.

Müller, R., and G. Krampitz (1955a). *Naturwissenschaften* **42,** 648–649.

Müller, R., and G. Krampitz (1955b). *Z. Tierzuecht. Zuechtungsbiol.* **65,** 187–198.

Muhrer, M. E., and E. J. Carroll (1964). *J. Animal Sci.* **23,** 885.

Mukherjee, D. B. (1960). Application of isotope technique in the study of fermentation in the rumen of sheep. Ph.D. Thesis, University of Aberdeen. Aberdeen, Scotland.

Munch-Petersen, E., and C. A. P. Boundy (1963). *Appl. Microbiol.* **11,** 190–195.

Munch-Petersen, E., and C. A. P. Boundy (1964). *Zentr. Bakteriol. Parasitenk. Abt. I. Orig.* **191,** 512–524.

Murray, M. G., R. S. Reid, and T. M. Sutherland (1962). *J. Physiol. (London)* **164,** 26P.

Muth, O. H., J. E. Oldfield, J. R. Schubert, and L. F. Remmert (1959). *Am. J. Vet. Res.* **20,** 231–234.

Muth, O. H., J. R. Schubert, and J. E. Oldfield (1961). *Am. J. Vet. Res.* **22,** 466–469.

Mylroie, R. L., and R. E. Hungate (1954). *Can. J. Microbiol.* **1,** 55–64.

Naga, M. M. A., and K. el-Shazly (1963). *J. Agr. Sci.* **61,** 73–79.

Nagy, J. G., H. W. Steinhoff, and G. M. Ward (1964). *J. Wildlife Management* **28,** 785–797.

Nanda, P. N., and G. Singh (1941). *Indian J. Vet. Sci.* **11,** 16–27.

Nangeroni, L. L. (1954). *Cornell Vet.* **44,** 403–416.

Nehring, K., ed. (1956). *Wiss. Abhandl. Deut. Akad. Landwirtschaftswiss. Berlin* **5,** 133–137.

Nelson, W. O., R. A. Opperman, and R. E. Brown (1958). *J. Dairy Sci.* **41**, 545–551.

Nelson, W. O., R. E. Brown, and R. G. Kingwill (1960). *J. Dairy Sci.* **43**, 1654–1655.

Neubauer, J. (1905). *Arch. Wiss. Prakt. Tierheilk.* **31**, 153–176.

Neumann, A. L., R. R. Snapp, and L. S. Gall (1951). *J. Animal Sci.* **10**, 1058–1059.

Newland, H. W., W. T. Magee, G. A. Branaman, and L. H. Blakeslee (1962). *J. Animal Sci.* **21**, 711–715.

Nichols, R. E., and K. Penn (1959). *Am. J. Vet. Res.* **20**, 445–447.

Nicholson, J. W. G., and H. M. Cunningham (1961). *Can. J. Animal Sci.* **41**, 134–142.

Nicholson, J. W. G., H. M. Cunningham, and D. W. Friend (1962a) *Can. J. Animal Sci.* **42**, 75–81.

Nicholson, J. W. G., H. M. Cunningham, and D. W. Friend (1962b). *Can. J. Animal Sci.* **42**, 82–87.

Nicholson, J. W. G., D. W. Friend, and H. M. Cunningham (1964). *Can. J. Animal Sci.* **44**, 39–44.

Nicolai, J. H., and W. E. Stewart (1965). *J. Dairy Sci.* **48**, 56–60.

Nikitin, V. I. (1939). *Biochem. J.* (*Ukraine*) **14**, 203–222.

Nisizawa, K., and W. Pigman (1960). *Biochem. J.* **75**, 293–298.

Niven, C. F., K. L. Smiley, and J. M. Sherman (1941). *J. Biol. Chem.* **140**, 105–109.

Niven, C. F., Z. Kiziuta, and J. C. White (1946). *J. Bacteriol.* **51**, 711–716.

Niven, C. F., M. R. Washburn, and J. C. White (1948). *J. Bacteriol.* **55**, 601–606.

Noffsinger, T. L., K. K. Otagaki, and C. T. Furukawa (1961). *J. Animal Sci.* **20**, 718–722.

Noirot-Timothée, C. (1956a). *Compt. Rend.* **242**, 1076–1078.

Noirot-Timothée, C. (1956b). *Compt. Rend.* **242**, 2865–2867.

Noirot-Timothée, C. (1956c). *Mem. Inst. Franc. Afrique Noire* **48**, 259–266.

Noirot-Timothée, C. (1956d). *Bull. Soc. Zool. France* **86**, 44–47.

Noirot-Timothée, C. (1956e). *Bull. Soc. Zool. France* **81**, 47–52.

Noirot-Timothée, C. (1957). *Compt. Rend.* **244**, 2847–2849.

Noirot-Timothée, C. (1958a). *Compt. Rend.* **246**, 1286–1289.

Noirot-Timothée, C. (1958b). *Compt. Rend.* **246**, 2293–2295.

Noirot-Timothée, C. (1958c). *Compt. Rend.* **247**, 692–695.

Noirot-Timothée, C. (1959a). *Ann. Sci. Nat., Zool.* [12] **1**, 265–281.

Noirot-Timothée, C. (1959b). *Ann. Sci. Nat., Zool.* [12] **1**, 331–337.

Noirot-Timothée, C. (1960). Etude d'une famille de cilies: les "ophroyscolecidae." Structures et ultrastructures. Ph.D. Thèse, University of Paris. Masson et Cie, Paris.

Noland, P. R., B. F. Ford, and M. L. Ford (1955). *J. Animal Sci.* **14**, 860–865.

Noller, C. H., M. C. Stillons, F. A. Martz, and D. L. Hill (1959). *J. Animal Sci.* **18**, 671–674.

Nordfeldt, S., W. Cagell, and M. Nordkvist (1961). *Kgl. Lantbrukshogskol. Statens Lantbruksforsok, Statens Husdjusforsok, Sartryck Forhandsmedd.* **151**, 16 pp. Cited from *Nutr. Abst. Rev.* **32**, abst. 6475.

Nottle, M. C. (1956). *Australian J. Biol. Sci.* **9**, 593–604.

Nuvole, P. (1961). *Boll. Soc. Ital. Biol. Sper.* **37**, 1128–1130.

Ocariz, J. (1963). *Rev. Patronato. Biol. Animal* **7**, 5–15.

Ogilivie, B. M., G. L. McClymont, and F. B. Shorland (1961). *Nature* **190,** 725–726.

Ogimoto, K., and T. Suto (1963). *Japan. J. Zootech. Sci.* **34,** 282–287.

Ogimoto, K., F. Shibata, and C. Furasaka (1962). *Japan. J. Zootech. Sci.* **32,** 330–334.

O'Halloran, M. W. (1962). *Proc. Australian Soc. Animal Prod.* **4,** 18–21.

Olson, T. M. (1940). *J. Dairy Sci.* **23,** 343–353.

Oltjen, R. R., E. F. Smith, B. A. Koch, and F. H. Baker (1959). *J. Animal Sci.* **18,** 1196–1200.

Oltjen, R. R., R. J. Sirny, and A. D. Tillman (1962a). *J. Nutr.* **77,** 269–277.

Oltjen, R. R., R. J. Sirny, and A. D. Tillman (1962b). *J. Animal Sci.* **21,** 277–283.

Oltjen, R. R., R. J. Sirny, and A. D. Tillman (1962c). *J. Animal Sci.* **21,** 302–305.

Oltjen, R. R., J. D. Robbins, and R. E. Davis (1964). *J. Animal Sci.* **23,** 767–770.

Onodera, R., T. Tsuda, and M. Umezu (1964). *Tohoku J. Agr. Res.* **15,** 99–104.

Oppermann, R. A., W. O. Nelson, and R. E. Brown (1957). *J. Dairy Sci.* **40,** 779–788.

Oppermann, R. A., W. O. Nelson, and R. E. Brown (1961). *J. Gen. Microbiol.* **25,** 103–111.

Orla-Jensen, S. (1919). "The Lactic Acid Bacteria." Høst, Copenhagen.

Ota, Y. (1958). *J. Japan Vet. Med. Assoc.* **11,** 356–366.

Otagaki, K. K., A. L. Black, H. Goss, and M. Kleiber (1955). *J. Agr. Food Chem.* **3,** 948–951.

Otagaki, K. K., A. L. Black, J. C. Bartley, M. Kleiber, and B. O. Eggam (1963). *J. Dairy Sci.* **46,** 690–695.

Owens, E. L. (1959). *New Zealand Vet. J.* **7,** 43–46.

Oxford, A. E. (1951). *J. Gen. Microbiol.* **5,** 83–90.

Oxford, A. E. (1955a). *J. Sci. Food Agr.* **6,** 413–418.

Oxford, A. E. (1955b). *Intern. Bull. Bacteriol. Nomenclature Taxonomy* **5,** 131–132.

Oxford, A. E. (1955c). *Exptl. Parasitol.* **4,** 569–605.

Oxford, A. E. (1958a). *New Zealand Sci. Rev.* **16,** 38–44.

Oxford, A. E. (1958b). *New Zealand J. Agr. Res.* **1,** 809–824.

Oxford, A. E. (1958c). *J. Gen. Microbiol.* **19,** 617–623.

Oxford, A. E. (1959). *New Zealand J. Agr. Res.* **2,** 365–374.

Oyaért, W., and J. H. Bouckaert (1960). *Zentr. Veterinaermed.* **7,** 929–935.

Oyaért, W., and J. H. Bouckaert (1961). *Res. Vet. Sci.* **2,** 41–52.

Oyaért, W., J. I. Quin, and R. Clark (1951). *Onderstepoort J. Vet. Res.* **25,** 59–65.

Packett, L. V., and R. W. McCune (1965). *Appl. Microbiol.* **13,** 22–27.

Packett, L. V., M. L. Plumlee, R. Barnes, and G. O. Mott (1965). *J. Nutr.* **85,** 89–101.

Paladines, O. L., J. T. Reid, B. D. H. van Niekerk, and A. Bensadoun (1964). *J. Nutr.* **83,** 49–59.

Palmquist, D. L., L. M. Smith, and M. Ronning (1964). *J. Dairy Sci.* **47,** 516–520.

Paloheimo, L., A. Mäkela, and M. L. Salo (1955). *Maataloustieteellinen Aikakauskirja* **27,** 70–76.

Pant, R. K., R. K. Bhatnagar, and A. Roy (1963a). *Indian J. Dairy Sci.* **16,** 22–28.

Pant, H. C., M. D. Pandey, J. S. Rawat, and A. Roy (1963b). *Indian J. Dairy Sci.* **16**, 29–33.

Parsons, A. R., G. Hall, and W. E. Thomas (1952). *J. Animal Sci.* **11**, 772–773.

Parsons, A. R., A. L. Neumann, C. K. Whitehair, and J. Sampson (1955). *J. Animal Sci.* **14**, 403–411.

Parthasarathy, D., and A. T. Phillipson (1953). *J. Physiol. (London)* **121**, 452–469.

Pasteur, L. (1861). *Compt. Rend.* **52**, 1260–1264.

Pasteur, L. (1863). *Compt. Rend.* **56**, 734–740.

Paul, J. I. (1961). *Australian J. Biol. Sci.* **14**, 567–579.

Payne, L. C. (1960). *Science* **131**, 611–612.

Paynter, M. J. B. (1962). Mechanism of propionate formation by *Selenomonas ruminantium*. M. S. thesis, Sheffield Univ., Sheffield, England.

Pazur, J. H., T. Budovich, E. W. Shuey, and C. E. Georgi (1957). *Arch. Biochem. Biophys.* **70**, 419–425.

Pazur, J. H., E. W. Shuey, and C. E. Georgi (1958). *Arch. Biochem. Biophys.* **77**, 387–394.

Pearce, G. R. (1965a). *Australian J. Agr. Res.* **16**, 635–648.

Pearce, G. R. (1965b). *Australian J. Agr. Res.* **16**, 649–660.

Pearce, G. R., and R. J. Moir (1964). *Australian J. Agr. Res.* **15**, 635–644.

Pearson, P. B., A. H. Winegar, and L. H. Schmidt (1940). *J. Nutr.* **20**, 551–563.

Pearson, P. B., L. Struglia, and I. L. Lindahl (1953). *J. Animal Sci.* **12**, 213–218.

Pearson, R. M., and J. A. B. Smith (1943). *Biochem. J.* **37**, 153–164.

Peel, J. L. (1955). *J. Gen. Microbiol.* **12**, ii.

Peel, J. L. (1960). *Biochem. J.* **74**, 525–541.

Peel, L. J., and A. D. Wilson (1964). *Australian J. Agr. Res.* **15**, 625–634.

Pennington, R. J. (1952). *Biochem. J.* **51**, 251–258.

Pennington, R. J. (1954). *Biochem. J.* **56**, 410–416.

Pennington, R. J., and J. M. Appleton (1958). *Biochem. J.* **69**, 119–125.

Pennington, R. J., and W. H. Pfander (1957). *Biochem. J.* **65**, 109–111.

Pennington, R. J., and T. M. Sutherland (1956a). *Biochem. J.* **63**, 353–361.

Pennington, R. J., and T. M. Sutherland (1956b). *Biochem. J.* **63**, 618–628.

Penny, R. H. C. (1954). *Vet. Record* **66**, 134.

Perrault, C. (1680). "Essais de physique, ou recueil de plusieurs traités touchant les choses naturelles," 4 vols. Coignard, Paris.

Perry, K. D., and C. A. E. Briggs (1955). *J. Pathol. Bacteriol.* **70**, 546.

Perry, K. D., and C. A. E. Briggs (1957). *J. Appl. Bacteriol.* **20**, 119–123.

Perry, K. D., M. K. Wilson, L. G. M. Newland, and C. A. E. Briggs (1955). *J. Appl. Bacteriol.* **18**, 436–442.

Peyer, J. C. (1685). "Merycologia, sive de ruminantibus et ruminatione commentarius," 288 pp. Koenig and Brandmyllerum, Basil. Cited from Colin (1886).

Pfander, W. H., and A. T. Phillipson (1953). *J. Physiol. (London)* **122**, 102–110.

Pfander, W. H., R. W. Kelley, C. C. Brooks, C. W. Gehrke, and M. E. Muhrer (1957a). *Missouri, Univ., Agr. Expt. Sta., Res. Bull.* **632**, 15 pp.

Pfander, W. H., G. B. Garner, W. C. Ellis, and M. E. Muhrer (1957b). *Missouri. Univ., Agr. Expt. Sta., Res. Bull.* **637**, 12 pp.

Phaneuf, L. P. (1961). *Cornell Vet.* **51**, 47–56.

Phatak, S. S., and V. N. Patwardhan (1953). *Nature* **172**, 456–457.

Phillips, G. D. (1960). *J. Agr. Sci.* **54**, 231–234.

Phillips, G. D. (1961a). *Res. Vet. Sci.* **2**, 202–208.

Phillips, G. D. (1961b). *Res. Vet. Sci.* **2**, 209–216.

Phillips, G. D., and G. W. Dyck (1964). *Can. J. Animal Sci.* **44**, 220–227.

Phillips, G. D., R. E. Hungate, A. MacGregor, and D. P. Hungate (1960). *J. Agr. Sci.* **54**, 416–420.

Phillips, R. W. (1964). The nature of the glucogenic effect of butyrate. Ph.D. thesis, University of California, Davis, California.

Phillipson, A. T. (1952a). *Proc. Roy. Soc.* **B139**, 196–202.

Phillipson, A. T. (1952b). *J. Physiol.* (*London*) **116**, 84–97.

Phillipson, A. T. (1952c). *Brit. J. Nutr.* **6**, 190–198.

Phillipson, A. T. (1958a). *Proc. 6th Ann. Res. Conf., Chicago, Illinois, 1958* pp. 125–135. Charles Pfizer Co., Terre Haute, Indiana.

Phillipson, A. T. (1958b). *Natl. Agr. Advisory Serv. Quart. Rev.* No. **40**.

Phillipson, A. T., and J. R. M. Innes (1939). *Quart. J. Exptl. Physiol.* **29**, 333–341.

Phillipson, A. T., and R. McAnally (1942). *J. Exptl. Biol.* **19**, 199–214.

Phillipson, A. T., and J. L. Mangan (1959). *New Zealand J. Agr. Res.* **2**, 990–1001.

Phillipson, A. T., and R. L. Mitchell (1952). *Brit. J. Nutr.* **6**, 176–189.

Phillipson, A. T., and R. S. Reid (1957). *Brit. J. Nutr.* **11**, 27–41.

Phillipson, A. T., and C. S. W. Reid (1958). *Nature* **181**, 1722–1723.

Phillipson, A. T., M. J. Dobson, T. H. Blackburn, and M. Brown (1962). *Brit. J. Nutr.* **16**, 151–166.

Piana, C. (1953a). *Arch. Vet. Ital.* **4**, 147–149.

Piana, C. (1953b). *Zootec. Vet.* **8**, 11–19.

Pigden, W. J., and J. E. Stone (1952). *Sci. Agr.* **32**, 502–506.

Pilgrim, A. F. (1948). *Australian J. Sci. Res.* **B1**, 130–138.

Pinsent, J. (1954). *Biochem. J.* **57**, 10–16.

Pittman, K. A., and M. P. Bryant (1964). *J. Bacteriol.* **88**, 401–410.

Pochon, J. (1935). *Ann. Inst. Pasteur* **55**, 676–697.

Poljansky, G., and A. Strelkow (1934). *Zool. Anz.* **107**, 215–220.

Poljansky, G., and A. Strelkow (1938). *Arch. Zool.* **80**, 1–123.

Popjak, G., T. H. French, and S. J. Folley (1951a). *Biochem. J.* **48**, 411–416.

Popjak, G., T. H. French, G. D. Hunter, and J. P. Martin (1951b). *Biochem. J.* **48**, 612–618.

Popoff, L. (1875). *Arch. Ges. Physiol.* **10**, 123–146.

Popow, N. A., A. A. Kudrjavcew, and W. K. Krasousky (1933). *Arch. Tierernaehr. Tierzucht* **9**, 243–252.

Porter, J. W. G. (1957). *Vet. Record* **69**, 250–253 and 253–259.

Porter, J. W. G., and A. M. Dollar (1958). *4th Intern. Congr. Biochem., Vienna, 1958* p. 96.

Portugal, A. V. (1963). Some aspects of amino acid and protein metabolism in the rumen of the sheep. Ph.D. Thesis, Aberdeen University, Aberdeen, Scotland.

Portway, B. (1957). *Australian Vet. J.* **33**, 210–212.

Postgate, J. R., and Campbell, L. L. (1963). *J. Bacteriol.* **86**, 274–279.

Potter, B. J. (1952). *Nature* **170**, 541.

Pounden, W. D. (1954). *Vet. Med.* **49**, 463–464.

Pounden, W. D., and J. W. Hibbs (1948a). *J. Dairy Sci.* **31**, 1041–1050.

Pounden, W. D., and J. W. Hibbs (1948b). *J. Dairy Sci.* **31**, 1051–1054.

Pounden, W. D., and J. W. Hibbs (1949a). *J. Dairy Sci.* **32**, 1025–1031.

Pounden, W. D., and J. W. Hibbs (1949b). *J. Am. Vet. Med. Assoc.* **114**, 33–35.

Pounden, W. D., and J. W. Hibbs (1949c). *Farm Home Res.* **34**, No. 257, 43–46.

Pounden, W. D., and J. W. Hibbs (1950). *J. Dairy Sci.* **33**, 639–644.

Pounden, W. D., L. C. Ferguson, and J. W. Hibbs (1950). *J. Dairy Sci.* **33**, 565–572.

Pounden, W. D., H. R. Conrad, and J. W. Hibbs (1954). *J. Am. Vet. Med. Assoc.* **124**, 394–396.

Powell, E. B. (1938). *Proc. Am. Soc. Animal Prod.* **31**, 40–53.

Prescott, J. M. (1961). *J. Bacteriol.* **82**, 724–728.

Prescott, J. M., and A. L. Stutts (1955). *J. Bacteriol.* **70**, 285–288.

Prescott, J. M., R. S. Ragland, and A. L. Stutts (1957). *J. Bacteriol.* **73**, 133–138.

Prescott, J. M., W. T. Williams, and R. S. Ragland (1959). *Proc. Soc. Exptl. Biol. Med.* **102**, 490–493.

Preston, R. L., and W. H. Pfander (1964). *J. Nutr.* **83**, 369–378.

Preston, T. R. (1958). *Proc. Brit. Soc. Animal Prod.* **1958**, 33–38.

Preston, T. R. (1963). *Ann. Rept. Animal Nutr. Allied Sci., Rowett Res. Inst.* **19**, 54–58.

Preston, T. R., F. G. Whitelaw, A. MacDearmid, N. A. MacLeod, and E. B. Charleson (1961). *Proc. Nutr. Soc. (Engl. Scot.)* **20**, xlii–xliii.

Prewitt, R. D., and C. P. Merilan (1958). *J. Dairy Sci.* **41**, 807–811.

Pringsheim, H. (1912). *Z. Physiol. Chem.* **78**, 266–291.

Pritchard, G. I., J. A. Newlander, and W. H. Riddell (1955). *J. Animal Sci.* **14**, 336–339.

Pritchard, G. I., W. J. Pigden, and D. J. Minson (1962). *Can. J. Animal Sci.* **42**, 215–217.

Provost, P. J., and R. N. Doetsch (1960). *Biochem. J.* **22**, 259–264.

Pugh, P. S., and R. Scarisbrick (1952). *Nature* **170**, 978–979.

Purdom, M. R. (1963). *Nature* **198**, 307–308.

Purser, D. B. (1961a). *Nature* **190**, 831–832.

Purser, D. B. (1961b). Factors affecting the concentration of the ciliate protozoal population of the rumen. Ph.D. Thesis, University of Western Australia, Perth, Australia.

Purser, D. B., and Cline, J. H. (1962). *Federation Proc.* **21**, 396.

Purser, D. B., and R. J. Moir (1959). *Australian J. Agr. Res.* **10**, 555–564.

Purser, D. P., and R. B. Tompkin (1965). *Life Science* **4**, 1493–1501.

Purser, D. B., and H. H. Weiser (1963). *Nature* **200**, 290.

Putman, P. A., J. Gutierrez, and R. E. Davis (1961). *J. Dairy Sci.* **44**, 1364–1365.

Quicke, G. V., and O. G. Bentley (1959). *J. Animal Sci.* **18**, 365–373.

Quicke, G. V., O. G. Bentley, H. W. Scott, and A. L. Moxon (1959a). *J. Animal Sci.* **18**, 275–287.

Quicke, G. V., O. G. Bentley, H. W. Scott, R. R. Johnson, and A. L. Moxon (1959b). *J. Dairy Sci.* **42**, 185–186.

Quin, A. H., J. A. Austin, and K. Ratcliff (1949). *J. Am. Vet. Med. Assoc.* **114**, 313–314.

Quin, J. I. (1933). *Onderstepoort J. Vet. Sci. Animal Ind.* **1**, 501–504.

Quin, J. I. (1943a). *Onderstepoort J. Vet. Sci. Animal Ind.* **18**, 91–112.

Quin, J. I. (1943b). *Onderstepoort J. Vet. Sci. Animal Ind.* **18**, 113–117.

Quin, J. I., and J. G. van der Wath (1938). *Onderstepoort J. Vet. Sci. Animal Ind.* **11**, 361–382.

Quin, J. I., C. Rimington, and G. C. S. Roets (1935). *Onderstepoort J. Vet. Sci. Animal Ind.* **4**, 463–478.

Quin, J. I., J. G. van der Wath, and S. Myburgh (1938). *Onderstepoort J. Vet. Sci. Animal Ind.* **11**, 341–360.

Quin, J. I., W. Oyaért, and R. Clark (1951). *Onderstepoort J. Vet. Res.* **25**, 51–58.

Quinn, L. Y. (1962). *Appl. Microbiol.* **10**, 580–582.

Quinn, L. Y., W. Burroughs, and W. C. Christiansen (1962). *Appl. Microbiol.* **10**, 583–592.

Quitteck, G. (1936). Ein Beitrag zur Frage der Fettbildung im Pansen des Schafes. Dissertation, Berlin.

Radeff, T., and I. Stojanoff (1955). *Arch. Tierernaehr.* **5**, 331–347.

Radloff, H. D., and L. H. Schultz (1963). *J. Dairy Sci.* **46**, 517–521.

Rakes, A. H., E. E. Lister, and J. T. Reid (1961). *J. Nutr.* **75**, 86–92.

Ralston, A. T., D. C. Church, and J. E. Oldfield (1962). *J. Animal Sci.* **21**, 306–308.

Rambe, L. (1938). *Skand. Arch. Physiol.* **78**, Suppl. 13, 1–157.

Rathnow, H. D. (1938). Über das Verhalten des Eisens der Nahrung während der Pansenverdauung des Schafes. Dissertation, Munich.

Raun, A., E. Cheng, and W. Burroughs (1956). *J. Agr. Food Chem.* **4**, 869–871.

Ray, S. N., W. C. Weir, A. L. Pope, G. Bohstedt, and P. H. Phillips (1948). *J. Animal Sci.* **7**, 3–15.

Raymond, W. F., C. E. Harris, and V. G. Harker (1953a). *J. Brit. Grassland Soc.* **8**, 301–314.

Raymond, W. F., C. E. Harris, and V. G. Harker (1953b). *J. Brit. Grassland Soc.* **8**, 315–320.

Raynaud, P. (1955). *Arch. Sci. Physiol.* **9**, 83–96.

Raynaud, P. (1961). *Ann. Nutr. Aliment.* **15**, 275–277.

Raynaud, P., and J. Bost (1957). *Arch. Ges. Physiol.* **264**, 306–313.

Read, B. E., R. Kellar, and V. J. Cabelli (1956). *Bacteriol. Proc.* (*Soc. Am. Bacteriologists*) **56**, 39.

Réaumur, R. E. F. (1752). *Mem. Acad. Sci.* (*Paris*) p. 461. Cited by Colin (1886).

Reed, F. M., R. J. Moir, and E. J. Underwood (1949). *Australian J. Sci. Res.* **B2**, 304–317.

Rees, C. W. (1931). *J. Morphol.* **52**, 195–215.

Reichenow, E. (1920). *Arch. Protistenk.* **41**, 1–33.

Reichl, J. (1960). *V.Ú.K. Čsazv* **4**, 37–65.

Reid, C. S. W. (1957a). *New Zealand J. Sci. Technol.* **A38**, 825–838.

Reid, C. S. W. (1957b). *New Zealand J. Sci. Technol.* **A38**, 853–856.

Reid, C. S. W. (1958). *New Zealand J. Agr. Res.* **1**, 349–364.

Reid, C. S. W. (1959). *Proc. Nutr. Soc.* (*Engl. Scot.*) **18**, 127–133.

Reid, C. S. W., and J. B. Cornwall (1959). *Proc. New Zealand Soc. Animal Prod.* **19**, 23–35.

Reid, C. S. W., and A. T. Johns (1957). *New Zealand J. Sci. Technol.* **38**, 908–927.

Reid, J. T. (1953). *J. Dairy Sci.* **36**, 955–996.

Reid, J. T., P. G. Woolfolk, C. R. Richards, R. W. Kaufmann, J. K. Loosli, K. L. Turk, J. I. Miller, and R. E. Blaser (1950). *J. Dairy Sci.* **33**, 60–71.

Reid, J. T., C. C. Balch, and M. J. Head (1957). *Nature* **179**, 1034.

Reid, R. L. (1950a). *Australian J. Agr. Res.* **1**, 182–199.

Reid, R. L. (1950b). *Nature* **165**, 448–449.

Reid, R. L. (1950c). *Australian J. Agr. Res.* **1**, 338–354.

Reid, R. L. (1951). *Australian J. Agr. Res.* **2**, 146–157.

Reid, R. L. (1953). *Australian J. Agr. Res.* **4**, 213–223.

Reid, R. L., M. C. Franklin, and E. G. Holdsworth (1947). *Australian Vet. J.* **23,** 136–140.

Reid, R. L., J. P. Hogan, and P. K. Briggs (1957). *Australian J. Agr. Res.* **8,** 691–710.

Reis, P. J., and Reid, R. L. (1959). *Australian J. Agr. Res.* **10,** 71–80.

Reiser, R. (1951). *Federation Proc.* **10,** 236.

Reiser, R., and H. Fu (1962). *J. Nutr.* **76,** 215–218.

Reiser, R., and H. G. R. Reddy (1956). *J. Am. Oil Chemists' Soc.* **33,** 155–156.

Reiset, J. (1863a). *Ann. Chim. Phys.* [3] **69,** 129–169.

Reiset, J. (1863b). *Compt. Rend.* **56,** 740–747.

Reiset, J. (1868a). *Compt. Rend.* **66,** 172–175.

Reiset, J. (1868b). *Compt. Rend.* **66,** 176–177.

Repp, W. W., W. H. Hale, and W. Burroughs (1955a). *J. Animal Sci.* **14,** 901–908.

Repp, W. W., W. H. Hale, E. W. Cheng, and W. Burroughs (1955b). *J. Animal Sci.* **14,** 118–131.

Rérat, A., and R. Jacquot (1954). *Compt. Rend.* **239,** 1693–1695.

Rérat, A., H. LeBars, and R. Jacquot (1956). *Compt. Rend.* **242,** 679–681.

Rérat, A., J. Mollé, and H. LeBars (1958a). *Compt. Rend.* **246,** 2051–2054.

Rérat, A., R. Boccard, and R. Jacquot (1958b). *Compt. Rend.* **247,** 1786–1788.

Rérat, A., O. Champigny, and R. Jacquot (1959). *Compt. Rend.* **249,** 1274–1276.

Rice, R. W., R. L. Salsbury, J. A. Hoeffer, and R. W. Luecke (1962). *J. Animal Sci.* **21,** 418–425.

Richards, C. R., and J. T. Reid (1953). *J. Dairy Sci.* **36,** 1006–1015.

Richards, C. R., H. G. Weaver, and J. D. Connelly (1958). *J. Dairy Sci.* **41,** 956–962.

Richardson, A., and A. C. Hulme (1957). *J. Sci. Food Agr.* **8,** 326–330.

Richter, K., and M. Becker (1956). *Z. Tierernaehr. Futtermittelk.* **11,** 289–295.

Richter, K., K. L. Cranz, and H. J. Oslage (1961). *Z. Tierphysiol. Tierernaehr. Futtermittelk.* **16,** 47–58.

Rickes, E. L., N. G. Brink, F. R. Konuiszy, T. R. Wood, and K. Folkers (1948). *Science* **108,** 134.

Riek, R. F. (1954). *Australian Vet. J.* **30,** 29–37.

Ritzman, E. G., L. E. Washburn, and F. G. Benedict (1936). *New Hampshire Agr. Expt. Sta. Tech. Bull.* **66.**

Robertson, A. (1963). *World Conf. Animal Prod.* **1,** 99–112.

Robertson, A., and C. Thin, (1953). *Brit. J. Nutr.* **7,** 181–195.

Robertson, J. A., and J. C. Hawke (1964). *J. Sci. Food Agr.* **15,** 274–282.

Robinson, R. Q., R. N. Doetsch, F. M. Sirotnak, and J. C. Shaw (1955). *J. Dairy Sci.* **38,** 13–19.

Robinson, T. J., D. E. Jasper, and H. R. Guilbert (1951). *J. Animal Sci.* **10,** 733–741.

Rodrique, C. B., and N. N. Allen (1960). *Can. J. Animal Sci.* **40,** 23–29.

Rodwell, A. W. (1953). *J. Gen. Microbiol.* **8,** 224–232.

Rogers, T. A., and M. Kleiber (1955). *Federation Proc.* **14,** No. 1.

Rogerson, A. (1955). *E. African Agr. J.* **20,** 254–255.

Rogerson, A. (1956). *E. African Agr. J.* **21,** 163.

Rogerson, A. (1958). *Brit. J. Nutr.* **12,** 164–176.

Rogerson, A. (1960). *J. Agr. Sci.* **55,** 359–364.

Rogosa, M. (1964). *J. Bacteriol.* **87,** 162–170.

Rogosa, M., and F. S. Bishop (1964a). *J. Bacteriol.* **87,** 574–580.

Rogosa, M., and F. S. Bishop (1964b). *J. Bacteriol.* **88,** 37–41.

Rogosa, M., M. I. Krichevsky, and F. S. Bishop (1965). *J. Bacteriol.* **90,** 164–171.

Rojahn, J. (1960). *Arch. Tierernaehr.* **10,** 438–471.

Rojas, M. A., and I. A. Dyer (1964). *J. Animal Sci.* **23,** 600.

Rollinson, D. H. L., K. W. Harker, and J. I. Taylor (1955). *J. Agr. Sci.* **46,** 124–129.

Rook, J. A. F. (1959). *Proc. Nutr. Soc. (Engl. Scot.)* **18,** 117–123.

Rook, J. A. F. (1964). *Proc. Nutr. Soc. (Engl. Scot.)* **23,** 71–80.

Rook, J. A. F., and C. C. Balch (1958). *J. Agr. Sci.* **51,** 199–207.

Rook, J. A. F., and C. C. Balch (1962). *J. Agr. Sci.* **59,** 103–108.

Rook, J. A. F., and R. C. Campling (1959). *J. Agr. Sci.* **53,** 330–332.

Rook, J. A. F., and C. Line (1961). *Brit. J. Nutr.* **15,** 109–119.

Rook, J. A. F., and M. Wood (1960). *J. Sci. Food Agr.* **11,** 137–143.

Rook, J. A. F., C. C. Balch, and C. Line (1958). *J. Agr. Sci.* **51,** 189–198.

Rook, J. A. F., C. C. Balch, R. C. Campling, and L. J. Fisher (1963). *Brit. J. Nutr.* **17,** 399–406.

Rook, J. A. F., R. C. Campling, and V. W. Johnson (1964). *J. Agr. Sci.* **62,** 273–279.

Rosen, W. G., H. Fassel, and R. E. Nichols (1956). *J. Animal Sci.* **15,** 1305.

Rosen, W. G., H. Fassel, and R. E. Nichols (1961). *Am. J. Vet. Res.* **22,** 117–122.

Roth, L. E., and Y. Shigenaka (1964). *J. Cell Biol.* **20,** 249–270.

Rothery, P., J. M. Bell, and J. W. T. Spinks (1953). *J. Nutr.* **49,** 173–181.

Rufener, W. H., W. O. Nelson, and M. J. Wolin (1963). *Appl. Microbiol.* **11,** 196–201.

Russell, E. L., W. H. Hale, and F. Hubbert (1962). *J. Animal Sci.* **21,** 523–526.

Ryan, R. K. (1964a). *Am. J. Vet. Res.* **25,** 646–651.

Ryan, R. K. (1964b). *Am. J. Vet. Res.* **25,** 653–659.

Ryś, R., L. Górski, and H. Styczński (1957). *Acta Biochim. Polon.* **4,** 147–164.

Ryś, R., H. Styczński, M. Kretowska, and H. Wcisto (1962). *Roczniki Nauk Rolniczych, Ser. B.* **79,** 195–211. Cited from *Nutr. Abst. Rev.* **32,** 1376.

Sabine, J. R., and B. C. Johnson (1964). *J. Biol. Chem.* **239.** 89–93.

Sahashi, Y., K. Iwamoto, M. Mikata, A. Nakayama, H. Sakai, J. Takahashi, J. Hayashi, N. Seno, T. Akatsuka, T. Miki, K. Harashima, and R. Matsumoto (1953). *J. Biochem. (Tokyo)* **40,** 227–244.

Salsbury, R. L., C. K. Smith, and C. F. Huffman (1956). *J. Animal Sci.* **15,** 863–868.

Salsbury, R. L., A. L. van der Kolk, B. V. Baltzer, and R. W. Luecke (1958a). *J. Animal Sci.* **17,** 293–297.

Salsbury, R. L., P. Elliott, and R. W. Luecke (1958b). *J. Animal Sci.* **17,** 1190.

Salsbury, R. L., J. A. Hoefer, and R. W. Luecke (1961). *J. Dairy Sci.* **44,** 1122–1128.

Sampson, J., and C. E. Hayden (1936). *Cornell Vet.* **26,** 183–199.

Sander, E. G., R. G. Warner, H. N. Harrison, and J. K. Loosli (1959). *J. Dairy Sci.* **42,** 1600–1605.

Sanford, J. (1963). *Nature* **199,** 829–830.

Sapiro, M. L., S. Hoflund, R. Clark, and J. I. Clark (1949). *Onderstepoort J. Vet. Sci. Animal Ind.* **22,** 357–372.

Sasaki, Y., and M. Umezu (1962). *Tohoku J. Agr. Res.* **13,** 211–219.

Satter, L. D., and B. R. Baumgardt (1962). *J. Animal Sci.* **21,** 897–900.

Satter, L. D., J. W. Suttie, and B. R. Baumgardt (1964). *J. Dairy Sci.* **47,** 1365–1370.

Sauer, F., W. M. Dickson, and H. H. Hoyt (1958). *Am. J. Vet. Res.* **19,** 567–574.

Sayama, K. (1953). Studies on the rumen ciliates from California deer. Ph.D. thesis, University of California, Berkeley, California.

Scaife, J. F. (1956a). *New Zealand J. Sci. Technol.* **A38,** 285–292.

Scaife, J. F. (1956b). *New Zealand J. Sci. Technol.* **A38,** 293–298.

Scardovi, V. (1963). *Ann. Microbiol. Enzymologia* **13,** 171–187.

Scarisbrick, R. (1954). *Vet. Record* **66,** 131–132.

Schad, G. A., R. Knowles, and E. Meerovitch (1964). *Can. J. Microbiol.* **10,** 801–804.

Schalk, A. F., and R. S. Amadon (1921). *J. Am. Vet. Med. Assoc.* **59,** 151–172.

Schalk, A. F., and R. S. Amadon (1928). *N. Dakota Agr. Expt. Sta., Bull.* **216,** 64 pp.

Schambye, P. (1951a). *Nord. Veterinarmed.* **3,** 555–574.

Schambye, P. (1951b). *Nord. Veterinarmed.* **3,** 748–762.

Schambye, P. (1951c). *Nord. Veterinarmed.* **3,** 1003–1014.

Schambye, P. (1951d). *Acta Physiol. Scand.* **25,** Supp. 89, 70–71.

Schambye, P. (1955a). *Nord. Veterinarmed.* **7,** 961–973.

Schambye, P. (1955b). *Nord. Veterinarmed.* **7,** 1001–1016.

Schambye, P., and A. T. Phillipson (1949). *Nature* **164,** 1094–1095.

Schellenberger, P. R., N. L. Jacobson, P. A. Hartman, and A. D. McGilliard (1964). *J. Animal Sci.* **23,** 196–202.

Scheuerbrandt, G., and K. Bloch (1962). *J. Biol. Chem.* **237,** 2064–2068.

Scheunert, A., and A. Trautmann (1921a). *Arch. Ges. Physiol.* **192,** 33–69.

Scheunert, A., and A. Trautmann (1921b). *Arch. Ges. Physiol.* **192,** 70–80.

Scheunert, A., and F. W. Krzywanek (1930). *Arch. Ges. Physiol.* **223,** 472–476.

Scheunert, A., F. W. Krzywanek, and K. Zimmermann (1929a). *Arch. Ges. Physiol.* **223,** 453–461.

Scheunert, A., F. W. Krzywanek, and K. Zimmermann (1929b). *Arch. Ges. Physiol.* **223,** 462–471.

Schlottke, E. (1936). *Sitzber. Naturforsch. Ges. Rostock, 3F* **6,** 59.

Schmid, H. (1939). *Z. Tierzuecht. Zuechtungsbiol.* **43,** 238–253.

Schmidt, G. H., and L. H. Schultz (1958). *J. Dairy Sci.* **41,** 169–175.

Schmidt, J., and J. Kliesch (1939). *Zuechtungskunde* **14,** 193–204.

Schmidt-Nielsen, B. (1959). *Sci Am.* **201,** 140–152.

Schmidt-Nielsen, B., and H. Osaki (1958). *Am. J. Physiol.* **193,** 657–661.

Schmidt-Nielsen, B., K. Schmidt-Nielsen, T. R. Houpt, and S. A. Jarnum (1957). *Am. J. Physiol.* **188,** 477–484.

Schmidt-Nielsen, B., H. Osaki, H. V. Murdaugh, Jr., and R. O'dell (1958). *Am. J. Physiol.* **194,** 221–228.

Schneider, B. H. (1947). "Feeds of the World." West Virginia Univ. Press, Morgantown, West Virginia.

Schoenemann, K. (1953). "The New Rheinau Wood Saccharification Process." Schön & Wetzel, Frankfurt-am-Main.

Schuberg, A. (1888). *Zool. Jahrb., Abt. System Oekol. Geograph. Tiere* **3,** 365–416.

Schultz, L. H. (1954). *J. Dairy Sci.* **37,** 664.

Schultz, L. H. (1958). *J. Dairy Sci.* **41,** 160–168.

Schultz, L. H., and Smith, V. R. (1951). *J. Dairy Sci.* **34,** 1191–1199.

Schulz, H. (1963). *Zool. Anz.* **170,** 306–310.

Schumacher, I. C. (1915). *Univ. Calif. Pub. Zool.* **16,** 95–106.

Schwarz, C., and H. Steinlechner (1925). *Biochem. Z.* **156,** 130–137.

Schwarz, K., *et al.* (1961). *Federation Proc.* **20**, 665–702.

Scott, H. W., and B. A. Dehority (1965). *J. Bacteriol.* **89**, 1169–1175.

Scott, T. W., P. F. V. Ward, and R. M. C. Dawson (1964). *Biochem. J.* **90**, 12–24.

Seekles, L. (1951). *Vet. Record* **63**, 494.

Seeley, H. W., and J. A. Dain (1960). *J. Bacteriol.* **79**, 230–235.

Sellers, A. F., and A. Dobson, (1960). *Res. Vet. Sci.* **1**, 95–102.

Setchell, B. P., and A. J. Williams (1962). *Australian Vet. J.* **38**, 58–62.

Seto, K., and M. Umezu (1959). *Tohoku J. Agr. Res.* **9**, 151–158.

Seto, K., I. Okabe, T. Tsuda, and M. Umezu (1959). *Tohoku J. Agr. Res.* **9**, 133–149.

Seto, K., I. Okabe, and M. Umezu (1964). *Tohoku J. Agr. Res.* **14**, 163–170.

Sharp, R. G. (1914). *Univ. Calif. (Berkeley) Publ. Zool.* **13**, 43–122.

Shattock, P. M. F. (1949). *J. Gen. Microbiol.* **3**, 80–92.

Shaw, J. C., B. C. Hatziolos, and A. C. Chung (1951). *Science* **114**, 574–576.

Shaw, J. C., A. C. Chung, R. A. Gessert, and G. Bajwa (1955). *Univ. Maryland Agr. Expt. Sta. Misc. Publ.* **238**, 15–20.

Shawver, C. B., and L. W. Williams (1960). *J. Animal Sci.* **19**, 656.

Shehata, A. M. E. (1958). *Appl. Microbiol.* **6**, 422–426.

Sheppard, A. J., R. E. Blaser, and C. M. Kincaid (1957). *J. Animal Sci.* **16**, 681–687.

Sheriha, G. M., R. J. Sirny, and A. D. Tillman (1962). *J. Animal Sci.* **21**, 53–56.

Shibata, F., K. Ogimoto, and C. Furusaka (1961). *Japan. J. Zootech. Sci.* **32**, 159–163.

Shinozaki, K. (1957). *Tohoku J. Agr. Res.* **8**, 149–154.

Shinozaki, K. (1959). *Tohoku J. Agr. Res.* **9**, 237–250.

Shinozaki, K., and H. Sugawara (1958). *Iwate Daigaku Nogakubu Hokoku* **4**, 88–95.

Shinozaki, K., H. Sugawara and H. Obonai (1957). *Iwate Daigaku Nogakubu Hokoku* **3**, 398–406.

Shoji, Y., K. Miyazaki, and M. Umezu (1964). *Tohoku J. Agr. Res.* **15**, 91–97.

Shor, A. L., W. P. Johnson, and A. Abbey (1959). *J. Dairy Sci.* **42**, 1203–1208.

Shorland, F. B. (1950). *Nature* **165**, 766.

Shorland, F. B. (1953). *J. Sci. Food Agr.* **4**, 497–503.

Shorland, F. B., and Hansen, R. P. (1957). *Dairy Sci. Abstr.* **19**, 168–189.

Shorland, F. B., R. O. Weenink, and A. T. Johns (1955). *Nature* **175**, 1129–1130.

Shorland, F. B., R. O. Weenink, A. T. Johns, and I. R. C. McDonald (1957). *Biochem. J.* **67**, 328–333.

Sijpesteijn, A. K. (1948). Cellulose-decomposing bacteria from the rumen of cattle. Ph.D. Thesis, Leiden.

Sijpesteijn, A. K. (1949). *Antonie van Leeuwenhoek, J. Microbiol. Serol.* **15**, 49–52.

Sijpesteijn, A. K. (1951). *J. Gen. Microbiol.* **5**, 869–879.

Sijpesteijn, A. K., and S. R. Elsden (1952). *Biochem. J.* **52**, 41–45.

Silver, I. A. (1954). *J. Physiol. (London)* **125**, 8P–9P.

Simesen, M. G., J. R. Luick, M. Kleiber, and C. Thin (1961). *Acta Vet. Scand.* **2**, 214–225.

Simonnet, H., and H. LeBars (1953a). *Rec. Med. Vet.* **129**, 401–423.

Simonnet, H., and H. LeBars (1953b). *Compt. Rend.* **237**, 751–753.

Simonnet, H., H. LeBars, and J. Mollé (1957). *Compt. Rend.* **244**, 943–945.

Sinclair, K. B., and D. I. H. Jones (1964). *J. Sci. Food Agr.* **15**, 717–721.

Singh, R. and C. P. Merilan (1957). *Missouri, Univ., Agr. Expt. Sta. Res. Bull.* **639**.

Sirotnak, F. M., R. N. Doetsch, R. Q. Robinson, and J. C. Shaw (1954). *J. Dairy Sci.* **37**, 531–537.

Sjollema, B. (1950). *J. Brit. Grassland Soc.* **5**, 179–194.

Slyter, L. L., W. O. Nelson, and M. J. Wolin (1964). *Appl. Microbiol.* **12**, 374–377.

Smart, W. W. G., Jr., T. A. Bell, N. W. Stanley, and W. A. Cope (1961). *J. Dairy Sci.* **44**, 1945–1946.

Smart, W. W. G., Jr., T. A. Bell, R. D. Machrie, and N. W. Stanley (1964). *J. Dairy Sci.* **47**, 1220–1223.

Smiles, J., and M. J. Dobson (1956). *J. Roy. Microscop. Soc.* **75**, 244–253.

Smith, A. D. (1957). *J. Range Management* **10**, 162–164.

Smith, A. D., R. B. Turner, G. A. Harris (1956). *J. Range Management* **9**, 142–145.

Smith, A. H., M. Kleiber, A. L. Black, and C. F. Baxter (1955). *J. Nutr.* **57**, 507–527.

Smith, A. H., M. Kleiber, A. L. Black, and G. P. Lofgreen (1956). *J. Nutr.* **58**, 95–111.

Smith, C., and Trexler, P. C. (1961). *Proc. 1961 Rumen Function Conf., Chicago* U.S. Dept. Agr. Mimeographed report.

Smith, C. K., J. R. Brunner, C. R. Huffman, and C. W. Duncan (1953). *J. Animal Sci.* **12**, 932.

Smith, E. E., G. D. Goetsch, and H. D. Jackson (1961). *Arch. Biochem. Biophys.* **95**, 256–262.

Smith, H. G. (1944). *Am. J. Vet. Res.* **5**, 234–242.

Smith, J. A. B., and F. Baker (1944). *Biochem. J.* **38**, 496–505.

Smith, J. A. B., and N. N. Dastur (1938). *Biochem. J.* **32**, 1868–1876.

Smith, K. J., and W. Woods (1962). *J. Animal Sci.* **21**, 798–803.

Smith, L. M. (1961). *J. Dairy Sci.* **44**, 607–622.

Smith, L. M., and E. L. Jack (1954a) *J. Dairy Sci.* **37**, 380–389.

Smith, L. M., and E. L. Jack (1954b) *J. Dairy Sci.* **37**, 390–398.

Smith, L. M., N. K. Freeman, and E. L. Jack (1954). *J. Dairy Sci.* **37**, 399–406.

Smith, P. H., and R. E. Hungate (1958). *J. Bacteriol.* **75**, 713–718.

Smith, P. H., H. C. Sweeney, J. R. Rooney, K. W. King, and W. E. C. Moore (1956). *J. Dairy Sci.* **39**, 598–609.

Smith, R. M., and K. J. Monty (1959). *Biochem. Biophys. Res. Communs.* **1**, 105–109.

Smith, S. E., B. A. Koch, and K. L. Turk (1951). *J. Nutr.* **44**, 455–464.

Smith, V. R. (1941). *J. Dairy Sci.* **24**, 659–665.

Smuts, D. B., B. A. du Toit, and J. G. van der Wath (1941). *Onderstepoort J. Vet. Sci. Animal Ind.* **16**, 181–190.

Somers, M. (1957). *Australian Vet. J.* **33**, 297–302.

Somers, M. (1961a). *Australian J. Exptl. Biol. Med. Sci.* **39**, 111–122.

Somers, M. (1961b). *Australian J. Exptl. Biol. Med. Sci.* **39**, 123–131.

Somers, M. (1961c). *Australian J. Exptl. Biol. Med. Sci.* **39**, 133–143.

Somers, M. (1961d). *Australian J. Exptl. Biol. Med. Sci.* **39**, 145–156.

Sorensen, H. (1955). *Nature* **176**, 74.

Sosnovskaja, E. A. (1959). *Zivotnovodstvo* **7,** 61–63. Cited from *Nutr. Abst. Rev.* **30,** 172.

Spallanzani, L. (1776). "Opuscoli di fisica animale e vegetabile." Vol. II, p. 639. Societa Tipografica, Modena. Cited in Colin (1886).

Sperber, I., and S. Hydén (1952). *Nature* **169,** 587.

Sperber, I., S. Hydén, and J. Ekman (1953). *Kgl. Lantbruks-Hogskol. Ann.* **20,** 337–345.

Sperber, I., S. Hydén, and J. Ekman (1956). *Proc. 7th Intern. Congr. Animal Husbandry, Madrid, 1956* pp. 131–135.

Spisni, D. (1952). *Atti Soc. Ital. Sci. Vet.* **6,** 144–149.

Sprengel, K. (1832). "Chemie für Landwirthe, Forstmänner und Cameralisten," Vol. I, 793 pp.; Vol. II, 699 pp. Vandenhouk and Ruprecht, Göttingen.

Squibb, R. L., C. Rivera, and R. Jarquin (1958). *J. Animal Sci.* **17,** 318–321.

Stalfors, H. (1926). *Arch. Wiss. Prakt. Tierheilk.* **54,** 519–524.

Stallcup, O. T., J. L. Cason, and B. J. Walker (1956). *Arkansas, Univ. (Fayetteville), Agr. Expt. Sta. Bull.* **572,** 16 pp.

Stanley, R. W., and E. M. Kesler (1959). *J. Dairy Sci.* **42,** 127–136.

Starks, P. B., W. H. Hale, U.S. Garrigus, and R. M. Forbes (1953). *J. Animal Sci.* **12,** 480–491.

Starks, P. B., W. H. Hale, U. S. Garrigus, R. M. Forbes, and M. F. James (1954). *J. Animal Sci.* **13,** 249–257.

Steiger, P. A. (1904). Bakterienbefunde bei der Euterentzündung von Kuh und Ziege. Dissertation, Bern. From Ankersmit (1905).

Stein, F. (1858a). *Z. Naturwiss.* **8,** 57–59.

Stein, F. (1858b). *Abhandl. K. Boehm. Ges. Wiss.* **10,** 69–70.

Stellmach-Helwig, R. (1961). *Arch. Mikrobiol.* **38,** 40–51.

Steurer, E., and K. Hess (1944). *Z. Physik. Chem.* **193,** 248–257.

Stevens, C. E., and A. F. Sellers (1959). *Am. J. Vet. Res.* **20,** 461–482.

Stevens, C. E., A. F. Sellers, and F. A. Spurrell (1960). *Am. J. Physiol.* **198,** 449–455.

Stewart, D. G., R. G. Warner, and H. W. Seeley (1961). *Appl. Microbiol.* **9,** 150–156.

Stewart, J. (1953). *Brit. J. Nutr.* **7,** 231–235.

Stewart, J., and E. W. Moodie (1956). *J. Comp. Pathol. Therap.* **66,** 10–12.

Stewart, W. E., and R. W. Dougherty (1958). *J. Animal Sci.* **17,** 1220.

Stewart, W. E., and J. H. Nicolai (1964). *J. Dairy Sci.* **47,** 654.

Stewart, W. E., and L. H. Schultz (1958). *J. Animal Sci.* **17,** 737–742.

Stewart, W. E., D. G. Stewart, and L. H. Schultz (1958). *J. Animal Sci.* **17,** 723–736.

Steyn, D. G. (1943). *Farming S. Africa* **18,** 313–318.

Stickland, L. H. (1934). *Biochem. J.* **28,** 1746–1759.

Stillings, B. R., J. W. Bratzler, L. F. Macriott, and R. C. Miller (1964). *J. Animal Sci.* **23,** 1148–1154.

Stoddard, G. E., N. N. Allen, and W. H. Peterson (1949). *J. Animal Sci* **8,** 630–631.

Stolk, K. (1959). *Verslag. Landbouwk. Onderzoek.* **65.13,** 20 pp. From *Nutr. Abst. Rev.* **30,** 639.

Stone, E. C. (1949). *Am. J. Vet. Res.* **10,** 26–29.

Storr, G. M. (1964). *Australian J. Biol. Sci.* **17,** 469–481.

Storry, J. E. (1961a). *J. Agr. Sci.* **57,** 97–102.

Storry, J. E. (1961b). *J. Agr. Sci.* **57,** 103–109.

Stranks, D. W. (1956). *Can. J. Microbiol.* **2**, 56–62.

Strelkow, A., G. Poljansky, and M. Issakowa-Keo (1933). *Arch. Tierernaehr. Tierzuecht* **9**, 679–697.

Strong, H. T., M. T. Clegg and J. H. Meyer (1960). *Calif. Agr.* **14**, 15.

Sugden, B. (1953). *J. Gen. Microbiol.* **9**, 44–53.

Sugden, B., and A. E. Oxford (1952). *J. Gen. Microbiol.* **7**, 145–153.

Sullivan, J. T. (1955). *J. Animal Sci.* **14**, 710–717.

Sullivan, J. T. (1961). *Proc. 1961 Rumen Function Conf., Chicago.* U.S. Dept. Agr. Mimeographed report.

Sullivan, J. T., and T. V. Hershberger (1959). *Science* **130**, 1252.

Summers, C. E., F. H. Baker, and R. B. Grainger (1957). *J. Animal Sci.* **16**, 781–786.

Sutherland, D. N. (1952). *Australian Vet. J.* **28**, 204–208.

Sutherland, T. M., W. C. Ellis, R. S. Reid, and M. G. Murray (1962). *Brit. J. Nutr.* **16**, 603–614.

Swift, R. W., and C. E. French (1954). "Energy Metabolism and Nutrition," 264 pp. Scarecrow Press, Washington, D.C.

Swift, R. W., E. J. Thacker, A. Black, J. W. Bratzler, and W. H. James (1947). *J. Animal Sci.* **6**, 432–444.

Swift, R. W., J. W. Bratzler, W. H. James, A. D. Tillman, and D. C. Meek (1948). *J. Animal Sci.* **7**, 475–485.

Swift, R. W., R. L. Cowan, G. P. Barron, K. H. Maddy, and E. C. Grose (1951). *J. Animal Sci.* **10**, 434–438.

Sym, E. A. (1938). *Acta Biol. Exptl. (Varsovie)* **12**, 192–210.

Synge, R. L. M. (1951). *Biochem. J.* **49**, 642–650.

Synge, R. L. M. (1952a). *Brit. J. Nutr.* **6**, 100–104.

Synge, R. L. M. (1952b). *Proc. Roy. Soc.* **B139**, 205.

Synge, R. L. M. (1953). *J. Gen. Microbiol.* **9**, 407–409.

Synge, R. L. M. (1957). *Collected Papers, Rowett Res. Inst.* **13**, 37–41.

Tagari, H., I. Ascarelli, and A. Bondi (1962). *Brit. J. Nutr.* **16**, 237–243.

Takahashi, N., T. Kasukawa, and Y. Nosaki (1963). *Japan J. Zootechn. Sci.* **34**, 380–386.

Takahashi, N., T. Kasukawa, and Y. Nosaki (1964). *Japan J. Zootechn. Sci.* **35**, 147–152.

Talapatra, S. K., S. C. Ray, and K. C. Sen (1948a). *Indian J. Vet. Sci.* **18**, 99–108.

Talapatra, S. K., S. C. Ray, and K. C. Sen (1948b). *J. Agr. Sci.* **38**, 163–173.

Tamate, H., K. Ishida, and O. Itakawa (1964). *Tohoku J. Agr. Res.* **14**, 195–207.

Taylor, J. I., D. H. L. Rollinson, and K. W. Harker (1955). *J. Agr Sci.* **45**, 257–263.

Teeri, A. E., and N. F. Colovos (1963). *J. Dairy Sci.* **46**, 864–865.

Teeri, A. E., M. Leavit, D. Josselyn, N. F. Colovos, and H. A. Keener (1950). *J. Biol. Chem.* **182**, 509–514.

Teeri, A. E., D. Josselyn, N. F. Colovos, and H. A. Keener (1951). *J. Dairy Sci.* **34**, 299–302.

Templeton, J. A., and I. A. Dyer (1964). *J. Animal Sci.* **23**, 601.

Ter-Karapetjan, M. A., and A. M. Ogandzanjan (1959). *Dokl. Akad. Nauk SSSR* **125**, 666–669.

Ter-Karapetjan, M. A., and A. M. Ogandzanjan (1960a). *Dokl. Akad. Nauk SSSR* **131**, 1187–1190.

Ter-Karapetjan, M. A., and A. M. Ogandzanjan (1960b). *Izv. Akad. Nauk Arm. SSR* **13**, 23–31.

Thaysen, A. C. (1945). *Proc. Nutr. Soc. (Engl. Scot.)* **3**, 195–199.

Theurer, B., W. Woods, and W. Burroughs (1963). *J. Animal Sci.* **22**, 146–149.

Thin, C., and A. Robertson (1953). *J. Comp. Pathol. Therap.* **63**, 184–194.

Thin, C., M. Kleiber, and D. S. Kronfeld (1962). *Am. J. Vet. Res.* **23**, 544–547.

Thomas, B. H., C. C. Culbertson, and F. Beard (1934). *Proc. Am. Soc. Animal Prod. 27th Ann. Meeting, 1934* pp. 193–199.

Thomas, G. J. (1960). *J. Agr. Sci.* **54**, 360–372.

Thomas, W. E. (1956). *J. Animal Sci.* **15**, 1295.

Thomas, W. E., J. K. Loosli, H. H. Williams, and L. A. Maynard (1951). *J. Nutr.* **43**, 515–523.

Tiedemann, F., and L. Gmelin (1831). "Die Verdauung nach Versuchen," 2nd ed., Vol. 1, 382 pp. Groos, Heidelberg and Leipzig.

Tillman, A. D., and J. R. Brethour (1958). *J. Animal Sci.* **17**, 104–112.

Tillman, A. D., and R. MacVicar (1956). *J. Animal Sci.* **15**, 211–217.

Tillman, A. D., C. F. Chappel, R. J. Sirny, and R. MacVicar (1954a). *J. Animal Sci.* **13**, 417–424.

Tillman, A. D., R. J. Sirny, and R. MacVicar (1954b). *J. Animal Sci.* **13**, 726–731.

Tillman, A. D., W. D. Gallup, L. S. Pope, G. A. McLaren, and W. Price (1957a). *J. Animal Sci.* **16**, 179–189.

Tillman, A. D., W. D. Gallup, and W. Woods (1957b). *J. Animal Sci.* **16**, 419–425.

Titchen, D. A. (1958). *J. Physiol.* **14**, 1–21.

Titchen, D. A. (1960). *J. Physiol. (London)* **151**, 139–153.

Topps, J. H., and R. C. Elliot (1964). *Animal Prod.* **6**, 91–96.

Tosic, J., and R. L. Mitchell (1948). *Nature* **162**, 502–504.

Toth, L. (1948). *Experientia* **4**, 395–396.

Toth, L. (1949). *Experientia* **5**, 474–475.

Toussaint, H. (1875). *Arch. Physiol. Norm. Pathol.* **7**, 141–176.

Trautmann, A. (1932). *Arch. Tiorernaehr. Tierzucht* **7**, 401–420.

Trautmann, A. (1933a). *Arch. Tierernaehr. Tierzucht* **9**, 19–30.

Trautmann, A. (1933b). *Arch. Tierernaehr. Tierzucht* **9**, 178–193.

Trautmann, A. (1933c). *Arch. Tierernaehr. Tierzucht* **9**, 575–584.

Trautmann, A., and H. Albrecht (1931). *Arch. Wiss. Prakt. Tierheilk.* **64**, 93–104.

Trautmann, A., and T. Asher (1939). *Z. Tierernaehr. Futtermittelk.* **3**, 45–52.

Trautmann, A., and T. Asher (1941a). *Arch. Wiss. Prakt. Tierheilk.* **76**, 317–328.

Trautmann, A., and T. Asher (1941b). *Deut. Tieraerztl. Wochschr.* **49**, 24–26.

Trautmann, A., and H. Hill (1949). *Arch. Ges. Physiol.* **252**, 30–39.

Trautmann, A., and J. Schmitt (1932). *Arch. Tierernaehr. Tierzucht* **7**, 421–435.

Trautmann, A., and J. Schmitt (1933a). *Arch. Tierernaehr. Tierzucht* **9**, 1–10.

Trautmann, A., and J. Schmitt (1933b). *Arch. Tierernaehr. Tierzucht* **9**, 11–18.

Trier, H. J. (1926). *Z. Vergleich. Physiol.* **4**, 305–330.

Tsuda, T. (1956a). *Tohoku J. Agr. Res.* **7**, 231–239.

Tsuda, T. (1956b). *Tohoku J. Agr. Res.* **7**, 241–256.

Tsuda, T. (1962). *Tohoku J. Agr. Res.* **13**, 29–42.

Tsuda, T. (1964). *Tohoku J. Agr. Res.* **15**, 83–90.

Turner, A. W., and V. E. Hodgetts (1949–1959). *Australia, Commonwealth Sci. Ind. Res. Organ., 1st to 11th Ann. Rept.* Cited from Ryan (1964b).

Turner, A. W., and V. E. Hodgetts (1953). *Australian J. Agr. Res.* **3**, 453–459.

Turner, A. W., and V. E. Hodgetts (1955a). *Australian J. Ag. Res.* **6**, 115–124.

Turner, A. W., and V. E. Hodgetts (1955b). *Australian J. Agr. Res.* **6**, 125–144.

Twigg, R. S. (1945). *Nature* **155**, 401–402.

Tyznik, W., and N. N. Allen (1951). *J. Dairy Sci.* **34**, 493.

Uesaka, S., R. Kawashima, Y. Hashimoto, and Y. Kamade (1959). *Bull. Res. Inst. Food Sci. Kyoto Univ.* No. 22, 17–25.

Uesaka, S., R. Kawashima, K. Namikawa, and M. Yagi (1962a). *Bull. Res. Inst. Food Sci. Kyoto Univ.* No. 25, 10–26.

Uesaka, S., R. Kawashima, Y. Tsubota, and S. Tabuchi (1962b). *Bull. Res. Inst. Food Sci. Kyoto Univ.* No. 26, 1–16.

Ueyama, E., and Y. Hirose (1964). *J. Faculty Agr. Hokkaido Univ.* **54**, 1–8.

Umezu, M., T. Tuda, M. Katuno, S. Omori, and S. Minagaki (1951). *Tohoku J. Agr. Res.* **2**, 73–81.

Underkofler, L. A., W. D. Kitts, and R. L. Smith (1953). *Arch. Biochem. Biophys.* **44**, 492–493.

Underwood, E. J. (1956). "Trace Elements in Human and Animal Nutrition." Academic Press, New York (2nd ed., 1962).

Underwood, E. J. (1957). *Australian Vet. J.* **33**, 283–286.

Underwood, E. J. (1959). *Ann. Rev. Biochem.* **28**, 499–526.

Underwood, E. J., and J. F. Filmer (1935). *Australian Vet. J.* **11**, 84–92.

Underwood, E. J., and R. J. Moir (1953). *J. Australian Inst. Agr. Sci.* **19**, 214–221.

U.S. Dept. Agr. (1956). *Agr. Res. ARS* **4**, 14.

Ushijima, R. N., and R. H. McBee (1957). *Proc. Montana Acad. Sci.* **17**, 33–35.

Ustjanzew, W. (1911). *Biochem. Z.* **37**, 457–476.

Usuelli, F. (1930a). *Wiss. Arch. Landwirtsch. Abt. B, Arch. Tierernaehr. Tierzucht* **3**, 4–19.

Usuelli, F. (1930b). *Wiss. Arch. Landwirtsch. Abt. B, Arch. Tierernaehr. Tierzucht* **3**, 368–382.

Usuelli, F. (1956). *Proc. 7th Intern. Congr. Animal Husbandry, Madrid, 1956* Subject 6, pp. 107–114.

Usuelli, F., and P. Fiorini (1938) *Boll. Soc. Ital. Biol. Sper.* **13**, 11–14.

Valentine, R. C., and R. S. Wolfe (1963). *J. Bacteriol.* **85**, 1114–1120.

Vallenas, G. A. (1956). *Am. J. Vet. Res.* **17**, 79–89.

Van Campen, D. R., and G. Matrone (1960). *J. Nutr.* **72**, 277–282.

Van Campen, D. R., and G. Matrone (1962). *Federation Proc.* **21**, 396.

Van Campen, D. R., and G. Matrone (1964a). *Biochim. Biophys. Acta* **85**, 400–409.

Van Campen, D. R., and G. Matrone (1964b). *Biochim. Biophys. Acta* **85**, 410–419.

van den Hende, C., W. Oyaért, and J. H. Bouckaert (1959). *Zentr. Veterinaermed.* **6**, 681–692.

van den Hende, C., W. Oyaért, and J. H. Bouckaert (1963a). *Res. Vet. Sci.* **4**, 77–88.

van den Hende, C., W. Oyaért, and J. H. Bouekaert (1963b). *Res. Vet. Sci.* **4**, 89–95.

van den Hende, C., W. Oyaért and J. H. Bouckaert (1963c). *Res. Vet. Sci.* **4**, 382–389.

van der Horst, C. J. G. (1961). *Nature* **191**, 73–74.

Vanderveen, J. E., and H. A. Keever (1964). *J. Dairy Sci.* **47**, 1224–1230.

van der Wath, J. G. (1948a). *Onderstepoort J. Vet. Sci. Animal Ind.* **23**, 367–383.

van der Wath, J. G. (1948b). *Onderstepoort J. Vet. Sci. Animal Ind.* **23**, 385–387.

van der Wath, J. G., and S. J. Myburgh (1941). *Onderstepoort J. Vet. Sci. Animal Ind.* **17**, 61–88.

Van Dyne, G. M., and J. H. Meyer (1964). *J. Animal Sci.* **23,** 1108–1115.

Van Horn, H. H., and E. E. Bartley (1961). *J. Animal Sci.* **20,** 85–85.

Van Horn, H. H., P. A. Hartman, N. L. Jacobson, A. D. McGilliard, P. R. Shellenberger, and S. M. Kassir (1961). *J. Animal Sci.* **20,** 751–758.

Van Horn, H. H., N. L. Jacobson, P. A. Hartman, A. D. McGilliard, and J. V. Debarthe (1963). *J. Animal Sci.* **22,** 399–409.

Van Soest, P. J., and N. N. Allen (1959). *J. Dairy Sci.* **42,** 1977–1985.

van Uden, N., L. C. Sousa, and M. Farinha (1958). *J. Gen. Microbiol.* **19,** 435–445.

Van Keuren, R. W., and W. W. Heinemann (1962). *J. Animal Sci.* **21,** 340–345.

Virtanen, A. I. (1963). *Biochem. Z.* **338,** 443–453.

Visek, W. J. (1964). *Proc. Cornell Nutr. Conf. Feed Manuf., 1963,* pp. 121–133.

Volcani, B. E., J. I. Toohey and H. A. Barker, (1961). *Arch. Biochem. Biophys.* **92,** 381–391.

Voltz, W. (1919). *Berlin. Klin. Wochschr.* **56,** 693.

Voltz, W. (1920). *Biochem. Z.* **102,** 151–227.

von Tappeiner, H. (1883). *Z. Biol.* **19,** 228–279.

von Tappeiner, H. (1884a). *Z. Biol.* **20,** 52–134.

von Tappeiner, H. (1884b). *Z. Biol.* **20,** 215–233.

von Tappeiner, H. (1886). *Z. Biol.* **22,** 236–240.

Vridnik, F. I. (1961). *Nauk. Pratsi, Ukr. Akad. Sil's' kogospodar. Nauk* **14,** 28–32.

Waentig, P., and W. Gierisch (1919). *Z. Physiol. Chem.* **107,** 213–225.

Waite, R., and J. Boyd (1953a). *J. Sci. Food Agr.* **4,** 197–204.

Waite, R., and J. Boyd (1953b). *J. Sci. Food Agr.* **4,** 257–261.

Waite. R., M. J. Johnston, and D. G. Armstrong (1964). *J. Agr. Sci.* **62,** 391–398.

Waldern, D. E., V. L. Johnson, and T. H. Blosser (1963). *J. Dairy Sci.* **46,** 327–332.

Waldo, D.R., and L. H. Schultz (1955). *J. Dairy Sci.* **38,** 605.

Waldo, D. R., and L. H. Schultz (1956). *J. Dairy Sci.* **39,** 1453–1460.

Walker, D. J. (1958). *Biochem. J.* **69,** 524–530.

Walker, D. J. (1961). *Australian J. Agr. Res.* **12,** 171–175.

Walker, D. J., and W. W. Forrest (1964). *Australian J. Agr. Res.* **15,** 299–315.

Walker, D. M. (1959a). *J. Agr. Sci.* **53,** 374–380.

Walker, D. M. (1959b). *J. Agr. Sci.* **53,** 381–386.

Walker, D. M. (1960). *Australian Vet. J.* **36,** 17–20.

Walker, D. M., and R. A. Simmonds (1962). *J. Agr. Res.* **59,** 375–379.

Walker, G. J., and P. M. Hope (1963). *Biochem. J.* **86,** 452–462.

Walker, G. J., and P. M. Hope (1964). *Biochem. J.* **90,** 398–408.

Wallace, V., A. E. Dracy, and R. K. Oines (1959). *Proc. S. Dakota Acad. Sci.* **38,** 146.

Walter, W. G. (1952). A study of microbial activities in the ovine rumen. Ph.D. thesis. Michigan State College, East Lansing, Michigan.

Ward, J. K., C. W. Tefft, A. J. Sirny, H. N. Edwards, and A. D. Tillman (1957). *J. Animal Sci.* **16,** 631–641.

Warner, A. C. I. (1955). Some aspects of the nitrogen metabolism of the microorganisms of the rumen with special references to proteolysis. Ph.D. Thesis, Univ. of Aberdeen, Aberdeen, Scotland.

Warner, A. C. I. (1956a). *J. Gen. Microbiol.* **14,** 733–748.

Warner, A. C. I. (1956b). *J. Gen. Microbiol.* **14,** 749–762.

Warner, A. C. I. (1962a). *J. Gen. Microbiol.* **28,** 119–128.

Warner, A. C. I. (1962b). *J. Gen. Microbiol.* **28,** 129–146.

Warner, A. C. I. (1964a). *Nutr. Abstr. Rev.* **34,** 339–352.

Warner, A. C. I. (1964b). *Australian J. Biol. Sci.* **17,** 170–182.

Warner, A. C. I., and B. D. Stacy (1965). *Quart. J. Exptl. Physiol.* **50,** 169–184.

Warth, F. J., and T. S. Krishnan (1935). *Indian J. Vet. Sci.* **5,** 319–331.

Washburn, L. E., and S. Brody (1937). *Missouri Univ. Agr. Expt. Sta. Res. Bull.* **263,** 40 pp.

Watanabe, Y., and M. Umezu (1963). *Tohoku J. Agr. Res.* **14,** 29–37.

Watson, C. J., J. W. Kennedy, W. M. Davidson, C. H. Robinson, and G. W. Muir (1949a). *Sci. Agr.* **29,** 173–184.

Watson, C. J., W. M. Davidson, and J. W. Kennedy (1949b). *Sci. Agr.* **29,** 185–188.

Watson, R. H. (1933). *Australian J. Exptl. Biol. Med. Sci.* **11,** 67–74.

Watson, R. H., and I. G. Jarrett (1941). *Australian Vet. J.* **17,** 137–142.

Watts, P. S. (1957). *Australian J. Agr. Res.* **8,** 266–270.

Watts, P. S. (1959a). *J. Agr. Sci.* **52,** 244–249.

Watts, P. S. (1959b). *J. Agr. Sci.* **52,** 250–255.

Weenink, R. O. (1959). *New Zealand J. Sci. Technol.* **A2,** 273.

Weenink, R. O. (1962). *Biochem. J.* **82,** 523–527.

Wegner, G. H., and E. M. Foster (1960). *J. Dairy Sci.* **43,** 566–568.

Wegner, G. H., and E. M. Foster (1963). *J. Bacteriol.* **85,** 53–61.

Wegner, M. I., A. N. Booth, G. Bohstedt, and E. B. Hart (1940a). *J. Dairy Sci.* **23,** 1123–1129.

Wegner, M. I., A. N. Booth, C. A. Elvehjem, and E. B. Hart (1940b). *Proc. Soc. Exptl. Biol. Med.* **45,** 769–771.

Wegner, M. I., A. N. Booth, G. Bohstedt, and E. B. Hart (1941a). *J. Dairy Sci.* **24,** 51–56.

Wegner, M. I., A. N. Booth, G. Bohstedt, and E. B. Hart (1941b). *J. Dairy Sci.* **24,** 835–844.

Wegner, M. I., A. N. Booth, C. A. Elvehjem, and E. B. Hart (1941c). *Proc. Soc. Exptl. Biol. Med.* **47,** 90–94.

Weineck, E. (1934). *Arch. Protistenk.* **82,** 169–202.

Weir, W. C., and D. T. Torell (1959). *J. Animal Sci.* **18,** 641–649.

Weir, W. C., J. H. Meyer, W. N. Garrett, G. P. Lofgreen, and N. R. Ittner (1959). *J. Animal Sci.* **18,** 805–814.

Weiske, H. (1879). *Landwirtsch. Jahrb.* **8,** 499–501.

Weiss, K. E. (1953a). *Onderstepoort J. Vet. Res.* **26,** 241–250.

Weiss, K. E. (1953b). *Onderstepoort J. Vet. Res.* **26,** 251–283.

Weldy, J. R., R. E. McDowell, P. J. van Soest, and J. Bond (1964). *J. Animal Sci.* **23,** 147–153.

Weller, R. A. (1957). *Australian J. Biol. Sci.* **10,** 384–389.

Weller, R. A., F. V. Gray, and A. F. Pilgrim (1958). *Brit. J. Nutr.* **12,** 421–429.

Weller, R. A., A. F. Pilgrim, and F. V. Gray (1962). *Brit. J. Nutr.* **16,** 83–90.

Wertheim, P. (1935). *Quart. J. Microscop. Sci.* [NS] **78,** 31–49.

Wester, J. (1926). "Die Physiologie und Pathologie der Vormagen beim Rinde," 110 pp. Schoetz, Berlin.

Westermarck, H. (1959). *Acta Vet. Scand.* **1,** 67–73.

Westhuizen, G. C. A. van der, A. E. Oxford and J. I. Quin (1950). *Onderstepoort J. Vet. Sci. Animal Ind.* **24,** 119–124.

Westphal, A. (1934a). *Zool. Anz.* **7,** Suppl., 207–210.

Westphal, A. (1934b). *Z. Parasitenk.* **7,** 71–117.

White, D. C., M. P. Bryant, and D. R. Caldwell (1962a). *Abstr. 8th Intern. Congr. Microbiol., Montreal, 1962* p. 42.

White, D. C., M. P. Bryant, and D. R. Caldwell (1962b). *J. Bacteriol.* **84**, 822–828.

White, E. A. (1955). *Vet. Med.* **50**, 199–202.

White, T. W., R. B. Grainger, F. H. Baker, and J. W. Stroud (1958). *J. Animal Sci.* **17**, 797–803.

Whitlock, J. H., and J. Fabricant (1947). *Cornell Vet.* **37**, 211–230.

Wieringa, G. W. (1956). *Tijdschr. Diergeneesk.* **81**, 242–249.

Wiese, A. C., B. C. Johnson, H. H. Mitchell, and W. B. Nevens (1947). *J. Nutr.* **33**, 263–270.

Wilckens, M. (1872). "Untersuchungen über den Magen der wiederkäuenden Hausthiere." H. and P. Wiegandt, Berlin.

Wildt, E. (1874). *Jahrb. Landwirtschaftsges.* **22**, 1–34.

Wildt, E. (1879). *Landwirtsch. Versuchsstationen* **23**, 54–58.

Williams, E. I. (1955a). *Vet. Record* **67**, 907–911.

Williams, E. I. (1955b). *Vet. Record* **67**, 922–927.

Williams, N. M., and D. E. Tribe (1957). *J. Agr.* (*Victoria*) **55**, 769–771.

Williams, P. P., and R. N. Doetsch (1960). *J. Gen. Microbiol.* **22**, 635–644.

Williams, P. P., J. Gutierrez and R. N. Doetsch (1960). *Proc. Soc. Am. Bacteriologists, Ann. Meeting, 1960* p. 32.

Williams, P. P., R. E. Davis, R. N. Doetsch, and J. Gutierrez (1961). *Appl. Microbiol.* **9**, 405–409.

Williams, P. P., J. Gutierrez, and R. E. Davis (1963). *Appl. Microbiol.* **11**, 260–264.

Williams, V. I., and K. R. Christian (1956). *New Zealand J. Sci. Technol.* **A38**, 194–202.

Williams, V. J., and R. J. Moir (1951). *Australian J. Sci. Res.* **B4**, 377–390.

Williams, V. J., M. C. Nottle, R. J. Moir, and E. J. Underwood (1953). *Australian J. Biol. Sci.* **6**, 142–151.

Williams, W. F., H. Hoernicke, D. R. Waldo, W. P. Flatt, and M. J. Allison (1963). *J. Dairy Sci.* **46**, 992–993.

Willing, E. (1933). *Arch. Tierernaehr. Tierzucht* **9**, 637–670.

Wilsing, H. (1885). *Z. Biol.* **21**, 625–630.

Wilson, A. D. (1963a). *Australian J. Agr. Res.* **14**, 680–689.

Wilson, A. D. (1963b). *Australian J. Agr. Res.* **14**, 808–814.

Wilson, A. D. (1964). *Brit. J. Nutr.* **18**, 163–172.

Wilson, A. D., and D. E. Tribe (1963). *Australian J. Agr. Res.* **14**, 670–679.

Wilson, A. I. (1962). *Outlook Agr.* **3**, 160–166.

Wilson, M. K., and C. A. E. Briggs (1954). *Vet. Record* **66**, 187–188.

Wilson, M. K., and C. A. E. Briggs (1955). *J. Appl. Bacteriol.* **18**, 294–306.

Wilson, P. N. (1961). *J. Agr. Sci.* **56**, 351–364.

Wilson, R. K., and W. J. Pigden (1964). *Can. J. Animal Sci.* **44**, 122–123.

Wilson, S. M. (1953). *J. Gen. Microbiol.* **9**, i-ii.

Winegar, A. H., P. B. Pearson, and H. Schmidt (1940). *Science* **91**, 508–509.

Winogradow, M., T. Winogradowa-Fedorowa, and A. Wereninow (1930). *Zentr. Bakteriol. Parasitenk. Abt. II* **81**, 230–244.

Winogradowa, T. (1936). *Z. Parasitenk.* **8**, 359–364.

Winogradowa-Fedorowa, T., and M. Winogradow (1929). *Zentr. Bakteriol. Parasitenk. Abt. II* **78**, 246–254.

Winter, A. J. (1962). *Am. J. Vet. Res.* **23**, 500–505.

Winter, K. A., R. R. Johnson, and B. A. Dehority (1964). *J. Dairy Sci*, **47**, 793–797.

Wise, M. B., S. E. Smith, and L. L. Barnes (1958). *J. Animal Sci.* **17**, 89–99.

Wiseman, H. G., and H. M. Irwin (1957). *J. Agr. Food Chem.* **5**, 213–215.

Wiseman, R. F., D. R. Jacobson, and W. M. Miller (1960). *Appl. Microbiol.* **8**, 76–79.

Woehler, F. (1824). *Tiedemann Treviranus Z.* **1**, 290–317.

Wolin, M. J. (1964). *Science* **146**, 775–776.

Wolin, M. J., and G. Weinberg (1960). *J. Dairy Sci.* **43**, 825–830.

Wolin, M. J., G. B. Manning, and W. O. Nelson (1959). *J. Bacteriol.* **78**, 147.

Wolin, M. J., E. A. Wolin, and N. J. Jacobs (1961). *J. Bacteriol.* **81**, 911–917.

Woodcock, H. M., and G. Lapage (1913). *Quart. J. Microscop. Sci.* **59**, 431–458.

Woodhouse, N. S., R. F. Davis, and G. H. Beck (1955). *Univ. Maryland Agr. Expt. Sta., Misc. Publ.* **238**, 8–9.

Woodman, H. E., and R. E. Evans (1938). *J. Agr. Sci.* **28**, 43–63.

Woodman, H. E., and R. E. Evans (1947). *J. Agr. Sci.* **37**, 81–93.

Woods, W., and R. Luther (1962). *J. Animal Sci.* **21**, 809–814.

Work, E. (1950). *Biochim. Biophys. Acta* **5**, 204–209.

Work, E. (1951). *Biochem. J.* **49**, 17–23.

Wrenn, T. R., J. Bitman, and J. F. Sikes (1961). *J. Dairy Sci.* **44**, 2077–2080.

Wright, D. E. (1959). *Nature* **184**, 875–876.

Wright, D. E. (1960a). *Nature* **185**, 540–547.

Wright, D. E. (1960b). *Arch. Biochem. Biophys.* **86**, 251–254.

Wright, D. E. (1960c). *J. Gen. Microbiol.* **22**, 713–725.

Wright, D. E. (1961a). *New Zealand J. Agr. Res.* **4**, 203–215.

Wright, D. E. (1961b). *New Zealand J. Agr. Res.* **4**, 216–223.

Wright, E. (1955a). *New Zealand J. Sci. Technol.* **A37**, 332–348.

Wright, E. (1955b). *Nature* **176**, 351–352.

Wright, P. I., A. L. Pope, and P. H. Phillips (1963). *J. Animal Sci.* **22**, 586–591.

Wynne, K. N., and G. L. McClymont (1956). *Australian J. Agr. Res.* **7**, 45–56.

Yadawa, I. S., and E. E. Bartley (1964). *J. Dairy Sci.* **47**, 1352–1358.

Yadawa, I. S., E. E. Bartley, L. R. Fina, C. L. Alexander, R. M. Meyer, and E. L. Sorensen (1964). *J. Dairy Sci.* **47**, 1346–1351.

Yarns, D. A., and P. A. Putnam (1962). *J. Animal Sci.* **21**, 744–745.

Zuntz, N. (1879). *Landwirtsch. Jahrb.* **8**, 65–117.

Zuntz, N. (1891). *Arch. Ges. Physiol.* **49**, 477–483.

Zuntz, N. (1913). *Landwirtsch. Versuchsstationen* **79/80**, 781–814.

Zuntz, N., and von Mehring (1883). *Arch. Ges. Physiol.* **32**, 173–221.

INDEX